Lecture Notes in Computer Science 4178

Commenced Publication in 1973
Founding and Former Series Editors:
Gerhard Goos, Juris Hartmanis, and Jan van Leeuwen

Editorial Board

David Hutchison
  Lancaster University, UK
Takeo Kanade
  Carnegie Mellon University, Pittsburgh, PA, USA
Josef Kittler
  University of Surrey, Guildford, UK
Jon M. Kleinberg
  Cornell University, Ithaca, NY, USA
Friedemann Mattern
  ETH Zurich, Switzerland
John C. Mitchell
  Stanford University, CA, USA
Moni Naor
  Weizmann Institute of Science, Rehovot, Israel
Oscar Nierstrasz
  University of Bern, Switzerland
C. Pandu Rangan
  Indian Institute of Technology, Madras, India
Bernhard Steffen
  University of Dortmund, Germany
Madhu Sudan
  Massachusetts Institute of Technology, MA, USA
Demetri Terzopoulos
  University of California, Los Angeles, CA, USA
Doug Tygar
  University of California, Berkeley, CA, USA
Moshe Y. Vardi
  Rice University, Houston, TX, USA
Gerhard Weikum
  Max-Planck Institute of Computer Science, Saarbruecken, Germany

Andrea Corradini   Hartmut Ehrig
Ugo Montanari   Leila Ribeiro
Grzegorz Rozenberg (Eds.)

Graph Transformations

Third International Conference, ICGT 2006
Natal, Rio Grande do Norte, Brazil
September 17-23, 2006
Proceedings

 Springer

Volume Editors

Andrea Corradini
Ugo Montanari
University of Pisa
Department of Computer Science
56127 Pisa, Italy
E-mail: {andrea,ugo}@di.unipi.it

Hartmut Ehrig
Technical University of Berlin
Department for Software Technology and Theoretical Informatics
10587 Berlin, Germany
E-mail: ehrig@cs.tu-berlin.de

Leila Ribeiro
Universidade Federal do Rio Grande do Sul
Instituto de Informática
91501-970 Porto Alegre, Brazil
E-mail: leila@inf.ufrgs.br

Grzegorz Rozenberg
Leiden University
Leiden Institute of Advanced Computer Science (LIACS)
2333 CA Leiden, The Netherlands
E-mail: rozenber@liacs.nl

Library of Congress Control Number: 2006931633

CR Subject Classification (1998): E.1, G.2.2, D.2.4, F.1, F.2.2, F.3, F.4.2-3

LNCS Sublibrary: SL 1 – Theoretical Computer Science and General Issues

ISSN   0302-9743
ISBN-10   3-540-38870-2 Springer Berlin Heidelberg New York
ISBN-13   978-3-540-38870-8 Springer Berlin Heidelberg New York

Springer is a part of Springer Science+Business Media

springer.com

© Springer-Verlag Berlin Heidelberg 2006
Printed in Germany

Typesetting: Camera-ready by author, data conversion by Scientific Publishing Services, Chennai, India
Printed on acid-free paper   SPIN: 11841883   06/3142   5 4 3 2 1 0

Preface

ICGT 2006 was the 3rd International Conference on Graph Transformation, following the previous two in Barcelona (2002) and Rome (2004), and a series of six international workshops between 1978 and 1998. ICGT 2006 was held in Natal (Rio Grande do Norte, Brazil) on September 17-23, 2006, co-located with the Brazilian Symposium on Formal Methods (SBMF 2006), under the auspices of the Brazilian Computer Society (SBC), the European Association of Software Science and Technology (EASST), the European Association for Theoretical Computer Science (EATCS) and the IFIP WG 1.3 on Foundations of Systems Specification. The conference obtained partial support from Formal Methods Europe and IFIP TC 1 on Foundations of Computer Science.

The scope of the conference concerned graphical structures of various kinds (like graphs, diagrams and visual sentences) that are useful when describing complex structures and systems in a direct and intuitive way. These structures are often enriched with formalisms that model their evolution via suitable kinds of transformations. The field of the conference was concerned with the theory, applications, and implementation issues of such formalisms. Particular emphasis was put on metamodels which can accommodate a variety of graphical structures within the same abstract theory.

The theory is strongly related to areas such as graph theory and graph algorithms, formal language and parsing theory, the theory of concurrent and distributed systems, formal specification and verification, logics, and semantics. The application areas include all those fields of computer science, information processing, engineering, biology and the natural sciences where static and dynamic modelling using graphical structures and graph transformations, respectively, play important roles. In many of these areas tools based on graph transformation technology have been implemented and used.

The proceedings of ICGT 2006 consist of two parts. The first part contains the contributions of the invited speakers followed by 28 accepted papers that were selected out of 62 carefully reviewed submissions. The topics of the papers range over a wide spectrum, including graph theory and graph algorithms, theoretic and semantic aspects, modelling, contributions to software engineering and global computing, applications to biology, and tool issues. The second part contains a short description of a tutorial on foundations and applications of graph transformations, and short presentations of the satellite events of ICGT 2006.

We would like to thank the members of the program committee and the secondary reviewers for their enormous help in the selection process. Moreover, we would like to express our gratitude to the local organizers who did a great job, in particular to the Organizing Committee chair Anamaria Martins Moreira.

July 2006 Andrea Corradini, Hartmut Ehrig, Ugo Montanari (PC co-chair)
Leila Ribeiro (PC co-chair), Grzegorz Rozenberg

Organization

Program Committee

Paolo Baldan	Venice (Italy)
Paolo Bottoni	Rome (Italy)
Bruno Courcelle	Bordeaux (France)
Andrea Corradini	Pisa (Italy)
Hartmut Ehrig	Berlin (Germany)
Gregor Engels	Paderborn (Germany)
Reiko Heckel	Leicester (UK)
Dirk Janssens	Antwerp (Belgium)
Gabor Karsai	Nashville (Tennessee, USA)
Hans-Jörg Kreowski	Bremen (Germany)
Barbara König	Duisburg-Essen (Germany)
Mercè Llabrés	Mallorca (Spain)
Anamaria Martins Moreira	Natal (Brazil)
Ugo Montanari (co-chair)	Pisa (Italy)
Manfred Nagl	Aachen (Germany)
Fernando Orejas	Barcelona (Spain)
Francesco Parisi-Presicce	Rome (Italy) and Fairfax (Virginia, USA)
Mauro Pezzè	Milan (Italy)
John Pfaltz	Charlottesville (Virginia, USA)
Rinus Plasmeijer	Nijmegen (The Netherlands)
Detlef Plump	York (UK)
Leila Ribeiro (co-chair)	Porto Alegre (RS, Brazil)
Grzegorz Rozenberg	Leiden (The Netherlands)
Andy Schürr	Darmstadt (Germany)
Gabriele Taentzer	Berlin (Germany)
Dániel Varró	Budapest (Hungary)
Daniel Yankelevich	Buenos Aires (Argentina)

Secondary Referees

András Balogh	Søren Christensen	Claudia Ermel
Benjamin Braatz	Giovanni Cignoni	Maribel Fernandez
Victor Braberman	Juan de Lara	Esteban Feuerstein
Chiara Braghin	Anne Dicky	Irene Finocchi
Clara Bertolissi	Dino Distefano	Fabio Gadducci
Laura Bocchi	Mike Dodds	Giorgio Ghelli
Roberto Bruni	Juan Echague	Luís Gomes
Alexey Cherchago	Karsten Ehrig	Annegret Habel

Tobias Heindel
Frank Hermann
Dan Hirsch
Kathrin Hoffmann
Nicolas Kicillof
Markus Klein
Manuel Koch
Radu Kopetz
Maciej Koutny
Georgios Lajios
Leen Lambers

Marc Lohmann
Ivan Lanese
Greg Manning
Leonardo Mariani
Tony Modica
Olaf Muliawan
Nikos Mylonakis
Julia Padberg
Jean-Guy Penaud
Ulrike Prange
Paola Quaglia

Guilherme Rangel
Christophe Ringeissen
Francesc Rosselló
Domenico Saccà
Hans Schippers
Gabriel Valiente
Niels Van Eetvelde
Pieter Van Gorp
Gergely Varró
Jessica Winkelmann

Sponsoring Institutions

Brazilian Computer Society (SBC)
European Association for Theoretical Computer Science (EATCS)
European Association of Software Science and Technology (EASST)
Formal Methods Europe (FME)
International Federation for Information Processing (IFIP) WG 1.3

Table of Contents

Borrowed Contexts and Adhesive Categories

Extensions for Distributed and Global Computing

Software Engineering Methods and Tools

Model-Driven Development

Efficient Implementation

Logics

Tutorial and Workshops

Nested Quantification in Graph Transformation Rules

Arend Rensink

Department of Computer Science, University of Twente
P.O. Box 217, 7500 AE, The Netherlands
rensink@cs.utwente.nl*

Abstract. In this paper we describe a way to integrate Taentzer's rule amalgamation with the recently proposed notions of nested graph conditions. The resulting so-called *quantified graph transformation rules* include (universally and existentially) quantified sub-structures in a flexible way. This can be used for instance to specify a larger-step operational semantics, thus improving the scalability of graph transformation as a technique for software verification.

1 Introduction

The idea presented in this paper is motivated by the goal to use graph transformation as a technique for specifying, and eventually verifying, the dynamic behaviour of software systems. In this setup, each transformation rule corresponds to a single computation step of the system, in which, for instance, a method is called or a variable is assigned. We have observed in previous work, e.g., [13], that such a computation step frequently involves acting upon a structure whose size is not *a priori* known, but instead involves sub-structures of which there may be arbitrarily many copies.

A typical example of this is the encoding of parameter transfer from a method caller to the called method. Obviously this is the same mechanism for all methods, and so we would like to have a single rule that captures it. Unfortunately, the number of parameters is not the same for all methods: in fact, the parameters indeed form a sub-structure with a varying and a priori unknown number of copies.[1] For this reason, it is not possible, using the standard graph transformation formalism, to capture the parameter transfer mechanism in a single rule. Fig. 1 shows an example rule for two parameters.

Although there are workarounds, typically involving the use of auxiliary edges which successively mark all copies of the substructure involved, these have undesirable consequences (besides being inelegant). In particular, such a solution results in a number of small steps that is linear in the number of substructures involved. In particular in a setting where the system under analysis has parallelism, these steps get interleaved with independent actions in other parts of the system, contributing to the state space blow-up (which is the most urgent problem in verification methods in the first place). This,

---

* The work reported in this paper was carried out in the context of the Dutch NWO project GROOVE (project number 612.000.314).

[1] To be more precise, the number of parameters is fixed and known for each individual method, but not from the more global perspective of our semantics, in which calls to all methods are to be treated as instances of the same mechanism.

A. Corradini et al. (Eds.): ICGT 2006, LNCS 4178, pp. 1–13, 2006.

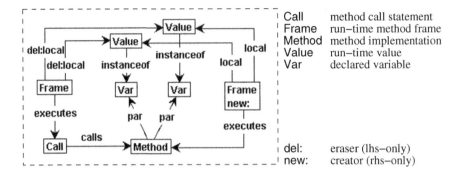

Fig. 1. Method call rule for two parameters

in turn, may be alleviated by further modifying the formalism, for instance by imposing priorities or other forms of control on the set of rules, but at the price of increased complexity of the formalism and hence of the verification task itself.

To make the example even more challenging: if we are interested in a data-flow analysis, we want the method call rule to be enabled if and only if all the arguments to the method are available (and not when control has explicitly reached the point in the program where the method is called). This involves a further condition on the rule, involving the existence of all relevant substructures — in this case, values for all method parameters. This is *not* a standard (positive or negative) application condition, since, once more, the number of substructures that are required to exist is not a priori known.

The problem described above has been studied before. On the practical side, many tools for graph transformation (for instance GREAT [6], FuJaBa [4]) have some notion of graph patterns that may be matched with cardinality greater than 1, i.e., that match to an a priori unknown number of sub-graphs in the host graph. Furthermore, Taentzer [18] has developed an elegant theoretical basis for this type of extension, called *rule amalgamation*. This is based on the concept of an *interaction scheme* which essentially imposes a sub-rule embedding on a set of rules, and a *covering condition* which imposes further conditions on the matches to be considered. Taking the above example, there would be a single so-called *elementary rule* that takes care of the hand-over of a single parameter, and a sub-rule that selects the caller Frame node and creates the called Frame node; the covering condition would be *local-all*, which gathers as many copies of the elementary rule as there are suitable combinations of Var- and Value-nodes attached to the selected Frame-node.

Elegant and natural though this solution is, it does not yet meet all demands. For instance, the data flow analysis rule proposed above cannot be captured by a simple sub-rule embedding, even in the presence of (standard) negative application conditions: the result would be that the rule as a whole is always enabled, with copies of the elementary rule precisely for those parameters for which an argument value is already available. More generally, the problem is to enforce, in a covering condition, that elementary rules are actually enabled for all sub-structures (in the host graph) of a particular kind, i.e., satisfying a certain application condition. In other words, there are further gains to be made in the appropriate combination of covering conditions and application conditions.

Recently there have been proposals, by ourselves in [16] and independently by Habel and Pennemann in [9], to extend the power of application conditions, by generalising the two-level structures originally introduced in [8] to trees of arbitrary nesting depth. As shown in [16], every further level of nesting effectively corresponds to an additional level of quantification in terms of logic.

The core contribution of this paper is to recognise that the two principles of subrule embedding on the one hand and condition nesting on the other can be fruitfully combined, giving rise to a notion of *quantified graph transformation rule* that is both natural and powerful, and complements the framework for rule amalgamation so as to solve the problem outlined above.

Summarising, in this paper we combine two pre-existing ideas:

Rule amalgamation, developed by Taentzer [18] and later applied for, e.g., refactoring [1], parallel graph transformation [3] and multi-formalism simulation [2].
Nested graph predicates, recently proposed by Rensink [16] and independently by Habel and Pennemann [9].

In terms of these techniques, a brief explanation of our proposal is that we merge the nesting structure of the nested graph predicates with the sub-rule embeddings of interaction schemes, as used in rule amalgamation, so that the left hand sides of the rules are part of a nested graph predicate that simultaneously acts as an application condition. Another way to put it (slightly more loosely) is to say that we present a way to use nested graph predicates as a language for covering conditions, in the sense of Taentzer [18].

The paper is structured as follows: Sect. 2 provides the necessary technical concepts, Sect. 3 illustrates their use on the basis of a number of examples, and Sect. 4 concludes the paper.

2 Definitions

We first recall the notion of graph predicates from [16] and in passing establish the connection to [9]. We assume some category of graphs Graph with an initial element $\emptyset$, objects G, H and morphisms f, g etc. For concrete examples, we will take the common edge-labelled graphs $\langle N, E, L, s, t, \ell \rangle$, with N as set of nodes, E a set of edges, L a set of labels, and $s, t: E \to N$ and $\ell: E \to L$ the source and target mapping and labelling function, respectively; morphisms will be the homomorphisms over this structure.

We characterise graph predicates as rooted diagrams in the category Graph. Given such a diagram d and an object G in d, $root_d$ will denote the object at the root of d, and $out_d(G)$ will denote the morphisms in d that originate in G. (Note that a rooted diagram has a well-defined root even if it has no arrows.) Furthermore, $sub_d(G)$ will denote the reachable sub-diagram rooted in G, and for arbitrary $f: G \to root_d$, $c = d \circ f$ is the diagram with $root_c = G$, $init_c = \{g \circ f \mid g \in init_d\}$ and $sub_c(H) = sub_d(H)$ for all graphs H in d except for $root_d$. (In other words, $d \circ f$ is obtained from d by using f's source as the new root, concatenating f with the initial arrows of d and leaving the remainder of d unchanged.

Definition 1 (graph predicate). *Let* $G \in$ Graph *be arbitrary. A graph predicate over* G *is a tree-shaped diagram* p *in* Graph *rooted in* G. p *is called* ground *if* $G = \emptyset$.

Predicate satisfaction *is a binary relation* $\models$ *between predicates and graph morphisms:* $p \models f$ *expresses that* f *satisfies* p. $\models$ *is defined as the* smallest *relation such that* $p \models f$ *whenever the following conditions hold:*

- *$root_p = H$;*
- *There are $g: G \to K \in init_p$ and $h: K \to H$ such that $f = h \circ g$ and $sub_p(K) \not\models h$.*

Moreover, if we say that a graph G satisfies p, denoted $p \models G$, if $p \models f: \emptyset \to G$. (Note that this implies that p is ground.)

It might take some getting used to that the subjects are morphisms f rather than graphs. The intuition is that the source of f, which corresponds to the root of the predicate diagram, only identifies the common context or pattern; typically a subgraph which is already known to be present. The predicate itself actually states something about the *target* graph of the morphism. This intuition is confirmed by the special case where the source of f is the empty graph: in that case f really contains only the information in its target.

Satisfaction as defined in Def. 1 is slightly tricky in that it seems to rely on a smallest fixpoint construction for a function that is not monotonic, but rather anti-monotonic, due to the negation in the second bullet above. However, this is only superficially true, since in the sub-clause the satisfaction predicate is applied to a strictly smaller diagram; hence we can conduct proofs on the depth of the predicate diagram. (Actually, the above definition and most of the developments of this paper would still work for dags, and even for diagrams with cycles, as long as for any arrow in the diagram, the length of every path from the root to that arrow has the same parity, i.e., either all paths have odd length or all paths have even length. This implies that, in particular, all cycles must have even length. However, we restrict to tree-shaped diagrams in this paper.)

Alternatively, and perhaps more understandably, predicate satisfaction can be formulated in terms of two distinct satisfaction relations, $\models_\exists$ and $\models_\forall$, as follows:

Definition 2. Existential *and* universal satisfaction *are the smallest pair of binary relations* $\models_\exists$ *and* $\models_\forall$ *between graph predicates p and graph morphisms $f: G \to H$ such that*

- *$p \models_\exists f$ whenever the following conditions hold:*
  - *$root_p = H$;*
  - *There are $g: G \to K \in init_p$ and $h: K \to H$ such that $f = h \circ g$ and $sub_p(K) \models_\forall h$.*
- *$p \models_\forall f$ whenever the following conditions hold:*
  - *$root_p = H$;*
  - *For all $g: G \to K \in init_p$ and all $h: K \to H$ such that $f = h \circ g$, $sub_p(K) \models_\exists h$.*

Note that the concept of "smallest pair of binary relations" is indeed well-defined, under a pairwise ordering on sets. Also note that the problem due to the negation in the inductive definition of Def. 1 is no longer present in this formulation; instead, we can mark every arrow of a predicate diagram as $\exists$ or $\forall$, depending on their distance from

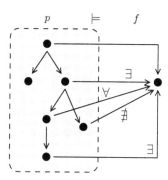

Fig. 2. Illustration of a satisfaction relation

the root, to indicate if the arrow should be satisfied existentially or universally. This is illustrated in Fig. 2, which shows a predicate diagram p and a morphism f, and the necessary morphisms from the objects of p to the target of f that establish satisfaction.

Obviously we can take this one step further by making the existentiality or universality a property of the diagram itself rather than of the satisfaction relation. In this view, graph predicates themselves are either always existentially satisfiable — roughly corresponding to *application conditions* in [9], though those may also be negative — or always universally satisfiable — roughly corresponding to *constraints*. Further variations, where individual objects or arrows of a predicate diagram are marked existential or universal, are also possible; when it comes to usability, rather than extensions to the theory, these are certainly worth investigating.

The following proposition states that the two definitions of satisfaction given above are indeed interchangeable.

Proposition 1. *For an arbitrary graph predicate p and graph morphism f, $p \models_\exists f$ iff $p \models f$, whereas $p \models_\forall f$ iff $p \not\models f$.*

So far we have presented satisfaction as a *relation* between predicate and subject; in some cases, we need a concrete *proof* of satisfaction. This is defined as follows.

Definition 3. *Let p be a graph predicate over G and let $f: G \to H$ be a graph morphism.*

- *A proof of existential satisfaction $\Phi : p \models_\exists f$ is a triple $\langle g: G \to K, h: K \to H, \Psi \rangle$, where $g \in init_p$, $f = h \circ g$ and $\Psi : sub_p(K) \models_\forall h$ is a proof of universal satisfaction.*
- *A proof of universal satisfaction $\Psi : p \models_\forall f$ is a partial function such that for all decompositions $f = h \circ g$ with $g: G \to K \in init_p$ and $h: K \to H$, the image $\Psi(g)(h) : sub_p(K) \models_\exists h$ is a proof of existential satisfaction.*

For instance, in Fig. 2, the proof of satisfaction consists (loosely speaking) of the (horizontal) arrows on the existential levels, and the mapping from arrows to the proof of the sub-diagram on the universal levels.

Where satisfaction establishes a relation between predicates and subjects (i.e., morphisms), we can also relate predicates among each other. This is formalised in the following notion of *implication*.

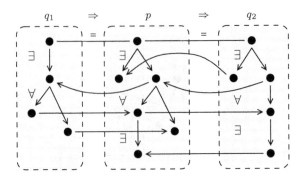

Fig. 3. Two illustrative implications

Definition 4. *Let p, q be graph predicates over G. A proof of implication $\phi : p \Rightarrow q$ is a mapping that associates to every initial arrow $f \colon G \to H \in init_p$ of p a triple $\langle i_f \colon G \to K, e_f \colon K \to H, \psi_f \rangle$, where*

- *$i_f \in init_q$ is an initial arrow of q;*
- *$f = e_f \circ i_f$, i.e., f is decomposed into i_f followed by e_f;*
- *$\psi_f : sub_q(K) \Rightarrow sub_p(H) \circ e_f$ is a proof of implication.*

This is a recursive definition; however, since graph predicates have finite depth there is no ambiguity in the interpretation. Proofs of implication compose and give rise to a category **Pred** with graph predicates as objects and proofs of implications as morphisms. An illustration is given in Fig. 3: this shows implications into and from the predicate p of Fig. 2.

For any predicate diagram p, since it is tree-shaped, every object G has a well-defined notion of *distance* from the root, which we will denote $dist_p(G)$. We call p *non-disjunctive* if $|out_p(G)| \leq 1$ whenever $dist_p(G)$ is even; in other words, when the tree does not branch at existential levels. For instance, q_1 in Fig. 3 is non-disjunctive but the other two predicates are not. (In terms of Def. 2, non-disjunctivity means that when $dist_p(G)$ is even, either $out_p(G)$ is empty, in which case it cannot give rise to a factorisation of the subject and the sub-diagram rooted at G is essentially equivalent to false, or the subject can be factored through the unique morphism in $out_p(G)$e. If, on the other hand, $|out_p(G)| > 1$ then there is a *choice* of morphisms starting in G, which acts as a disjunction.)

If q is non-disjunctive, then a proof $\phi : p \Rightarrow q$ closely follows the structure of q and essentially selects, for every morphism f in the diagram of q, at most one morphism e_f incident with the target of f: $tgt(e_f) = tgt(f)$ if f is existential in q (i.e., $dist_q(src(f))$ is even) or $src(e_f) = tgt(f)$ otherwise. The other end of e_f, i.e., the end not incident with f, is an object in p; in this way ϕ establishes a mapping from a prefix of the tree structure of q into the tree structure of p.

A proof of implication $\phi : p \Rightarrow q$ can be used to modify proofs of existential and universal satisfaction, in the following way: for arbitrary $\Phi : p \models_\exists f$ and $\Psi : q \models_\forall f$

$$\phi(\Phi) = \langle i_g, h \circ e_g, \psi_g(\Psi') \rangle \qquad \text{if } \Phi = \langle g, h, \Psi' \rangle$$
$$\phi(\Psi) = (g_1, h_1) \mapsto \psi_g(\Psi(i_{g_1}, h_1 \circ e_{g_1})) \quad \text{for } g_1 \in init_p \text{ and } h_1 \text{ with } f = h_1 \circ g_1.$$

This gives rise to the following relation between satisfaction and implication proofs:

Proposition 2. *Let $\phi : p \Rightarrow q$, $\Phi : p \models_\exists f$ and $\Psi : q \models_\forall f$.*

- *$\phi(\Phi) : q \models_\exists f$ is a proof of existential satisfaction;*
- *$\phi(\Psi) : p \models_\forall f$ is a proof of universal satisfaction.*

Hence we have the following corollary, which is a property one would expect, essentially stating that implication is sound w.r.t. satisfaction:

Corollary 1. *If $p \Rightarrow q$ then $p \models_\exists f$ implies $q \models_\exists f$ and $q \models_\forall f$ implies $p \models_\forall f$.*

Note that we do not have the dual *completeness* property; i.e., if $p \models_\exists f$ implies $q \models_\exists f$ for arbitrary f then it does not follows that there exists a proof of implication $p \Rightarrow q$. However, for our purposes Def. 4 suffices.

We now come to the core definition of this paper, namely that of quantified transformation rules:

Definition 5 (quantified rule). *A quantified rule $\mathcal{R} = (p_L \Rightarrow p_I \Leftarrow p_R)$ is a cospan of implication proofs in the category* Pred, *where p_L, p_I, p_R are non-disjunctive predicates rooted in $\emptyset$.*

A quantified rule R gives rise to an interaction scheme in a sense slightly extended from [18]: namely, for every morphism k occurring existentially in p_I (i.e., such that $dist_{p_I}(src(k))$ is even), due to the fact that p_I is non-disjunctive, there is at most one span of morphisms $r_k = (L_k \leftarrow I_k \rightarrow R_k)$ with $I_k = tgt(k)$, where the two morphisms in the span are part of the proofs of implication in $\mathcal{R}$. Each such r_k is a transformation rule of a more ordinary kind. Moreover, $\mathcal{R}$ induces a tree structure on the r_k, where the branches of the tree correspond to sub-rule embeddings.

The *match* of a quantified rule $\mathcal{R}$ for a host graph G is a proof of (existential) satisfaction $\Phi : p_L \models G$ (recall that p_L is ground). Φ includes a (possibly empty) set of matches for each L_k, which (due to the constraints on Φ) overlap in the matches of sub-rules. By gluing together these individual matches, we obtain an amalgamated rule in the sense of [18], which transforms G, via some intermediate K, to a target graph H. It can be shown that (automatically) $p_I \models K$. We consider the transformation *valid* and denote $G \xrightarrow{\mathcal{R}, \Phi} H$ if, in addition $p_R \models H$. Together, this gives rise to a transformation according to the following schema:

$$(\mathcal{R}) \qquad \begin{array}{ccccc} L & \Longrightarrow & p_I & \Longleftarrow & p_R \\ \big\| \Phi & & \big\| & & \big\| \\ G & \longleftarrow & K & \longrightarrow & H \end{array}$$

Note that this is not a diagram in any category: the object and arrows on top form a cospan in Pred and those on the bottom form a span in Graph, whereas the vertical relations are proofs of satisfaction. We do have some form of commutativity because (due to Prop. 2) $p_I \models G$ and $p_I \models H$, but we currently do not see any universal characterisation arising out of the above diagram. In other words, quantified rules are a way to program amalgamated rules. We argue in Sect. 4 that there may be further interest in extending the notion theoretically.

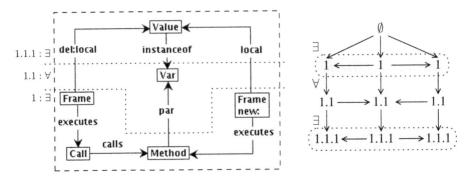

Fig. 4. Quantified parameter transfer rule

3 Examples

In this section we demonstrate the use of quantified rules by showing a number of illustrative examples.

Parameter transfer. First of all we come back to the example mentioned in the introduction (see Fig. 1). A rule that captures parameter transfer for an arbitrary number of parameters is given in Fig. 4. This involves three levels of quantification: the usual existentially quantified rule (corresponding to the sub-rule of an interaction scheme), universal quantification over all Var-nodes connected to the signature of the method in question, and existential quantification over the Value-node that the caller Frame has for this method. Thus, the application condition specifies that the rule is enabled if and only if there is a value for every parameter. The effect of the rule is then to create a single new Frame (since this is done by the sub-rule) and deleter, respectively create local-edges to the individual parameter Values.

The figure should be read as follows: the graphs and their hierarchical structure are identified by vectors of natural numbers, of the form $n_1.n_2\ldots$ — in this case, 1, 1.1 and 1.1.1. All elements appearing in a box (delineated by a dotted line) labelled n are considered to be present in all graphs with an identifier extending n. For instance, since the Method-node of Fig. 4 occurs in the box labelled 1, it is considered to be part of graphs 1, 1.1 and 1.1.1. The Var-node, on the other hand, only appears in 1.1 and 1.1.1. Arrows are considered to be part of the graph with the most deeply nested identifier that they cross — so, the arrows from the Frame-nodes to the Value-node only occur in graph 1.1.1.

On the right hand side of the figure the structure of the nested predicates is shown explicitly, with both the nesting morphisms (vertical) and the implication morphisms (horizontal). The shaded areas are the actual rules of the resulting interaction scheme — in this case, a single sub-rule and a single elementary rule.

Petri Net transitions. A second example we present is inspired by [2, 3]: the firing rule of a Petri net. We do this without giving the full definition of Petri nets or their encoding in terms of graphs; let it suffice that there are three types of nodes, encoding

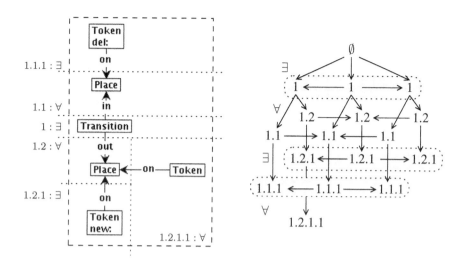

Fig. 5. Quantified condition/event net firing rule

(respectively) places, transitions and tokens, with edges expressing that a place is in the in- or out-set of a transition, and that a token is on a place.

To make the example more challenging, we will take a condition/event net, which carry as an additional restriction that a transition is disabled if one or more of the out-places already have a token. The corresponding rule is given in Fig. 5. In this case the sub-rule (1) does nothing; there are two elementary rules (1.1.1 and 1.2.1) to take care of the in-places and out-places, respectively. The fourth level, 1.2.1.1, encodes the negative condition on the out-places: it is universally quantified but has no further sub-conditions, which due to Def. 2 means that, if the graph matches (meaning that the corresponding out-place already has a token), the LHS predicate is always violated.

Note that, in this example, it is crucial that the token on the in-place is existentially quantified (on level 1.1.1) rather than universally (on level 1.1): otherwise the rule would not require that all in-places actually have a token, but rather remove all existing tokens from all in-places.

It is also noteworthy that the *only* difference with the firing rule for P/T nets is in the negative condition, 1.2.1.1. In particular, if we remove this condition and apply the rule to a transition with an in-place with more than one token, the (amalgamated) rule would have two different matches, each of which removes a single token from that in-place. In fact, this corresponds to the *individual token semantics* of Petri nets (cf. [5]).

Gossipping Girls. Finally, we present an example for which we have carried out a small experiment. It concerns a puzzle described (and solved analytically) in [12], which gives rise to state spaces so large that no model checking approach without good symmetry reduction can check cases of size greater than 6. In GROOVE (see [7]) we had so far been able to tackle cases up to size 8; by using universal quantification we achieve an improvement of an order of magnitude, so that we are now able to go one higher and check size 9.

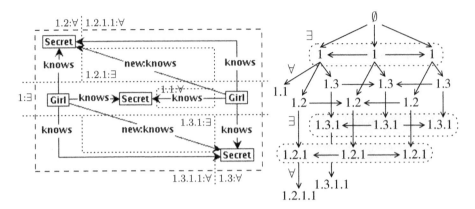

Fig. 6. Half of the quantified gossip rule

The puzzle is the following: given a number of girls, each of which has her own secret, and given a protocol whereby girls repeatedly call each other, at which point both girls divulge (to each other) all the secrets they know, what is the minimal number of calls after which all girls can know all secrets? For those interested in studying the puzzle without having the answer, we have put it in a footnote.[2] To simulate this problem in a model checker, it is necessary to do a breadth-first search where all states are generated; the number of those is roughly in the order of the number of partitionings of a set of size n, which grows super-exponentially with n. However, if the basic step, where one girl calls another and they exchange all secrets, cannot be modelled atomically, the simulation problem becomes much worse still. This is a clear case where universal quantification is required.

The problem can be modelled simply by having Girl and Secret nodes, and knows-edges linking each Girl to the Secrets she knows. However, due to the fact that our graphs can have parallel edges, we must take care explicitly that there is at most one edge between each given Girl and Secret: it would make no sense (and make the state space infinite) if we would allow girls to know secrets "more than once".

The rule required is shown in Fig. 6. Here the graph 1.1 is once more a negative application condition; apart from this, as in the condition/event rule there are two universally quantified parts, this time completely symmetric, which represent the two directions of communication.

Table 7 reports the results of the experiments. It clearly shows the gain achieved by using quantification in the gossip rule: the "plain" version of the problem involves a smaller-step protocol where each production just divulges a single secret, whereas in the "quantified" version this is done atomically, using the quantified rule above. The improvement is by an order of magnitude, both in the number of states and in the time needed for exploration, although the gain in time is less (relatively speaking) than the gain in number of states — which was to be expected since, naturally, the quantified rule itself is more complex to evaluate.

---

[2] The answer is $2n - 4$ where n is the number of girls.

Table 7. Comparison: plain versus quantified gossiping

# girls	plain # states	# sec	quantified # states	# sec
5	381	2	52	1
6	4,448	11	353	4
7	80,394	240	3,684	45
8	2,309,763	13,308	60,990	1,400
9	–	–	2,132,210	87,302

4   Conclusions

We briefly summarise the contribution of this paper and discuss related and future work.

In the context of existing work, the theory developed in this paper, and illustrated on the examples in Sect. 3, can be presented in different ways.

- It is a combination of the ideas of nested graph predicates (or generalised application conditions) and rule amalgamation;
- It provides a specification language for the covering conditions of rule amalgamation.

In any case, we think to have found a powerful and new combination of rule amalgamation and application conditions that is both theoretically justified and usable in practice. As discussed below, we intend to provide tool support in the (near) future. The experiment reported in Table 7 has shown that the ability to specify universal elements in rules may have a large impact on verification performance.

At the same time, the work reported here raises a new question of a theoretical nature. We have characterised quantified rules as cospans of morphisms in the category of (nested) graph predicates, but the application of the rules does not use this characterisation — instead it goes via the existing technique for rule amalgamation. Ideally, to fit into the algebraic approach, we should establish a category where rule application corresponds to a double pullback (rather than a double pushout because the direction of the arrows in the category of nested graph predicates is reversed with respect to graph morphisms). A problem here, as pointed out in Sect. 2, is that the cospans that make up the rules live in a different category than the objects we want to transform, i.e., the graphs, and rather than one type of morphism from which we can build a pullback diagram, we have three: implication morphisms (between predicates), satisfaction morphisms (from predicates to graphs) and graph morphisms. As far as we are concerned, this issue is entirely open.

Related work. We have already put this paper in the context of some important related work in the introduction. Here we restrict ourselves to mentioning some more places where the desire to formulate rules with more general or "programmable" matching conditions has been noted or addressed.

Outside the algebraic graph rewriting approach, Schürr in [17] presents a different type of formalisation for graph rewriting, which also offers the possibility to include "set-oriented pattern matching", as it is called there. The relation between the expressive power of the approach described there and the one in this paper is not easy to establish, due to the differences in presentation and the different thrust of the approaches: Schürr's work aims at programming on the basis of graph transformation; in contrast, our interest in modelling and verification imposes different requirements, where a powerful and uniform theoretical foundation is important even at the price of the heavier machinery of the algebraic approach.

More recently, Hoffmann in [11] proposes to use *variables* to capture part of the unknown structure of a host graph and so make rules more generic. In this way, too, a single finite structure can stand for a rule schema with a potentially unbounded number of instances, which in this case differ by the concrete instantiation of the variables. Variables are proposed for different purposes; one of them, called *clone variables*, allows to encode a limited form of universal matching. We conjecture that this corresponds to a special case of the technique proposed here. This proposal is worked out in a more application-oriented context in [15].

Also recently, Hausmann in his thesis [10] analyses the problem and proposes a limited solution based on so-called *universally quantified structures*, essentially consisting of specially marked nodes and their adjacent edges. These give rise to an interaction scheme in the sense of Taentzer [18]. As Hausmann observes, after listing a number of different scenarios in which universal quantification is useful, "up to date no graph transformation approach takes [all] these possibilities into account and provides distinguishing notations". We concur with that statement; where we believe that the theory presented here takes a step towards addressing the first part of this observation, the second is still open.

Future work. We intend to extend the tool set GROOVE [7] to universally quantified rules, along the lines described here. As described in Sect. 3, we have already carried out an experiment, using an *ad hoc* implementation of a single quantified rule, showing a speedup, for this particular (selected) case, of an order of magnitude.

For this purpose, as pointed in the above quote from [10], we need "distinguishing notations": the input to the GROOVE tools is (currently only) through a graph editor, so the notation needs to be visual and understandable. In our case this first of all involves developing a notation for nested graph predicates. The initial proposal of [16] is not usable in practice. An alternative is visually separating the (elements residing on) different levels of a nested graph condition, as in of Figs. 4–6, but this certainly also has its limitations.

Acknowledgement. Many thanks to Theo Ruys for suggesting the Gossipping Girls example, which has been a source of inspiration.

References

[1] P. Bottoni, F. Parisi-Presicce, and G. Taentzer. Specifying integrated refactoring with distributed graph transformations. In J. L. Pfaltz, M. Nagl, and B. Böhlen, editors, *Applications of Graph Transformations with Industrial Relevance (AGTIVE)*, volume 3062 of *LNCS*, pages 220–235. Springer, 2004.

[2] J. de Lara, C. Ermel, G. Taentzer, and K. Ehrig. Parallel graph transformation for model simulation applied to timed transition petri nets. In *Graph Transformation and Visual Modelling Techniques (GTVMT)*, volume 109 of *ENTCS*, pages 17–29, 2004.

[3] C. Ermel, G. Taentzer, and R. Bardohl. Simulating algebraic high-level nets by parallel attributed graph transformation. In Kreowski et al. [14], pages 64–83.

[4] FuJaBa tool suite: From UML to Java and back again. URL: www.cs.uni-paderborn.de/cs/fujaba/downloads, 2005.

[5] R. V. Glabbeek and G. Plotkin. Configuration structures. In *Tenth Annual Symposium on Logic in Computer Science*, pages 199–209. IEEE Computer Society, 1995.

[6] The graph rewrite and transformation (GReAT) tool suite. URL: http://escher.isis.vanderbilt.edu/tools/get_tool?GReAT, 2006.

[7] Graphs for object-oriented verification (GROOVE). URL: groove.sf.net, 2006.

[8] A. Habel, R. Heckel, and G. Taentzer. Graph grammars with negative application conditions. *Fundamenta Informaticae*, 26(3/4):287–313, 1996.

[9] A. Habel and K.-H. Pennemann. Nested constraints and application conditions for high-level structures. In Kreowski et al. [14], pages 293–308.

[10] J. H. Hausmann. *DMM: Dynamic Meta Modelling — A Semantics Description Language for Visual Modeling Languages*. PhD thesis, University of Paderborn, 2006.

[11] B. Hoffmann. Graph transformation with variables. In Kreowski et al. [14], pages 101–115.

[12] C. A. J. Hurkens. Spreading gossip efficiently. *Nieuw Archief voor Wiskunde*, 1(2):208, 2000.

[13] H. Kastenberg, A. Kleppe, and A. Rensink. Defining object-oriented execution semantics using graph transformations. In R. Gorrieri and H. Wehrheim, editors, *Formal Methods for Open Object-Based Distributed Systems (FMOODS)*, volume 4037 of *Lecture Notes in Computer Science*, pages 186–201. Springer-Verlag, 2006.

[14] H.-J. Kreowski, U. Montanari, F. Orejas, G. Rozenberg, and G. Taentzer, editors. *Formal Methods in Software and Systems Modeling — Essays Dedicated to Hartmut Ehrig*, volume 3393 of *LNCS*. Springer, 2005.

[15] M. Minas and B. Hoffmann. An example of cloning graph transformation rules for programming. In R. Bruni and D. Varró, editors, *Fifth International Workshop on Graph Transformation and Visual Modeling Techniques (GT-VMT)*, pages 235–244, 2006.

[16] A. Rensink. Representing first-order logic using graphs. In H. Ehrig, G. Engels, F. Parisi-Presicce, and G. Rozenberg, editors, *International Conference on Graph Transformations (ICGT)*, volume 3256 of *LNCS*, pages 319–335. Springer, 2004.

[17] A. Schürr. Logic based programmed structure rewriting systems. *Fundamenta Informaticae*, 26(3/4):363–385, 1996.

[18] G. Taentzer. *Parallel and Distributed Graph Transformation: Formal Description and Application to Communication-Based Systems*. PhD thesis, Technische Universität Berlin, 1996.

Idioms of Logical Modelling

Daniel Jackson

Computer Science and Artificial Intelligence Laboratory
Massachusetts Institute of Technology
Cambridge, Massachusetts, USA

Most modeling languages embody a particular idiom: the state/invariant/operations idiom for VDM and Z; the variable-update/temporal-logic idiom for SMV and Murphi; the imperative-programming idiom for Promela and Zing; and so on. Fixing an idiom makes tools easier to build, and helps novice modellers. But it also makes the language less flexible.

Alloy is a modelling language that was designed, in contrast, to support multiple idioms. Its core is a simple but expressive relational logic, whose semantics consists of a set of bindings of relations to global variables. In other words, an Alloy model is a constraint, and its meaning is a set of graphs with labelled edges of varying arity. The Alloy Analyzer is a constraint solver that can find a graph satisfying a given constraint.

A variety of idioms can be readily expressed in Alloy, and analyzed using the Alloy Analyzer. You can structure the state as a global state variable with multiple components, or follow an object-oriented style, where the state is collection of objects, each with components whose values vary of time. You can express and analyze individual operation executions, using the inductive approach of languages like Z, or introduce traces and check them against linear temporal logic properties. Frame conditions can be written as conventional equalities in each operation, or in the style invented by Reiter.

In my talk, I'll explain the basics of Alloy, show how to express a variety of idioms, and describe some case studies using these idioms, including most recently an analysis by Tahina Ramanandro of the Mondex electronic purse (developed by NatWest Bank, and originally modelled in Z by Susan Stepney, David Cooper and Jim Woodcock).

A. Corradini et al. (Eds.): ICGT 2006, LNCS 4178, p. 14, 2006.

New Algorithms and Applications of Cyclic Reference Counting

Rafael Dueire Lins

Departamento de Eletrônica e Sistemas
CTG, Universidade Federal de Pernambuco, Recife, PE, Brazil
rdl@ufpe.br

Abstract. Reference counting is a simple and efficient way of performing graph transformation and management in which each graph node stores the number of pointers to it. Graph operations are performed in such a way to keep this property invariant. The major drawback of standard reference counting is its inability to work with cyclic structures, which appear ever so often in real applications. The author of this talk developed a series of cyclic reference counting algorithms whose applicability goes far beyond the implementation of garbage collectors in programming languages. This paper presents the milestones in the history of cyclic reference counting followed by two new applications: the consistent management of Web pages in the Internet and the correctly handling of processes in clusters and grids.

Keywords: cyclic graphs, reference counting, webpage management, process management.

1 Background

Reference counting is a management technique for oriented graphs in which each node has a counter (RC) that stores the number of arcs or edges pointing at it. A node B is *connected* to a node A, $(A \to B)$, if and only if there is an oriented edge $<A, B>$, with source A and target B. A cell B is *transitively connected* to a cell A $(A^* \to B)$, if and only if there is a chain of oriented edges from A to B. Graph operations must keep the validity of the reference counter and are generally performed in small steps interleaved with computation. The graph has a *root* to which all nodes in use are transitively connected to. Unused nodes (RC=0) are in a *free-list*. Depending on the application, nodes are also known as *cells* or *objects* and oriented arcs or edges are called *pointers* or *references*. These terms are now on used as synonyms.

Graph transformation and management are generally described by three operations:

New(R) gets a cell U from the free-list and links it to the graph:

```
New (R) = select U from free-list
          make_pointer <R, U>
```

Copy(R, <S,T>) gets a cell R and a pointer <S, T> to create a pointer <R, T>, incrementing the counter of the target cell:

A. Corradini et al. (Eds.): ICGT 2006, LNCS 4178, pp. 15 – 29, 2006.

```
Copy(R, <S,T>) = make_pointer <R, T>
                 Increment RC(T)
```

Delete performs pointer removal:

```
Delete (R,S) = Remove <R,S>
               If (RC(S) == 1) then
                   for T in Sons(S) do Delete(S, T);
                   Link_to_free_list(S);
               else Decrement_RC(S);
```

A cell T belongs to the bag Sons(S) iff there is a pointer <S,T>.

Reference counting was developed by G.E.Collins [9] in the context of the implementation of programming languages to avoid the suspensions in LISP mark-scan algorithm [32], the first automatic dynamic memory management algorithm. It has several advantages over mark-scan garbage collection (see [16] for a survey of the field). The most important of them is that it is a local and non-suspending algorithm It is the memory management technique of most widespread use today [1].

The major drawback of standard reference counting is its inability to reclaim cyclic graphs, as reported by J.H.McBeth in [33]. Figure 01 presents the deletion of the last pointer that links an island of objects to root in a graph, introducing a *space-leak* [16].

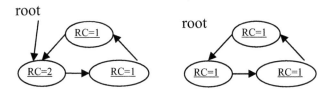

Fig. 1. Isolating a cycle from *root* causes a space-leak

In real applications cyclic structures appear very often. For instance, recursion is frequently represented by a cyclic graph and web pages have hyperlinks to other web pages that frequently point back to themselves. These are two examples that may give an account of the importance of being able to handle cycles in reference counting. Several researchers looked for solutions for this problem. Friedman and Wise [11] present an algorithm which can recover cyclic structures that are created in one operation, and never modified thereafter. This is the case for the cyclic representation of recursive functions in LISP and functional languages. Bobrow [6] gives an algorithm which can, in principle, recover all cyclic structures. His method relies on explicit information provided by the programmer. Bobrow collects nodes of the graph together to form groups and associates a reference count with a group rather than an individual data structure. Hughes' algorithm [15] is based on Bobrow's. It has the major advantage of not needing extra information provided by the programmer. Hughes' algorithm is suitable for the implementation of referentially transparent (pure) functional languages such as Haskell [14]. Another algorithm for cyclic reference counting was presented by Brownbridge [7], which as explained by Salkild in his thesis [36], was not correct. Corrections to Browbridge's algorithm were suggested independently by Salkild [36], Pepels et al. [35], and Thompson and Lins [39], yielding tremendously inefficient similar algorithms.

Reference [34] describes the first widely acknowledged general solution for cyclic reference counting. The Martinez-Wachenchauzer-Lins algorithm performs a local mark-scan whenever a pointer to a shared data structure is deleted. Lins largely improved the performance of the algorithm in two different ways. The first optimization [23] postpones the mark-scan, as much as possible. The benchmarking published in reference [23] showed that delaying the mark-scan by storing a reference to a shared deleted pointer in a control structure for later analysis largely increases the performance of the algorithm as a whole. This may even reduce the number of calls to the mark-scan to only collect cycles in the case of an empty free-list. The second optimization [24], presented in the Generational cyclic reference counting algorithm, relies on a creation-time stamp to help in cycle detection.

Another important optimization to cyclic reference counting   [25] was made possible by   introducing a data structure, called the Jump_stack, which stores a reference to the "critical points" in the graph while performing the local marking (after the deletion of a pointer to a shared cell). These nodes are revisited directly, saving a whole scanning phase in [23]. The work reported in reference [31] makes the use of the Jump_stack more efficient, as the constant of linearity of the algorithm in [23] was reduced from $3\Theta(n)$ to $2\Theta(n)$, where n is the size of the sub-graph below the deleted (shared) pointer. This gain in efficiency was obtained without the need for creation-time stamps and this lower complexity bound is obtained for all sub-graphs, being more general and lower cost in time and space than the generational solution [24].

The introduction of the Jump_stack and of the control structure has drastic impact in the performance of cyclic reference counting. A recent paper merges the Jump_stack and the control structure together in a single data structure called the Status_analyser. The introduction of the Status_analyser not only makes the description of the algorithm simpler and more uniform, but also brings gains in space-time performance in relation to the preceding algorithms.

The initial attempts to develop concurrent garbage collection architectures [38, 3] were based on mark-scan algorithms. Intel even implemented such algorithms in hardware, as described in reference [37]. The aforementioned cyclic reference counting algorithms served as the basis for several parallel ones suitable for multi-threaded uniprocessors and strongly couple architectures. Before those, Wise proposed an on-board reference count architecture [41] and Kakuta-Nakamura-Iida [17] presented a complex architecture based on reference counting, both unable to deal with cyclic data structures. The first general concurrent architecture for reference counting was presented by Lins in [21], which worked with two processors: one in charge of graph rewritings, called *mutator*, and another dedicated to garbage collection, the *collector*. Reference [22] generalized the previous architecture *à la* Lamport [18] allowing any number of mutators and collectors to work concurrently. Recent work, developed at IBM T.J.Watson Research Center, aimed at the efficient implementation of concurrent garbage collection [1] in the context of the Jalapeño Java virtual machine [2]. That architecture is based on Lins' concurrent strongly-coupled algorithms [21, 22], which on their turn are based on the cyclic reference counting algorithms presented in [23, 34]. Reference [28] introduces a new concurrent algorithm for cyclic reference counting which parallelizes the sequential algorithm presented in reference [27]. The architecture introduced is generalised to work with any number of mutators and collectors. Besides that, the collaboration between collectors is increased removing a criticism made by Blelloch and Cheng [5] to Lins' previous concurrent architectures.

2 Distributed Reference Counting

Distributed reference counting is a simple extension to uniprocessor reference counting. On a loosely coupled system the creation of a new reference to an object requires that a message be sent to it in order to increment its reference count. Likewise, if a remote reference is discarded then a decrement message must be sent. Special care must be taken to avoid an object being reclaimed while references to it still exist. This may happen if messages which refer to a certain object arrive in a different order than expected, for instance, if the message deleting the last reference to an object arrives at it before a copying message. The discard message will set the object's count to zero allowing it to be recycled. A solution to these problems appears in [19], in which a communication protocol provides a correct distributed reference count scheme at the cost of three messages per inter-processor reference.

Weighted Reference Counting [4, 40] makes reference counting suitable for use in loosely-coupled multiprocessor architectures. It has low communication overhead, namely one message per inter-processor reference and extra space associated with each reference. It is not able to cope with cyclic data structures, however. Reference [30] extends Weighted Reference Counting with general algorithms for cyclic reference counting for uniprocessors described in references [23]. A new distributed concurrent algorithm for cyclic weighted reference counting is presented in reference [26]. In what follows we detail the features of weighted reference counting and one of its cyclic version taken as the basis for the new applications presented further on in his paper.

2.1 Weighted Reference Counting

Here, the original weighted reference counting algorithm as presented in references [4, 38] is presented. Each *object* or *cell* has a count in which its weight is stored. An object has *fields* or *slots* which store pointers. These pointers reference objects. A *weight* is associated with each pointer. The reference count field of a cell contains the total weight of all pointers that refer to it. A pointer from a cell R to a cell S is denoted by <R,S>, its weight by Weight(<R,S>) (or W(<R,S>), for short) and the reference count of cell S by RC(S). For all cells N, X the following invariant is maintained:

$$RC(N) = \sum_{N} Weight(<X,N>)$$

For simplicity, it is assumed that the graph formed by the objects in use has a starting point, which is called *root*. All cells in use (active or non-garbage cells) are transitively connected to root. For similar reasons all cells not in use are assumed to be linked together forming a free-*list*. Any cell that is not reachable either from the free-list or by tracing pointers from root is garbage.

The algorithm is described as in standard reference counting, in terms of three primitive operations on the graph:

New(R) gets a cell *U* from the *free-list* and creates the pointer <R,U>, where *R* is a cell transitively connected to *root*. The reference count of *U* and the weight of the pointer <R,U> are both equal to the maximum weight, say *w*. This can be expressed as:

```
New (R) = if free_list not empty then
                select U from free_list
                set RC (U) := w
                make pointer <R,U>
                set W(<R,U>) := w
                    else write_out "No cells available"
```

Copy (R,<S,U>) creates the pointer <R,U>, where R and S are cells transitively connected to *root* and the pointer <S,U> exists. The weight of each pointer <R,U> and <S,U> is equal to half of the original weight of <S,U>. No communication with T takes place.

Algorithmically, one has:

```
Copy (R, <S,U>) = make pointer <R,U>
                    set W(<R,U>) := W(<S,U>)/2
                    set W(<S,U>) := W(<R,U>)
```

Delete (<R,S>) removes the pointer <R,S> from the graph and re-adjusts the graph. Only now does inter-processor communication take place. Object R will send S the weight of the deleted pointer. This weight is subtracted from the reference count of S. If its reference count is zero then S is free and its sons can also be reclaimed by recursive calls to Delete. A cell T belongs to the bag *Sons(S)* if and only if there is a pointer <S,T>.

```
Delete (<R,S>) = send Message_Delete(<R,S>) to S
    (in processor R)
                    remove <R,S>

Handle_Delete(<R,S>) = set RC(S) := RC(S) - W(<R,S>)
    (in processor S)
                        if RC(S) = 0  then
                            for T in Sons (S) do
                                    Delete (<S,T>)
                            link S to free_list
```

From this description it can be seen that pointer weights are powers of 2. This permits a practical technique for implementation: each pointer stores the *logarithm* of its weight. Indirection cells are used when copying pointers of weight one: to execute Copy(R<S,T>) when Weight(<S,T>)=1, an indirection cell U is created (in the same processing element as S so that no communication is necessary). The indirection cell simply contains a pointer to the target T - the pointer's weight is one so need not be stored. R and S are both set to refer to the indirection cell, each pointer weight W/2. Notice that the reference count of T need not be changed --- no communication is necessary.

2.2 Cyclic Weighted Reference Counting

The algorithm presented in reference [30] merges together weighted reference counting and the cyclic reference counting algorithm presented in [23]. In what follows a new optimized version of the algorithm is presented based on the sequential algorithm presented in [27]. In addition to the weight of a cell, two extra fields are needed. The first field holds the color of the cell. Two colors are used: green and red. The second field is a secondary weight count. The usual initial condition that every cell except root is on the *free-list* is assumed. For simplicity, all operations are atomic, i.e. once an operation is started a processor does not perform any other operation until its conclusion.

New(R) behaves as in Weighted Reference counting, but also sets the color of the new cell to *green*.

```
New (R) = if free-list not empty then
              select U from free_list set RC (U) := w
              set colour (U) := green
              make pointer <R,U>
              set W(<R,U>) := w
          else write_out "No cells available"
```

Copy (R,<S,T>) creates the pointer <R,T>, where R and S are cells transitively connected to *root* and the pointer <S,T> exists. The weight of each pointer <R,T> and <S,T> is equal to half of the original weight of <S,T>. No communication with T takes place.

```
Copy (R, <S,T>) = make pointer <R,T>
                  set W(<R,T>) := W(<S,T>)/2
                  set W(<S,T>) := W(<R,T>)
```

As in the weighted reference counting algorithm Delete forces communication to take place. The general idea of the cyclic algorithm here is to perform a local mark-scan whenever a pointer to a shared structure is deleted. Again for simplicity, the condition that all processors suspend computation while the local mark-scan is performed is imposed. In fact, this condition can be relaxed substantially below. The algorithm works in two phases. First, the graph below the deleted pointer is marked; rearranging counts due to internal references and possible connecting points to root are stored in a data structure, called the *Jump_stack*. In phase two, the cells in the Jump_stack are visited directly and if external references are found the sub-graph below that point is remarked as ordinary cells (*green*), and have their counts reset. All other nodes are garbage cells, thus collected and returned to the free-list.

Delete (<R,S>) extends the Weighted Reference Counting algorithm by invoking a local mark-scan on the sub-graph if S has other references.

```
Delete (<R,S>) = send Message_Delete(<R,S>) to S      (in processor R)
                 remove <R,S>
Handle_Delete(<R,S>) = set RC(S) := RC(S) - W(<R,S>)  (in processor S)
                       if RC(S) = 0  then
                           for T in Sons (S) do
                               Delete (<S,T>)
                               link S to free_list
```

```
                        else broadcast Suspend
                            mark_red(S)
                            scan(S)
                            collect(S)
                            broadcast Continue
```

mark_red(S) is an auxiliary function which paints all the cells in the sub-graph below *S* as red. This indicates that these cells may be garbage. Each cell visited has its reference count decremented, leaving only weights which refer to pointers external to the sub-graph.

```
mark_red (S) = if colour (S) is green then        (in processor S)
                    set colour (S) := red
                    for T in Sons (S) do
                        send Message_mark_red(<S,T>) to T

Handle_mark_red (<S,T>) = set RC(T) := RC(T) - W(<S,T>)
    (in processor T)
                        if (RC(T)>0 && T not in Jump_stack) then
                            Jump_stack := T
                            mark_red (T)
```

Scan(S) makes processors verify whether the Jump_stack is empty. If so, the algorithm sends cells hanging from S to the free-list. If the Jump-stack is not empty then there are nodes in the graph to be analysed. If their reference count is greater than zero, there are external pointers linking the cell under observation to root and counts should be restored from that point on, by calling Scan_green(T).

```
scan (S) = If (Colour(S) == red && RC(S)>0) then
        (in processor S)
                    scan_green(T);
            else For T in Jump_stack do
                        send Message_scan(T)

Handle_scan(T) = scan(T)    (in processor T)
```

scan_green (S) paints green all the sub-graph below *S*.

```
scan_green (S) = set colour (S) := green        (in processor S)
                    for T in Sons (S) do
                        send Message_scan_green(<S,T>) to T

Handle_scan_green(<S,T>) = set RC(T) := RC(T) - W(<S,T>)
    (in processor T)
                        if colour (T) is not green then
                            scan_green (T)
```

collect(S) recovers all cells in the sub-graph below *S* (garbage) and links them to the *free-list*.

```
collect(S) = if colour (S) is red then        (in processor S)
                for T in Sons(S) do
                    send Message_collect(T) to T
                    set back_of_control_queue := S
```

3 Webpage Management

This section explains the scheme for the consistent management of web pages. The algorithm is an extension of the algorithm presented in [29] and is based on cyclic weighted reference counting as presented in the last section. Some assumptions are made to start with:

- Each webpage at creation has a count in which its weight is stored.
- Webpages may store hyperlinks, pointers to other webpages.
- Each hyperlink has also a weight associated with.
- A hyperlink from a webpage R to a webpage S is denoted by <R,S>, its weight by Weight(<R,S>) (or W(<R,S>), for short) and the reference count of webpage S by RC(S).
- A webpage's reference count field contains the total weight of all hyperlinks that refer to the webpage.
- For all webpages N, X the following invariant is maintained:

$$RC(N) = \sum_N Weight(<X,N>)$$

- For simplicity, it is assumed that the graph formed by the webpages in use has a starting point, which is called root.
- All webpages in use (active or non-garbage webpages) are transitively connected to root.
- Any webpage that is not reachable by tracing hyperlinks from root are garbage and may be deleted.

The algorithm is described in terms of three primitive operations on the graph:

New(R) creates a webpage U from makes the hyperlink <R,U>, where R is a webpage transitively connected to *root*. The weighted reference count of U and the weight of the hyperlink <R,U> are both equal to the maximum weight, say w. This can be expressed as:

New (R) = create webpage U
                set RC (U) := w
                make hyperlink <R,U>
                set W(<R,U>) := w

Copy (R,<S,T>) creates the hyperlink <R,T>, where R and S are webpages transitively connected to *root* and the hyperlink <S,T> exists. The weight of each hyperlink <R,T> and <S,T> is half of the original weight of <S,T>. No communication with T takes place.

Copy (R, <S,T>) = make hyperlink <R,T>
                      set W(<R,T>) := W(<S,T>)/2
                      set W(<S,T>) := W(<R,T>)

Delete (<R,S>) removes the hyperlink <R,S> from the graph and re-adjusts the graph. Only now does interhost communication take place. Webpages R will send S the weight of the deleted hyperlink. This weight is subtracted from the reference count of S. If its reference count is zero then S is free and its Sons can also be reclaimed by recursive calls to Delete. A webpage T belongs to the bag *Sons(S)* if and only if there is a hyperlink <S,T>.

```
Delete (<R,S>) = send Message_Delete(<R,S>) to S
                    (in the host that stores webpage R)
                    remove <R,S>
Handle_Delete(<R,S>) = set RC(S) := RC(S) - W(<R,S>)
                    (in the host that stores webpage S)
                          if RC(S) = 0  then
                              for T in Sons (S) do Delete (<S,T>)
                              remove webpage S
```

From this description it can be seen that hyperlink weights are powers of 2. This permits a practical technique for implementation: each hyperlink stores the *logarithm* of its weight. Indirection webpages are used when copying hyperlinks of weight one: to execute Copy(R<S,T>) when Weight(<S,T>)=1, an indirection webpage U is created (in the same processing element as S so that no communication is necessary). The indirection webpage simply contains a hyperlink to the target T - the hyperlink's weight is one so need not be stored. R and S are both set to refer to the indirection webpage, each hyperlink weight W/2. Notice that the reference count of T need not be changed -- - no communication is necessary. One can also observe that the weight of a page lying on a cycle never drops to zero, similarly to the situation depicted on Figure 01.

3.1 Allowing Self-references

The algorithm presented here adapts the distributed algorithm presented in reference [29] for managing self-references in web pages. It may be seen as a cyclic extension of the algorithm presented in the last section. In addition to the weight of a webpage, two extra fields are needed. The first field holds the color of the webpage. Two colors are used: green and red. The second field is a secondary weight count. The usual initial condition that every webpage except root is on the *free-list* is assumed. For simplicity, all operations are atomic, i.e. once an operation is started a host does not perform any other operation until its conclusion.

New(R) behaves as in Weighted Reference counting, but also sets the color of the new webpage to *green*.

```
New (R) = create webpage
             set RC (U) := w
             set color (U) := green
             make hyperlink  <R,U>
             set W(<R,U>) := w
```

Copy remains unchanged as in the original weighted reference counting algorithm:

```
Copy (R, <S,T>) = make hyperlink <R,T>
                      set W(<R,T>) := W(<S,T>)/2
                      set W(<S,T>) := W(<R,T>)
```

As in the weighted reference counting algorithm Delete forces communication to take place. The general idea of the cyclic algorithm here is to perform a local mark-scan whenever a pointer to a shared structure is deleted. Again for simplicity, the condition that all hosts suspend computation while the local mark-scan is performed is imposed. In fact, this condition can be relaxed substantially.

The algorithm works in two phases. First, the graph below the deleted pointer is marked; rearranging counts due to internal references and possible connecting points to root are stored in a data structure, called the *Jump_stack*. In phase two, the webpages in the Jump_stack are visited directly and if external references are found the sub-graph below that point is remarked as ordinary webpages (*green*), and have their counts reset. All other nodes are garbage webpages, thus removed.

Delete (<R,S>) extends the conservative scheme above by invoking a local mark-scan on the sub-graph if S has other references.

```
Delete (<R,S>) =
     send Message_Delete(<R,S>) to S          (in host R)
     remove hyperlink  <R,S>

Handle_Delete(<R,S>) =
     set RC(S) := RC(S) - W(<R,S>)             (in host S)
                     if RC(S) = 0  then
                         for T in Sons (S) do
                             Delete (<S,T>)
                                 remove webpage S
                         else broadcast Suspend
                             mark_red(S)
                             scan(S)
                             collect(S)
                             broadcast Continue
```

mark_red(S) is an auxiliary function which paints all the webpages in the sub-graph below S as red. This indicates that these webpages may be garbage. Each webpage visited has its reference count decremented, leaving only weights which refer to hyperlinks external to the sub-graph.

```
mark_red (S) =                                       (in host S)
                 if colour (S) is green or black then
                     set colour (S) := red
                     for T in Sons (S) do
                     send Message_mark_red(<S,T>) to host T

Handle_mark_red (<S,T>) =
                 set RC(T) := RC(T) - W(<S,T>)     (in host T)
                     if (RC(T)>0 && T not in Jump_stack) then
                             Jump_stack := T
                             mark_red (T)
```

Scan(S) makes hosts verify whether the Jump_stack is empty. If so, the algorithm disposes the web pages hanging from S. If the Jump-stack is not empty then there are nodes in the graph to be analyzed. If their reference count is greater than zero, there are external hyperlinks linking the webpage under observation to root and counts should be restored from that point on, by calling Scan_green(T).

```
    scan (S) =                                       (in host S)
                 If (Colour(S) == red && RC(S)>0)
               then  scan_green
                     else For T in Jump_stack do
                     send Message_scan(T)

    Handle_scan(T) = scan(T)                          (in host T)
```

scan_green (S) paints green all the sub-graph below *S*.

```
scan_green (S) = set colour (S) := green           (in host S)
                 for T in Sons (S) do
                     send Message_scan_green(<S,T>) to T

Handle_scan_green(<S,T>) =                          (in host T)
              set RC(T) := RC(T) - W(<S,T>)
              if colour (T) is not green then
              scan_green (T)
```

collect(S) disposes all garbage webpages in the sub-graph below S.

```
collect(S) = if colour (S) is red then             (in host S)
             for T in Sons(S) do
                 send Message_collect(T) to host  T
```

The algorithm presented is robust with respect to the loss of messages.

3 Distributed Speculative Process Management

Computer networks have opened a wide range of new possibilities in parallel computation. The recent technological advances have made possible to distribute tasks in parallel over processors spread in a network [8]. In general, if these processors are either a few meters to kilometers apart forming a LAN (Local Area Network) the parallel architecture is called a *cluster*. In the case of the processors being from several kilometers to even continents apart one has a WAN or internet. Exploiting parallelism in such network is called a *grid*. Both clusters and grids are part of the technological reality of today [12, 10].

Speculative parallelism is a control strategy often used in different architectures. Tasks are generated and start running and later on the evaluation process decide whether needed or not. Unnecessary tasks either terminated or not, are aborted and the resources are allocated into another task. The simplest example of speculative parallelism is provided by the parallel execution of an if-then-else statement where the conditional clause and both branches are sparkled simultaneously. Once finished the evaluation of the clause the process knows that only one of the results is needed, and the other should be automatically discarded. This section presents how the cyclic weighted algorithm may be used to control processes in a distributed environment such as a grid.

Hudak and Keller [13] were the first to mention the possibility of using a garbage collection mechanism for doing process management. They did not go any further than saying that cells in memory allocation play a similar role to a process, however. Inter-process and inter-processor communication should be kept as low as possible. Load balancing and process (or task) granularity are two other issues of paramount importance to improve the cluster or grid throughput. These are considered out of the scope of the process management.

The distributed process management algorithm proposed here is based on weighted reference counting. In general a processor (or process) may generate another (new) process. The algorithm links a weighted reference count to each process and also to each inter-process reference. Process creation of a new process U by a process P is done by:

```
New_process (P) = create process U
                  set RC (U) := w
                  make_reference <P,U>
                  set W(<P,U>) := w
```

In process management reference copying is an operation of seldom use. Instead of copy, *fork* or *process replication* is more often used. The simplest way to fork a process is to make the original process to work as an indirection node to two new processes exactly equal, thus:

```
Fork_process (P) = create process U (copy_of_P)
                   create process V (copy_of_P)
                   set RC (U) := w
                   set RC (V) := w
                   make_reference <P,U>
                   set W(<P,U>) := w/2
                   make_reference <P,V>
                   set W(<P,V>) := w/2
```

Processes are *killed* or *deleted* only whenever the information they are processing is no longer needed. In reality, the global parent process (root) "looses interest" in the information the process (or processor) is evaluating and that means that the reference is discarded.

```
Delete (<R,S>) = process R sends Message_Delete(<R,S>) to S
                 Remove_reference <R,S>

Receive_Delete(<R,S>) = set RC(S) := RC(S) - W(<R,S>)
                        if RC(S) = 0  then
                           for T in Sons (S) do Delete (<S,T>)
                           kill_process_ S
```

In theory, dynamic process creation may form cycles through some form of general recursion. No real application so far has made use of such complex process setting. The process management algorithm as presented above is unable to work with cycles. One may observe that the algorithm above work consistently with process migration in a simple way.

4 Conclusions

Whenever an application may be modeled by an oriented graph, reference counting is an elegant and efficient way of performing the management of operations. The problem of being unable to work with cyclic graphs has already been overcome.

Weighted reference counting is a simple way of working with reference counting in a distributed environment. Its cyclic version, opens a wide range of new applications, two of them are presented here: webpage and process management in distributed environments.

The webpage management algorithm presented in this paper consistently avoids the burden of following hyperlinks to non-existent web pages. Its space overhead is minimal: one count to store the number of references to a page and also a similar count in each hyperlink. The communication cost of the proposed scheme is only one inter-host message per each non-local hyperlink deleted. The algorithm is robust in relation to the order messages are dispatch and arrive. The algorithm imposes no suspension at any time and the operations performed to the connectivity of the graph involve only repositories that are transitively connected to the webpage under analysis. Thus, that is by far more efficient than stop-the-world alternatives that would suspend all changes to the connectivity of the World-Wide-Web to be able to globally check the status of web pages. The consistent management scheme proposed herein is completely orthogonal to web browsing, thus it is transparent to users.

Very often web pages form cycles as related pages point at each other. This severely complicates any scheme for distributed memory management and, in particular the management of web pages. Although, they do not form dangling references they case space leaks as the web pages are kept unnecessarily. The algorithm presented herein is able to recycle web pages that form cycles removing such burden.

Process management in distributed systems is an area of rising importance due to the dissemination of cluster and grid computing. A scheme for distributed process management based on weighted reference counting is proposed.

Acknowledgements

This work was sponsored by CNPq to whom the author is grateful.

References

1. D.F.Bacon and V.T.Rajan. Concurrent Cycle Collection in Reference Counted Systems, Proceedings of European Conference on Object-Oriented Programming, June, 2001, Springer Verlag, LNCS vol 2072.
2. D.F.Bacon, C.R.Attanasio, H.B.Lee, R.T.Rajan and S.Smith. Java without the Coffee Breaks: A Nonintrusive Multiprocessor Garbage Collector, Proceedings of the SIGPLAN Conference on Programming Language Design and Implementation, June, 2001 (SIGPLAN Not. 36,5).
3. M.Ben-Ari. Algorithms for on-the-fly garbage collection. ACM Transactions on Programming Languages and Systems, 6(3):333--344, July 1984.
4. D.I.Bevan. Distributed garbage collection using reference counting. *In Proceedings of PARLE'87*, LNCS 259:176--187, Springer Verlag.
5. G.Blelloch and P.Cheng, On Bonding Time and Space for Multiprocessor Garbage Collection. In Proc. of ACM SIGPLAN Conference on Programming Languages Design and Architecture, March 1999.
6. D.G. Bobrow, Managing reentrant structures using reference counts, *ACM Trans. On Prog. Languages and Systems* **2** (3) (1980).
7. D.R. Brownbridge, Cyclic reference counting for combinator machines, in J.P.Jouannaud (ed.), Record of 1985 Conf. on Func. Programming and Computer Architecture, LNCS, Springer Verlag, 1985.
8. R.Buyya (ed.). High Performance Cluster Computing: Architectures and Systems. Prentice Hall, 1999.
9. G.E. Collins, A method for overlapping and erasure of lists, Comm. of the ACM, 3(12):655—657, Dec.1960.
10. J.Dongarra, I.Foster, G.Fox, W.Gropp, K.Kennedy, L.Torczon, and A.White. Sourcebook of Parallel Computing .Morgan Kauffman Publishers, 2003.

11. D.P. Friedman and D.S. Wise, reference counting can manage the circular environments of mutual recursion, IPL 8(1) 41—45, 1979.
12. I.Foster and C.Kesselman. The Grid 2: Blueprint for a New Computing Infrastructure. Morgan Kauffman Publishers, 2004.
13. P. Hudak and R.M. Keller. Garbage collection and task deletion in distributed applicative processing systems. Proceedings of the Conference on Lisp and Functional Programming Languages, 168-178, ACM (Aug. 1982).
14. P.R. Hudak et al., Report on the programming language Haskell, ACM SIGPLAN Notices, 27(5), May 1992.
15. J.R.M.Hughes, Managing reduction graphs with reference counts, Departamental research report CSC/87/R2, University of Glasgow, 1987.
16. R.E. Jones and R.D. Lins, Garbage Collection Algorithms for Dynamic Memory Management, John Wiley & Sons, 1996. (Revised edition in 1999).
17. Kakuta, Nakamura and Iida, Information Proc. Letters, 23(1):33-37, 1986.L.Lamport. Garbage collection with multiple processes: an exercise in parallelism. IEEE Conference on Parallel Processing, p50--54. IEEE, 1976.
19. L-M.Lermen and D.Maurer. A protocol for distributed reference counting. *Proc.s of ACM Conference on Lisp and Functional Prog*ramming, 343--350.
20. D.Gries. An exercise in proving parallel programs correct. Communications of ACM, 20(12):921--930, December 1977.
21. R.D.Lins. A shared memory architecture for parallel cyclic reference counting, Microprocessing and microprogramming, 34:31—35, Sep. 1991.
22. R.D.Lins. A multi-processor shared memory architecture for parallel cyclic reference counting, Microprocessing and microprogramming, 35:563—568, Sep. 1992.
23. R.D.Lins. Cyclic Reference counting with lazy mark-scan, IPL 44(1992) 215—220, Dec. 1992.
24. R.D.Lins. Generational cyclic reference counting, IPL 46(1993) 19—20, 1993.
25. R.D.Lins. An Efficient Algorithm for Cyclic Reference Counting, Information Processing Letters, vol 83 (3), 145-150, North Holland, August 2002.
26. R.D.Lins. Efficient Cyclic Weighted Reference Counting. In Proceedings of SBAC-PAD'2002. IEEE Press, 2002.
27. R.D.Lins. Efficient Lazy Reference Counting. Journal Of Universal Computer Science, Springer Pub. Co. - Austria, v. 9, n. 8, p. 813-828, 2003.
28. R.D.Lins. A New Multi-Processor Architecture for Parallel Lazy Cyclic Reference Counting. In Proceedings of SBAC-PAD 2005. IEEE Press, 2005.
29. R.D.Lins and A.S.de França. A Scheme for the Consistent Management of Web Pages. In: IADIS - INTERNATIONAL CONFERENCE ON COMPUTER APPLICATIONS, 2005, Algarve. IADIS Press, 2005. v. 2, p. 425-428.
30. R.D.Lins and R.E.Jones, Cyclic weighted reference counting, in K. Boyanov (ed.), Intern. Workshop on Parallel and Distributed Processing, NH, May 1993.
31. R.D.Lins and J.A.Salzano Filho, Optimising the Jump_Stack, Proceedings of SBLP'2002, September 2002.
32. J.MacCarthy. Recursive functions of symbolic expressions and their computation by machine. CACM, April 1960.
33. J.H. McBeth, On the reference counter method, Comm. of the ACM, 6(9):575, Sep. 1963.
34. A.D. Martinez, R. Wachenchauzer and R.D. Lins, Cyclic reference counting with local mark-scan, IPL 34(1990) 31—35, North Holland, 1990.
35. E.J.H. Pepels, M.C.J.D. van Eekelen and M.J. Plasmeijer, A cyclic reference counting algorithm and its proof, Internal Rept. 88-10, U. of Nijmegen, 1988.

36. J.D. Salkild. Implementation and Analysis of two Reference Counting Algorithms. Master thesis, University College, London, 1987.
37. F.J.Pollack, G.W.Cox, D.W.Hammerstein, K.C.Kahn, K.K.Lai, and J.R.Rattner. Supporting Ada memory management in the iAPX-432, Proceedings of the Symposium on Architectural Support for Programming Languages and Operating Systems, pages 117--131. SIGPLAN Not. (ACM) 17,4, 1982.
38. G.L.Steele. Multiprocessing compactifying garbage collection. Communications of ACM, 18(09):495--508, September 1975.
39. S.J.Thompson and R.D.Lins. A correction to Brownbridge´s algorithm. 1987.
40. Watson, P. and Watson, I. 1987. An efficient garbage collection scheme for parallel computer architectures. *PARLE'87,* LNCS 259:432--443, 1987.
41. D.S.Wise. Design for a multiprocessing heap with on-board reference counting, In J.P. Jouannaud, editor, Functional Programming Languages and Computer Architecture, volume LNCS 201, pages 289--304. Springer-Verlag, 1985.

Sesqui-Pushout Rewriting[*]

Andrea Corradini[1], Tobias Heindel[1],
Frank Hermann[2], and Barbara König[3]

[1] Dipartimento di Informatica, Università di Pisa, Italy
[2] Institut für Softwaretechnik und Theoretische Informatik, TU Berlin, Germany
[3] Institut für Informatik und interaktive Systeme, Univ. Duisburg-Essen, Germany

Abstract. *Sesqui-pushout* (SqPO) *rewriting*—"sesqui" means "one and a half" in Latin—is a new algebraic approach to abstract rewriting in any category. SqPO rewriting is a deterministic and conservative extension of double-pushout (DPO) rewriting, which allows to model "deletion in unknown context", a typical feature of single-pushout (SPO) rewriting, as well as cloning.

After illustrating the expressiveness of the proposed approach through a case study modelling an access control system, we discuss sufficient conditions for the existence of final pullback complements and we analyze the relationship between SqPO and the classical DPO and SPO approaches.

1 Introduction

In the area of graph transformation the two main categorical approaches used to describe the effect of applying a rule to a graph are the double-pushout approach (DPO [7,2]) and the single-pushout approach (SPO [17,5]). Both approaches use concepts of category theory to obtain an elegant and compact description of graph rewriting, but they differ with respect to the kind of morphisms under consideration, the form of the rules, and the diagrams the rewriting steps are based on. The aim of this paper is to propose a new categorical approach to rewriting that combines the good properties of both approaches and improves them by allowing to model cloning of structures in a natural way.

In the DPO approach [7,2] a rule q is a span $q = L \xleftarrow{\alpha} K \xrightarrow{\beta} R$ of arrows in a category of graphs and *total* graph morphisms. Given an *occurrence* of q in a graph G, i.e., a *match morphism* $m: L \to G$ from the *left-hand side* L to G, to apply q to G one first deletes from G the part of the occurrence of L which is not present in the *interface* K, and then one adds to the resulting graph those parts of the *right-hand side* R which are not in the image of K.

This construction is described by a double-pushout diagram as in (1) which, given q and m, can be constructed if there exists a *pushout complement* of α and m, i.e., arrows $A \xleftarrow{\gamma} D \xleftarrow{i} K$ making the resulting square a pushout.

$$
\begin{array}{ccccc}
L & \xleftarrow{\alpha} & K & \xrightarrow{\beta} & R \\
{\scriptstyle m}\downarrow & \_ & {\scriptstyle i}\downarrow & \_ & \downarrow{\scriptstyle c} \\
A & \xleftarrow{\gamma} & D & \xrightarrow{\delta} & B
\end{array}
\qquad (1)
$$

[*] Research partially supported by EPSRC grant GR/T22049/01, DFG project SANDS, EC RTN 2-2001-00346 SEGRAVIS, EU IST-2004-16004 SENSORIA, and MIUR PRIN 2005015824 ART.

A. Corradini et al. (Eds.): ICGT 2006, LNCS 4178, pp. 30–45, 2006.

Since the pushout complement is not characterised by a universal property, the DPO approach is in general non-deterministic: given a rule and a match, there could be several (possibly non-isomorphic) resulting graphs. To guarantee determinism, one usually sticks to *left-linear* rules, i.e., α must be injective. In this case, it is known that a pushout complement of α and m exists if and only if m satisfies the so-called *dangling* and *identification condition* with respect to α.

In the SPO approach [17,5], instead, a rule is an arrow $q \colon L \rightharpoonup R$ in a category of graphs and *partial* graph morphisms. The application of the rule q to a match m is modelled by a single pushout in this category and thus, by the universal property of pushouts, SPO rewriting is instrinsically deterministic. It is well known that a *partial* map $q \colon L \rightharpoonup R$ can be represented, in a category with *total* maps as arrows, as a span $L \hookleftarrow dom(q) \xrightarrow{q} R$, where $dom(q)$ is the domain of definition of q. Given an SPO rule and a match, the result of SPO rewriting is isomorphic to the result of DPO rewriting using the corresponding span and the same match, provided that the pushout complement exists. Thus, as explained in [17,5], SPO rewriting on graphs (or similar structures) subsumes DPO rewriting: this fact is exploited in practice by the AGG system[1], which implements SPO rewriting but offers to developers both SPO and DPO.

Unlike DPO, SPO rewriting is possible even if the match does not satisfy the dangling or identification condition w.r.t. the rule. The dangling condition requires that if a node of G is to be deleted, then any arc incident to it is deleted as well. If this condition does not hold, both the node and all incident arcs are deleted in the graph resulting from SPO rewriting. Thus SPO rewriting allows to model *deletion in unknown context*, and this is recognized as a useful feature in several applications. The identification condition does not allow the match to identify in G an item to be preserved by the rule with one to be deleted. If this condition does not hold, the SPO construction deletes that item from G, and thus the morphism from R to the resulting graph is partial. Most often this feature (called "precedence of deletion over preservation") is ruled out by restricting the class of allowed matches.

The use of categorical machinery made possible, along the years, the generalization of basic definitions and main results of both the DPO and the SPO approach to a more abstract setting, where the structures on which rewriting is performed are objects of a generic category satisfying suitable properties. The characterization of such properties has been the main topic of the theory of High Level Replacement (HLR) systems [4,6], and recently the definition of adhesive categories [16] (and their variants) provided a more manageable definition of them for the DPO case. The generalization of the SPO approach to suitable categories of spans has been elaborated in [14,18].

Here we propose a new categorical approach to rewriting, called *sesqui-pushout* (SqPO) *rewriting*. Following the trend described in the last paragraph, the approach is presented abstractly in an arbitrary category. Rules are DPO-like spans of arrows, and a rewriting step is defined as for DPO rewriting, but the left pushout is replaced by a pullback satisfying a certain universal property: the

---

[1] See http://tfs.cs.tu-berlin.de/agg/

role of the pushout complement is played now by the so-called *final pullback complement* (PBC).

Since final pullback complements are unique up to isomorphism, SqPO rewriting is deterministic. For *left-linear* rules, the final pullback complement coincides with the pushout complement, if the latter exists, and in this sense SqPO rewriting subsumes DPO rewriting. When the pushout complement does not exist but the final pullback complement does, SqPO rewriting models faithfully deletion in unknown context, like the SPO: dangling edges are removed. Strictly speaking, however, sesqui-pushout rewriting does not subsume SPO rewriting fully. In fact, by construction, there is always a total morphism from the right-hand side of the rule to the result of a sesqui-pushout rewriting step; thus if the match identifies items to be deleted with items to be preserved then the final pullback complement does not exist and rewriting is not allowed.

Interestingly, the final pullback complement is unique (if it exists) even for rules which are not left-linear, unlike the pushout complement. In this case, the final pullback complement *is not* a pushout complement in general, but it models faithfully the effect of *cloning*, at least for some concrete structures where the details have been worked out.

Based on the above discussion, we can explain the name we chose for the proposed approach. "Sesqui" is the latin word for "one and a half" and suggests that our approach is placed halfway between the single-pushout and the double-pushout approach. In fact, metodologically, it is based on a construction similar to the DPO, but it captures essential features of the SPO as well.

After introducing the basic definitions and properties of SqPO rewriting in Sec. 2, we demonstrate the expressiveness of the approach in Sec. 3 by modelling the access control problem described in [11]. In Sec. 4 we show how to construct the final pullback complement in some concrete categories, and we discuss its existence in general. Sec. 5 is dedicated to the comparison of SqPO rewriting with DPO and SPO rewriting, and also presents a Local Church Rosser theorem for parallel independent direct derivations. A concluding section summarizes the results of the paper and discusses further topics of investigation.

2   Defining Sesqui-Pushout Rewriting

In this section we present only the fundamentals of sesqui-pushout rewriting: an example illustrating the expressiveness of this new approach is presented in the next section.

Let   be a category, about which we do not make any assumption, for the time being: objects and arrows belong to   if not specified otherwise. As in the DPO approach, a *rule* q is a span of arrows $q = L \xleftarrow{\alpha} K \xrightarrow{\beta} R$. Now the general idea of SqPO-rewriting is to replace the left square of a DPO rewriting diagram by a "most general" pullback complement with respect to the rule and the matching arrow, called a *final pullback complement* [18,3]. We will discuss in Section 4 how the universal property characterizing this construction is related to the right adjoint to the pullback functor.

After the general definition, we list some basic properties of final pullback complements and we present a simpler characterization which is applicable when the rule q is left-linear, i.e., α is mono.

Definition 1 (Sesqui-pushout rewriting). *Let $q = L \xleftarrow{\alpha} K \xrightarrow{\beta} R$ be a rule, and $m: L \to A$ be an arrow, called a* match. *Then we write $A \xRightarrow{\langle m, q \rangle} B$, and we say that there is a* direct derivation *from A to B (using m and q) if we can construct a diagram such as (2) where the following conditions hold:*

- *the right square is a pushout, and*
- *the arrows $A \xleftarrow{\gamma} D \xleftarrow{i} K$ form a final pull-back complement of $A \xleftarrow{m} L \xleftarrow{\alpha} K$, (this is indicated by the sign $\angle$ in Diagram (2))*

$$
\begin{array}{ccccc}
L & \xleftarrow{\alpha} & K & \xrightarrow{\beta} & R \\
{\scriptstyle m}\downarrow & {\scriptstyle \angle} & \downarrow{\scriptstyle i} & {\scriptstyle -} & \downarrow{\scriptstyle c} \\
A & \xleftarrow{\gamma} & D & \xrightarrow{\delta} & B
\end{array} \quad (2)
$$

where a final pullback complement *of $A \xleftarrow{m} L \xleftarrow{\alpha} K$ is defined to be a pair of arrows $A \xleftarrow{\gamma} D \xleftarrow{i} K$ such that*

1. *the square $K \xrightarrow{\alpha} L \xrightarrow{m} A \xleftarrow{\gamma} D \xleftarrow{i} K$ is a pullback, and*
2. *for each pullback $K' \xrightarrow{\alpha'} L \xrightarrow{m} A \xleftarrow{\gamma'} D' \xleftarrow{i'} K'$, and for each $f: K' \to K$ such that $\alpha \circ f = \alpha'$, there exists a unique $\widehat{f}: D' \to D$ such that $\gamma \circ \widehat{f} = \gamma'$ and $i \circ f = \widehat{f} \circ i'$ (see the right hand diagram).*

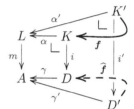

It immediately follows from the defining properties that the final pullback complement of any pair of composable arrows is unique up to isomorphism, if it exists. Additionally, if the rule is left-linear, i.e., $L \xleftarrow{\alpha} K$ is mono, then also γ is mono, and it can be characterized as the largest pullback complement, where *largest* is interpreted in the poset of subobjects of A. These facts are formalized in the following lemma.

Lemma 2 (Properties of final pullback complements)

In the square on the right let $A \xleftarrow{\gamma} D \xleftarrow{i} K$ be a final pullback complement. Then the following facts hold.

$$
\begin{array}{ccc}
L & \xleftarrow{\alpha} & K \\
{\scriptstyle m}\downarrow & {\scriptstyle \angle} & \downarrow{\scriptstyle i} \\
A & \xleftarrow{\gamma} & D
\end{array}
$$

1. *If $A \xleftarrow{\gamma'} D' \xleftarrow{i'} K$ is another final pullback complement of $A \xleftarrow{m} L \xleftarrow{\alpha} K$, then there is a unique isomorphism $\phi: D' \to D$ such that $\phi \circ i' = i$ and $\gamma \circ \phi = \gamma'$. Thus final pullback complements are unique up to iso.*

$$
\begin{array}{ccc}
L & \xleftarrow{\alpha} & K \\
{\scriptstyle m}\downarrow & {\scriptstyle -} & \downarrow{\scriptstyle i'} \quad {\scriptstyle i} \\
A & \xleftarrow{\gamma'} & D' \xdashrightarrow{\;=\;} D \\
& {\scriptstyle \gamma} &
\end{array}
$$

2. *If additionally α is monic then*
   (a) *the arrow γ is monic*
   (b) *the arrow γ can be characterized as the largest among the subobjects of A that provide a pullback complement of $A \xleftarrow{m} L \xleftarrowtail{\alpha} K$, i.e.*

   for every pullback complement $A \xleftarrowtail{\delta} E \xleftarrow{j} K$ there exists a unique arrow $\ell: E \to D$ such that $i = \ell \circ j$ and $\delta = \gamma \circ \ell$.

$$
\begin{array}{ccc}
L & \xleftarrowtail{\alpha} & K \\
{\scriptstyle m}\downarrow & {\scriptstyle -} & \downarrow{\scriptstyle j} \\
A & \xleftarrowtail{\delta} & E
\end{array}
\quad \Rightarrow \quad
\begin{array}{ccc}
L & \xleftarrowtail{\alpha} & K \\
{\scriptstyle m}\downarrow & & \downarrow{\scriptstyle i} \quad {\scriptstyle j} \\
A & \xleftarrowtail{\gamma} & D \xleftarrow{\ell} E \\
& {\scriptstyle \delta} &
\end{array}
$$

It is worth stressing that, by the first point of Lemma 2, uniqueness of final pullback complements holds in any category, even if α is not monic: this fact guarantees that the result of SqPO rewriting is determined (up to iso), also in situations where DPO rewriting is "ambiguous" because of the existence of several pushout complements.

3   Modelling the Access Control Problem

To show the expressive power of SqPO rewriting, we model the basic Access Control systems of [11]; for this we use *simple graphs*, i.e., graphs with at most one edge of each type between two nodes. This category has all pullbacks and pushouts, and hence a rule is applicable at a match if the relevant final pullback complement exists. Unlike pullbacks, pushouts are not computed componentwise on nodes and edges, because multiple parallel edges are not allowed. As discussed in [16], the category of simple graphs is *quasi-adhesive*; this implies, among other things, that pushout complements along *regular monos* are unique (if they exists).[2]

The Discrete Access Control model [11] considers a *protection system*, which controls the access of a set of *subjects* to a set of *objects*. Moreover suitable commands can change the state of the system. The corresponding decision problem consists in deciding whether a subject can obtain a certain right after applying a sequence of commands to a given initial state. Commands are sequences of *primitive operations* guarded by a Boolean condition: such operations model elementary changes of the system. Here we shall model the configurations of a system and the primitive operations. We also introduce a new operation called `clone`, which allows us to show a non-left-linear rule at work.

Definition 3 (Protection system). *A protection system $P = (R, C)$ consists of a finite set of* rights R *and a finite set of* commands C. *A configuration of a protection system is a triple $c = (S, O, A)$, where S is a set of* current subjects, O *is a set of* current objects *and A is an* access matrix $A[s, o] \subseteq R$, *with $s \in S, o \in O$.*

We model the configurations of a protection system as simple graphs, and the primitive operations as SqPO-rules. We depict subjects by shaded boxes $\square$, objects by rounded boxes $\bigcirc$, and if a subject possesses a right $i \in R$ to an object, we draw a labelled arc between them: $\square\!\!-i\!\!\rightarrow\!\!\bigcirc$. In the examples we use the common rights for "read" and "write" access, which are labelled by r and w, respectively.

The transformations of a configuration are defined by six primitive operations. They are `create subject` X_s and `create object` X_o for creating subjects and objects, `destroy subject` X_s and `destroy object` X_o for destroying subjects and objects, and finally `enter` i `into`(X_s, X_o) and `delete` i `from`(X_s, X_o) for

---

[2] A *regular* mono is an arrow which is an equalizer of some pair of parallel arrows. In the category of simple graphs, an injective morphism is regular if it reflects edges.

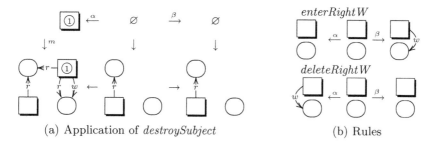

(a) Application of *destroySubject*   (b) Rules

Fig. 1. Basic rules for transforming an Access Control configuration

entering and deleting rights. Figure 1(a) shows the destruction of a subject by the application of the rule *destroySubject*, i.e., how the corresponding node and its incident edges in the graph are deleted. The morphisms are defined by mappings according to the numbers within the boxes. The left square is clearly a final pullback with the bottom right graph being its final pullback complement object. In fact, the square is a pullback, and it satisfies Condition *2(b)* of Lemma 2. Note that the effect of SqPO rewriting is similar to SPO, while DPO rewriting would not be applicable here.

Figure 1(b) depicts the rules which correspond to the operations of establishing and deleting the "write" access to a subject.

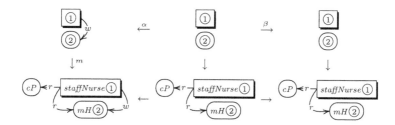

Fig. 2. Application of *deleteRightW*

Figure 2 shows the application of *deleteRightW* using SqPO rewriting: the left square is clearly a final pullback. Notice that in this case the DPO approach would be non-deterministic, as there are two non-isomorphic pushout complements for the given α and $L \xrightarrow{m} G$: the shown final pullback complement and G itself. Indeed, the category of simple graphs is quasi-adhesive, and uniqueness of pushout complements is guaranteed along *regular* monos only, i.e., morphisms reflecting edges: but α is not regular.

A new and interesting aspect of non-linear rules is used in Figure 3, where a subject is *cloned*. The final pullback complement construction automatically generates copies of the adjacent edges (cf. Construction 5). The rule *cloneSubject* is applied to a configuration representing a staff nurse, which has access to two objects, namely cP (representing contact information of patients) and mH (denoting medical history information of patients). When a hospital employs a new

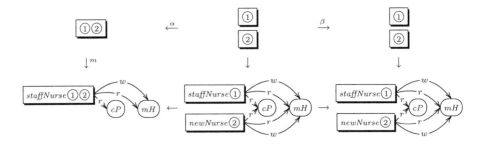

Fig. 3. Application of *cloneSubject*

nurse, the administrator might not want to define all rights separately again. In general, in systems with complex configurations, operations which model cloning are of great help.

The role based model for Access Control RBAC described in [8] and [20] is widely used, and it can be considered as an extension of the model in [11]. Graph transformation is used in [15] for defining and verifying an RBAC model: unlike in our approach, negative application conditions are needed there to avoid multiple edges.

4 Existence and Construction of Final Pullback Complements

In this section we first give, as it is usual for algebraic approaches to rewriting, a concrete set-based description of SqPO-rewriting steps in the category of graphs; since pushouts are treated as usual, we only provide a construction for final pullback complements. In the sequel we address the question under which conditions final pullback complements exist and how they may be constructed "abstractly" in categories where right adjoints to pullback functors exist (most of the categories used in practice are of this kind).

4.1 Constructing Final Pullback Complements in Graph

Since in practice one usually works with concrete objects, i.e., with objects that are representable by structured sets, it is useful to present a set-theoretical construction of final pullback complements in a sample category of this kind. We consider here directed (multi-)graphs, but the construction can be generalized easily to algebras over an arbitrary *graph structure*, i.e., a signature with unary operator symbols only [17]. We present explicit constructions of the final pullback complement in **Graph** for the case in which either the left morphism of the rule or the match are monic, i.e., injective. For left-linear rules, we also provide a necessary and sufficient condition for its existence.

Recall that a *graph* is a tuple $G = \langle \mathsf{V}_G, \mathsf{E}_G, \mathsf{src}_G \colon \mathsf{E}_G \to \mathsf{V}_G, \mathsf{tgt}_G \colon \mathsf{E}_G \to \mathsf{V}_G \rangle$ where V_G and E_G are disjoint sets, which are called *vertices* and *edges*, respectively; the latter are connected according to the *source* and *target* functions

src_G and tgt_G, respectively. A *graph morphism* $f: G \to H$ is a pair $\langle f_V: V_G \to V_H, f_E: E_G \to E_H \rangle$ such that $src_H \circ f_E = f_V \circ src_G$ and $tgt_H \circ f_E = f_V \circ tgt_G$. The category of graphs and graph morphisms is denoted by **Graph**.

Left-linear rules. Given a rule $q = L \xleftarrow{\alpha} K \xrightarrow{\beta} R$ with monic α (that we assume w.l.o.g. to be an inclusion) and a match $m: L \to A$, it is easy to show that there does not exist any pullback complement (and thus, *a fortiori*, any final PBC) of $A \xleftarrow{m} L \xleftarrow{\alpha} K$ if the match m is not *conflict-free* with respect to α.

Definition 4 (Conflict freeness). *A match $m: L \to A$ is* conflict-free *with respect to $L \xleftarrow{\supseteq} K$ if $m(L \setminus K) \cap m(K) = \quad$.*

For example, for the non-conflict-free match m shown to the right, any graph closing the square and making it commutative should contain at least one node (the image of ① under the vertical arrow), but in this case the resulting square would not be a pullback.

$$\{①, ②\} \xleftarrow{\quad} \{①\}$$
$$m \Big\downarrow {\scriptstyle 1 \mapsto 0 \atop 2 \mapsto 0} \qquad \Big\downarrow$$
$$\{⓪\} \xleftarrow{\quad} \ ?$$

It is worth observing that conflict-freeness is weaker than other conditions that are often imposed on matches in the framework of algebraic graph rewriting. For example, if the match m is monic, *d-injective* [17], or it satisfies the *identification condition* of DPO rewriting, then it is conflict-free.

Assuming conflict-freeness, the final pullback complement exists in **Graph**, and it can be described as follows as a subgraph of A (according to Condition *2(b)* of Lemma 2).

Construction 5 (Final PBC for left-linear rules in Graph). *Let be given a rule $L \xleftarrow{\supseteq} K \xrightarrow{\beta} R$ and a conflict-free match $m: L \to A$. Then the final PBC for $A \xleftarrow{m} L \xleftarrow{\supseteq} K$ is given by $A \xleftarrow{\supseteq} D \xleftarrow{m|_K} K$, where D is defined as*

$$V_D = V_A \setminus m(V_L \setminus V_K)$$
$$E_D = \{e \in E_A \setminus m(E_L \setminus E_K) \mid src_A(e) \in V_D \wedge tgt_A(e) \in V_D\}$$

It is evident from the construction that all the edges of A that are connected to deleted nodes are deleted as well, thus D is a well-defined graph; it is easily shown that it is indeed a pullback complement of the given arrows, and that no larger subgraph of A would be a PBC.

General rules, monic matches. If the left-hand side of the rule is not monic but the match is, the final pullback complement exists in **Graph** and it can be described as follows.

Construction 6 (Final PBC for monic matches in Graph). *Let be given a rule $L \xleftarrow{\alpha} K \xrightarrow{\beta} R$ and a monic match $m: L \rightarrowtail A$. Then a final pullback complement $A \xleftarrow{\gamma} D \xleftarrow{i} K$ can be constructed as follows.*

$$\mathsf{V}_D = \big(\mathsf{V}_A \setminus m(\mathsf{V}_L)\big) \cup \mathsf{V}_K, \qquad\qquad \gamma_{\mathsf{V}}(u) = \begin{cases} m(\alpha_{\mathsf{V}}(u)) & \textit{if } u \in \mathsf{V}_K \\ u & \textit{if } u \in \mathsf{V}_A \end{cases}$$

$$\mathsf{E}_D = \left\{ \begin{array}{c} \langle e, u, v\rangle \mid e \in \mathsf{E}_A \setminus m(\mathsf{E}_L) \wedge u, v \in \mathsf{V}_D \wedge \\ \mathsf{src}_A(e) = \gamma_{\mathsf{V}}(u) \wedge \mathsf{tgt}_A(e) = \gamma_{\mathsf{V}}(v) \end{array} \right\} \cup \mathsf{E}_K$$

$$\gamma_{\mathsf{E}}(e) = \begin{cases} e' & \textit{if } e = \langle e', u, v\rangle \\ m(\alpha_{\mathsf{E}}(e)) & \textit{otherwise} \end{cases}$$

$$\mathsf{src}_D(e) = \begin{cases} u & \textit{if } e = \langle e', u, v\rangle \\ \mathsf{src}_K(e) & \textit{otherwise} \end{cases} \qquad\qquad \mathsf{tgt}_D(e) = \begin{cases} v & \textit{if } e = \langle e', u, v\rangle \\ \mathsf{tgt}_K(e) & \textit{otherwise} \end{cases}$$

In words, the resulting graph D contains a copy of K, a copy of the largest subgraph of A which is not in the image of m, and a suitable number of copies of each arc of A incident to a node in $m(\alpha(K))$: this has the effect of "cloning" part of A. The proof that D is indeed a final PBC is omitted for space reasons.

Example 7 (Final PBC of a non-left-linear rule). According to Construction 6, the final pullback complement for $\bigcirc \xleftarrow{\ m\ } \cdot \xleftarrow{\ \alpha\ }$ in **Graph** is $\bigcirc\!\bigcirc\!\bigcirc$. Notice that there are four pushout complements of the given arrows: $\bigcirc$, $\smile$, $\frown$, and $\circlearrowright$; hence in this case the final PBC is not a pushout complement. Incidentally, it can be shown that also in the category of *simple* graphs $\bigcirc\!\bigcirc\!\bigcirc$ is a final PBC of the given arrows, but in this case it is a pushout complement as well.

Interestingly, note that one can derive from $\bigcirc$ a clique with n nodes by $n-1$ consecutive applications of the rule $\cdot \xleftarrow{\ \alpha\ } \cdot \xrightarrow{\ id\ } \cdot$.

General rules, general matches. In the case of non-left-linear rules and non-injective matches, the exact conditions for the existence of final pullback complements in **Graph** and the details of its construction are rather involved, and go beyond the scope of this paper; the interested reader is encouraged to use the constructions of [9] to specialize the results in Section 4.2 to the category of graphs. One of the main issues of this general case is that the final pullback construction cannot be performed componentwise on nodes and edges.

Recall that in the case of DPO-rewriting, restricting to monic matches actually enhances expressiveness [10], in the sense of modelling power. It is left as future work to check if a similar result holds for SqPO-rewriting as well.

4.2   Final Pullback Complements in Arbitrary Categories

In this section we provide sufficient conditions for the existence of final pullback complements in a category. We first need to introduce some categorical concepts. We assume a fixed category , to which all mentioned objects and arrows belong unless we say otherwise.

Definition 8 (Slice category and pullback functor). *Let A be an object. The* slice category over A, *denoted by* $\downarrow A$, *has all -arrows $(B \xrightarrow{\ \beta\ } A)$ with*

codomain A as objects, and given two objects $(B \xrightarrow{\beta} A)$ and $(C \xrightarrow{\gamma} A)$ of $\downarrow A$
each   -arrow $\boldsymbol{f}\colon B \to C$ satisfying the equality $\gamma \circ \boldsymbol{f} = \beta$ is an arrow $\boldsymbol{f}\colon \beta \to \gamma$
in $\downarrow A$.

A pullback functor along an arrow $m\colon L \to A$ is a
functor $m^{\quad}\colon\ \downarrow A \to\ \downarrow L$ which maps each object

$$\beta \in \mathsf{ob}(\ \downarrow A) \quad to \quad m^{\quad}(\beta) \in \mathsf{ob}(\ \downarrow L)$$

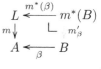

and provides an additional arrow $m'_\beta\colon m^{\quad}(B) \to B$
such that the right hand diagram is a pullback.
Further each arrow $\boldsymbol{f}\colon \beta \to \gamma$ in $\downarrow A$ is mapped to the
unique arrow $m^{\quad}(\boldsymbol{f})\colon m^{\quad}(\beta) \to m^{\quad}(\gamma)$ such that the
following is true, by the universal property of pullbacks:

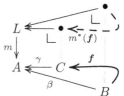

$$m^{\quad}(\gamma) \circ m^{\quad}(\boldsymbol{f}) = m^{\quad}(\beta) \land m'_\gamma \circ m^{\quad}(\boldsymbol{f}) = \boldsymbol{f} \circ m'_\beta.$$

Given a category   and an arrow m such that the pullback functor $m^{\quad}\colon\ \downarrow A \to$
$\downarrow L$ exists, the right adjoint to $m^{\quad}$, if it exists, is usually denoted by $\varPi_m\colon\ \downarrow L \to$
$\downarrow A$. Even if $\varPi_m$ does not exist, it might exist *partially* at an object $\alpha \in$
$\mathsf{ob}(\ \downarrow L)$. In this case $\varPi_m(\alpha)$ satisfies a univeral property which can be described
as follows.

Definition 9 (Right adjoints (partial))

Let $m\colon L \to A$ be an arrow, let $m^{\quad}\colon\ \downarrow A \to$
$\downarrow L$ be a pullback functor, and let $(K \xrightarrow{\alpha} L) \in$
$\mathsf{ob}(\ \downarrow L)$ be an object. Then the right adjoint
$\varPi_m\colon\ \downarrow L \to\ \downarrow A$ to $m^{\quad}$ exists partially at α if
there is an object $\varPi_m(\alpha) \in\ \downarrow A$ and an arrow
$\varepsilon_\alpha\colon m^{\quad}(\varPi_m(\alpha)) \to \alpha$ in $\downarrow L$ such that for ev-
ery $(D \xrightarrow{\delta} A) \in \mathsf{ob}(\ \downarrow A)$ and each $\boldsymbol{f}\colon m^{\quad}(\delta) \to$
α there exists a unique $\widehat{\boldsymbol{f}}\colon \delta \to \varPi_m(\alpha)$ such that
$\varepsilon_\alpha \circ m^{\quad}(\widehat{\boldsymbol{f}}) = \boldsymbol{f}$.

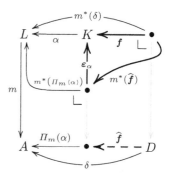

To illustrate these definitions we give an example, based on [9], where we talk
about the simpler right adjoint to the preimage functor in **Set**.

Example 10 (The adjunction $m^{\quad -1} \dashv \forall_m$) Consider a function $m\colon L \to A$ and the
pre-image functor $m^{\quad -1}\colon \langle \wp(A), \subseteq \rangle \to \langle \wp(L), \subseteq \rangle$ (recall that every poset gives rise
to a category). The functor $m^{\quad -1}$ is essentially the restriction of a pullback functor
$m^{\quad}\colon \mathbf{Set}{\downarrow}A \to \mathbf{Set}{\downarrow}L$, since given a subset $D \in \wp(A)$, $m^{\quad}$ maps the inclusion
morphism $D \xhookrightarrow{\subseteq} A$ to some mono $m^{\quad}(D) \rightarrowtail L$, such that $m^{\quad}(D) \cong m^{\quad -1}(D)$.
For each subset $K \in \wp(L)$ we define the set $\forall_m(K) \subseteq A$ by

$$\forall_m(K) = \{a \in A \mid \forall \ell \in m^{\quad -1}(\{a\}).\, \ell \in K\}.$$

In fact, this definition of $\forall_m$ makes it a functor $\forall_m\colon \langle \wp(L), \subseteq \rangle \to \langle \wp(A), \subseteq \rangle$.
Note that $\forall_m$ can be seen as the restriction of $\varPi_m\colon \mathbf{Set}{\downarrow}L \to \mathbf{Set}{\downarrow}A$ to the
subcategory $\langle \wp(L), \subseteq \rangle$, since $\varPi_m$ maps monos into L to monos into A.

Further one verifies that for all subsets $D \in \wp(A)$

$$m^{-1}(D) \subseteq K \quad \text{if and only if} \quad D \subseteq \forall_m(K). \tag{3}$$

To make the link to Definition 9 more precise, note that the co-unit for K corresponds to the inclusion $m^{-1}(\forall_m(K)) \subseteq K$; further the Equivalence (3) implies that for all sets $D \in \wp(A)$, if the inclusion $m^{-1}(D) \subseteq K$ holds then $D \subseteq \forall_m(K)$ and hence also $m^{-1}(D) \subseteq m^{-1}(\forall_m(K))$ hold.

The above definitions provide a sufficient condition for the existence of final pullback complements in an arbitrary category, as stated by the following lemma.

Lemma 11 (Existence and construction of final PBC). *Let* $A \xleftarrow{m} L \xleftarrow{\alpha} K$ *be a pair of composable arrows. Assume that the pullback functor* $m : \quad \downarrow A \rightarrow \downarrow L$ *exists, that the right adjoint* Π_m *to it exists partially at* α, *and that the arrow* $\varepsilon_\alpha : m(\Pi_m(\alpha)) \rightarrow \alpha$ *satisfies the conditions of Definition 9. Then*

1. *There exists a final pullback complement for* $A \xleftarrow{m} L \xleftarrow{\alpha} K$ *iff* ε_α *is iso.*
2. *If* ε_α *is iso, then the pair of composable arrows* $\langle \Pi_m(\alpha), m'_{\Pi_m(\alpha)} \circ \varepsilon_\alpha^{-1} \rangle$ *is a final pullback complement.*

5   Putting SqPO into Context

This section is dedicated to the relation of sesqui-pushout rewriting to the double- and single-pushout approach, which are the most widely used categorical approaches to rewriting. It should be mentioned that SqPO rewriting can also be seen as a "conceptual instance" of the very general categorical approach proposed by Wolfram Kahl [13], which is based on fibred categories, but space limitations prevent us to discuss the relationship to the latter.

For the case of left-linear rules, we will, in a certain sense, locate SqPO in between SPO and DPO. In fact we will see that SqPO rewriting coincides with DPO rewriting under mild assumptions, but its deletion mechanism is more general and closer to the one of SPO rewriting.

5.1   Relation Between the SqPO and the DPO Approach

The definition of SqPO rewriting differs from that of DPO rewriting only in the construction of the left square, which is a final pullback in the former case, and a pushout in the latter. Therefore, whenever the pushout complement of a match with respect to (the left-hand side of) a rule exists and it is also a final pullback complement, then the results of both constructions is the same. This holds in a very general case, namely for *left-regular rules* in *quasi-adhesive categories*.[3]

---

[3] Hence it also holds for *left-linear rules* in *adhesive categories*, which include **Set**, **Graph**, and several categories of graph-like objects. In fact, an adhesive category is a quasi-adhesive one where all monos are regular (see [16]).

Proposition 12 (DPO vs. SqPO). *Let   be a quasi-adhesive category, let $q =$ $L \xleftarrow{\alpha} K \xrightarrow{\beta} R$ be a left-regular rule (i.e., such that α is regular mono) and let $A \xleftarrow{m} L$ be a match in   . Then any pushout complement $A \xleftarrow{\gamma} D \xleftarrow{i} K$ for $A \xleftarrow{m} L \xleftarrow{\alpha} K$ is a final pullback complement. As a consequence, the following hold.*

1. *If $A \xRightarrow[\text{DPO}]{\langle m,q \rangle} B$ then also $A \xRightarrow{\langle m,q \rangle} B$.*

2. *If $A \xRightarrow{\langle m,q \rangle} B$ and a pushout complement of $A \xleftarrow{m} L \xleftarrow{\alpha} K$ exists, then also $A \xRightarrow[\text{DPO}]{\langle m,q \rangle} B$.*

Proof. In [16] it is shown (Lemma 2.3) that in a quasi-adhesive category pushouts along regular monos are pullbacks.

Furthermore, it is proved (Lemma 2.8) that if the square to the right is a pushout and α is regular, then $\gamma\colon D \to A$ enjoys the universal property of the right adjoint to the pullback functor m at α, i.e., $\gamma \cong \Pi_m(\alpha)$.

$$
\begin{array}{ccc}
L & \xleftarrow{\ \alpha\ } & K \\
{\scriptstyle m}\downarrow & - & \downarrow{\scriptstyle i} \\
A & \xleftarrow[\ \gamma\ ]{} & D
\end{array}
$$

Thus by Lemma 11 $A \xleftarrow{\gamma} D \xleftarrow{i} K$ is a final pullback complement.   □

For non-left-regular rules, as shown by Example 7, there exist in general several pushout complements and hence DPO rewriting is ambiguous. In contrast, SqPO rewriting is always deterministic, and its result models cloning, which cannot be obtained with DPO.

5.2   Relation Between the SqPO and the SPO Approach

We discuss now the relation between the SPO and the SqPO approach. First we concentrate on algebras for a graph structure, where the SPO approach coincides with the SqPO approach when we restrict the first to conflict-free matches and the latter to left-linear rules. Then we briefly discuss that a similar result holds for non-left-linear rules, in the context of the categorial generalization of the SPO approach presented in [18].

SPO **over graph structures.** Single-pushout rewriting has been defined in [17,5] for categories of algebras over *graph structures*, i.e., over signatures with unary operator symbols only.[4] For example, **Graph** can be seen as the category of algebras for the signature including two sorts, V and E, and two operator symbols, src, tgt: E → V.

For the rest of this subsection let   be the category of algebras and total homomorphisms of an arbitrary but fixed graph structure, and let   P be the category having the same objects and *partial morphisms* as arrows: that is, an arrow $f\colon X \to Y$ of   P is a total homomorphism $f\colon dom(f) \to Y$ from a subalgebra $dom(f) \subseteq X$.

---

[4] Note that all such categories can be seen as categories of set-valued functors, and therefore they are adhesive (see [16]).

As recalled in the introduction, according to the SPO approach a rule is an arrow $q\colon L \rightharpoonup R$ of ^{P}, and it is applied to a *total* match $m\colon L \to A$ by constructing a pushout in ^{P}. This is always possible, because ^{P} is co-complete. To simulate such a direct derivation using the SqPO approach, we consider the rule as a span $\hat{q} = L \hookleftarrow dom(q) \xrightarrow{q} R$ in , and look for the final pullback complement of $A \xleftarrow{m} L \hookleftarrow dom(q)$ in . Then, as summarized by the next proposition, it is possible to show that the results of the two constructions are equal if and only if the final pullback complement exists, i.e., by Construction 5, if and only if match m is conflict-free with respect to $L \hookleftarrow dom(q)$.

Proposition 13 (SPO vs. SqPO). *Let* L, K, R *and* A *be objects; let* $q\colon L \rightharpoonup R$ *be an arrow of* ^{P}, *and* $\hat{q} = L \xleftarrow{\supseteq} dom(q) \xrightarrow{q} R$ *be the corresponding span in* , *and let* $m\colon L \to A$ *be a total match morphism. Then the following are true*

1. *If* $A \xRightarrow{\langle m,\hat{q}\rangle} B$ *then* $A \xRightarrow[\text{SPO}]{\langle m,q\rangle} B$.

2. *If* $A \xRightarrow[\text{SPO}]{\langle m,q\rangle} B$ *and* m *is conflict-free then* $A \xRightarrow{\langle m,\hat{q}\rangle} B$.

The square to the right shows the result of SPO rewriting with a rule q and a non-conflict-free match m in category $\mathbf{Set}^{\mathrm{P}}$. Note that the function from the right-hand side of the rule to the resulting set is partial: this effect is often considered as unintuitive, and it is ruled out by imposing suitable constraints on the matches.

$$
\begin{array}{ccc}
\{①, ②\} & \xrightarrow[1\mapsto 1]{q} & \{①\} \\
m \downarrow {\scriptstyle \begin{smallmatrix} 1\mapsto 0 \\ 2\mapsto 0 \end{smallmatrix}} & & \downarrow \\
\{⓪\} & \xrightarrow{} &
\end{array}
$$

As shown in [17] the morphism from the right-hand side to the resulting object is total *if and only if* the match is conflict-free, thus SqPO rewriting rules out exactly the SPO direct derivations where this unintuitive effect shows up.

SPO over arbitrary categories. The SPO approach has been lifted to an abstract, categorical setting in [14,18]. Following the approach of [19], in [14] a *partial morphism* in a category is defined as an equivalence class of spans of , where the left arrows are monic. Generalizing even further, in [18] rules are defined as spans like $L \xleftarrow{m} K \xrightarrow{h} R$, where $m \in \mathcal{M}$ and $h \in \mathcal{H}$ are required to belong to two classes of arrows of satisfying suitable properties: in particular, it is not required that arrows in $\mathcal{M}$ are mono. Even if the technical details of this analysis are beyond the scope of the present paper, it turns out that for these classes of rules, every SqPO-derivation is a SPO-derivation. Moreover the reverse holds, whenever there exists a final pullback complement for the involved matching morphism. In other words, the statement of Proposition 13 holds true in the more general framework of [18] by replacing in point *2* the condition of conflict-freeness with that of existence of final pullback complements.

5.3   Parallelism

After having discussed which fragments of the classical algebraic approaches are subsumed by the new one, we present the local Church-Rosser theorem as evidence that (part of) the existing parallelism theory can be transferred to the realm of SqPO rewriting. Also a theorem concerning sequential commutativity holds true for SqPO rewriting, but we do not present it because of space limitations. We assume here that objects and arrows belong to a fixed quasi-adhesive category , and that rules are spans of regular monos.

Definition 14 (Parallel Independence). *Let there be two direct derivations* $G \xrightarrow{\langle m_1, p_1 \rangle} H_1$ *and* $G \xrightarrow{\langle m_2, p_2 \rangle} H_2$. *Then they are* parallel independent *if there exist morphisms* $u \colon L_1 \to D_2$ *and* $v \colon L_2 \to D_1$, *such that* $\gamma_2 \circ u = m_1$ *and* $\gamma_1 \circ v = m_2$.

$$
\begin{array}{ccccccccc}
R_1 & \xleftarrow{\beta_1} & K_1 & \xrightarrow{\alpha_1} & L_1 & & L_2 & \xleftarrow{\alpha_2} & K_2 & \xrightarrow{\beta_2} & R_2 \\
n_1 & & k_1 \,\Big\downarrow\, v & & m_1 & m_2 & & u \, k_2 & & & n_2 \\
H_1 & \xleftarrow{\delta_1} & D_1 & \xrightarrow{\gamma_1} & G & \xleftarrow{\gamma_2} & D_2 & \xrightarrow{\delta_2} & H_2
\end{array}
$$

Definition 14 can be seen as a conservative extension of the definitions given in the literature for SPO and DPO. More precisely, if two SPO direct derivations are also SqPO-derivations, then they are parallel independent in the SPO sense if and only if they are so according to Definition 14. The same holds for DPO-parallel independence as well, obviously.

Theorem 15 (Local Church-Rosser)
Given two parallel independent direct transforma-tions $G \xrightarrow{\langle m_1, p_1 \rangle} H_1$ *and* $G \xrightarrow{\langle m_2, p_2 \rangle} H_2$, *there are an object* G' *and direct transformations* $H_1 \xrightarrow{\langle m_2', p_2 \rangle} G'$ *and* $H_2 \xrightarrow{\langle m_1', p_1 \rangle} G'$.

$$
\begin{array}{ccc}
 & G & \\
{}^{m_1,p_1} & & {}^{m_2,p_2} \\
H_1 & & H_2 \\
{}^{m_2',p_2} & G' & {}^{m_1',p_1}
\end{array}
$$

The proof of this theorem is very similar to the one given in [16], the difference being that we need some additional SqPO-specific lemmas.

6 Conclusion

We have proposed a new algebraic approach to rewriting in arbitrary categories, called *sesqui-pushout rewriting*, and we discussed its basic properties. In the classical case of graphical structures and left-linear rules, its relation to the SPO and DPO approaches is summarized by the following table, where application conditions are listed below the features of the approaches.

	DPO	$\lesssim$	SqPO	$\lesssim$	SPO
deletion in unknown context	-		✓		✓
precedence of deletion over preservation	-		-		✓
indentification and dangling condition	✓		-		-
conflict-free matches	✓		✓		-

We have a chain of simulations (indicated by the symbol $\lesssim$): every DPO derivation is a SqPO derivation, and every SqPO derivation is a SPO derivation, by seeing left-linear rules as partial morphisms. Furthermore, when DPO rewriting is not possible because the dangling condition is not satisfied, if SqPO rewriting is possible then SPO is possible as well, and both model deletion in unknown context. Finally, when SqPO rewriting is not possible because the match is not conflict-free, then DPO rewriting is not possible because the identification condition is not satisfied, but SPO rewriting is possible and the conflict is resolved in favour of deletion. However, in this case there is no total morphism from the right-hand side of the rule to the resulting graph: an effect that is often considered as undesirable, and that is ruled out automatically by the new approach.

Probably the most original and interesting feature of sesqui-pushout rewriting is the fact that it can be applied to non-left-linear rules as well, and in this case it models the cloning of structures.

We presented a Local Church Rosser theorem for the new approach. We are confident that most of the parallelism and concurrency theory of the DPO and SPO approaches can be lifted smoothly to sesqui-pushout rewriting: this is a topic of ongoing research. Concluding, let us remark that we compared the new approach only with SPO and DPO because they are the most widely used categorical approach to rewriting, but there are several others to which sesqui-pushout rewriting has to be related as well, including the fibred approach by Kahl [13], the double-pullback approach by Heckel [12], and the pullback approach by Bauderon [1].

Acknowledgements. We would like to thank Paolo Baldan and Paweł Sobociński for enlightening discussions about the topic of the paper.

References

1. Michel Bauderon. A uniform approach to graph rewriting: The pullback approach. In M. Nagl, editor, *Graph-Theoretic Concepts in Computer Science*, volume 1017 of *Lecture Notes in Computer Science*, pages 101–115. Springer, 1995.
2. Andrea Corradini, Ugo Montanari, Francesca Rossi, Hartmut Ehrig, Reiko Heckel, and Michael Löwe. Algebraic approaches to graph transformation—part I: Basic concepts and double pushout approach. In G. Rozenberg, editor, *Handbook of Graph Grammars and Computing by Graph Transformation, Vol. 1: Foundations*, chapter 3. World Scientific, 1997.
3. Roy Dyckhoff and Walter Tholen. Exponentiable morphisms, partial products and pullback complements. *Journal of Pure and Applied Algebra*, 49(1-2):103–116, 1987.
4. Hartmut Ehrig, Annegret Habel, Hans-Jörg Kreowski, and Francesco Parisi-Presicce. Parallelism and concurrency in high level replacement systems. *Mathematical Structures in Computer Science*, 1:361–404, 1991.
5. Hartmut Ehrig, Reiko Heckel, Martin Korff, Michael Löwe, Leila Ribeiro, Annika Wagner, and Andrea Corradini. Algebraic approaches to graph transformation—part II: Single pushout approach and comparison with double pushout approach. In G. Rozenberg, editor, *Handbook of Graph Grammars and Computing by Graph Transformation, Vol.1: Foundations*, chapter 4. World Scientific, 1997.

6. Hartmut Ehrig and Michael Löwe. Categorical principles, techniques and results for high-level-replacement systems in computer science. *Applied Categorical Structures*, 1:21–50, 1993.

7. Hartmut Ehrig, Michael Pfender, and Hans-Jürgen Schneider. Graph grammars: An algebraic approach. In *Proc. 14th IEEE Symp. on Switching and Automata Theory*, pages 167–180, 1973.

8. David F. Ferriolo, D. Richard Kuhn, and Ramaswamy Chandramouli. *Role-Based Access Control*. Artech House computer security series. Artech House, 2003. ISBN: 1-580-53370-1.

9. Robert Goldblatt. *Topoi: The Categorial Analysis of Logic*. Studies in Logic and the Foundations of Mathematics. North-Holland Publishing Company, Amsterdam, The Netherlands, 1984.

10. Annegret Habel, Jürgen Müller, and Detlef Plump. Double-pushout graph transformation revisited. *Mathematical Structures in Computer Science*, 11(5):637–688, 2001.

11. Michael A. Harrison, Walter L. Ruzzo, and Jeffrey D. Ullman. Protection in operating systems. *Commun. ACM*, 19(8):461–471, 1976.

12. Reiko Heckel, Hartmut Ehrig, Uwe Wolter, and Andrea Corradini. Double-pullback transitions and coalgebraic loose semantics for graph transformation systems. *Applied Categorical Structures*, 9(1), 2001.

13. Wolfram Kahl. A fibred approach to rewriting — how the duality between adding and deleting cooperates with the difference between matching and rewriting. Technical Report 9702, Fakultät für Informatik, Universität der Bundeswehr München, May 1997.

14. Richard Kennaway. Graph Rewriting in Some Categories of Partial Morphisms. In Hartmut Ehrig, Hans-Jörg Kreowski, and Grzegorz Rozenberg, editors, *Proc. 4th. Int. Workshop on Graph Grammars and their Application to Computer Science*, volume 532 of *Lecture Notes in Computer Science*, pages 490–504. Springer-Verlag, 1991.

15. Manuel Koch, Luigi V. Mancini, and Francesco Parisi-Presicce. A formal model for role-based access control using graph transformation. In Frédéric Cuppens, Yves Deswarte, Dieter Gollmann, and Michael Waidner, editors, *ESORICS*, volume 1895 of *Lecture Notes in Computer Science*, pages 122–139. Springer, 2000.

16. Stephen Lack and Paweł Sobociński. Adhesive and quasiadhesive categories. *Theoretical Informatics and Applications*, 39(2):511–546, 2005.

17. Michael Löwe. Algebraic approach to single-pushout graph transformation. *Theoretical Computer Science*, 109:181–224, 1993.

18. Miquel Monserrat, Francesc Rosselló, Joan Torrens, and Gabriel Valiente. Single pushout rewriting in categories of spans I: The general setting. Technical Report LSI-97-23-R, Departament de Llenguatges i Sistemes Informàtics, Universitat Politècnica de Catalunya, 1997.

19. Edmund Robinson and Giuseppe Rosolini. Categories of partial maps. *Information and Computation*, 79(2):95–130, 1988.

20. Ravi S. Sandhu, Edward J. Coyne, Hal L. Feinstein, and Charles E. Youman. Role-based access control models. *IEEE Computer*, 29(2):38–47, 1996.

Automata on Directed Graphs: Edge Versus Vertex Marking[*]

Dietmar Berwanger and David Janin

LaBRI, Université Bordeaux 1
351, Cours de la Libération, 33405 Talence cedex, France
{dwb, janin}@labri.fr

Abstract. We investigate two models of finite-state automata that operate on rooted directed graphs by marking either vertices (V-automata) or edges (E-automata). Runs correspond to locally consistent markings and acceptance is defined by means of regular conditions on the paths emanating from the root. Comparing the expressive power of these two notions of graph acceptors, we show that E-automata are more expressive than V-automata. Moreover, we prove that E-automata are at least as expressive as the μ-calculus. Our main result implies that every MSO-definable tree language can be recognised by E-automata with uniform runs, that is, runs that do not distinguish between isomorphic subtrees.

Introduction

Extending the formal language theory of words and trees to general classes of graphs is a very challenging endeavour. During the last two decades, this topic has attracted much attention, and several notions of graph-language recognisability have been developed [15, 4, 25, 7, 10, 8, 6].

Over the domain of arbitrary finite graphs, Courcelle [4] proposes a powerful algebraic theory of languages recognisable via interpretations of tree-shaped terms. This characterisation inherits many features from the well-established theory of tree automata [24]. Thus, the notion of recognisability is closed under Boolean operations and projection, and its expressive power reaches beyond Monadic Second-Order Logic (MSO).

For directed graphs of bounded degree, Thomas [25] develops an automata-theoretic approach in terms of *tiling systems*, or more generally *graph acceptors*. These are devices that proceed by marking graph vertices according to local constraints, tailored to match the expressive power of the existential fragment of MSO (monadic Σ_1). The associated notion of recognisability is closed under union, intersection, and projection. In general it is not closed under complement since, on many classes of graphs, monadic Σ_1 is not closed under complement [27].

The most substantial research is, however, focused on the special case of finite directed acyclic graphs (see [26] for a survey). Robust notions of recognisability

---

[*] This research has been partially supported by the EU RTN GAMES: "Games and Automata for Synthesis and Validation."

A. Corradini et al. (Eds.): ICGT 2006, LNCS 4178, pp. 46–60, 2006.
c Springer-Verlag Berlin Heidelberg 2006

are available, in particular, for partially ordered sets which serve as models for concurrent computation [7, 17].

In this paper, we are concerned with arbitrary directed graphs of unbounded degree that may be infinite or contain cycles. Aside from the generic interest, this framework is fundamental for modelling the behaviour of state-transition systems. Our point of departure is the notion of a graph acceptor, introduced by Thomas in [25]. Adapting this notion to graphs of unbounded degree, we define two kinds of finite-state automata that operate on rooted directed graphs by marking either vertices (V-automata) or edges (E-automata), starting from the root and proceeding according to transitions specified by local first-order formulae. Because we are interested in the infinite behaviour of models, we equip these automata with ω-regular acceptance conditions over the marking of paths emanating from the designated root.

The question whether to label edges or vertices is subject to a fundamental choice in the design of automata that may revisit vertices of their input. The most common option is to mark vertices, but the alternative to mark edges also has some tradition, going back to the early 80ies and Kamimura and Slutzki's variant of tree-walking automata over planar graphs [15]. In [22], Potthoff, Seibert, and Thomas discuss the expressive power of graph acceptors that mark edges compared to those that mark vertices and show that, in their specific framework restricted to ranked acyclic graphs of bounded degree, edge and vertex marking lead to the same notion of recognisability.

Taking up an analogue investigation for our extended setting, we find that the situation is radically different over arbitrary graphs, even if global acceptance conditions are not involved. In a comparative study, we separate the expressive power of vertex and edge-marking automata and relate it to Monadic Second-Order Logic and to its bisimulation-invariant fragment, the μ-calculus. Our main result establishes a correspondence between runs of edge-marking automata over arbitrary graphs and vertex-marking automata over trees obtained by unravelling these graphs. Besides showing that edge-marking automata capture the μ-calculus over arbitrary graphs, as vertex-marking automata do on trees, this result opens a perspective on *uniform* recognisability of graph languages.

Outline. The paper is structured as follows. After fixing our notation in Section 1, we introduce V-automata and E-automata in Section 2, and point out some elementary properties. Thus, recognisable classes of graphs are closed under conjunction, disjunction and projection, but not under complement. In terms of MSO, E-automata and V-automata define graph properties in the level Σ_3 of the monadic quantifier-alternation hierarchy, respectively MSO_2, the variant of MSO augmented with edge quantifiers [5].

In Section 3, we show that, over arbitrary directed graphs, E-automata are strictly more expressive than V-automata. Actually, E-automata can describe properties like perfect matching that are not yet definable in MSO. However, we show that even when we restrict to MSO-definable properties, E-automata are more expressive than V-automata. A separating property is directed reachability.

In Section 4, we investigate the relation between our automata model and the (counting) μ-calculus [16, 2], a logic which captures the MSO-definable properties that are invariant under (counting) bisimulation [14]. These properties are particularly relevant for the specification of state-transition models, because they do not distinguish between a model and its behaviour, understood as the unravelling of its possible computations in a tree-like manner.

Our main technical result is a simulation theorem relating recognisability of graphs and their unravelling. It states that, for every V-automaton $\mathcal{A}$, there exists an equivalent E-automaton that recognises precisely the class of graphs whose unravelling is accepted by $\mathcal{A}$. Intuitively, this means that E-automata can simulate on their input graph the behaviour of a V-automaton on the unravelling of this graph. Notice that every element of the input graph may have infinitely many copies in its unravelling, such that the V-automaton has potentially infinite "space" to apply his marking. However, we argue that all these copies can be encoded into a single marking of the original input element.

We discuss three consequences of this theorem. First, it implies that E-automata subsume the μ-calculus over arbitrary directed graphs, yielding an operative model that differs substantially from previous automata-theoretic characterisations of the μ-calculus (see, e.g., [20, 9, 3, 13]), where automata essentially run on a tree unravelling rather than on the input graph. The model of V-automata is, however, not strong enough to capture the μ-calculus over arbitrary graphs, since it cannot express directed reachability, which is μ-definable. Secondly, it follows that our definition of E-automata, with a universal linear-time condition on infinite computation paths, is fairly robust, as far as expressive power is concerned. In particular, adding branching-time acceptance condition over the computation trees of automata, i.e., the unravelling of runs, would not increase their expressiveness. Finally, when rephrased in terms of automata over infinite trees, our main result shows that every MSO-definable language of infinite trees can be recognised by a non-deterministic tree automata with uniform runs, i.e., runs that do not distinguish between isomorphic subtrees. In other words, shared substructure of the input can also be shared by the run. This uniformisation result is particularly surprising as it does not incur a decrease in expressiveness, as it is usually the case for such normalisations [11, 28].

1   Background

1.1   Words

A *word* over an alphabet A is a partial function $\alpha :\quad \to A$ with prefix-closed domain. We say that α is finite, when $\mathrm{dom}(\alpha)$ is so. The set of finite words over the alphabet A is denoted by $A\quad$, whereas the set of infinite words is denoted by A^ω; the union of these two sets is $A\quad$. The concatenation of a word $\alpha \in A\quad$ with a word $\beta \in A\quad$ is denoted by $\alpha\beta$. This notation naturally extends to sets of words. We sometimes refer to the element $\alpha(i)$ of a word α by α_i.

In the sequel, we will use sets $L \subseteq A^\omega$ as infinitary acceptance conditions. We say that L is a *ω-regular* condition when L a finite union of languages of the form LU^ω where $L \subseteq A$ and $U \subseteq A$ are nonempty and regular. The set L is called a *parity condition*, if there exist a *priority* mapping $\Omega : A \to$ of finite image, such that $L = \{\alpha \in A^\omega \mid \liminf \Omega(\alpha) \equiv 0 \pmod{2}\}$, i.e., L is the set of infinite sequences where the least priority that occurs infinitely often is even.

1.2 Graphs

A *graph* is a structure $\mathcal{G} = (V, E)$ over a domain V of *vertices* with a binary *edge* relation $E \subseteq V \times V$. A *rooted* graph $\mathcal{G}, u$ is a graph with a distinguished *root* vertex u. Given an edge $(v, w) \in E$, we refer to v as its *source* and to w as its *target*. A *(directed) path* in the graph $\mathcal{G}$ is a finite or infinite non-empty sequence $v_1, v_2, \cdots \in V$ of vertices such that for any two consecutive elements v_i and v_{i+1}, we have $(v_i, v_{i+1}) \in E$. An *undirected path* in $\mathcal{G}$ is a sequence $v_1, v_2 \cdots \in V$ where, for any two consecutive elements v_i and v_{i+1}, either $(v_i, v_{i+1}) \in E$ or $(v_{i+1}, v_i) \in E$. The *distance* $d(v, v')$ between two vertices v, v' of V, is the least number n such that there exists an undirected path $v_1, \ldots, v_n$ in $\mathcal{G}$ with $v_1 = v$ and $v_n = v'$. If no such path exists, we set $d(v, v') = \infty$.

Fix an alphabet C of colours. For a graph $\mathcal{G} = (V, E)$, a *vertex colouring* over C is a function $\lambda : V \to C$; likewise, an *edge colouring* over C is a function $\gamma : E \to C$. We refer to the expansion of a graph by edge and/or vertex colourings as a *coloured* graph. When, instead of total functions, we consider partial edge or vertex colourings, we refer to them as *markings*. The elements in the domain of such a partial colouring are said to be *marked* by the respective function. In addition to this, we say that a vertex $v \in V$ is *involved* in an edge marking $\gamma : E \to C$, if it is either the source or the target of an edge marked by γ. A vertex colouring $\lambda : V \to C$ is naturally extended to a path π in $\mathcal{G}$ by setting $\lambda(\pi) = \lambda \circ \pi \in C$. Likewise, for an edge colouring $\gamma : E \to C$, we define the edge colouring of a path $\pi = v_1, v_2, \ldots$ in $\mathcal{G}$ to be the word $\gamma(\pi) \in C$, such that $\gamma(\pi)(i) = \gamma(\pi(i), \pi(i+1))$ for all indices i with $i + 1 \in \mathrm{dom}(\pi)$. For markings, the corresponding definitions are restricted to paths in $\mathcal{G}$ that involve only marked elements.

Besides the functional notation $\mathcal{G} = (V, E, \lambda)$ for vertex-marked graphs, it is sometimes convenient to use a relational notation $\mathcal{G} = (V, E, (P_c)_{c\ C})$, with monadic symbols P_c interpreted by $P_c := \{v \in V \mid \lambda(v) = c\}$, for every $c \in C$. Similarly, for edge-marked graphs $\mathcal{G} = (V, E, \gamma)$, we use the notation $\mathcal{G} = (V, E, (R_c)_{c\ C})$ with binary relational symbols R_c interpreted by $R_c := \{(v, w) \in E \mid \gamma(v, w) = c\}$, for every colour $c \in C$.

Bisimulation. The main part of this paper is concerned with devices taking as input vertex-coloured graphs. For the sake of clarity, the following definitions are formulated for this setting, the generalisation to edge-coloured graphs being straightforward.

A *counting bisimulation* between two vertex-coloured graphs $\mathcal{G} = (V, E, \lambda)$ and $\mathcal{G}' = (V', E', \lambda')$ is a relation $Z \subseteq V \times V'$ such that, if $(v, v') \in Z$, then $\lambda(v) = \lambda'(v')$ and Z contains a bijection between the sets $\{w \mid (v, w) \in E\}$ and $\{w' \mid (v', w') \in E'\}$. Two rooted graphs $\mathcal{G}, u$ and $\mathcal{G}', u'$ are counting *bisimilar*, if there exists a counting bisimulation Z between them with $(u, u') \in Z$. Two vertices v, v' of a graph $\mathcal{G}$ are counting bisimilar, if $\mathcal{G}, v$ and $\mathcal{G}, v'$ are so.

The *unravelling* of a graph $\mathcal{G} = (V, E, \lambda)$ from a vertex $u \in V$ is the graph $\mathcal{T}(\mathcal{G}, u)$ with domain V consisting of all directed paths π through $\mathcal{G}$ that start from u, edge relation E containing all the pairs $(\pi, \pi v) \in V \times V$, and vertex colouring λ defined by $\lambda(\pi v) = \lambda(v)$. A rooted graph $\mathcal{G}, u$ is a *tree*, if it is isomorphic to its unravelling $\mathcal{T}(\mathcal{G}, u), u$. For trees, and in particular for unravellings, we will generally not specify the root explicitly. Obviously, the natural projection which sends every path $v_1, \ldots, v_\ell \in V$ to its last node v_ℓ defines a counting bisimulation between $\mathcal{G}$ and $\mathcal{T}(\mathcal{G}, u)$. It is well-known (see, e.g., [12]), that two graphs $\mathcal{G}, u$ and $\mathcal{G}', u$ are counting bisimilar if, and only if, their unravelling $\mathcal{T}(\mathcal{G}, u)$ and $\mathcal{T}(\mathcal{G}, u)$ are isomorphic.

1.3 Logic

We consider standard predicate logics, in particular First-Order Logic (FO) and Monadic Second-Order Logic (MSO) interpreted over coloured graphs. Given an alphabet C of vertex colours, we write P_C for the collection of monadic symbols $(P_c)_c$ $_C$. When using edge colours from an alphabet D, we write R_D for the collection of binary relational symbols $(R_d)_d$ $_D$. Thus, the vocabulary of formulae is typically (a subset of) $E \cup P_C \cup R_D$. We refer to any formula $\varphi(x)$ with precisely one free first-order variable as a *predicate*, and to a formula without free first-order variables as a *sentence*. For any integer k, the k-*sphere* around a vertex v of a graph is the set of vertices w such that $d(v, w) \leq k$. An FO-predicate $\varphi(x)$ is called k-*local* around x, if it is equivalent to the predicate $\varphi'(x)$ obtained by relativising every quantifier in φ to elements of the k-sphere around x. We say that a predicate $\varphi(x)$ is local around x if it is k-local for some k.

We also consider MSO_2, the extension of MSO where quantification over sets of edges is provided [5]. The syntax of MSO_2 allows binary second-order variables and quantification over them. Given a graph $\mathcal{G} = (V, E)$, the semantics of this quantification is however relativised to subsets of E. It is well known that, on arbitrary graphs, MSO_2 is strictly more expressive than MSO. Though, on trees and on graphs of bounded degree, hence in particular on grids, MSO and MSO_2 are equally expressive [5].

The monadic quantifier-alternation hierarchy is defined as follows. The first level, monadic Π_0, also called monadic Σ_0, is the set of FO-formulae. Then, for every n, the level monadic Σ_{n+1} (resp. monadic Π_{n+1}) is the closure of the set of monadic Π_n-formulae (respectively monadic Σ_n-formulae) under existential quantification (respectively universal quantification). This hierarchy is known to be strict over arbitrary graphs, i.e., for every $n \in$, there exists a property φ_n definable in MSO that is not definable in the level Σ_n of the monadic hierarchy. More recently [18], it has also been shown that the monadic hierarchy is strict already over finite grids. Since MSO and MSO_2 are equally expressive over grids, and because the translation between the two preserves the quantification structure, this strictness result carries over to MSO_2.

A class $\mathcal{K}$ of rooted graphs is counting-bisimulation *closed* if, for any graph $\mathcal{G}, u$, we have $\mathcal{G}, u \in \mathcal{K}$ if, and only if, there exists a counting-bisimilar graph

$\mathcal{G}', u' \in \mathcal{K}$. A sentence φ of FO or MSO is counting-bisimulation *invariant*, if its model class is counting-bisimulation closed, that is, for any two counting-bisimilar graphs $\mathcal{G}, u$ and $\mathcal{G}, u'$ we have $\mathcal{G}, u \models \varphi$ if, and only if, $\mathcal{G}', u' \models \varphi$. The (counting) μ-calculus [16, 2, 12] is an extension of (counting) modal logic with fixed-point operators that provides an effective syntax for the (counting) bisimulation-invariant fragment of MSO [14, 12].

2 Vertex and Edge-Marking Automata

Traditionally, automata are finite-state devices that produce a marking of their input objects with states. The process of marking starts from designated input elements and propagates locally, depending on the local properties of the input structure and of the previously produced marking. When the input structures are homogeneous, these propagation transitions can often be described pictorially. However, as we are concerned with coloured directed graphs of unbounded branching which are not homogeneous, we choose a more abstract way to describe transitions using local FO-formulae that refer to both the input structure and the produced marking.

We introduce automata that take as input graphs with vertex colourings and produce either edge or vertex markings. Whenever we speak of a Σ-coloured graph, we mean a graph with a vertex colouring over a finite alphabet Σ.

Definition 1 (V-automaton). *Let Σ be a finite alphabet of vertex colours. A* vertex-marking automaton (V-automaton) *for Σ-coloured graphs is a tuple*

$$\mathcal{A} = (Q, \Sigma, \delta_0, \delta, Acc)$$

with a finite set Q of states, two local formulae $\delta_0(x), \delta(x) \in$ FO, called root constraint respectively transition specification, over the vocabulary $E \cup P_\Sigma \cup P_Q$ of Σ-coloured graphs augmented with unary symbols associated to the states of Q, and an ω-regular acceptance condition $Acc \subseteq Q^\omega$.

Given a Σ-coloured graph $\mathcal{G} = (V, E, \lambda)$ with a designated root u, a *run* of the V-automaton $\mathcal{A}$ on $\mathcal{G}, u$ is a vertex marking $\rho : V \to Q$ with the following properties:

(i) *initial condition:* $\mathcal{G}, \rho \models \delta_0(u)$, and
(ii) *local consistency:* for every vertex $v \in V$ marked by ρ, we have $\mathcal{G}, \rho \models \delta(v)$.

A run ρ is *accepting* if, for every infinite path π in $\mathcal{G}$ that starts from u and consists of vertices marked by ρ, we have $\rho(\pi) \in Acc$. A graph $\mathcal{G}, u$ is *accepted* by $\mathcal{A}$, if there exists an accepting run of $\mathcal{A}$ on $\mathcal{G}, u$. We define $\mathcal{L}_V(\mathcal{A})$ to be the class of all rooted graphs accepted by the V-automaton $\mathcal{A}$. A class $\mathcal{K}$ of rooted graphs is V-*recognisable*, if there exists a V-automaton $\mathcal{A}$ with $\mathcal{L}_V(\mathcal{A}) = \mathcal{K}$.

We observe that V-automata generalise most of the classical nondeterministic automata models, such as top-down or bottom-up automata over finite trees, but also, e.g., Muller-automata over infinite trees.

Remark 2. Notice that the run of an automaton is independent of the part of the input graph that is unreachable from the designated root, since local consistency is enforced just at marked vertices and extends only to a neighbourhood of the current vertex, so that no marking of an unreachable vertex can be required. However we may assume, without loss of generality, that accepting runs of a V-automaton $\mathcal{A}$ are total functions. To achieve this, we can add an extra dummy state $\bot$ and modify the transition specification to be $P \vee \delta$ and the acceptance condition to include the set $Q \{\bot\}(Q \cup \{\bot\})^{\omega}$.

Lemma 3. *Every V-recognisable class of graphs is definable in the level Σ_3 of MSO.*

Proof. Let $\mathcal{A} = (Q, \Sigma, \delta_0, \delta, Acc)$ be a V-automaton with state set $\{1, \ldots, n\}$. We construct an MSO-formula φ of the form $\exists P_1 \cdots \exists P_n (\forall x \psi(x) \wedge \varphi_{Acc})$ with $\varphi_{Acc} \in \Pi_2$ and $\psi \in \Sigma_0$ such that, for every graph $\mathcal{G}$ with a designated root u, we have $\mathcal{G}, u \models \varphi$ if, and only if, $\mathcal{G}, u \in \mathcal{L}_V(\mathcal{A})$. In this formula, the block of existential quantifiers $\exists P_1 \cdots \exists P_n$ guesses a vertex marking, the subformula $\forall x \psi(x)$ expresses the local constraints,

$$\psi(x) := \big(x = u \rightarrow \delta_0(x)\big) \wedge \big(\bigvee_{q\ Q} P_q \rightarrow \delta(x)\big),$$

and φ_{Acc} checks the infinitary path condition. To see that φ_{Acc} can be described in Σ_2, notice that its negation $\neg\varphi_{Acc}$ expresses the property that there exists a marked path starting at u that does not satisfy the ω-regular acceptance condition Acc. Using the representation of $Q^{\omega} \setminus Acc$ as a non-deterministic Büchi word automaton, this property can be defined by a monadic Σ_2-formula (more precisely, a $\mu\nu$-formula of the μ-calculus). □

Our second automata model differs from V-automata only by its way of applying state labels to edges rather than to vertices.

Definition 4 (E-automaton). *Let Σ be a finite alphabet of vertex colours. An* edge-marking automaton *(E-automaton) for Σ-coloured graphs is a tuple*

$$\mathcal{A} = (Q, \Sigma, \delta_0, \delta, Acc),$$

with a finite set Q of states, *two local predicates $\delta_0(x), \delta(x) \in$ FO, called* root constraint *and* transition specification, *over the vocabulary $E \cup P_{\Sigma} \cup R_Q$ of Σ-coloured graphs augmented with binary relational symbols associated to the states of Q, and an ω-regular acceptance condition $Acc \subseteq Q^{\omega}$.*

Given a Σ-coloured graph $\mathcal{G} = (V, E, \lambda)$ with a designated root u, an accepting run of the E-automaton $\mathcal{A}$ on $\mathcal{G}, u$ is now an edge marking $\rho : E \rightarrow Q$ with the following properties:

(i) *initial condition:* $\mathcal{G}, \rho \models \delta_0(u)$, and
(ii) *local consistency:* for every vertex $v \in V$ involved in ρ, we have $\mathcal{G}, \rho \models \delta(v)$.

A run ρ is *accepting* if, for every infinite path $\pi = v_0, v_1, \ldots$ in $\mathcal{G}$ that starts from the root $u = v_0$ and proceeds along edges (v_i, v_{i+1}) marked by ρ, we have $\rho(\pi) \in Acc$. As in the case of V-automata, we say that the graph $\mathcal{G}, u$ is *accepted* by $\mathcal{A}$ if there exists an accepting run of $\mathcal{A}$ on $\mathcal{G}, u$, and we define $\mathcal{L}_E(\mathcal{A})$ to be the class of rooted graphs accepted by the E-automaton $\mathcal{A}$. A class $\mathcal{K}$ of rooted graphs is E-*recognisable*, if there exists an E-automaton $\mathcal{A}$ with $\mathcal{L}_E(\mathcal{A}) = \mathcal{K}$.

We remark that there are E-recognisable classes of graphs that cannot be described in MSO. An example is the class of graphs that allow a perfect matching between the vertices reachable from the root. Essentially, this is because edge marking corresponds to a quantification over sets of edges which is not available in MSO. Nevertheless, for every E-automaton, the class $\mathcal{L}_E(\mathcal{A})$ is definable in MSO_2. The proof is a straightforward adaptation of the proof of Lemma 3.

Lemma 5. *Every E-recognisable class of graphs is definable in the level Σ_3 of MSO_2.*

2.1 Elementary Properties

We survey some elementary properties of our graph automata. An essential feature is that we can specify grid properties, even without marking edges, by simulating a grid vocabulary consisting of two functional edge symbols, say R_N and R_E, standing for North and East. Towards this, we use two extra monadic symbols P_N and P_E, and we require that the root is in both P_N and P_E, and every vertex v has exactly two outgoing edges (v, v_E) and (v, v_N) such that either both or none of v and v_E belong to P_E whereas they never belong together to P_N and, similarly, either both or none of v and v_N belong to P_N whereas they never belong together to P_E. The intended grid relations can now be defined by $R_N := \{(x, y) \in E \mid P_N(x) \rightarrow P_N(y)\}$ and $R_E := \{(x, y) \in E \mid P_E(x) \rightarrow P_E(y)\}$.

Lemma 6. *E-automata and V-automata with $k+1$-local transition specifications are strictly more expressive than E-automata respectively V-automata with k-local transition specification.*

Proof. A corresponding statement for tiling systems is proved in [26]. The argument carries over to our automata. □

Lemma 7. *Both V-recognisable and E-recognisable classes of graphs are closed under union, intersection, and projection.*

Proof. These properties follow directly from the definition of our automata model. To show, for instance, closure under union for V-automata, consider two V-automata $\mathcal{A} = (Q, \Sigma, \delta_0, \delta, Acc)$ and $\mathcal{A}' = (Q', \Sigma, \delta_0', \delta', Acc')$. Then the automaton over the state set $Q \cup Q'$, with root constraint $\delta_0 \vee \delta_0'$, transition specification $\delta \vee \delta'$, and infinitary condition $Acc \cup Acc'$ recognises $\mathcal{L}_V(\mathcal{A}) \cup \mathcal{L}_V(\mathcal{A}')$. □

Lemma 8. *Neither V-recognisable nor E-recognisable classes of graphs are closed under complement. The statement also holds for classes of finite graphs.*

Proof. We have already seen that FO-definable classes of finite grids can be recognised by V-automata and also by E-automata. If, in addition to being closed under projection, recognisable classes were closed under complement, any MSO-definable class of finite grids would be recognisable, and hence Σ_3-definable, by Lemma 3 respectively Lemma 5. This contradicts the infiniteness of the monadic hierarchy over grids [18]. $\qquad\square$

3   E-Automata Versus V-Automata

In this section, we compare the expressive power of the two notions of automata. It turns out that E-automata are, even on finite graphs, strictly more expressive than V-automata.

3.1   Encoding V-Automata into E-Automata

Proposition 9. *Every V-recognisable class of graphs is also E-recognisable.*

Proof. For simplicity, we assume here that automata are normalised so that accepting runs are total functions. Given a V-automaton $\mathcal{A} = (Q, \Sigma, \delta_0, \delta, Acc)$, we construct an E-automaton $\mathcal{B} = (Q \times Q, \Sigma, \delta'_0, \delta', Acc')$ that marks edges of its input graph with pairs of (vertex) states from Q, in such a way that the marking of a vertex v with a state q in a run of $\mathcal{A}$ corresponds to the marking of all incoming and outgoing edges from v by pairs of the form (q', q) respectively (q, q') in a run of $\mathcal{B}$.

The following one-local formula expresses that the edge marking around a vertex z encodes a vertex marking of z by q:

$$\varphi_q(z) = \forall y \Big[\Big(E(y, z) \to \bigvee_{q' \; Q} R_{(q', q)}(y, z) \Big) \wedge \Big(E(z, y) \to \bigvee_{q' \; Q} R_{(q, q')}(z, y) \Big) \Big].$$

Now, we define the root constraint $\delta'_0(x)$ and the transition specification $\delta'(x)$ for $\mathcal{B}$ to be the conjunction of $\bigvee_{q \; Q} \varphi_q(x)$ with the formula obtained from $\delta_0(x)$ respectively $\delta(x)$ by replacing every atom $P_q z$ with the subformula $\varphi_q(z)$. The acceptance condition Acc' consists of all infinite words $\beta \in (Q \times Q)^\omega$ for which there exists a word $\alpha \in Acc$, such that $\beta_i = (\alpha_i, \alpha_{i+1})$, for all indices i.

To verify that the construction is correct in the case of graphs with no isolated vertices (these need to be treated separately, but pose no great difficulty), let us consider a rooted graph $\mathcal{G}, u \in \mathcal{L}_V(\mathcal{A})$ and let $\rho : V \to Q$ be an accepting run of the V-automaton $\mathcal{A}$. Then, the marking $\rho' : E \to Q \times Q$ defined by $\rho'(v, w) = (\rho(v), \rho(w))$, for every edge $(v, w) \in E$, is an accepting run of the E-automaton $\mathcal{B}$ on $\mathcal{G}, u$. Conversely, for an accepting run $\rho' : E \to Q \times Q$ of $\mathcal{B}$ on a graph $\mathcal{G}, u$, the conditions δ'_0 and δ' ensure that, for every vertex $v \in V$, there exists a unique state $q \in Q$ such that $\mathcal{G}, \rho' \models \varphi_q(v)$; the vertex-marking $\rho : V \to Q$ defined by associating to every vertex $v \in V$ this unique state, is an accepting run of the V-automaton $\mathcal{A}$ on $\mathcal{G}$. $\qquad\square$

3.2 E-Automata Are More Expressive Than V-Automata

Next, we prove that edge-marking yields a strict increase in expressiveness over vertex-marking. Instead of relying on properties from $MSO_2 \setminus MSO$ we show, moreover, that separating properties exist already in MSO.

Theorem 10. *There exists an E-recognisable class of directed graphs that is MSO-definable, but not V-recognisable.*

Proof. We show that directed reachability is definable by an E-automaton, but not by a V-automaton. Let $\mathcal{K}$ be the class of graphs $\mathcal{G}, u$ over the alphabet $\Sigma = \{a, b\}$ in which there exists a finite directed path from u to a vertex v with $\lambda(v) = a$. Clearly, this class is definable in MSO, in fact, already in the μ-calculus by the formula $\mu X.(P_a \vee \ \ X)$.

To see that $\mathcal{K}$ is E-recognisable, consider the automaton $\mathcal{B} = (\{q\}, \Sigma, \delta_0, \delta, \emptyset)$ with only one state q. The root constraint δ_0 states that, if the root is not coloured with a, precisely one outgoing edge is marked,

$$\delta_0(x) := \neg P_a \rightarrow \exists y \left(Exy \wedge Rxy \wedge \forall z(Exy \wedge Rxz \rightarrow z = y) \right).$$

The transition specification requires that, if a vertex is not coloured with a and has an incoming marked edge, then precisely one outgoing edge is marked,

$$\delta(x) := \left(\neg P_a \wedge \exists y(Eyx \wedge Ryx) \right) \rightarrow \exists y \left(Exy \wedge Rxy \wedge \forall z(Exz \wedge Rxz \rightarrow z = y) \right).$$

In this way, accepting runs of $\mathcal{B}$ correspond to markings of a directed path from the root to some vertex of colour a. Therefore, $\mathcal{L}_V(\mathcal{B}) = \mathcal{K}$.

To show that $\mathcal{K}$ is not V-recognisable, we use an idea of Ajtai and Fagin [1]. Towards a contradiction, let $\mathcal{A} = (Q, \Sigma, \delta_0, \delta, Acc)$ be a V-automaton with $\mathcal{L}_V(\mathcal{A}) = \mathcal{K}$. We fix real number p between 0 and 1. For every integer n, we construct a random Σ-coloured graph $\mathcal{G}_p^n = (V_n, E_n, \lambda_n)$ over the domain $V_n = \{0, \ldots, n\}$ with the following edge relation:

– for all $i < n$, the *forward-edge* $(i, i+1)$ is contained in E_n, and
– for every (i, j) with $1 \leq i < j \leq n$, the *back-edge*(j, i) is contained in E_n with probability p.

The vertices of $\mathcal{G}_p^n$ are coloured by $\lambda_n(i) = b$, for all $i < n$, and $\lambda_n(n) = a$.

Clearly, for every n, the graph $\mathcal{G}_p^n$ with root 0 belongs to $\mathcal{K}$. For every element $i < n$, let now $\mathcal{G}_{p,i}^n$ be the graph obtained from $\mathcal{G}_p^n$ by removing the forward-edge $(i, i+1)$. Obviously, $\mathcal{G}_{p,i}^n, 0 \notin \mathcal{K}$. A technical theorem of [1] implies that, for every size of a vertex-alphabet C and every quantifier rank r, there exists an integer n and a value for p, such that the following property holds with positive probability: for every marking $\rho : V_n \rightarrow Q$ of $\mathcal{G}_p^n$, there exists an index $i < n$ such that, for every FO-formula of quantifier rank at most r, if $\mathcal{G}_p^n, \rho \models \varphi$ then $\mathcal{G}_{p,i}^n, \rho \models \varphi$.

Applying this statement to an accepting run ρ of $\mathcal{A}$ on $\mathcal{G}_p^n, 0$, it implies that there exist values $n \in \ \ , p \in (0, 1)$, and $i < n$ such that, with positive probability, ρ is also an accepting run of $\mathcal{A}$ on $\mathcal{G}_{p,i}^n$ since the initial condition and the local consistency of $\mathcal{A}$ cannot distinguish between the graphs $\mathcal{G}_{p,i}^n, 0$ and $\mathcal{G}_p^n, 0$ when they are marked in the same way. Notice that the path condition cannot discriminate between these graphs either, since any infinite path in the former is also an infinite path in the latter. Hence, we have $\mathcal{G}_{p,i}^n, 0 \in \mathcal{L}_V(\mathcal{A})$, in contradiction to our assumption that $\mathcal{L}_V(\mathcal{A}) = \mathcal{K}$.   □

4   Simulation and Uniform Recognisability

In this section, we establish a relation between runs of automata on graphs with runs on their unravelling. This allows us, on the one hand, to conclude that over arbitrary graphs E-automata capture the counting μ-calculus and, on the other hand, that V-automata over trees can be normalised to mark isomorphic subtrees identically.

4.1   E-Automata and the μ-Calculus

Observe that, on trees, the marking of edges can be simply moved to their targets, and hence the notions of E-recognisability and V-recognisability coincide. Furthermore, V-automata –with one-local root constraint and transition specification– generalise MSO tree-automata [29, 12]. Accordingly, for classes of trees, recognisability equals MSO-definability.

Theorem 11 (Rabin [23], Muchnik and Walukiewicz [19, 29]). *For every* MSO *formula* φ *there exists a one-local V-automaton* $\mathcal{A}_\varphi$ *such that, for every tree* $\mathcal{T}$, *we have* $\mathcal{T} \models \varphi$ *if, and only if,* $\mathcal{T} \in \mathcal{L}_V(\mathcal{A}_\varphi)$.

Proof. This follows from Walukiewicz's automata-theoretic characterisation of MSO on trees. As already observed in [12], Walukiewicz's automata are, in this case, V-automata with transition specification definable by means of counting one-local formulae.   □

The following theorem is our main result. Informally, it states that every V-automaton on trees can be simulated by an E-automaton on graphs that is equivalent in the sense that a graph is accepted by the E-automata if, and only, if its unravelling is accepted by the V-automaton.

Theorem 12. *For every V-automaton* $\mathcal{A}$ *we can construct an E-automaton* $\mathcal{B}$ *such that a rooted graph* $\mathcal{G}, u$ *is accepted by* $\mathcal{B}$ *if, and only if, its tree unravelling* $\mathcal{T}(\mathcal{G}, u)$ *is accepted by* $\mathcal{A}$.

Proof. Let $\mathcal{A} = (Q, \Sigma, \delta_0, \delta, Acc)$ be a V-automaton. Without loss of generality [29], we may assume that Acc is a parity condition, and that the root constraint and the transition specification are one-local formulae.

As in the proof of Proposition 9, we construct an E-automaton that operates by encoding a vertex-marking run into an edge-marking one. However, in that setting, the two markings were defined on the same graph. Here, we need to overlay all the (counting bisimilar) copies of a graph vertex that occur in the unravelling. To handle this, we proceed by a power-set construction. Essentially, we intend to mark every edge (v, w) in $\mathcal{G}$ by the set of all pairs of states (q, q'') that label copies of v and w connected in $\mathcal{T}(\mathcal{G}, u)$.

Formally, let $\mathcal{B} = (Q', \Sigma, \delta'_0, \delta', Acc')$ be an E-automaton with set of states $Q' = \mathscr{P}(Q \times Q)$. The following formula expresses that the edge marking around a vertex z encodes that a copy of z in the unravelling tree is marked by q:

$$\varphi_q(z) := \forall y \left[\left(E(z, y) \to \bigvee_{\substack{q'\ Q' \\ (q, q'')\ q'}} R_{q'}(z, y) \right) \wedge \left(E(y, z) \to \bigvee_{\substack{q'\ Q' \\ (q'', q)\ q'}} R_{q'}(y, z) \right) \right].$$

Observe that the corresponding formulae in the proof of Proposition 9 are mutually exclusive for different states $q \in Q$. Here, several formulae φ_q may hold at one vertex, as it corresponds to the overlay of several vertices in the unravelling.

The root constraint $\delta'_0(x)$ and the transition specification $\delta'(x)$ for $\mathcal{B}$ is obtained from $\delta_0(x)$ respectively $\delta(x)$ by replacing every atom $P_q z$ with the subformula $\varphi_q(z)$. The infinitary path condition is defined in terms of traces. Given an infinite word $\beta \in (\mathscr{P}(Q \times Q))^\omega$, we say that a word $\alpha \in Q^\omega$ is a *trace* of β, if it is the case that $(\alpha_i, \alpha_{i+1}) \in \beta_i$ for every index i. The acceptance condition Acc' consists of all infinite words $\beta \in Q'^\omega$ for which every trace α belongs to Acc. Clearly, if Acc is regular, then Acc' is regular as well. Notice however, that even when Acc is a parity condition, Acc' is not a prefix-invariant property.

To verify that the construction is correct, let $\mathcal{G} = (V, E, \lambda)$ be a coloured graph with a distinguished root u, and let $\mathcal{T}(\mathcal{G}, u) = (V\,, E\,, \lambda\,)$ be its unravelling from u. Recall that $\mathcal{T}(\mathcal{G}, u)$ is built from $\mathcal{G}$ by taking as vertices the finite paths in $\mathcal{G}$ starting from u. Let $f : V\ \to V$ be the projection that maps every finite path in $V\,$ to its last vertex. Assuming that $\mathcal{T}(\mathcal{G}, u) \in \mathcal{L}_V(\mathcal{A})$, let $\rho : V\ \to Q$ be an accepting run of the automaton $\mathcal{A}$ on this unravelling. We define the edge marking $\rho' : E \to Q'$ by setting, for every $(v, w) \in E$,

$$\rho'(v, w) = \left\{ (\rho(v'), \rho(w')) \in Q \times Q \mid (v', w') \in E \quad \text{with } f(v') = v \text{ and } f(w') = w \right\}.$$

It is not difficult to check that ρ' defined in such a way is an accepting run of the automaton $\mathcal{B}$ on $\mathcal{G}, u$.

Conversely, assume $\mathcal{G}, u \in \mathcal{L}_E(\mathcal{B})$ and let $\rho' : E \to Q'$ be an accepting run of the E-automaton $\mathcal{B}$ on the graph $\mathcal{G}, u$. By induction on the length of paths in $V\,$, and exploiting the memoryless determinacy of parity games [9], one can verify that there exists a vertex marking $\rho : V\ \to Q$ such that $\mathcal{T}(\mathcal{G}, u), \rho \models \delta_0(u)$ and, for every path of the form $\pi v w$ in $V\,$, we have $(\rho(\pi v), \rho(\pi v w)) \in \rho'(v, w)$ and $\mathcal{T}(\mathcal{G}, u), \rho \models \delta(\pi v w)$. By definition of $\mathcal{B}$, it follows that ρ is an accepting run of automaton $\mathcal{A}$ on the unravelling $\mathcal{T}(\mathcal{G}, u)$. $\square$

Corollary 13. *Every class of graphs definable in the counting μ-calculus is recognisable by an E-automaton.*

Proof. Let φ be a formula in the counting μ-calculus. By Theorem 11 there exists a V-automaton $\mathcal{A}_\varphi$ such that for every tree, $\mathcal{T} \in \mathcal{L}_V(\mathcal{A})$ if, and only if $\mathcal{T} \models \varphi$.

Since φ is counting-bisimulation invariant, a graph $\mathcal{G}, u$ satisfies φ if, and only if, its unravelling $\mathcal{T}(\mathcal{G}, u)$ also satisfies φ, that is, if $\mathcal{T}(\mathcal{G}, u) \in \mathcal{L}_V(\mathcal{A})$. Now, the E-automaton $\mathcal{B}$ that simulates $\mathcal{A}$ according to Theorem 12 recognises the model class of φ, as $\mathcal{G}, u \in \mathcal{L}_E(\mathcal{B})$ if, and only if, $\mathcal{T}(\mathcal{G}, u) \in \mathcal{L}_V(\mathcal{A})$.   □

4.2   Application to Expressiveness

As a consequence of the above results, it follows that our ω-regular path conditions for E-automata are optimal in the sense that a more general model, where global acceptance conditions are given by MSO-formulae interpreted over the unravelling of locally consistent markings, would not be more expressive.

Proposition 14. *Consider an E-automaton $\mathcal{A} = (Q, \delta_0, \delta, Acc)$, and an MSO-formula φ over infinite trees with edges marked by Q. Then, there exists an E-automaton $\mathcal{A}_\varphi$ such that $\mathcal{G}, u \in \mathcal{A}_\varphi$ if, and only if, there is a run ρ of $\mathcal{A}$ on $\mathcal{G}, u$ such that $\mathcal{T}((\mathcal{G}, \rho), u) \models \varphi$.*

Proof. We refer to a generalisation of our automata model over input graphs where both edges and vertices are coloured. According to (straighforward generalisations of) Theorem 11 and 12, there exists an E-automaton $\mathcal{B}$ running on graphs $\mathcal{H}, u$ with Q-coloured edges such that, for every such input graph $\mathcal{T}(\mathcal{H}, u) \models \varphi$ if, and only if, $\mathcal{H}, u \in L_E(\mathcal{B})$. The desired automaton $\mathcal{A}_\varphi$ is then obtained by combining the automata $\mathcal{A}$ and $\mathcal{B}$ as a wreath product in which $\mathcal{B}$ reads runs of $\mathcal{A}$.

4.3   Weakly Uniform Tree Automata

A consequence of Gurevich and Shelah's Non-Uniformisation Theorem [11] for Monadic Second-Order Logic on the binary tree is that there exist MSO-definable languages that are not recognised by unambiguous tree-automata [28]. This negative result suggests that developing a notion of uniform recognisability [27] for MSO-definable languages of infinite trees could be very difficult. On the other hand, a success in achieving a notion of uniform recognisability may bear with itself many decision and classification results as those obtained, e.g., for languages of infinite words [21]. Our result on E-automata shows that at least a weak notion of uniformity is available without sacrificing expressiveness.

Definition 15. *Given a V-automaton $\mathcal{A}$ and a tree $\mathcal{T}$, we say that a run ρ of $\mathcal{A}$ on $\mathcal{T}$ is uniform, if any two isomorphic subtrees of $\mathcal{T}$ are marked by ρ in the same way. We say that a V-automaton $\mathcal{A}$ is weakly uniform if, for every tree $\mathcal{T} \in L_V(\mathcal{A})$, there exists an accepting run of $\mathcal{A}$ on $\mathcal{T}$ that is uniform.*

Observe that, in contrast to Thomas' notion of uniform recognisability, our assertion of weak uniformity is constrained to the particular input tree; runs over isomorphic subtrees of different trees may be different.

Theorem 16. *For every MSO-formula φ on trees, there exists a weakly uniform V-automaton that recognises the models of φ.*

Proof. Let $\mathcal{A}$ be a V-automaton equivalent to φ according to Theorem 11. Further, let $\mathcal{B} = (Q, \delta_0, \delta, Acc)$ be the E-automaton that simulates $\mathcal{A}$ according to Theorem 12, such that, for every graph, $\mathcal{G}, u \in \mathcal{L}_E(\mathcal{B})$ if, and only if, $\mathcal{T}(\mathcal{G}, u) \models \varphi$. For every graph $\mathcal{G}, u$ and every accepting run ρ of $\mathcal{B}$ on $\mathcal{G}, u$, let ρ be the marking induced by ρ on the unravelling $\mathcal{T}(\mathcal{G}, u)$. Since $\mathcal{L}_E(\mathcal{B})$ is closed under counting bisimulation, we can modify the transition specification δ, without modifying $\mathcal{L}_E(\mathcal{B})$, in such a way that, whenever ρ is an accepting run of $\mathcal{B}$ on $\mathcal{G}, u$, the unravelling ρ of the marking $(\mathcal{G}, \rho), u$ is also an accepting run of $\mathcal{B}$ on $\mathcal{T}(\mathcal{G}, u)$.

We conclude by proving that $\mathcal{B}$ modified in such a way is weakly uniform. Given any tree $\mathcal{T} \in L_E(\mathcal{B})$, consider the quotient $\mathcal{G}, u_\varepsilon$ of $\mathcal{T}$ under counting bisimulation, with u_ε corresponding to the class of the root. Then there exists an accepting run ρ of $\mathcal{B}$ on $\mathcal{G}, u_\varepsilon$ (since $\mathcal{T}$ and $\mathcal{G}, u_\varepsilon$ are counting bisimilar). Now, one can observe that ρ is a uniform accepting run of $\mathcal{B}$ on $\mathcal{T}$. We obtain the desired V-automaton over trees, by pushing the state-marking of $\mathcal{B}$ from edges towards their target. □

References

[1] M. AJTAI AND R. FAGIN, *Reachability is harder for directed rather than undirected finite graphs*, Journal of Symbolic Logic, 55 (1990), pp. 113–150.

[2] A. ARNOLD AND D. NIWIŃSKI, *Rudiments of mu-calculus*, vol. 146 of Studies in Logic and the Foundations of Mathematics, North-Holland, 2001.

[3] O. BERNHOLTZ AND O. GRUMBERG, *Branching time temporal logic and amorphous tree automata*, in Conf. on Concurrency Theory (CONCUR), vol. 715 of LNCS, Springer-Verlag, 1993, pp. 262–277.

[4] B. COURCELLE, *The monadic second order logic of graph I: Recognizable sets of finite graphs*, Inf. and Comp., 85 (1990), pp. 12–75.

[5] ———, *The monadic second-order logic of graphs VI: On several representations of graphs by logical structures*, Discrete Applied Mathematics, 54 (1994), pp. 117–149.

[6] B. COURCELLE AND P. WEIL, *The recognizability of sets of graphs is a robust property*, Theoretical Computer Science, 342 (2005), pp. 173–228.

[7] V. DIEKERT, *The Book of Traces*, World Scientific Publishing Co., Inc., 1995.

[8] M. DROSTE, P. GASTIN, AND D. KUSKE, *Asynchronous cellular automata for pomsets*, Theoretical Computer Science, 247 (2000), pp. 1–38.

[9] E. A. EMERSON AND C. S. JUTLA, *Tree automata, mu-calculus and determinacy*, in IEEE Symp. on Logic in Computer Science (LICS), 1991, pp. 368–377.

[10] D. GIAMMARESI AND A. RESTIVO, *Two-dimensional languages*, in Handbook of Formal Languages, G. Rozenberg and A. Salomaa, eds., vol. III, Springer-Verlag, 1997, pp. 215–268.

[11] Y. GUREVICH AND S. SHELAH, *Rabin's uniformization problem*, J. Symb. Log., 48 (1983), pp. 1105–1119.

[12] D. JANIN AND G. LENZI, *Relating levels of the mu-calculus hierarchy and levels of the monadic hierarchy*, in IEEE Symp. on Logic in Computer Science (LICS), IEEE Computer Society, 2001, pp. 347–356.

[13] D. JANIN AND I. WALUKIEWICZ, *Automata for the modal mu-calculus and related results*, in Mathematical Found. of Comp. Science (MFCS), vol. 969 of LNCS, Springer-Verlag, 1995.

[14] ——, *On the expressive completeness of the modal mu-calculus with respect to monadic second order logic*, in Conf. on Concurrency Theory (CONCUR), vol. 1119 of LNCS, Springer-Verlag, 1996, pp. 263–277.

[15] T. KAMIMURA AND G. SLUTZKI, *Parallel and two-way automata on directed ordered acyclic graphs*, Information and Control, 49 (1981), pp. 10–51.

[16] D. KOZEN, *Results on the propositional μ-calculus*, Theoretical Comp. Science, 27 (1983), pp. 333–354.

[17] K. LODAYA AND P. WEIL, *Series-parallel languages and the bounded-width property*, Theoretical Computer Science, 237 (2000), pp. 347–380.

[18] O. MATZ AND W. THOMAS, *The monadic quantifier alternation hierarchy over finite graphs is infinite*, in IEEE Symp. on Logic in Computer Science (LICS), 1997, pp. 236–244.

[19] A. A. MUCHNIK, *Games on infinite trees and automata with dead-ends: a new proof for the decidability of the monadic second order theory of two successors*, Bull. EATCS, 42 (1992), pp. 220–267. (traduction d'un article en Russe de 1984).

[20] D. NIWIŃSKI, *Fixed points vs. infinite generation*, in IEEE Symp. on Logic in Computer Science (LICS), 1988, pp. 402–409.

[21] D. PERRIN AND J.-E. PIN, *Infinite Words*, Academic Press, 2002.

[22] A. POTTHOFF, S. SEIBERT, AND W. THOMAS, *Nondeterminism versus determinism of finite automata over directed acyclic graphs*, Bull. Belg. Math. Soc. Simon Stevin, 1 (1994), pp. 285–298.

[23] M. O. RABIN, *Decidability of second order theories and automata on infinite trees*, Trans. Amer. Math. Soc., 141 (1969), pp. 1–35.

[24] J. W. THATCHER AND J. B. WRIGHT, *Generalized finite automata theory with an application to a decision problem of second–order logic*, Math. Systems Theory, 2 (1968), pp. 57–81.

[25] W. THOMAS, *On logics, tilings, and automata*, in Proceedings of the 18th international colloquium on Automata, languages and programming, New York, NY, USA, 1991, Springer-Verlag New York, Inc., pp. 441–454.

[26] ——, *Automata theory on trees and partial orders*, in TAPSOFT'97, M. D. M. Bidoit, ed., no. 1214 in LNCS, Springer-Verlag, 1997, pp. 20–38.

[27] ——, *Uniform and nonuniform recognizability.*, Theor. Comput. Sci., 292 (2003), pp. 299–316.

[28] I. WALUKIEWICZ, *On the ambiguity problem*, 1994. Notes non publiées.

[29] ——, *Monadic second order logic on tree-like structures*, Theoretical Comp. Science, 275 (2002), pp. 311–346.

Conflict Detection for Graph Transformation with Negative Application Conditions

Leen Lambers[1], Hartmut Ehrig[2], and Fernando Orejas[3]

[1] Institut für Softwaretechnik und Theoretische Informatik
Technische Universität Berlin
Germany
leen@cs.tu-berlin.de,
[2] ehrig@cs.tu-berlin.de
[3] Dept. L.S.I. - Technical University Catalonia
Barcelona, Spain
orejas@lsi.upc.edu

Abstract. This paper introduces a new theory needed for the purpose of conflict detection for graph transformation with negative application conditions (NACs). Main results are the formulation of a conflict notion for graph transformation with NACs and a conflict characterization derived from it. A critical pair definition is introduced and completeness of the set of all critical pairs is shown. This means that for each conflict, occuring in a graph transformation system with NACs, there exists a critical pair expressing the same conflict in a minimal context. Moreover a necessary and sufficient condition is presented for parallel independence of graph transformation systems with NACs. In order to facilitate the implementation of the critical pair construction for a graph transformation system with NACs a correct construction is formulated. Finally, it is discussed how to continue with the development of conflict detection and analysis techniques in the near future.

1 Introduction

Several applications using graph transformation need or already use negative application conditions (NACs) to express that certain structures at a given time are forbidden, e.g., [1,2,3,4,5]. In order to allow conflict detection and analysis for these applications, the theory already worked out for graph transformation systems (gts) without NACs should be generalized to gts with NACs. The notion of critical pairs is central in this theory, allowing for conflict detection and analysis. It was developed at first in the area of term rewriting systems (e.g., [6]) and, later, introduced in the area of graph transformation for hypergraph rewriting [7,8] and then for all kinds of transformation systems fitting into the framework of adhesive high-level replacement categories [9].

This paper now generalizes the critical pair notion and some first important related results to gts with NACs. We tailored the theory presented in this paper for gts with NACs and not on other kind of constraints or application conditions, since NACs are already widely used in practice. It would be subject of future

A. Corradini et al. (Eds.): ICGT 2006, LNCS 4178, pp. 61–76, 2006.

work to develop also a critical pair theory for graph transformation with other
kind of constraints as presented in [9]. Subject of future work as well and more
directly related to the subject of this paper is the formulation of a critical pair
lemma which gives a sufficient condition for local confluence of a gts with NACs.

The structure of this paper is as follows. In the first paragraph we repeat the
necessary definitions for graph transformation in the double pushout approach
[10] with NACs. Then we explain carefully what new types of conflicts can occur
because of the NACs by means of a new conflict characterization. This conflict
characterization leads in the next paragraph to a critical pair definition for gts
with NACs. A critical pair describes a conflict in a minimal context. Since now
there occur new types of conflicts we also distinguish other types of critical
pairs. Afterwards we show completeness for this critical pair definition i.e. each
conflict is expressed at least by one critical pair. Moreover we demonstrate,
that if there are no critical pairs at all in the graph transformation system
with NACs then this system is locally confluent or, more exactly, each pair of
direct transformations is parallel independent. In the conclusion and outlook we
explain how to continue with the development of critical pair theory to enable
manageable conflict detection and analysis techniques for gts with NACs.

2   Graph Transformation with NACs

Definition 1 (graph and graph morphism). *A graph $G = (G_E, G_V, s, t)$
consists of a set G_E of edges, a set G_V of vertices and two mappings $s, t :
G_E \to G_V$, assigning to each edge $e \in G_E$ a source $q = s(e) \in G_V$ and target
$z = t(e) \in G_V$. A graph morphism $f : G_1 \to G_2$ between two graphs $G_i =
(G_{i,E}, G_{i,V}, s_i, t_i)$, $(i = 1, 2)$ is a pair $f = (f_E : G_{E,1} \to G_{E,2}, f_V : G_{V,1} \to G_{V,2})$
of mappings, such that $f_V \circ s_1 = s_2 \circ f_E$ and $f_V \circ t_1 = t_2 \circ f_E$. A graph morphism $f :
G_1 \to G_2$ is injective (resp. surjective) if f_V and f_E are injective (resp. surjective)
mappings. Two graph morphisms $m_1 : L_1 \to G$ and $m_2 : L_2 \to G$ are jointly
surjective if $m_{1,V}(L_{1,V}) \cup m_{2,V}(L_{2,V}) = G_V$ and $m_{1,E}(L_{1,E}) \cup m_{2,E}(L_{2,E}) = G_E$.
A pair of jointly surjective morphisms (m_1, m_2) is also called an overlapping of
L_1 and L_2. The category having graphs as objects and graph morphisms as arrows
is called* **Graph**.

Definition 2 (rule). *A graph transformation rule $p : L \xleftarrow{l} K \xrightarrow{r} R$ consists of a
rule name p and a pair of injective graph morphisms $l : K \to L$ and $r : K \to R$.
The graphs L, K and R are called the left-hand side (lhs), the interface, and the
right-hand side (rhs) of p, respectively.*

Definition 3 (match). *Given a rule $p : L \xleftarrow{l} K \xrightarrow{r} R$ and a graph G, one can
try to apply p to G if there is an occurence of L in G i.e. a graph morphism,
called* match *$m : L \to G$.*

A negative application condition or NAC as introduced in [11] forbids a certain
graph structure to be present before or after applying the rule.

Definition 4 (negative application condition)

Let M be the set of all injective graph morphisms.

- *A* negative application condition *or* $NAC(n)$ *on* L *is a graph morphism* $n : L \rightarrow N$. *A graph morphism* $g : L \rightarrow G$ *satisfies* $NAC(n)$ *on* L *i.e.* $g \models NAC(n)$ *if and only if* $\nexists\, q : N \rightarrow G \in M$ *such that* $q \circ n = g$.

$$
\begin{array}{ccc}
L & \xrightarrow{\ n\ } & N \\
{\scriptstyle g}\big\downarrow & {\scriptstyle X} & \Big\downarrow{\scriptstyle q} \\
G & &
\end{array}
$$

- *A* $NAC(n)$ *on* L *(resp. R) for a rule* $p : L \xleftarrow{l} K \xrightarrow{r} R$ *is called* left *(resp. right) NAC on* p. $NAC_{p,L}$ *(resp. $NAC_{p,R}$) is a set of left (resp. right) NACs on* p. $NAC_p = (NAC_{p,L}, NAC_{p,R})$, *consisting of a set of left and a set of right NACs on* p *is called a set of NACs on* p.

Definition 5 (graph transformation with NACs)

- *A graph transformation system with NACs* is a set of rules where each rule $p : L \xleftarrow{l} K \xrightarrow{r} R$ has a set $NAC_p = (NAC_{p,L}, NAC_{p,R})$ of NACs on p.
- *A direct graph transformation* $G \overset{p,g}{\Rightarrow} H$ *via a rule* $p : L \xleftarrow{l} K \xrightarrow{r} R$ *with* $NAC_p = (NAC_{p,L}, NAC_{p,R})$ *and a match* $g : L \rightarrow G$ *consists of the double pushout [10] (DPO)*

$$
\begin{array}{ccc}
L & \xleftarrow{\ l\ } K \xrightarrow{\ r\ } & R \\
{\scriptstyle g}\big\downarrow \quad \big\downarrow \quad & & \big\downarrow{\scriptstyle h} \\
G & \longleftarrow D \longrightarrow & H
\end{array}
$$

where g satisfies each NAC in $NAC_{p,L}$, written $g \models NAC_{p,L}$, and $h : R \rightarrow H$ satisfies each NAC in $NAC_{p,R}$, written $h \models NAC_{p,R}$. Since pushouts in **Graph** *always exist, the DPO can be constructed if the pushout complement of $K \rightarrow L \rightarrow G$ exists. If so, we say that, the match m satisfies the gluing condition of rule p. A graph transformation, denoted as $G_0 \Rightarrow G_n$ is a sequence $G_0 \Rightarrow G_1 \Rightarrow \cdots \Rightarrow G_n$ of direct graph transformations.*

In the example in Fig. 1 a pair of direct transformations via the rules $p_1 : L_1 \leftarrow K_1 \rightarrow R_1$, $p_2 : L_2 \leftarrow K_2 \rightarrow R_2$ and matches m_1 resp. m_2 is depicted. The match m_1 fullfills the negative application condition $NAC(n_1)$ since there is no ingoing edge into node 1 in graph G. The morphism $e_2 \circ m_2$ though doesn't fullfill the negative application condition $NAC(n_1)$ since now there is an edge from node 7 to node 1 in graph H_2.

Remark: From now on we consider only gts with rules having an empty set of right NACs. This is without loss of generality, because each right NAC can be translated into an equivalent left NAC as explained in [9], where Theorem 7.17 can be specialized to NACs.

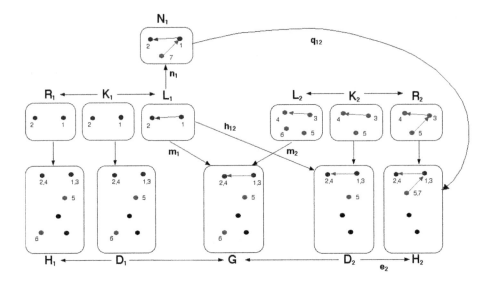

Fig. 1. forbid-produce/delete-use-conflict

3   Conflicts for Graph Transformation with NACs

Confluence conflicts in term rewriting or graph transformation can typically occur when two rules are applied to the same term or graph in such a way that the corresponding redexes (i.e. for graph transformation the images of the corresponding matches) overlap. In particular, the conflict appears when one of the rules can *delete* part of the redex of the other rule. We call these conflicts *delete-use* (or *use-delete*) conflicts. As a consequence, this kind of conflicts are detected by computing the critical pairs of the given system, i.e., such delete-use or use-delete conflicts induced by the overlappings between any two rules. However, when dealing with graph transformation with NACs some new forms of conflict may be present. For instance, an otherwise harmless overlapping (e.g., if no deletion happens) may cause a conflict as the following example shows (for simplicity, in the examples below we will display the rules just in terms of their left and right-hand sides, leaving the context implicit). Suppose that we have two rules p_1 and p_2 with exactly the same left-hand side:

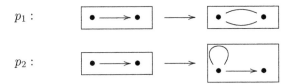

It should be clear that if we apply the rule p_1 to a given graph, then we can apply afterwards the rule p_2 at the same redex. And, conversely, if we apply p_1

we can apply afterwards p_2 at the same redex. However, suppose that the rule p_2 has a left NAC which coincides with the right-hand side of the rule p_1. That is, suppose that the NAC of rule p_2 is defined by the inclusion:

$NAC(p_2)$:

then, obviously, after applying the rule p_1 we would be unable to apply the rule p_2 at the same location because the associated NAC would forbid it. The problem here is that the application of the first rule produces some additional structure that is forbidden by the NAC of the second rule. For this reason we call these new kind of conflicts *produce-forbid* (or *forbid-produce*) conflicts. Actually, these new conflicts may arise even when the possible application of two rules do not overlap. For instance suppose that p_1 and p_2 are the rules below:

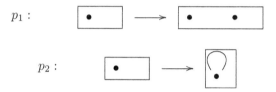

and suppose that the rule p_2 includes the left NAC:

$NAC(p_2)$:

meaning that the rule p_2 cannot be applied to a graph including at least three nodes. Now, suppose that we have a graph G including just two nodes a and b. Obviously, we can apply rule p_1 to G at node a and rule p_2 at node b without any overlapping. However, if we first apply rule p_1 this causes the creation of a new node that would now forbid the application of rule p_2 at node b.

In what follows, we will first look at the concept of parallel independence of two direct transformations with NACs, which expresses the condition to be fulfilled in order to apply two different rules to the same graph in any order with the same result. This is proven in Theorem 1, the Local Church-Rosser Theorem with NACs. Afterwards we will provide the conflict notion for gts with NACs and a characterization of the conflicts as described above.

Definition 6 (parallel independence). *Two direct transformations* $G \overset{(p_1,m_1)}{\Longrightarrow} H_1$ *with* NAC_{p_1} *and* $G \overset{(p_2,m_2)}{\Longrightarrow} H_2$ *with* NAC_{p_2} *are parallel independent if*

$$\exists h_{12} : L_1 \rightarrow D_2 \text{ s.t. } (d_2 \circ h_{12} = m_1 \text{ and } e_2 \circ h_{12} \models NAC_{p_1})$$

and

$$\exists h_{21} : L_2 \rightarrow D_1 \text{ s.t. } (d_1 \circ h_{21} = m_2 \text{ and } e_1 \circ h_{21} \models NAC_{p_2})$$

as in the following diagram:

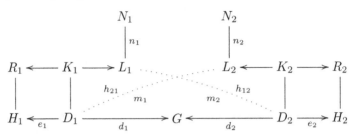

Theorem 1 (Local Church-Rosser Theorem with NACs). *If a pair of direct transformations $H_1 \overset{p_1}{\Leftarrow} G \overset{p_2}{\Rightarrow} H_2$ with NACs is parallel independent, then there are two direct transformations $H_1 \overset{p_2}{\Rightarrow} H$ and $H_2 \overset{p_1}{\Rightarrow} H$ with NACs s.t.*

$$
\begin{array}{ccc}
 & H_1 & \\
\scriptstyle p_1 & & \scriptstyle p_2 \\
G & & H \\
\scriptstyle p_2 & & \scriptstyle p_1 \\
 & H_2 &
\end{array}
$$

Proof. Because of the Local Church-Rosser Theorem for rules without NACs all necessary pushouts in $H_1 \overset{p_2}{\Rightarrow} H$ and $H_2 \overset{p_1}{\Rightarrow} H$ can be constructed and moreover the matches $e_2 \circ h_1$ and $e_1 \circ h_2$ satisfy the NACs of rule p_1 resp. p_2 by the definition of parallel independence for graph transformation with NACs.

The following lemma describes that, if a match for the potential second transformation exists, it is unique. Moreover this lemma will allow an elegant conflict characterization in Lemma 2.

Lemma 1 (unique match). *Given two direct transformations $G \overset{(p_1,m_1)}{\Longrightarrow} H_1$ with NAC_{p_1} and $G \overset{(p_2,m_2)}{\Longrightarrow} H_2$ with NAC_{p_2}, then the following holds:*

- *if $\exists h_{12} : L_1 \to D_2$ s.t. $d_2 \circ h_{12} = m_1$ then h_{12} is unique*
- *if $\exists h_{21} : L_2 \to D_1$ s.t. $d_1 \circ h_{21} = m_2$ then h_{21} is unique.*

Proof. Since each rule consists of two injective morphisms and pushouts are closed under injective morphisms, d_1 and d_2 are injective morphisms as well. If there would exist $h'_{12} : L_1 \to D_2 : d_2 \circ h'_{12} = m_1$ then because of d_2 injective and $d_2 \circ h'_{12} = d_2 \circ h_{12} = m_1$ it follows that $h'_{12} = h_{12}$. Analogously one can prove that h_{21} is unique.

Definition 7 (conflict). *Two direct transformations $G \overset{(p_1,m_1)}{\Longrightarrow} H_1$ with NAC_{p_1} and $G \overset{(p_2,m_2)}{\Longrightarrow} H_2$ with NAC_{p_2} are in conflict if they are not parallel independent i.e. if*

$$\nexists h_{12} : L_1 \to D_2 \text{ s.t. } (d_2 \circ h_{12} = m_1 \text{ and } e_2 \circ h_{12} \models NAC_{p_1})$$

or

$$\nexists h_{21} : L_2 \to D_1 \text{ s.t. } (d_1 \circ h_{21} = m_2 \text{ and } e_1 \circ h_{21} \models NAC_{p_2}).$$

The following lemma characterizes this conflict notion for graph transformation with NACs s.t. the difference with the conflict notion for graph transformation without NACs becomes more clear. As described in the introduction of this section new types of conflicts can occur and the lemma in fact characterizes four different types of conflicts that can occur partly simultaneously.

Two direct transformations $G \overset{(p_1,m_1)}{\Rightarrow} H_1$ and $G \overset{(p_2,m_2)}{\Rightarrow} H_2$ are in delete-use-conflict (resp.use-delete-conflict) if rule p_1 (resp. p_2) deletes part of the graph G, which is used by rule p_2 (resp.p_1) in the second (resp. first) direct transformation. This kind of conflict occurs also in gts without NACs [12]. In contrast a produce-forbid-conflict (resp. forbid-produce-conflict) occurs only in gts with NACs. Namely, if rule p_1 (resp.p_2) produces a graph structure which is forbidden by the NAC of rule p_2 (resp. p_1).

Lemma 2 (conflict characterizaton). *Two direct transformations* $G \overset{(p_1,m_1)}{\Rightarrow}$ H_1 *with* NAC_{p_1} *and* $G \overset{(p_2,m_2)}{\Rightarrow} H_2$ *with* NAC_{p_2} *are in conflict if and only if:*

1. (a) $\not\exists h_{12} : L_1 \to D_2 : d_2 \circ h_{12} = m_1$ *(use-delete-conflict)*
     or
   (b) *there exists a unique* $h_{12} : L_1 \to D_2 : d_2 \circ h_{12} = m_1$, *but* $e_2 \circ h_{12} \not\models NAC_{p_1}$
       (forbid-produce-conflict)

   or
2. (a) $\not\exists h_{21} : L_2 \to D_1 : d_1 \circ h_{21} = m_2$ *(delete-use-conflict)*
     or
   (b) *there exists a unique* $h_{21} : L_2 \to D_1 : d_1 \circ h_{21} = m_2$, *but* $e_1 \circ h_{21} \not\models NAC_{p_2}$
       (produce-forbid-conflict).

Proof. $G \overset{(p_1,m_1)}{\Rightarrow} H_1$ *with* NAC_{p_1} *and* $G \overset{(p_2,m_2)}{\Rightarrow} H_2$ *with* NAC_{p_2} *are in conflict if*

$$\not\exists h_{12} : L_1 \to D_2 \text{ s.t. } (d_2 \circ h_{12} = m_1 \text{ and } e_2 \circ h_{12} \models NAC_{p_1})$$

or

$$\not\exists h_{21} : L_2 \to D_1 \text{ s.t. } (d_1 \circ h_{21} = m_2 \text{ and } e_1 \circ h_{21} \models NAC_{p_2})$$

We consider at first the first line of this disjunction. Let $A(h_{12}) := d_2 \circ h_{12} = m_1$, $B(h_{12}) := e_2 \circ h_{12} \models NAC_{p_1}$, $P(h_{12}) := (A(h_{12}) \wedge B(h_{12}))$ and M_{12} be the set of all morphisms from L_1 to D_2. Then the first line is equivalent to

$$\not\exists h_{12} \in M_{12} : (A(h_{12}) \wedge B(h_{12})) \equiv \not\exists h_{12} \in M_{12} : P(h_{12})$$

This is equivalent to

$$\forall h_{12} \in M_{12} : \neg P(h_{12}) \equiv (M_{12} = \emptyset) \vee (M_{12} \neq \emptyset \wedge \forall h_{12} \in M_{12} : \neg P(h_{12}))$$

Moreover $P \equiv A \wedge B \equiv A \wedge (A \Rightarrow B)$ and thus $\neg P \equiv \neg(A \wedge B) \equiv \neg(A \wedge (A \Rightarrow B)) \equiv \neg A \vee \neg(A \Rightarrow B) \equiv \neg A \vee \neg(\neg A \vee B) \equiv \neg A \vee (A \wedge \neg B)$. This implies that $(M_{12} = \emptyset) \vee (M_{12} \neq \emptyset \wedge \forall h_{12} \in M_{12} : \neg P(h_{12})) \equiv$

$$(M_{12} = \emptyset) \vee (M_{12} \neq \emptyset \wedge \forall h_{12} \in M_{12} : \neg A(h_{12}) \vee (A(h_{12}) \wedge \neg B(h_{12})))$$

Because of Lemma 1 and because the disjunction holding for each morphism in M_{12} is an exclusive one this is equivalent to

$$(M_{12} = \emptyset) \vee (M_{12} \neq \emptyset \wedge \forall h_{12} \in M_{12} : \neg A(h_{12})) \vee (\exists ! h_{12} \in M_{12} : (A(h_{12}) \wedge \neg B(h_{12})))$$

Now $(M_{12} = \emptyset) \vee (M_{12} \neq \emptyset \wedge \forall h_{12} \in M_{12} : \neg A(h_{12})) \equiv \forall h_{12} \in M_{12} : \neg A(h_{12}) \equiv \nexists h_{12} \in M_{12} : A(h_{12})$. This implies finally that $\nexists h_{12} : L_1 \rightarrow D_2$ s.t. $(d_2 \circ h_{12} = m_1$ and $e_2 \circ h_{12} \models NAC_{p_1})$ is equivalent to

$$(\nexists h_{12} \in M_{12} : d_2 \circ h_{12} = m_1) \vee (\exists ! h_{12} \in M_{12} : (d_2 \circ h_{12} = m_1 \wedge e_2 \circ h_{12} \not\models NAC_{p_1}))$$

is equivalent to

1. (a) $\nexists h_{12} : L_1 \rightarrow D_2 : d_2 \circ h_{12} = m_1$ (use-delete-conflict)
   or
   (b) there exists a unique $h_{12} : L_1 \rightarrow D_2 : d_2 \circ h_{12} = m_1$, but $e_2 \circ h_{12} \not\models NAC_{p_1}$ (forbid-produce-conflict)

Analogously we can proceed for the second part of the disjunction.

Note that a use-delete-conflict (resp. delete-use-conflict) cannot occur simultaneously to a forbid-produce-conflict (resp. produce-forbid-conflict), since $(1.a) \Rightarrow \neg(1.b)$ (resp. $(2.a) \Rightarrow \neg(2.b)$). The following types of conflicts can occur simultaneously though: use-delete/delete-use-, use-delete/produce-forbid-, forbid-produce/delete-use-, forbid-produce/produce-forbid-conflict. In the example in Fig. 1 a pair of direct transformations in forbid-produce/delete-use-conflict is shown. In this case the first rule forbids an additional edge pointing to node (1,3) which is added by the second rule and the first rule deletes the edge (1,3)-(2,4) which is used by the second rule. Note that the labels express how nodes and edges are mapped to each other.

4   Critical Pairs for Graph Transformation with NACs

Now that we have a detailed conflict characterization we can look at a conflict in a minimal context i.e. a *critical pair*. Basically we exclude from the conflict all the graph parts that in no way can be responsible for the occurence of the conflict. In the case of a *delete-use-conflict* this would be the graph context which is not reached by any of the matches of the lhs's of the rules. This is because these graph parts can not be used nor deleted by any of the rules anyway. Therefore we consider only jointly surjective matches or overlappings of the lhs's of both rules in part $(1a)$ and $(2a)$ of the following critical pair definition. In the case of a *produce-forbid-conflict* we can leave out the graph parts which are not affected by any negative application condition of one rule and reached by a match of the rhs of the other rule. This is because these graph parts are not forbidden by a NAC of one rule and can not have been produced by the other rule anyway. Therefore we only consider overlappings or jointly surjective mappings of the NAC of one rule with the rhs of the other rule in part $(1b)$ and $(2b)$ of the following critical pair definition. Thus in fact in the example in Fig. 1 we obtain the critical pair by ignoring all unlabelled graph nodes. Remember that in the following critical pair definition M is the set of all injective graph morphisms as defined in Def. 4.

Definition 8 (critical pair). *A critical pair is a pair of direct transformations* $K \overset{(p_1,m_1)}{\Rightarrow} P_1$ *with* NAC_{p_1} *and* $K \overset{(p_2,m_2)}{\Rightarrow} P_2$ *with* NAC_{p_2} *such that:*

1. *(a)* $\nexists h_{12} : L_1 \rightarrow D_2 : d_2 \circ h_{12} = m_1$ *and* (m_1, m_2) *jointly surjective (use-delete-conflict)*

   or

   (b) there exists a $h_{12} : L_1 \rightarrow D_2$ *s.t.* $d_2 \circ h_{12} = m_1$, *but for one of the NACs* $n_1 : L_1 \rightarrow N_1$ *of* p_1 *there exists a morphism* $q_{12} : N_1 \rightarrow P_2 \in M$ *s.t.* $q_{12} \circ n_1 = e_2 \circ h_{12}$ *and* (q_{12}, h_2) *jointly surjective (forbid-produce-conflict)*

   or

2. *(a)* $\nexists h_{21} : L_2 \rightarrow D_1 : d_1 \circ h_{21} = m_2$ *and* (m_1, m_2) *jointly surjective (delete-use-conflict)*

   or

   (b) there exists a $h_{21} : L_2 \rightarrow D_1$ *s.t.* $d_1 \circ h_{21} = m_2$, *but for one of the NACs* $n_2 : L_2 \rightarrow N_2$ *of* p_2 *there exists a morphism* $q_{21} : N_2 \rightarrow P_1 \in M$ *s.t.* $q_{21} \circ n_2 = e_1 \circ h_{21}$ *and* (q_{21}, h_1) *jointly surjective (produce-forbid-conflict)*

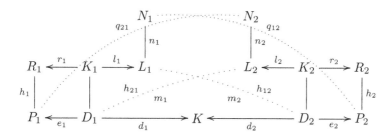

Remarks to related work: Note that the definition in this paper for parallel independence and conflict as well as the Local Church Rosser Theorem for graph transformation with NACs coincide with their equivalents as introduced in [11]. Moreover, if the gts doesn't hold any NAC, then the definition of parallel independence, conflict and critical pair as given in this paper correspond to the respective definition in the context of graph transformation without NACs [9]. Leadoff ideas to capture the critical pair notion for graph transformation with NACs were described in [13] and coincide with the formalization in this paper. Furthermore in [5] so-called critical conflict pairs for single pushout graph transformation with NACs are defined. The correspondence between this notion and the critical pair notion as introduced in this paper should be investigated in more detail.

Now we prove that Definition 8 of critical pairs leads to completeness. This means, that each occuring conflict in the graph transformation system with NACs can be expressed by a critical pair i.e. the same kind of conflict but in a minimal context. Therefore at first we need the following definition and lemma.

Definition 9 (extension diagram). *An* extension diagram *is a diagram (1),*

$$G_0 \overset{t}{\Longrightarrow} G_n$$

$$k_0 \downarrow \quad (1) \quad \downarrow k_n$$

$$G'_0 \overset{t'}{\Longrightarrow} G'_n$$

where, $k_0 : G_0 \to G'_0$ *is a morphism, called extension morphism, and* $t :$ $G_0 \Rightarrow G_n$ *and* $t' : G'_0 \Rightarrow G'_n$ *are transformations via the same productions* $(p_0, \cdots, p_{n-1})$ *and matches* $(m_0, \cdots, m_{n-1})$ *and* $(k_0 \circ m_0, \cdots, k_{n-1} \circ m_{n-1})$ *respectively, defined by the following DPO diagrams :*

$$p_i : \quad L_i \overset{l_i}{\longleftarrow} K_i \overset{r_i}{\longrightarrow} R_i$$

$$m_i \downarrow \qquad j_i \downarrow \qquad n_i \downarrow$$

$$G_i \overset{f_i}{\longleftarrow} D_i \overset{g_i}{\longrightarrow} G_{i+1}$$

$$k_i \downarrow \qquad d_i \downarrow \qquad k_{i+1} \downarrow$$

$$G'_i \overset{f'_i}{\longleftarrow} D'_i \overset{g'_i}{\longrightarrow} G'_{i+1}$$

Remark: Since t and t' are transformations for a gts with NACs, the matches $(m_0, \cdots, m_{n-1})$ and $(k_0 \circ m_0, \cdots, k_{n-1} \circ m_{n-1})$ have to satisfy the NACs of the rules $(p_0, \cdots, p_{n-1})$.

Lemma 3 (induced direct transformation). *Given a direct transformation* $G \overset{p}{\Rightarrow} H$ *with NACs via the rule* $p : L \overset{l}{\leftarrow} K \overset{r}{\to} R$ *and match* $m : L \to G$ *and given an object* K' *with two morphisms* $L \overset{m_{lk}}{\to} K' \overset{m_{kg}}{\to} G$ *s.t.* $m = m_{kg} \circ m_{lk}$, *with* $m_{kg} \in M$, *then there exists a direct transformation, the so called* induced direct transformation $K' \overset{p}{\Rightarrow} P$ *via the same rule* p *and the match* m_{lk}, *satisfying the NACs of* p *as in the following diagram :*

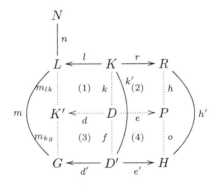

Proof. Given $G \overset{p}{\Rightarrow} H$ with NAC n as shown above. Since $d' \in M$ we can take pullback (3) of m_{kg} and d'. Since $m_{kg} \circ m_{lk} \circ l = m \circ l = d' \circ k'$ then there exists a morphism $k : K \rightarrow D$ with $k' = f \circ k$ and $d \circ k = m_{lk} \circ l$ because of the pullback property of (3). Because of the pushout-pullback-decomposition lemma [9], $l \in M$ and $m_{kg} \in M$ diagrams (1) and (3) are both pushouts. Now we can construct pushout (2) of $D \leftarrow K \rightarrow R$ because of $r \in M$. Since $e' \circ f \circ k = e' \circ k' = h' \circ r$ there exists a morphism $o : P \rightarrow H$ with $o \circ h = h'$ and $o \circ e = e' \circ f$ because of the pushout-property of (2). Because of the pushout-decomposition property also diagram (4) is a pushout.

It remains to show that m_{lk} satisfies the NACs of p. Suppose that m_{lk} doesn't fullfill some $NAC(n)$ of p, then there exists a morphism $q : N \rightarrow K' \in M$ s.t. $q \circ n = m_{lk}$, but this implies $m_{kg} \circ q \circ n = m_{kg} \circ m_{lk} = m$ with $m_{kg} \circ q \in M$ and this is a contradiction.

Theorem 2 (completeness of critical pairs). *For each pair of direct trans-formations* $H_1 \overset{(p_1, m_1')}{\Leftarrow} G \overset{(p_2, m_2')}{\Rightarrow} H_2$ *in conflict there is a critical pair with exten-sion diagrams (1) and (2) and* $m \in M$.

$$
\begin{array}{ccccc}
P_1 & \Longleftarrow & K & \Longrightarrow & P_2 \\
\big| & & \big| & & \big| \\
& (1)\ m & & (2) & \\
H_1 & \Longleftarrow & G & \Longrightarrow & H_2
\end{array}
$$

Proof. According to Lemma 2 the following reasons are responsible for a pair of direct transformations $G \overset{(p_1, m_1')}{\Rightarrow} H_1$ with NAC_{p_1} and $G \overset{(p_2, m_2')}{\Rightarrow} H_2$ with NAC_{p_2} to be *in conflict*:

1. (a) $\nexists h_{12}' : L_1 \rightarrow D_2' : d_2' \circ h_{12}' = m_1'$ (use-delete-conflict)
    or
   (b) there exists a unique $h_{12}' : L_1 \rightarrow D_2' : d_2' \circ h_{12}' = m_1'$, but $e_2' \circ h_{12}' \not\models NAC_{p_1}$ (forbid-produce-conflict)

    or
2. (a) $\nexists h_{21}' : L_2 \rightarrow D_1 : d_1' \circ h_{21}' = m_2'$ (delete-use-conflict)
    or
   (b) there exists a unique $h_{21}' : L_2 \rightarrow D_1 : d_1' \circ h_{21}' = m_2'$, but $e_1' \circ h_{21}' \not\models NAC_{p_2}$ (produce-forbid-conflict)

It is possible, that (1.b) and (2.b) are both false. In this case, (1.a) or (2.a) have to be true which corresponds to the usual use-delete-conflict (resp. delete-use-conflict) and in [9] it is described how to embed a critical pair into this pair of direct transformations. In the other case (1.b) or (2.b) are true. Let at first (1.b) be true. This means that there exists a unique $h_{12}' : L_1 \rightarrow D_2' : d_2' \circ h_{12}' = m_1'$, but $e_2' \circ h_{12}' \not\models NAC_{p_1}$. Thus for one of the NACs $n_1 : L_1 \rightarrow N_1$ of p_1 there exists a morphism $q_{12}' : N_1 \rightarrow H_2 \in M$ such that $q_{12}' \circ n_1 = e_2' \circ h_{12}'$. For each pair of graph morphisms with the same codomain, there exists an $E - M$ pair factorization [9] with E the set of all jointly surjective morphisms and M the set of injective graph morphisms as defined in Def. 4. Thus for $q_{12}' : N_1 \rightarrow H_2$ and $h_2' : R_2 \rightarrow H_2$ we obtain an object P_2 and morphisms $h_2 : R_2 \rightarrow P_2$,

$q_{12} : N_1 \to P_2$ and $o_2 : P_2 \to H_2$ with (h_2, q_{12}) jointly surjective and $o_2 \in M$ such that $o_2 \circ h_2 = h'_2$ and $o_2 \circ q_{12} = q'_{12}$. Because of Lemma 3 pushouts (5) - (8) can be constructed, if we consider the fact that also $H_2 \Rightarrow G$ is a direct transformation via the inverse rule of p_2. Since $o_2 \in M$ and (7) and (8) are pushouts also $f_2 \in M$ and $m \in M$. Because of the same argumentation as in Lemma 3, since m'_2 fullfills all the NACs of p_2 also m_2 fullfills them. Now we have the first half $K \Rightarrow P_2$ of the critical pair under construction.

We still have to check if this critical pair is in forbid-produce-conflict. Since (8) is a pullback and $o_2 \circ q_{12} \circ n_1 = q'_{12} \circ n_1 = e'_2 \circ h'_{12}$ there exists a morphism $h_{12} : L_1 \to D_2$, with $e_2 \circ h_{12} = q_{12} \circ n_1$ and $f_2 \circ h_{12} = h'_{12}$. Because $q'_{12} = o_2 \circ q_{12} \in M$ and $o_2 \in M$ we have $q_{12} \in M$. This means, that $e_2 \circ h_{12}$ doesn't fullfill the NAC $n_1 : L_1 \to N_1$.

Now we can start constructing the second half of the critical pair. Let m_1 be the morphism $d_2 \circ h_{12}$, then the following holds $m \circ m_1 = m \circ d_2 \circ h_{12} = d'_2 \circ f_2 \circ h_{12} = d'_2 \circ h'_{12} = m'_1$.

Because of Lemma 3 and $m \in M$ pushouts (1) - (4) can be constructed and m_1 satisfies the NACs of p_1. Thus finally we obtain a critical pair according to Def. 8 of type (1.b) because we have h_{12} with $d_2 \circ h_{12} = m_1$. Moreover there is $q_{12} \in M$ with (q_{12}, h_2) jointly surjective and $e_2 \circ h_{12} = q_{12} \circ n_1$.

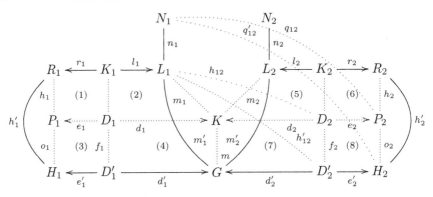

We can proceed analogously for the case of (2.b) being true leading to a critical pair of type (2.b) according to Def. 8.

In the example in Fig. 1 the critical pair, obtained by ignoring all unlabelled nodes, can be embedded into the forbid-produce-delete-use-conflict depicted in this figure in a bigger context (i.e. two extra nodes).

Fact 3 (necessary and sufficient condition for parallel independence). *Each pair of direct transformations $H_1 \Leftarrow G \Rightarrow H_2$ in a gts with NACs is parallel independent if and only if there are no critical pairs for this gts with NACs. A gts with NACs is locally confluent if there are no critical pairs for this gts with NACs.*

Proof. – Given a gts with NACs with an empty set of critical pairs and let $H_1 \Leftarrow G \Rightarrow H_2$ be a pair of non parallel independent direct graph transformations for this gts with NACs. This is a contradiction, since then there would exist a critical pair which can be embedded into this pair of direct transformations as in Theorem 2.

- Given a gts with NACs with only parallel independent pairs of direct transformations $H_1 \Leftarrow G \Rightarrow H_2$. Then the set of critical pairs has to be empty, otherwise a critical pair would be a pair of non parallel independent direct transformations.
- If each pair of direct transformations $H_1 \Leftarrow G \Rightarrow H_2$ in a gts with NACs is parallel independent then each pair is also locally confluent and in consequence this gts with NACs is locally confluent.

5   Conflict Detection for Graph Transformation with NACs

5.1   Construction of Critical Pairs

Critical pairs allow for static conflict detection. Each conflict, occuring at some moment in the graph transformation, is represented by a critical pair. Thus it is possible to foresee each conflict by computing the set of all critical pairs before running the gts as implemented in the graph transformation tool AGG [14]. Each pair of rules of the gts induces a set of critical pairs. Computing this set for each pair of rules delivers us in the end the complete set of critical pairs for a gts. Here a straightforward construction is given to compute the set of critical pairs for a given pair of rules of the gts with NACs. Note that there exists already a more efficient construction for critical pairs in delete-use- or use-delete-conflict in step 1 described in [12]. For lack of space we only refer to it here and give instead the straightforward construction. Moreover we think that, using similar techniques, we could provide also a more efficient construction for the produce-forbid and forbid-produce critical pairs, but this is current work.

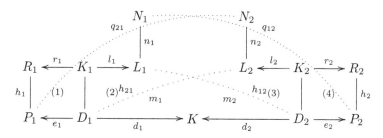

Given a pair of rules $(p_1 : L_1 \leftarrow K_1 \rightarrow R_1, p_2 : L_2 \leftarrow K_2 \rightarrow R_2)$ with NACs:

1. Consider any jointly surjective pair $(m_1 : L_1 \rightarrow K, m_2 : L_2 \rightarrow K)$.
   (a) Check gluing condition for (l_1, m_1) and (l_2, m_2). If it is satisfied then construct PO-complements D_1, D_2 in (2),(3) and PO's P_1, P_2 in (1) and (4).
   (b) Check if the pair of direct transformations $P_1 \Leftarrow K \Rightarrow P_2$ is in delete-use or use-delete-conflict, leading to critical pair $P_1 \Leftarrow K \Rightarrow P_2$.

2. Consider for each NAC $n_1 : L_1 \to N_1$ of p_1 any jointly surjective pair of morphisms $(h_2 : R_2 \to P_2, q_{12} : N_1 \to P_2)$ with q_{12} injective.
   (a) Check gluing condition for (h_2, r_2). If it is satisfied, then construct PO-complement D_2 in (4).
   (b) Construct PO K in (3) and abort, if $m_2 \not\models NAC_{p_2}$.
   (c) Check existence of $h_{12} : L_1 \to D_2$ s.t. $e_2 \circ h_{12} = q_{12} \circ n_1$ (e_2 injective implies uniqueness of h_{12}). If not existent, then abort.
   (d) Define $m_1 = d_2 \circ h_{12} : L_1 \to K$ and abort if $m_1 \not\models NAC_{p_1}$.
   (e) Check gluing condition for (m_1, l_1). If it is satisfied, then construct PO-complement D_1 in (2).
   (f) Construct P_1 as PO in (1) leading to critical pair $P_1 \Leftarrow K \Rightarrow P_2$.
3. Consider for each NAC $n_2 : L_2 \to N_2$ of p_2 any jointly surjective pair of morphisms $(h_1 : R_1 \to P_1, q_{21} : N_2 \to P_1)$ with q_{21} injective and continue analog to step 2.

5.2   Correctness of This Construction

The construction in the last paragraph is derived quite straightforwardly from Definition 8 and we are able to show that in fact it yields all critical pairs of a pair of rules of the gts with NACs.

Theorem 4. *The critical pair construction in paragraph 5.1 yields the set of all critical pairs for a pair of rules (p_1, p_2) of a gts with NACs.*

Proof.   – At first we prove that the pair of direct transformations constructed in steps 1,2 and 3 is really a critical pair. Step 1: Since the matches (m_1, m_2) of $P_1 \Leftarrow K \Rightarrow P_2$ are jointly surjective and this pair is in delete-use- or use-delete-conflict this is a critical pair. Step 2: Since there exists a morphism $h_{12} : L_1 \to D_2$ with $m_1 = d_2 \circ h_{12}$ and an injective morphism $q_{12} : N_1 \to P_2$ with $e_2 \circ h_{12} = q_{12} \circ n_1$ and (h_2, q_{12}) jointly surjective, this is a critical pair in forbid-produce-conflict. Step 3: Analog to Step 2.

– Secondly we prove that each critical pair is constructed by step 1, 2 or 3. Looking at Definition 8 there are three different types of critical pairs. Given a critical pair $P_1 \Leftarrow K \Rightarrow P_2$ of type 1a or 2a it is constructed by step 1. This is because the matches (m_1, m_2) are jointly surjective, (l_1, m_1) and (l_2, m_2) satisfy the gluing condition, because (2) and (3) are pushouts, (1) and (4) are also pushouts, pushouts are unique up to isomorphy and $P_1 \Leftarrow K \Rightarrow P_2$ are in delete-use- or use-delete-conflict. Given a critical pair $P_1 \Leftarrow K \Rightarrow P_2$ of type (1b) it is constructed by step 2. This is because (h_2, q_{12}) are jointly surjective, the gluing condition for (h_2, r_2) is satisfied because (4) is a pushout, (3) is a pushout, $m_2 \models NAC_{p_2}$, $h_{12} : L_1 \to D_2$ exists s.t. $e_2 \circ h_{12} = q_{12} \circ n_1$, $m_1 \models NAC_{p_1}$, the gluing condition for (m_1, l_1) holds since (2) is a pushout, (1) is a pushout and pushouts are unique up to ismorphy. Given a critical pair $P_1 \Leftarrow K \Rightarrow P_2$ of type (2b) it is constructed by step 3 analogously to a critical pair of type (1b).

6   Conclusion and Outlook

We presented the first foundations for a critical pair theory for gts with NACs which in the end should lead to good conflict detection and analysis algorithms for all kinds of systems described with means of gts with NACs. Main results in this paper are a conflict notion and conflict characterization for gts with NACs. The definition of a critical pair for gts with NACs for which we could prove completeness. We provided a straightforward and correct construction of the set of all critical pairs.

The theory presented in this paper can be generalized to adhesive HLR systems [9] with NACs. It is subject of future work to reformulate in detail all results and proofs mentioned in this paper on this more abstract level. Note that we tuned most reasonings in this paper already for this generalization such that it will be a relatively straightforward step. Once formulated the theory for adhesive HLR systems with NACs it is possible to instantiate it in particular for typed attributed graph transformation systems with NACs. This more general kind of graph transformation technique is most significant for modeling and metamodeling in software engineering and visual languages.

The theory of critical pairs consists of an other important part not mentioned yet in this paper. In gts without NACs the critical pair lemma holds. It gives a sufficient condition for the gts to be confluent. This is the case if all critical pairs are strictly confluent [9]. Thus, the critical pair lemma enables us to infer confluence behaviour of the whole graph transformation system by investigating the confluence behaviour of the set of all critical pairs. A similar result should be obtained for gts with NACs. This is work in progress and we are confident to be on the right path to complete it with the critical pair definition presented in this paper.

Moreover the results in this paper build a necessary theoretical foundation to continue with investigations on how to design conflict detection and analysis for typed, attributed gts with NACs as manageable as possible. For gts without NACs in [12] a rule analysis was proposed in order to obtain a more efficient conflict detection as the straightforward one. In [15] this efficiency investigation was continued by designing the so-called essential critical pairs. They build a subset of all critical pairs and represent each conflict not only in a minimal context, but also in a unique way. It should be possible to formulate also for critical pairs with NACs such a subset of essential critical pairs, by analyzing the rules and defining the exact conflict reason for each conflict. Finally future work is not only concerned with optimizations for conflict detection, but also for conflict analysis or finding a manageable way to investigate the resolvability of each conflict.

References

1. Hausmann, J., Heckel, R., Taentzer, G.: Detection of Conflicting Functional Requirements in a Use Case-Driven Approach. In: Proc. of Int. Conference on Software Engineering 2002, Orlando, USA (2002)

2. Mens, T., Taentzer, G., Runge, O.: Detecting Structural Refactoring Conflicts using Critical Pair Analysis. In Heckel, R., Mens, T., eds.: Proc. Workshop on Software Evolution through Transformations: Model-based vs. Implementation-level Solutions (SETra'04), Satellite Event of ICGT'04), Rome, Italy, ENTCS (2004)

3. Taentzer, G., Ehrig, K., Guerra, E., de Lara, J., Lengyel, L., Levendovsky, T., Prange, U., Varro, D., Varro-Gyapay, S.: Model Transformation by Graph Transformation: A Comparative Study. In: Proc. Workshop Model Transformation in Practice, Montego Bay, Jamaica (2005)

4. Bottoni, P., Schürr, A., Taentzer, G.: Efficient Parsing of Visual Languages based on Critical Pair Analysis and Contextual Layered Graph Transformation. In: Proc. IEEE Symposium on Visual Languages. (2000) Long version available as technical report SI-2000-06, University of Rom.

5. Koch, M., Mancini, L., Parisi-Presicce, F.: Graph-based Specification of Acces Control Policies. In: JCSS 71. (2005) 1–33

6. Huet, G.: Confluent reductions: Abstract properties and applications to term rewriting systems. JACM **27,4** (1980) 797–821

7. Plump, D.: Hypergraph Rewriting: Critical Pairs and Undecidability of Confluence. In Sleep, M., Plasmeijer, M., van Eekelen, M.C., eds.: Term Graph Rewriting. Wiley (1993) 201–214

8. Plump, D.: Confluence of graph transformation revisited. In Middeldorp, A., van Oostrom, V., van Raamsdonk, F., de Vrijer, R., eds.: Processes, Terms and Cycles: Steps on the Road to Infinity:Essays Dedicated to Jan Willem Klop on the Occasion of His 60th Birthday. Volume 3838 of Lecture Notes in Computer Science. Springer (2005) 280,308

9. Ehrig, H., Ehrig, K., Prange, U., Taentzer, G.: Fundamentals of Algebraic Graph Transformation. EATCS Monographs in Theoretical Computer Science. Springer (2006)

10. Corradini, A., Montanari, U., Rossi, F., Ehrig, H., Heckel, R., Löwe, M.: Algebraic approaches to graph transformation I : Basic Concepts and Double Pushout Approach. In Rozenberg, G., ed.: Handbook of Graph Grammars and Computing by Graph Transformation, Volume 1: Foundations. World Scientific (1997) 163–245

11. Annegret Habel, R.H., Taentzer, G.: Graph grammars with negative application conditions. Fundamenta Informaticae **26** (1996) 287–313

12. Lambers, L., Ehrig, H., Orejas, F.: Efficient detection of conflicts in graph-based model transformation. In: Proc. International Workshop on Graph and Model Transformation (GraMoT'05). Electronic Notes in Theoretical Computer Science, Tallinn, Estonia, Elsevier Science (2005)

13. Schultzke, T.: Entwicklung und implementierung eines parsers für visuelle sprachen basierend auf kritischer paaranalyse. Master's thesis, Technische Universität Berlin (2001)

14. Taentzer, G.: AGG: A Graph Transformation Environment for Modeling and Validation of Software. In Pfaltz, J., Nagl, M., Boehlen, B., eds.: Application of Graph Transformations with Industrial Relevance (AGTIVE'03). LNCS 3062, Springer (2004) 446 – 456

15. Lambers, L., Ehrig, H., Orejas, F.: Efficient conflict detection in graph transformation systems by essential critical pairs. In: Proc. Workshop GTVMT. (2006)

Adaptive Star Grammars[*]

Frank Drewes[1], Berthold Hoffmann[2], Dirk Janssens[3],
Mark Minas[4], and Niels Van Eetvelde[3,**]

[1] Umeå universitet, Sweden
[2] Universität Bremen, Germany
[3] Universiteit Antwerpen, Belgium
[4] Universität der Bundeswehr München, Germany

Abstract. We propose an extension of node and hyperedge replacement grammars, called adaptive star grammars, and study their basic properties. A rule in an adaptive star grammar is actually a rule schema which, via the so-called cloning operation, yields a potentially infinite number of concrete rules. Adaptive star grammars are motivated by application areas such as modeling and refactoring object-oriented programs. We prove that cloning can be applied lazily. Unrestricted adaptive star grammars are shown to be capable of generating every type-0 string language. However, we identify a reasonably large subclass for which the membership problem is decidable.

1 Introduction

Software engineering tools for model transformation or refactoring do often represent models and programs by graphs. Our earlier research in this area [1] revealed that the structure of such graphs cannot be captured by graph schemas, because models and programs have a recursive syntactical structure. Graph grammars are among the most natural candidates for specifying recursively structured graphs. For example, a graph grammar could be designed to generate the set of all program graphs as defined in [1].

The purpose of this paper is to introduce adaptive star grammars and to study their basic properties. Being context-free devices with nice computational properties, hyperedge and node replacement grammars [2,3,4] have proven particularly useful for defining graph languages. Unfortunately, these types of graph grammars turn out to be too weak to generate program graphs in a reasonable way. Therefore, we propose an extension, called adaptive star grammar, that is not only able to capture the context-free structure of object-oriented programs, but also aspects such as scope rules, overriding of methods, and references of variable and parameter uses to their definitions.

A star rule is a rule which replaces a nonterminal node together with its outgoing edges – a *star* – with another graph. This graph is glued to the *border*

---

[*] Supported by SEGRAVIS (www.segravis.org), a European research training network.
[**] On leave to Universität Bremen on a SEGRAVIS grant (October 2005–January 2006).

A. Corradini et al. (Eds.): ICGT 2006, LNCS 4178, pp. 77–91, 2006.

nodes of the star, i.e., to the nodes pointed to by the outgoing edges of the nonterminal node. The replacement process is similar to the well-known notion of hyperedge replacement, where the nonterminal node corresponds to the hyperedge being replaced. To increase the generative power of the device, border nodes of the left-hand side of a star rule may be designated as so-called multiple nodes. These nodes can be *cloned* prior to the application of the star rule. Cloning simply replicates a multiple node together with its incident edges any number of times (including 0). Thus, a star rule containing multiple nodes is actually a rule schema. In fact, even the host graph may contain multiple nodes, and these can be cloned as well in order to make a rule applicable.

We note here that the set nodes of PROGRES [5] and FUJABA [6] are similar to our multiple nodes. In the model transformation language GMORPH [7], a more general notion of cloning is provided whose *collection containers* correspond to the notion of a *multiple subgraph*. A similar concept is addressed in [8].

As our first main result, we show that cloning can be applied in both an eager and a lazy manner. Thus, derivations can be carried out effectively. Our second and third results concern the generative power of adaptive star grammars and the membership problem. Unrestricted adaptive star grammars can generate all recursively enumerable string languages (encoded as chain graphs in the usual way). Thus, these grammars are too powerful if given structures need to be parsed. However, in our third main result, we identify a reasonably large class of adaptive star grammars for which membership is decidable.

The structure of this paper is as follows. In the next section, we define the basic notions regarding stars and star replacement. Section 3 introduces the cloning operation. Based on this, adaptive star grammars are introduced in Section 4. In this section we also discuss a nontrivial example that applies adaptive grammars to generate program graphs. Two derivation strategies, eager and lazy cloning, are studied in Section 5 and demonstrated on the example. In Section 6, the generative power and the membership problem of adaptive star grammars are investigated. Section 7 concludes the paper.

2   Star Replacement

We start by defining the type of graphs considered in this paper. Throughout the paper, let Σ be a set of labels which is partitioned into two disjoint, countably infinite sets $\dot{\Sigma}$ and $\bar{\Sigma}$ of node and edge labels, resp. A finite subset Σ of Σ is called a labeling alphabet. Its two components are $\dot{\Sigma} = \Sigma \cap \dot{\Sigma}$ and $\bar{\Sigma} = \Sigma \cap \bar{\Sigma}$.

Intuitively, in the type of grammars to be defined later on, stars are the nonterminal items to be replaced. Therefore, we reserve an infinite supply $\mathcal{N} \subseteq \dot{\Sigma}$ of node labels called nonterminals. (We assume that the remaining set $\dot{\Sigma} \setminus \mathcal{N}$ of terminal labels is infinite as well.) In the following definition of graphs, we prohibit edges that point to nonterminal nodes. In particular, nonterminal nodes cannot be connected by edges. In this way, stars become a generalised version of the hyperedges known from hyperedge replacement grammars [2,3].

Definition 1 (Graph). A *graph* $G = \langle \dot{G}, \bar{G}, s_G, t_G, \dot{\ell}_G, \bar{\ell}_G \rangle$ consists of finite sets $\dot{G}$ of *nodes* and $\bar{G}$ of *edges*, of *source* and *target functions* $s_G, t_G \colon \bar{G} \to \dot{G}$, and of node and edge labeling functions $\dot{\ell}_G \colon \dot{G} \to \Sigma$ and $\bar{\ell}_G \colon \bar{G} \to \bar{\Sigma}$. For all edges $e \in \bar{G}$, it is required that $\dot{\ell}(t_G(e)) \notin \mathcal{N}$.

The set of all graphs labeled over a labeling alphabet Σ is denoted by $\mathcal{G}_\Sigma$.

We use common terminology regarding graphs. For instance, an edge is said to be *incident* with its source and target nodes, and makes these nodes *adjacent* to each other. For $A \subseteq \dot{G}$, $G \setminus A$ denotes the subgraph of G induced by $\dot{G} \setminus A$. Morphisms and isomorphisms are defined as usual. The notation $G \cong_m H$ denotes the fact that graphs G and H are isomorphic via the isomorphism m.

Next, we define a central notion of this paper, the *star*.

Definition 2 (Star). For a graph G and a node $x \in \dot{G}$, $G(x)$ denotes the subgraph of G consisting of x, all its incident edges, and all its adjacent nodes. A graph of the form $G(x)$ is a *star* if $\dot{\ell}_G(x) \in \mathcal{N}$. In this case, $G(x)$ is also called a *star occurrence in* G.

Thus, a star is a graph S that consists of a nonterminal node x and its adjacent nodes. In the following, these will be called the *center node* of S and the *border nodes* of S, resp. The edges are called the *arms* of x. By Definition 1, each arm points from the center node to a border node. A star is *straight* if the target nodes of its arms are pairwise distinct.

Definition 3 (Star Rule). A *star rule* $S ::= R$ consists of a star S, called its *left-hand side*, and a graph R, called its *right-hand side*, that share precisely the border nodes of S. When we modify such a rule, it is considered to be a single graph, namely the union of S and R.

Example 1 (Star Rules). Two examples of star rules are shown below:

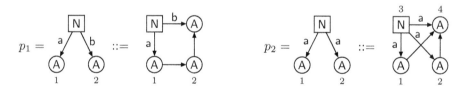

Nonterminal nodes are drawn as boxes; they have two border nodes labeled A, and two arms. For both rules, the border nodes 1 and 2 are drawn twice: they belong both to the right-hand side and to the left-hand side.

Definition 4 (Star Replacement). Let G be a graph, and $p = (S ::= R)$ a star rule so that $S \cong_m G(x)$ for some node $x \in \dot{G}$. The *replacement* of x by R yields the graph H which is obtained from the disjoint union of G and R by removing x and its arms, and identifying every border node b of S in R with its image $m(b) \in \dot{G}$. In this situation, we also write $G \Rightarrow_{x,p,m} H$.

Obviously, star replacement is a restricted form of DPO graph transformation [9] (with injective occurrence morphisms). In fact, star replacement is more or less equivalent to hyperedge replacement [2,3], because the center node of a star together with its arms can be seen as a hyperedge.

Star replacement does not cover node replacement [4], as the left-hand side of a star rule has a fixed number of arms, whereas nonterminals in node-replacement grammars can be replaced independently of the number of edges incident with them. The notion of cloning introduced in the next section is a formal mechanism that makes it possible to overcome this limitation of star replacement. Rules are specified in a generic way so that they adapt to several contexts of a nonterminal, but not necessarily to all. Next, we formalize the adaptation process, which we call cloning, and then we use it to define adaptive star grammars.

3   Cloning

In this section, we formalize the notion of cloning. We use a special set of labels designating so-called multiple nodes. A similar mechanism can be found in the PROGRES graph transformation language [5].

Formally, we assume from now on that $\Sigma \setminus \mathcal{N}$ contains a subset $\ddot{\Sigma}$ of *multiple node labels*. The remaining node labels are said to be *singular* ones. Further, we assume that there is a bijection $\ddot{}: \dot{\Sigma} \setminus (\mathcal{N} \cup \ddot{\Sigma}) \to \ddot{\Sigma}$. Thus, every singular node label l has a copy $\ddot{l}$ among the multiple node labels. A node is said to be singular or multiple depending on its label. The set of multiple nodes in a graph G is denoted by $\ddot{G}$, i.e., $\ddot{G} = \{v \in \dot{G} \mid \dot{\ell}_G(v) \in \ddot{\Sigma}\}$. In figures, we draw multiple nodes as nodes with a "shadow", as is seen in Definition 6.

We can now define the cloning operation. Using this operation, a multiple node can be turned into any number of singular nodes, its clones. However, we also want to be able to create clones that are multiple nodes. Thus, we define $G\frac{x}{(m,n)}$ to be obtained from G by replacing the multiple node x with m clones which are still multiple, and n singular clones.

Definition 5 (Cloning Operation). Let G be a graph, $x \in \ddot{G}$ a multiple node, and $m, n \geq 0$. The *clone* $G\frac{x}{(m,n)}$ is the graph constructed as follows. Let $G'(x)$ be obtained from $G(x)$ by replacing the label $\ddot{l}$ of x by l. Then take the disjoint union of the graph $G \setminus \{x\}$, m copies of $G(x)$, and n copies of $G'(x)$. Finally, identify the $m + n + 1$ copies of each node in $\dot{G}(x) \setminus \{x\}$ with each other.

The $m + n$ copies of x in $G\frac{x}{(m,n)}$ are called the clones of x. Obviously, $G\frac{x}{(m,n)}$ is defined only up to isomorphism. However, $G \setminus \{x\}$ is of course isomorphic to the subgraph of $G\frac{x}{(m,n)}$ induced by the nodes that are not clones of x.

The process of cloning can be described by graph transformation. The rules in the following definition should be considered as rules in the DPO approach, where the interface graph is the discrete graph consisting of the nodes $1, \ldots, p+q$.

Definition 6 (Cloning Rules). The set Δ of *cloning rules* consists of all rules of the form

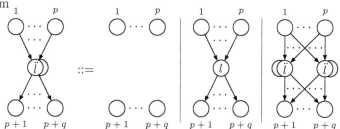

for all $\ddot{l} \in \ddot{\Sigma}$ and all $p, q \geq 0$.[1] The three rule schemas will be denoted by $rem(\ddot{l})$, $sing(\ddot{l})$, and $copy(\ddot{l})$, respectively.

The application of a cloning rule performs a cloning operation, in which a multiple node is either removed, turned into a singular node, or copied. It should be clear that the cloning rules suffice to describe all clonings. More precisely, let G be a graph containing a multiple node x. For all $m, n \geq 0$, $G\frac{x}{(m,n)}$ is derived from G using the cloning rules, as follows: $G\frac{x}{(0,0)}$ is obtained by an application of $rem(\ddot{l})$. Moreover, for $m + n > 0$, $G\frac{x}{(m,n)}$ is obtained by $m + n - 1$ applications of $copy(\ddot{l})$ and n applications of $sing(\ddot{l})$.

The result obtained by cloning a number of nodes is independent of the order in which those nodes are treated.

Lemma 1 (Cloning is Commutative). *For a graph G with distinct multiple nodes x and y, and for $m, n, m', n' \geq 0$,*

$$\left(G\frac{x}{(m,n)} \right) \frac{y}{(m',n')} \cong \left(G\frac{y}{(m',n')} \right) \frac{x}{(m,n)}.$$

Proof. Obviously, if two rules in Δ are applied to distinct multiple nodes of G, the result does not depend on the order of these rule applications. As argued above, Δ describes cloning correctly. This yields the statement. □

We define a cloning operation for a set of multiple nodes in a graph. For each multiple node in the set, the necessary information about the number of desired clones is given by a so-called multiplicity function.

Definition 7 (Iterated Cloning). Let G be a graph. A *multiplicity function* for G is a function $\mu \colon \ddot{G} \to \mathbb{N}^2$. If $\ddot{G} = \{x_1, x_2, \dots, x_k\}$ (where $x_1, \dots, x_k$ are pairwise distinct), then G^μ is the graph defined by

$$G^\mu = \left(\dots \left(\left(G\frac{x_1}{\mu(x_1)} \right) \frac{x_2}{\mu(x_2)} \right) \dots \frac{x_k}{\mu(x_k)} \right).$$

By Lemma 1, G^μ is defined uniquely up to isomorphism. In the following, when defining a multiplicity function μ, we will specify only those multiplicities $\mu(x)$ which are not equal to $(1, 0)$.

---

[1] The labels of the nodes $1, \dots, p + q$ as well as the edge labels have been omitted to avoid cluttering the figure. They carry over from the left-hand side to the right-hand sides in the obvious way. Note also that the nodes $1, \dots, p + q$ may be multiple.

4   Adaptive Star Grammars

In this section, we define adaptive star grammars. The rules of these grammars are star rules which may contain multiple nodes that can be cloned before a rule is applied. The graphs being derived may contain multiple nodes as well, and so they may also be cloned in order to make a rule applicable. Let us first define the cloning of (nodes in) star rules.

Definition 8 (Star Rule Clone). Let $p = (S ::= R)$ be a star rule. A *star rule clone* of p is a star rule p'' such that $p \Rightarrow_\Delta p'$ for some p' from which p'' can be obtained by taking a *quotient*, i.e., identifying pairs of border nodes (that have the same label). The set of all star rule clones of a set P of star rules is denoted by P^Δ.

Note that neither edges nor non-border nodes are identified by taking quotients. Clearly, every star rule clone is a star rule. We can now define adaptive star grammars and the graph languages they generate.

Definition 9 (Adaptive Star Grammar). An *adaptive star grammar* $\Gamma = \langle \Sigma, N, P, Z \rangle$ consists of

- a labeling alphabet Σ containing only terminal labels,
- a finite set $N \subseteq \mathcal{N}$ of nonterminals,
- a finite set P of star rules over $\Sigma \cup N$ with straight left-hand sides, and
- an *initial nonterminal* $Z \in N$.

The *language generated by* Γ is $\mathcal{L}(\Gamma) = \{G \in \mathcal{G}_{\Sigma\ \ddot{\Sigma}} \mid Z \Rightarrow^+_{\Delta P} G\}$. Here, Z denotes the graph consisting of a single node labeled Z, and $\Delta P = \Delta \cup P^\Delta$.

Thus, derivation steps in adaptive star grammars can be of two different types: On the one hand, multiple nodes in the host graph can be cloned, and, on the other hand, star rule clones can be applied.

We now discuss a particular application of star grammars.

Example 2 (A grammar for program graphs). As a nontrivial example we now discuss a star grammar modeling the structure of object-oriented programs by generating graphs called *program graphs*. This type of graphs has been developed for studying refactoring in [1]. Due to space restrictions, only a simplified method body specification is considered here, where method bodies contain only assignments and method calls. The grammar is shown in Fig. 1. A more complete specification based on star grammars can be found in [10].

We use terminal node labels B, E, V, M that correspond to method body root, entity occurrence, variable, and method, respectively. Furthermore we have non-terminal node labels $BODY, STS, ST, EXP, ACC, ASS, CALL, APS$. The label $BODY$ is the initial nonterminal. It generates an STS star (statement sequence), connected to a B node, and to a multiple N node. The latter is a shorthand covering both V and M. From the modeling point of view, it can be seen as a

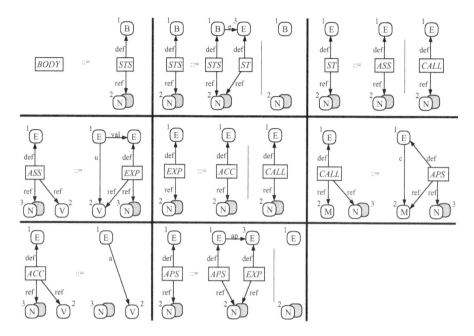

Fig. 1. A star grammar method body syntax tree specification

supertype of nodes of type V and M. Any singular node cloned out of a multiple N node becomes a node of type V or M. Alternatively, to stay within the formalism of adaptive star grammars, the rules of the grammar can be modified by applying $copy(N)$ to each of the N nodes and relabeling the two clones into a V node and an M node.

The STS star generates recursively a number of statements (ST), each of which can be rewritten into an assignment (ASS) or a method call ($CALL$). The right-hand side of an assignment is an expression (EXP). Expressions are either calls or variable accesses (ACC). Calls can have actual parameters (generated by APS) which are expressions.

The edge labels are e, a, u, c, val, ap, def, ref. The first six of these stand for syntax tree expression, variable access, variable update, method call, assignment value and actual parameter respectively. The edge labels def and ref are used for the arms of nonterminal nodes. The body root node B groups a set of E nodes, connected by e edges (cf. Fig. 2). Each of these nodes represents an occurrence of a variable, or a method call in the syntax tree. In the first case it is connected by an outgoing a or u edge to a variable and in the second case by a c edge to a method. Assignment and call occurrences may have additional val or ap edges to other E nodes.

Every nonterminal has a ref arm. It is always connected to a multiple node of type N representing all the *referable* symbols (visible methods, variables, formal parameters, types) that can be used in the program part derived from the nonterminal. The ACC star has three arms: the def arm shows the start node of the

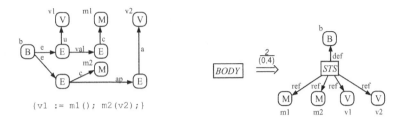

Fig. 2. Method body graph example   **Fig. 3.** Eager cloning derivation

statement. The two others are ref arms. This has the effect of selecting out of the complete set of referred elements one particular variable that is of particular relevance, all the other referable elements are represented by the multiple node. The nonterminal node disappears and an a edge is created between the E node and the V node (cf. Fig 1). The *CALL* and *ASS* rules are similar but create c and u edges and an additional *EXP* nonterminal.

An example of a method body graph that can be generated by the grammar is given in Figure 2, together with its textual equivalent. The node identifiers relate the program entities with their graphical representations.

By using adaptive star grammars instead of context-free string grammars for generating models of object-oriented programs, typical properties of object-oriented languages, such as the fact that every use of an identifier has a matching declaration, can be modeled. This is realized by the ref arms, which record the existing entities that can be used in a function. It can also be shown that the use of adaptive star grammars makes it possible to enforce the visibility constraints for class attributes, i.e. to ensure that method bodies in derived program graphs never contain accesses to private attributes of other classes. However, adaptive star grammars seem to be unable to cope with more complex constraints like the parameter correspondence. It does not seem to be possible to generate exactly as many actual parameters as formal parameters and make their types match.

5   Eager and Lazy Cloning

In this section, we will study an important aspect regarding derivations, namely the interplay between cloning and rule application. Cloning can be performed eagerly, where cloning on the host graph is done as early as possible. The following lemma shows that when a star replacement is followed by a cloning step in which one of the nodes of the host graph is cloned, then this results in the same graph as the one obtained by performing the cloning operation before the star replacement. If the cloned node is a border node of the star occurrence that is replaced, then an appropriate cloning of the rule used is needed. The straightforward proof is omitted.

Lemma 2 (Eager Cloning). *Let G, H be graphs, let p be a star rule, and let $G \Rightarrow_{x,p,m} H$ be a star replacement. Let $y \in \ddot{G}$ and let $k, l \geq 0$.*

1. *If y is not a border node of $G(x)$, then $G \frac{y}{(k,l)} \Rightarrow_{x,p,m} H \frac{y}{(k,l)}$.*
2. *If y is a border node of $G(x)$, then let $\tilde{y} = m^{-1}(y)$ and $\tilde{p} = p \frac{\tilde{y}}{(k,l)}$. Let $\tilde{m}$ be an extension of m, mapping the $k + l$ clones of $\tilde{y}$ bijectively to the $k + l$ clones of y. Then $G \frac{y}{(k,l)} \Rightarrow_{x,\tilde{p},\tilde{m}} H \frac{y}{(k,l)}$.*

Consequently, we obtain a normalform $nf(P)$ of a set P of star rules, such that $nf(P)$ is the subset of all star rule clones in P^Δ that do not contain multiple nodes. In particular, rule application creates only singular nodes.

Example 3 (Eager cloning in program graphs). The example program of Figure 2 can be derived using eager cloning. We consider the *STS* star in Figure 3. It is obtained by executing the initial rule being cloned by $\frac{2}{(0,4)}$ to generate all the needed variables and methods in advance. From that point, the star rules have to be cloned by $\frac{3}{(0,3)}$ for the *CALL*, *ACC* and *ASS* rules and $\frac{2}{(0,4)}$ for the other rules.

Corollary 1. *For every adaptive star grammar $\Gamma = \langle \Sigma, N, P, Z \rangle$, it holds that $\mathcal{L}(\Gamma) = \{ G \in \mathcal{G}_{\Sigma \ \ddot{\Sigma}} \mid Z \Rightarrow^+_{nf(P)} G \}$.*

When constructing a derivation of an adaptive star grammar, it would obviously be desirable to postpone cloning as much as possible: we use incremental cloning rules to construct derivations so that cloning is kept at a minimum. In order to characterize which clonings can be postponed until after a given star replacement, the following auxiliary notion is useful.

Definition 10 (Indistinguishability). Let $\tilde{p} = (\tilde{S} ::= \tilde{R}) \in P^\Delta$ be a quotient of a rule p' that is obtained by cloning a rule $p = (S ::= R) \in P$. For a border node y of $\tilde{S}$, its set of *precursors* is the set of nodes x of S such that y is the image of either x or a clone of x under the quotient. Two border nodes y_1, y_2 of $\tilde{S}$ are *indistinguishable* (in $\tilde{p}$) if they have the same set of precursors in S and there exists an automorphism of $\tilde{p}$ that interchanges them while leaving the other nodes invariant.

Definition 11 (Lazy Cloning). Let $\tilde{p} = (\tilde{S} ::= \tilde{R}) \in P^\Delta$. A derivation $G \Longrightarrow_\Delta \tilde{G} \Longrightarrow_{x,\tilde{p},m} H$ constitutes a *lazy step* if only border nodes of $G(x)$ are cloned in $G \Longrightarrow_\Delta \tilde{G}$, and, moreover, there do not exist distinct border nodes y_1, y_2 of $\tilde{S}$ such that y_1 and y_2 are indistinguishable in $\tilde{p}$, and $m(y_1)$ and $m(y_2)$ are clones of the same node of $G(x)$.

Note that $\tilde{p}$ can in general be obtained in different ways from a rule $p \in P$, and hence the notion of a lazy step is defined only with respect to a fixed choice of p, p' and a quotient map. However for our purposes it is sufficient to consider a step as lazy if there exists such a choice.

The next result shows that lazy cloning is correct: every derivation can be rearranged into a sequence of lazy steps followed by a number of cloning steps.

Theorem 1 (Correctness of Lazy Cloning). *Let P be a set of star rules. For every derivation $G \Longrightarrow_{\Delta P} H$ there is a graph $\hat{H}$ that can be derived from G by a sequence of lazy steps, such that $\hat{H} \Longrightarrow_{\Delta} H$.*

Proof. It suffices to consider a derivation of the form $G \Longrightarrow_{\Delta}^{n} \tilde{G} \Longrightarrow_{x,\tilde{p},m} H$ which is not a lazy step, and to prove that one of its cloning steps can be postponed until after the star replacement. For this purpose, assume that $\tilde{p} = (\tilde{S} ::= \tilde{R})$ is a quotient of some rule obtained from $p = (S ::= R) \in P$ by a sequence of cloning steps.

If $G \Longrightarrow_{\Delta}^{n} \tilde{G}$ clones a node that is not a border node of $G(x)$, then we may assume that the corresponding step is the last step of $G \Longrightarrow_{\Delta}^{n} \tilde{G}$ (see Lemma 1). This step is parallel independent of the star replacement, and thus the desired result follows from well-known results about DPO graph rewriting. So assume that all steps in $G \Longrightarrow_{\Delta}^{n} \tilde{G}$ clone border nodes of $G(x)$. By assumption, $G \Longrightarrow_{\Delta}^{n} \tilde{G} \Longrightarrow_{x,\tilde{p},m} H$ is not a lazy step. Hence, there exist distinct nodes y_1, y_2 of $\tilde{S}$, a multiple border node z of $G(x)$ and two clones z_1, z_2 of z in $\tilde{G}$ such that y_1 and y_2 are indistinguishable, $m(y_1) = z_1$, and $m(y_2) = z_2$. If at least one of z_1, z_2 is singular, then it is obtained by an application of $sing(\ddot{l})$, and obviously this step can be postponed until after the star replacement. So assume that both z_1 and z_2 are multiple. Again by Lemma 1 one may assume that the cloning step that produces z_2 is the last step of $G \Longrightarrow_{\Delta}^{n} \tilde{G}$. Moreover, since both z_1 and z_2 are clones of z, z_2 can be obtained as a clone of z_1. Thus, $G \Longrightarrow_{\Delta}^{n-1} G' \Longrightarrow_{\Delta} \tilde{G}$, where G' is the graph obtained from $\tilde{G}$ by deleting z_2 and its incident edges, and $\tilde{G} = G'\frac{z_1}{(2,0)}$. Now let $p' = (S' ::= R')$ and H' be obtained from $\tilde{p}$ and H by deleting y_2, z_2 and their incident edges, respectively. Then $G' \Longrightarrow_{x,p',m'} H'$, where m' is the restriction of m to S'. Moreover, $\tilde{p} = p'\frac{y_1}{(2,0)}$, because y_1 and y_2 are indistinguishable, and it follows from the definition of a star replacement that $H = H'\frac{z_1}{(2,0)}$. The result follows.   □

Example 4 (Lazy cloning in program graphs). A lazy derivation of the first statement `v1 := m1()` of the method body example would, after applying the initial rule, immediately apply the *STS* and *ST* rule without cloning to arrive at the *ASS* nonterminal. The rest of the derivation is shown in Fig. 4. To clarify which of the possible rules is used for star replacement, rule names carry an index. Note that all star replacements together with the preceding cloning steps are lazy steps.

6   The Membership Problem

This section consists of two parts. In the first part, we show that adaptive star grammars can generate every recursively enumerable string language. We will do this by sketching how to simulate a slightly modified version of the well-known counter machines. Hence, in particular, the membership problem is unsolvable. In the second part of the section, a restriction is studied under which this problem becomes decidable.

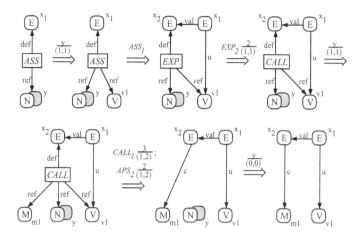

Fig. 4. A lazy derivation for a call statement

Let us first define a variant of (nondeterministic) counter machines that turns out to be particularly suitable for our situation. An *offline counter machine* (OCM, for short) with $k \geq 1$ counters is a system $M = (Q, \mathcal{A}, I, q_0, F)$ consisting of a finite set Q of states, a finite input alphabet $\mathcal{A}$, a finite set I of instructions, an initial state $q_0 \in Q$, and a set $F \subseteq Q$ of final states. Each instruction has the form $(q, i, z) \mapsto (q', j)$, where $q, q' \in Q$, $1 \leq i \leq k$, and $(z, j) \in \{(\text{ZERO}, +1), (\text{NONZERO}, -1), (\text{NONZERO}, +1)\}$.

A configuration $(q, c_1 \cdots c_k) \in Q \times \mathbb{N}^k$ consists of a state q and k counter values $c_1, \ldots, c_k$. There is a computation step $(q, c_1 \cdots c_k) \mapsto_M (q', c'_1 \cdots c'_k)$ if I contains an instruction $(q, i, z) \mapsto (q', j)$ with $z = \text{ZERO} \iff c_i = 0$ and

$$c'_l = \begin{cases} c_l + j & \text{if } l = i \\ c_l & \text{otherwise.} \end{cases}$$

Suppose $\mathcal{A} = \{a_1, \ldots, a_{m-1}\}$ (using an arbitrary but fixed order on the symbols in $\mathcal{A}$). The initial configuration for an input string $w = a_{i_1} \cdots a_{i_n}$ is given by $initial_M(w) = (q_0, c0 \cdots 0)$. Here, c is obtained by interpreting $i_1 \cdots i_n$ as a number written in base-m notation. The OCM M accepts w if a configuration $(q, c_1 \cdots c_k)$ with $q \in F$ (called a *final configuration*) can be reached from $initial_M(w)$. As usual, the *recognized language* is the set of all strings accepted by M. It is well known that counter machines recognize all recursively enumerable languages (see, e.g., [11]). As the reader may easily check, this holds also for the variant defined above.

Let us now see how star rules can simulate an OCM. For this, consider an OCM M as above. We use the states in Q as nonterminals. There is only one further node label in $\dot{\Sigma} \setminus \ddot{\Sigma}$. The corresponding nodes and their multiple counterparts are considered to be unlabelled in the following.

A configuration $C = (q, c_1 \cdots c_k)$ is represented by the non-straight star $gr(C)$ shown on the right. It consists of a nonterminal center node v labelled q, a terminal border node v', and $1 + \sum_{i=1}^{k} c_i$ parallel edges from v to v'. One of these edges is labelled with aux, whereas c_i edges are labelled with

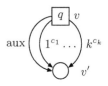

i, for $1 \le i \le k$. Here, an edge label carrying an exponent abbreviates the respective number of parallel edges. The edge labelled aux ensures that $gr(C)$ is a star even if $c_1 = \cdots = c_k = 0$.

It is now rather easy to define a set P_M of star rules which simulate the instructions of M by removing or adding the appropriate number of arms in each step. For example, if $k = 3$ and the instruction in question is $(q, 3, \text{NONZERO}) \mapsto (q', +1)$, the resulting star rule looks like this:

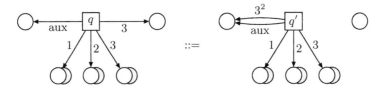

To see that this rule has the desired effect, note that its application to a graph of the form $gr(C)$ requires taking a quotient which identifies all border nodes. Intuitively, this means that the arms in the rule are parallel edges in disguise. Hence, the rule applies to $gr(C)$ if counter 3 has a nonzero value and will in this case increase the number of edges labelled 3 by one.

By adding terminating rules (which remove the nonterminal node if the nonterminal is a final state of M), we get the following lemma.

Lemma 3. *For every configuration C of M, there is a derivation $gr(C) \Rightarrow^{+}_{\Delta P_M} G$ for some terminal graph G if and only if there exists a computation of M that turns C into a final configuration. Furthermore, in this case, G is the graph consisting of a single node and no edges.*

Using Lemma 3, we can now prove the promised result. For this, we identify a string $b_1 \cdots b_n \in \mathcal{A}$ with the graph consisting of unlabelled nodes $v_0, \ldots, v_n$ and edges $e_1, \ldots, e_n$, where e_i points from v_{i-1} to v_i and is labelled with b_i.

Theorem 2. *Every recursively enumerable string language can be generated by an adaptive star grammar.*

Sketch of Proof. Consider any recursively enumerable string language L, and let M be an OCM recognizing L. Without loss of generality, we may assume that L does not contain the empty string. An adaptive star grammar generating L may work as follows. In a preprocessing phase, it generates an arbitrary string $w \in \mathcal{A}^{+}$ in a nondeterministic fashion. At the same time, the subgraph $gr(initial_M(w))$ is built. In the second phase, the star rules in P_M are used to simulate M.

To see that the first phase can really be implemented, note that we can turn n edges labeled with 1 into $n^m + j$ such edges using the rule

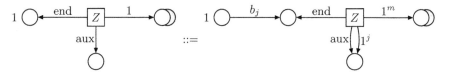

Using such rules, we can generate exactly the graphs G of the form

with $b_1, \ldots, b_n \in \mathcal{A}$ and s the representation of $b_1 \cdots b_n$ in base-m notation for some $n \geq 1$. Thus, the star occurrence $G(v)$ equals $gr(initial_M(b_1 \cdots b_n))$. Together with Lemma 3, this proves the theorem. $\qquad\qquad\qquad\qquad\quad \Box$

If adaptive star grammars shall be practically used, it is necessary to come up with restrictions that guarantee the decidability of the membership problem. Therefore, we now study a reasonably restricted class of star grammars that allows to decide this question. We consider *simple adaptive star grammars* first. In the following, let us call an edge e in a graph G terminal if it is not incident with a nonterminal node (i.e., if $\ell_G \circ s_G(e) \notin N$); otherwise, e is *nonterminal*.

Definition 12 (Simple adaptive star grammar). An adaptive star grammar $\Gamma = \langle \Sigma, N, P, Z \rangle$ and its set P of rules are called *simple* if P does not contain any rule whose right-hand side is either just its set of border nodes or contains a non-straight star.

Following Corollary 1 (p. 85), we restrict derivations to the set $nf(P)$ of rules without multiple nodes and to graphs without multiple nodes. Simple adaptive star grammars cannot produce parallel nonterminal edges as right-hand sides do not contain non-straight stars. Hence, in the following, we can ignore rules in $nf(P)$ that are obtained by taking quotients. Let $\widetilde{nf}(P)$ be the corresponding set of rules.

Lemma 4. *There is an algorithm that decides whether $G \Rightarrow_{\widetilde{nf}(P)}^* G'$ for every finite set P of simple star rules and all graphs G and G' without multiple nodes.*

Proof. We measure the size of a graph G by $\tau(G) = |\dot{G}| + |\{e \in \bar{G} \mid e \text{ terminal}\}|$. Each derivation $H \Rightarrow_{\widetilde{nf}(P)} H'$ removes a nonterminal node, but adds at least one other node or a terminal edge, i.e., $\tau(H) \leq \tau(H')$. We prove the lemma by showing that the set of all graphs G' with $G \Rightarrow_{\widetilde{nf}(P)}^* G''$ and $\tau(G'') \leq \tau(G')$ is finite. The number of graphs $\tilde{G}$ that can be derived from another graph in a single step such that $\tau(\tilde{G}) \leq \tau(G)$ is finite. We, therefore, have to show that there is no infinite derivation sequence $G = G_0 \Rightarrow_{p_0} G_1 \Rightarrow_{p_1} \ldots$ such that $G_i \not\cong G_j$ for all $i \neq j$ and $p_i \in \widetilde{nf}(P)$, $\tau(G_i) \leq \tau(G')$ for each i. If we assume that there is such an infinite derivation sequence, there must be an index s such that, for each $i \geq s$, $\tau(G_i) = \tau(G_s)$ and $p_i \in \widetilde{nf}(P)$ is obtained from a rule whose right-hand side is a straight star as Γ is simple. Hence, each graph G_i for $i \geq s$ has the same number of nodes resp. edges and, as a consequence, there must exist two indices $i, j \geq s$, $i \neq j$ such that $G_i \cong G_j$.

Next, we consider *straight adaptive star grammars*. An adaptive star grammar is called straight if its rules are straight, meaning that their right-hand sides do not contain non-straight stars. Hence, each simple adaptive star grammar is a straight one, but the converse does not hold. The method body grammar in Figure 1 is an example of a straight grammar that is not simple. In order to show that membership is decidable for straight adaptive star grammars, we will use the following lemma:

Lemma 5. *There is an algorithm that decides whether $G \Rightarrow_{\tilde{nf}(P)} G \setminus \{x\}$ for every finite set P of straight star rules and all straight stars G without multiple nodes, where x is the center node of G.*

Proof. The existence of a derivation $G \Rightarrow_{\tilde{nf}(P)} G \setminus \{x\}$ requires that no applied rule adds either a new terminal node or a terminal edge. Hence, the number of terminal nodes remains constant in the derivation, and the derivation does not contain graphs containing terminal edges.

Let P_n be the (finite) subset of all rules $(S ::= R) \in \tilde{nf}(P)$ such that $|\dot{S}| \quad n$ and R contains neither terminal edges nor terminal nodes that are not border nodes of S. Obviously, $G \Rightarrow_{\tilde{nf}(P)} G \setminus \{x\}$ iff $G \Rightarrow_{P_n} G \setminus \{x\}$ where $n = |\dot{G}|$.

Now, $G \Rightarrow_{P_n} G \setminus \{x\}$ is equivalent to the (decidable) question whether an appropriately constructed context-free Chomsky grammar G' generates the empty string. To see this, construct G' by using as nonterminals the set of all isomorphism classes S such that S is a star occurring in one of the rules in P_n. Now, let G' contain the rule $[S] \rightarrow [S_1] \cdots [S_k]$ if P_n contains a rule $S ::= R$, where $S_1, \ldots, S_k$ are the stars occurring in R. It should be clear that G' generates the empty string if and only if there is a derivation $G \Rightarrow_{P_n} G \setminus \{x\}$. □

This result allows to prove the following theorem:

Theorem 3. *The (uniform) membership problem is decidable for straight adaptive star grammars, i.e., there is an algorithm deciding whether $G \in \mathcal{L}(\Gamma)$ for every straight adaptive star grammar Γ and every graph G.*

Proof. For $\Gamma = \langle \Sigma, N, P, Z \rangle$, we construct a new rule set P' iteratively, as follows. Initially, P' is the (finite) set of all rules in $\tilde{nf}(P)$ whose left-hand sides consist of not more than $|\dot{G}| + 1$ nodes. Now, if P' contains a rule $S ::= R$ such that there occurs a star with center node x in R, then the rule $S ::= R \setminus \{x\}$ is added to P' provided that $R(x) \Rightarrow_{\tilde{nf}(P)} R(x) \setminus \{x\}$ (which can be decided by Lemma 5). This process is repeated until no new rule can be added to P'. Finally, each rule is removed from P' whose right-hand side is just its set of border nodes. Obviously, for every nonempty graph G, we have $G \in \mathcal{L}(\Gamma)$ iff $Z \Rightarrow_{P'}^+ G$. The result follows by Lemma 4 if G is not empty (as P' is simple), and from Lemma 5 otherwise. □

7 Conclusions

Adaptive star grammars are more expressive than context-free graph grammars while retaining a context-free flavour. The extended expressive power is indispensable for generating structures such as object-oriented program models. In this paper the authors have joined their earlier work: The mechanisms presented here are much simpler than those proposed in [12]. Future work will investigate how star grammars can be used in graph transformation rules to define parts of a rule that may be variable, but have a fixed shape. This is useful for modeling refactorings rules, in which complex, variable structures like syntax trees are manipulated as atomic parts. All these concepts will be implemented in the graph transformation language and tool Diaplan [13].

References

1. Tom Mens, Serge Demeyer, and Dirk Janssens. Formalising behaviour-preserving transformation. In Andrea Corradini, Hartmut Ehrig, Hans-Jörg Kreowski, and Grzegorz Rozenberg, editors, *First International Conference on Graph Transformation (ICGT'02)*, number 2505 in LNCS, pages 286–301. Springer, 2002.
2. Annegret Habel. *Hyperedge Replacement: Grammars and Languages*. Number 643 in LNCS. Springer, 1992.
3. Frank Drewes, Annegret Habel, and Hans-Jörg Kreowski. Hyperedge replacement graph grammars. In Rozenberg [14], chapter 2, pages 95–162.
4. Joost Engelfriet and Grzegorz Rozenberg. Node replacement graph grammars. In Rozenberg [14], chapter 1, pages 1–94.
5. Andy Schürr, Andreas Winter, and Albert Zündorf. The PROGRES approach: Language and environment. In Gregor Engels, Hartmut Ehrig, Hans-Jörg Kreowski, and Grzegorz Rozenberg, editors, *Handbook of Graph Grammars and Computing by Graph Transformation. Vol. II: Applications, Languages, and Tools*, chapter 13, pages 487–550. World Scientific, Singapore, 1999.
6. Jörg Niere and Albert Zündorf. Using fujaba for the development of production control systems. *LNCS*, 1779:181–191, 2000.
7. Shane Sendall. Combining generative and graph transformation techniques for model transformation: An effective alliance? In *Proc. OOPSLA'03-Workshop on Generative Techniques in the Context of MDA*, 2003. URL: www.softmetaware.com/oopsla2003/mda-workshop.html.
8. Berthold Hoffmann, Dirk Janssens, and Niels Van Eetvelde. Cloning and expanding graph transformation rules for refactoring. *ENTCS*, 152(4), 2006. Proc. Graph and Model Transformation Workshop (GRAMOT'05).
9. Hartmut Ehrig. Introduction to the algebraic theory of graph grammars. In V. Claus, Hartmut Ehrig, and Grzegorz Rozenberg, editors, *Graph Grammars and Their Application to Computer Science and Biology*, number 73 in LNCS, pages 1–69. Springer, 1979.
10. Berthold Hoffmann and Niels Van Eetvelde. A graph grammar for program graphs. Technical report, University of Antwerp, March 2006. UA WIS/INF 2006.
11. John E. Hopcroft and Jeffrey D. Ullman. *Introduction to Automata Theory, Languages and Computation*. Addison-Wesley, Reading, Massachusetts, 1979.
12. Berthold Hoffmann. Graph transformation with variables. In H.-J. Kreowski et al., editors, *Formal Methods in Software and System Modeling*, volume 3393 of *LNCS*, pages 101–115. Springer, 2005.
13. Frank Drewes, Berthold Hoffmann, Raimund Klein, and Mark Minas. Rule-based programming with diaplan. *ENTCS*, 117(1), 2005.
14. Grzegorz Rozenberg, editor. *Handbook of Graph Grammars and Computing by Graph Transformation, Vol. I: Foundations*. World Scientific, Singapore, 1997.

Narrowing Data-Structures with Pointers*

Rachid Echahed and Nicolas Peltier

LEIBNIZ-IMAG, CNRS
46, avenue Félix Viallet
38031 Grenoble Cedex, France
Rachid.Echahed@imag.fr, Nicolas.Peltier@imag.fr

Abstract. We investigate the narrowing relation in a wide class of (cyclic) term-graph rewrite systems. We propose a new sound and complete narrowing-based algorithm able to solve goals in presence of data structures with pointers (e.g., circular lists, doubly linked lists etc.). We first define the class of rewrite systems we consider. Our rules provide features such as pointer (edge) redirections, relabeling of existing nodes, in addition to the creation of new nodes. Moreover, we split the set of nodes of term-graphs in two (possibly empty) subsets: (i) variables and (ii) names. Variable nodes can be mapped against any other node whereas names act as constants and thus they are supposed to match themselves. This distinction between nodes allows us to synthesize, through the narrowing process, data-structures with circular shapes. In a second step, we define the rewriting and narrowing relations. We then show the soundness and completeness of narrowing.

1 Introduction

Narrowing is the heart of the operational semantics of declarative languages which integrate functional and logic programming paradigms [10]. Programs in these languages are term rewrite systems. Their operational semantics consists in solving goals. For example, let us consider the following program which defines the length of a sequence:

$$length(nil) \rightarrow 0 \qquad\qquad length(cons(x, u)) \rightarrow s(length(u)).$$

To solve the goal $length(cons(x, nil)) = z$, one may normalize the term $length(cons(x, nil))$ and gets the unique solution $z = s(0)$. But the following goal $length(u) = s(s(0))$ cannot be solved by simple normalization ; instead narrowing can be used to synthesize the answer $u = cons(x_1, cons(x_2, nil))$ where x_1 and x_2 are *fresh* variables.

Narrowing has been widely investigated in the framework of first order term rewrite systems and optimal strategies have been proposed (e.g. [2]). In this paper, we propose to extend narrowing to a large class of term-graph rewrite

* This work has been partly funded by the project ARROWS of the French *Agence Nationale de la Recherche*.

A. Corradini et al. (Eds.): ICGT 2006, LNCS 4178, pp. 92–106, 2006.

systems. There are at least two reasons that motivate our work. First, efficient implementation techniques of declarative languages use dags (directed acyclic graphs) to implement terms. Second, in recent developments of graph transformations [3,4,6], it is shown that data-structures with pointers could be handled by using graph rewrite systems, and thus rule-based languages such as declarative ones could benefit from such results in order to fully integrate cyclic term-graphs (with pointers), such as circular lists or doubly linked lists, as first class objects.

In this paper, we consider term-graph rewrite systems composed of rules of the following shape (see Section 4 for details): $L \rightarrow A$, where L is a (constrained) term-graph and A is a sequence of actions the aim of which is the construction of the right hand side. Roughly speaking, A could be split into two parts, say R and D, where R is a term-graph and D is a sequence of edge (pointer) redirections. For example, the following rule inserts an element in a circular list (we use the classical linear [5] notation of term-graphs, where non-connected graphs are separated by ","):

$$\gamma{:}insert(a, \alpha{:}cons(b, u)), \beta{:}cons(c, \alpha) \rightarrow \gamma{:}cons(a, \alpha); \beta.2 \gg \gamma.$$

Applying this rule to the term-graph $\gamma_1{:}insert(a, \alpha_1{:}cons(b, \beta_1{:}cons(c, \alpha_1)))$ we get the intermediate term-graph $\gamma_1{:}cons(a, \alpha_1{:}cons(b, \beta_1{:}cons(c, \alpha_1)))$ before we perform the action $\beta_1.2 \gg \gamma_1$. The aim of this action is to redirect the second edge outgoing of the node β_1 in order to point the node γ_1. The final result of the application of the rule above is then the term $\gamma_1{:}cons(a, \alpha_1{:}cons(b, \beta_1{:}cons(c, \gamma_1)))$.

The rewrite rules we consider define a large class of term-graph rewrite systems (formally defined in Section 4). It includes several useful features. Left-hand sides could be cyclic with some constraints (disequations) on the nodes. Actions building the right hand side can execute redirections of pointers (edges) either locally as in the example above or globally as it happens when rewriting rooted term graphs. We have no restriction over the cyclic term-graphs to be rewritten.

Solving goals with cyclic term-graphs is certainly not an easy task. Consider for instance the operation $\#$ which computes the number of elements of a circular list (the complete definition of this operation is given in Section 5). If we consider the goal $\#(u) = s(s(0))$ then we should get a solution such as $u = \alpha{:}cons(x, \beta{:}cons(y, \alpha))$ with the constraint $\alpha \not\approx \beta$. Note that there is no published algorithm which is able to synthesize such a solution. Nodes α, β, x and y are supposed to be fresh variable nodes. The distinction between (constant) nodes and variable nodes is essential in our setting. Variable nodes behave as classical first order variables in the unification process for example, while the remaining nodes (constants) could be seen as global variables in imperative languages.

Defining narrowing in our setting turns out to be trickier than in the previous works. This is mainly due to the actions we perform on term-graphs such as pointer redirections and also to the fact that graphs are not considered equal up to bisimulation. Consider for instance the following term-graph $f(\delta{:}a, \gamma{:}a)$ where δ and γ are variable nodes, to be narrowed by using the rule $f(\alpha{:}a, \beta) \rightarrow \alpha{:}b$

(β denotes a variable). We can get two different narrowing steps

$$f(\delta{:}a, \gamma{:}a) \quad _{\delta=\gamma} \quad f(\delta{:}b, \delta) \qquad \text{or} \qquad f(\delta{:}a, \gamma{:}a) \quad _{\delta \quad \gamma} \quad f(\delta{:}b, \gamma{:}a).$$

From this simple example, we can see that instantiation of variable nodes during the narrowing process is not usual. Indeed, in contrast to the usual case, the computed solutions may include disequations, such as $\delta \not\approx \gamma$ in the second derivation above.

There are very few results in the literature on term-graph narrowing. In [12,9,11], acyclic term-graph narrowing have been studied and basic narrowing strategies have been proposed in [11,9]. Cyclic term-graph narrowing was first studied in [7] in the context of weakly-orthogonal term-graph rewrite systems. Its extension with graph collapsing could be found in [8]. Optimal term-graph narrowing strategies have been proposed in [7,8]. Very recently, [1] extended [7] and proposed efficient term-graph narrowing strategies in the presence of non-deterministic functions (i.e. non-confluent rewrite systems).

In this paper, we go beyond these results and tackle cyclic term graph narrowing in a very large class of term-graph rewrite systems that subsumes by far the weakly-orthogonal graph rewrite systems studied in [7]. We define the narrowing relation induced by the considered term-graph rewrite systems and prove its soundness and completeness.

The paper is organised as follows. Section 2 gives the precise definition of the term-graphs we consider as well as some basic definitions we need in the paper. In Section 3 we give the definitions of different actions we operate on graphs such as node creation and redefinition, pointer redirections etc. Section 4 defines the rewrite rules, rewrite steps and the term-graph rewrite systems we consider. Section 5 is dedicated to the definition of narrowing relation. The soundness and completeness of the narrowing relation are investigated in Section 6. Finally, Section 7 concludes the paper. Due to length restrictions, proofs are omitted.

2   Term-Graph

In this section, we describe the class of data structures (i.e. term-graphs) considered in the paper. The definitions are close to the ones of [5], but some of the notations are slightly adapted in order to better suit our purposes.

We assume given a set of *names* $\mathcal{A}$, a set of *variables* $\mathcal{V}$ and a set of *function symbols* Σ. We denote by $\mathcal{N}$ the set $\mathcal{N} \stackrel{\text{def}}{=} \mathcal{A} \cup \mathcal{V}$. $\mathcal{N}$ is the set of *nodes*.

Definition 1. *(Term-Graph) A reference on a set of nodes $N \subseteq \mathcal{N}$ is an expression of the form $f(\alpha_1, \ldots, \alpha_n)$ where $f \in \Sigma$, $n \geq 0$ and $\alpha_1, \ldots, \alpha_n \in N$ (if $n = 0$ then $f(\alpha_1, \ldots, \alpha_n)$ should be written f). The set of references on a set of nodes N is denoted by $\mathcal{T}(N)$. A term-graph G is defined by a set of nodes $\mathcal{N}(G) \subseteq \mathcal{N}$ and a partial function ref_G from $\mathcal{N}(G)$ to $\mathcal{T}(\mathcal{N}(G))$.*

For instance, a term-graph consisting in a variable node α without reference may be seen as a *variable* (in the usual sense), i.e. denotes an arbitrary term-graph.

If α is a name, then the graph is *partially instantiated*: the name of one of its nodes is known, but its reference and its other nodes remain to be specified.

We denote by $head_G(\alpha)$ the head symbol of $ref_G(\alpha)$ (if it exists, otherwise $head_G(\alpha)$ is undefined). We denote by $dom(G)$ the set of nodes α s.t. $ref_G(\alpha)$ is defined. Note that we may have $dom(G) \neq \mathcal{N}(G)$. A term-graph G is said to be *ground* if $\mathcal{N}(G) \subseteq \mathcal{A}^1$. We write $G \subseteq H$ iff $\mathcal{N}(G) \subseteq \mathcal{N}(H)$ and if for any $\alpha \in dom(G)$ we have $\alpha \in dom(H)$ and $ref_H(\alpha) = ref_G(\alpha)$. Intuitively, $G \subseteq H$ if G is a subgraph of H. The notion of subgraph is the analogue of the notion of subterm for usual terms. In what follows, we always denote nodes (variables and names) by Greek letters $\alpha, \beta, \ldots$, function symbols by $f, g, \ldots$ and constant symbols by $a, b, \ldots$.

Although Definition 1 is useful from a theoretical point of view, in the forthcoming examples, we adopt a more convenient and readable (commonly used, see for instance [5]) linear notation for term-graphs. We write a term-graph as a standard term, but we prefix some of the subterms (those occurring several times in the considered term-graph) by nodes. Obviously, naming (i.e. prefixing) subterms with nodes allows one to share subterms and to denote infinite (rational) terms. For instance, the expression $\alpha{:}f(a, g(\alpha))$ denotes a (cyclic) term-graph s.t.: $dom(G) = \{\alpha, \beta, \gamma\}$, $ref_G(\alpha) = f(\beta, \gamma)$, $ref_G(\beta) = a$, $ref_G(\gamma) = g(\alpha)$ (β, γ are arbitrarily chosen nodes distinct from α). Depending on the context the unnamed nodes β, γ could be constants or variables. Note that the above term-graph could also be written $\alpha{:}f(\beta{:}a, \gamma{:}g(\alpha))$, but for the sake of clarity, we prefer to skip useless names. Two distinct names necessarily correspond to distinct nodes, whereas two distinct variables can be made identical by instantiation. For instance, let us consider the following term-graph $G = \alpha{:}cons(1, \beta{:}cons(1, \alpha))$. If α, β are variables, then β may be instantiated by α. Thus the term-graph $\delta{:}cons(1, \delta)$ is an *instance* of G. More precisely, G denotes a circular list of length either 1 or 2. In contrast, if α, β are distinct names, G denotes a (specific) circular of length 2. The possibility of handling abstract nodes allows one to handle *partially defined data-structures*, which is absolutely essential for defining narrowing algorithms. It also allows the programmer to define more general rules, which is capital from a practical point of view (for instance we could compare two lists of integers without knowing whether they are *physically* equal or not).

A *substitution* σ is a function mapping each variable x in $\mathcal{V}$ to a node $x\sigma \in \mathcal{N}$. The *domain* of a substitution σ is denoted by $dom(\sigma)$ and defined as the set of variables x s.t. $x\sigma \neq x$. A substitution is said to be *ground* iff $x\sigma \in \mathcal{A}$ for any $x \in dom(\sigma)$. If σ, θ are two substitutions, then $\sigma\theta$ denotes the composition of σ and θ (i.e. $x\sigma\theta = \theta(\sigma(x))$). σ is said to be *more general* than θ if there is a substitution σ' s.t. $\sigma\sigma' = \theta$.

The image of a standard term by a substitution is always a term. However, in our setting, the image of a term-graph by a substitution is not necessarily a term-graph. For instance if $G = f(\alpha{:}a, \beta{:}b)$ is a term-graph where α, β are variables, then the image of G by a substitution $\sigma : \{\alpha \rightarrow \gamma, \beta \rightarrow \gamma\}$ is not a

---

[1] This is not equivalent to the usual notion of "ground term" because the nodes do not need to be associated to a reference.

term-graph. Thus, we can instantiate a term-graph G by a substitution σ only if σ is *compatible* with the term-graph in the sense that if two variables are mapped to the same node then the corresponding references (if they exist) must be the same. Formally, a substitution σ is said to be *compatible* with a graph G iff for any $\alpha, \beta \in dom(G)$ s.t. $\alpha\sigma = \beta\sigma$ we have $ref_G(\alpha)\sigma = ref_G(\beta)\sigma$.

If σ is compatible with G, then we denote by $G\sigma$ the graph H s.t.: $\mathcal{N}(H) = \{\alpha\sigma \mid \alpha \in \mathcal{N}(G)\}$ and for any $\alpha \in dom(G)$, $ref_H(\alpha\sigma) \stackrel{\text{def}}{=} ref_G(\alpha)\sigma$. Note that $G\sigma$ is well-defined if σ is compatible with G, since by definition $\alpha\sigma = \beta\sigma \Rightarrow ref_G(\alpha)\sigma = ref_G(\beta)\sigma$. H is called an *instance* of G iff there exists a substitution σ compatible with G s.t. $G\sigma \subseteq H$.

3   Graph Transformation

We introduce some basic operations on term-graphs: creation of a new node, node redefinition (i.e. replacement of the reference associated to an existing node by a new reference) and global redirection (i.e. redirection of all edges pointing to a node α to a node β). Node redefinition subsumes in particular edge redirection (i.e. redirection of an existing edge). For every action a, we shall denote by $G_{[a]}$ the result of the application of the action a on the term-graph G. The actions and their applications are defined in the following sections.

A **node creation** is an expression of the form α^+ where α is a node in $\mathcal{V}$. Applying a node creation to a term-graph simply adds a new node in the term-graph (with no reference).

We assume given an (infinite) subset of $\mathcal{A}$, denoted by $\mathcal{C}$ and a total precedence $\prec$ among elements of $\mathcal{C}$. Every created node is associated to a name in $\mathcal{C}$. If G be a graph and $\alpha \in \mathcal{V}$ then $G_{[\alpha+]}$ denotes the term-graph H s.t.:

- $\mathcal{N}(H) \stackrel{\text{def}}{=} \mathcal{N}(G) \cup \{NewNode(G)\}$, where $NewNode(G)$ denotes the smallest (according to $\prec$) node in $\mathcal{C}$ not occurring in G.
- For every node $\beta \in dom(G)$, $ref_H(\beta) \stackrel{\text{def}}{=} ref_G(\beta)$.
- $ref_H(NewNode(G))$ is undefined.

Note that H does not depend on α. As we shall see, α will be instantiated by $NewNode(G)$ which is useful only when applying a *sequence* of actions.

A **node redefinition** is a pair $\alpha{:}r$ where α is a node in $\mathcal{V}$ and r a reference. We denote by $G_{[\alpha:r]}$ the term-graph H defined as follows:

- $\mathcal{N}(H) \stackrel{\text{def}}{=} \mathcal{N}(G)$.
- For every node $\beta \in dom(G)$, if $\beta \neq \alpha$ then $ref_H(\beta) \stackrel{\text{def}}{=} ref_G(\beta)$.
- $ref_H(\alpha) \stackrel{\text{def}}{=} r$.

For instance, $\beta{:}f(\alpha, \delta{:}a)_{[\alpha:f(\delta,\alpha)]} = \beta{:}f(\alpha{:}f(\delta{:}a, \alpha), \delta)$. Note that we may have $\alpha \in dom(G)$ (in this case α is redirected) or $\alpha \notin dom(G)$ (in this case new edges and label are created). Note that a node redefinition does not introduce new nodes in the term-graph(this has to be done before by the node creation action).

An **edge redirection** may be seen as a particular case of node redefinition in which a unique edge is redirected. It is an expression of the form $\alpha.i \gg \beta$, where $i \in$, $\alpha \in \mathcal{N}$ and $\beta \in \mathcal{N}$. Applying an edge redirection to a term-graph consists in redirecting the i-th argument of the node α to point to the node β.

If G is a term-graph and $\alpha, \beta \in \mathcal{N}(G)$, where $\alpha \in dom(G)$, then $G_{[\alpha.i\ \beta]}$ denotes the graph H defined as follows: $H \stackrel{\mathrm{def}}{=} G_{[\alpha:f(\beta_1,\ldots,\beta_{i-1},\beta,\beta_{i+1},\ldots,\beta_n)]}$ where $f(\beta_1,\ldots,\beta_n) = ref_G(\alpha)$. Note that if $n < i$ then by convention $f(\beta_1,\ldots,\beta_{i\ 1},\beta,\beta_{i+1},\ldots,\beta_n) = f(\beta_1,\ldots,\beta_n)$ thus $H = G$.

For instance, $\beta{:}f(\alpha{:}f(\delta : a, \delta), \delta)_{[\alpha.1\ \beta]} = \beta{:}f(\alpha{:}f(\beta,\delta{:}a),\delta)$.

A **global redirection** is an expression of the form $\alpha \gg \beta$, where $\alpha \in \mathcal{N}$ and $\beta \in \mathcal{N}$. Applying a global redirection to a term-graph consists in redirecting any edge pointing to α to the node β, i.e. in replacing any occurrence of α in a reference in G by β.

If G is a term-graph and $\alpha, \beta \in \mathcal{N}(G)$ then $G_{[\alpha\ \beta]}$ denotes the graph H defined as follows: $\mathcal{N}(H) \stackrel{\mathrm{def}}{=} \mathcal{N}(G)$ and for every node $\gamma \in dom(G)$ s.t. $ref_G(\gamma) = f(\beta_1,\ldots,\beta_n)$ then $ref_H(\gamma) \stackrel{\mathrm{def}}{=} f(\beta_1',\ldots,\beta_n')$ where for every $i \in [1..n]$ we have $\beta_i' \stackrel{\mathrm{def}}{=} \beta_i$ if $\beta_i \neq \alpha$ and $\beta_i' \stackrel{\mathrm{def}}{=} \beta$ otherwise ($ref_H(\gamma)$ is undefined if $ref_G(\gamma)$ is undefined). This action is said to be "global" because it may affect any node in the term-graph (in the worst case all nodes may be affected). Global redirections are necessary to express easily collapsing rules of the form $f(x) \to x$ (any occurrence of $f(x)$ in the term-graph should be replaced by x).

For instance, $\beta{:}h(\delta{:}g(\alpha{:}a,\delta),\alpha)_{[\alpha\ \beta]} = \beta{:}h(\delta{:}g(\beta,\delta),\beta),\alpha$.

An **action** is either a node creation, or an edge redirection or a node definition or a global redirection. Substitutions can be extended to sequences of actions using the following definitions (where ϵ denotes the empty sequence and $\tau.\tau'$ denotes the concatenation of τ and τ').

- $\epsilon\sigma \stackrel{\mathrm{def}}{=} \epsilon$, $(a.\tau)\sigma \stackrel{\mathrm{def}}{=} a\sigma.\tau\sigma$.
- $(\alpha^+)\sigma \stackrel{\mathrm{def}}{=} \alpha^+$ (α is not instantiated since α is a variable denoting the new node).
- $(\alpha{:}f(\boldsymbol{\beta}))\sigma \stackrel{\mathrm{def}}{=} \alpha\sigma{:}f(\boldsymbol{\beta}\sigma)$, $(\alpha \gg \beta)\sigma \stackrel{\mathrm{def}}{=} \alpha\sigma \gg \beta\sigma$, $(\alpha.i \gg \beta)\sigma \stackrel{\mathrm{def}}{=} \alpha\sigma.i \gg \beta\sigma$.

If τ is a sequence of actions, and G is a term-graph, then $G_{[\tau]}$ denotes the term-graph defined as follows:

- $G_{[\epsilon]} \stackrel{\mathrm{def}}{=} G$
- If $a = \alpha^+$, $G_{[a.\tau]} \stackrel{\mathrm{def}}{=} G_{[a]_{[\tau\ \alpha\ NewNode(G)\ ]}}$. Note that α is instantiated by the new created node in the rest of the sequence (this allows one to "reuse" this node, hence to create edges starting from or pointing to this node).
- $G_{[a.\tau]} \stackrel{\mathrm{def}}{=} G_{[a]_{[\tau]}}$, if a is not a node creation.

Informally, $G_{[\tau]}$ is obtained from G by applying the actions in τ, in the corresponding order. For instance $\alpha{:}f(\alpha,\alpha)_{[\delta^+,\delta{:}a,\alpha.1\ \delta]} = \alpha{:}f(\alpha,\alpha),\alpha'_{[\alpha'{:}a,\alpha.1\ \alpha']} = \alpha{:}f(\alpha,\alpha),\alpha'{:}a_{[\alpha.1\ \alpha']} = \alpha{:}f(\alpha'{:}a,\alpha)$, where $\alpha' = NewNode(\alpha{:}f(\alpha,\alpha))$ (commas are used to separate the actions in the sequence).

Note that $G_{[\tau]}$ is not defined if τ is an action $\alpha.i \gg \beta$ s.t. $\alpha \notin dom(G)$. For instance $\alpha{:}a_{[\beta.1\ \ \alpha]}$ is undefined because β is not a node in $\alpha{:}a$. Otherwise, $G_{[\tau]}$ is always defined.

If τ is a sequence of actions then we denote by $r(\tau)$ the set of nodes α s.t. τ contains an action of the form $\alpha \gg \beta$ or $\alpha.i \gg \beta$ or $\alpha{:}f(\boldsymbol{\beta})$. Intuitively, $r(\tau)$ denotes the set of nodes that are affected by the sequence of actions τ.

4   Rewrite Rules

Obviously, rewrite rules operating on term-graphs should be able to check whether two nodes are equal or not. This is useful for instance when traversing a circular list: in order to avoid looping, we need to compare the current node with the initial one before proceeding to the tail of the list. These conditions correspond to *disequality constraints* between nodes, that need to be "attached" to the left-hand side of the rule. More precisely, a *node constraint* is a finite conjunction of (possibly none) disequations of the from $\alpha \not\approx \beta$, where $\alpha, \beta \in \mathcal{N}$. The empty node constraint is denoted by $\top$.

A disequation $\alpha \not\approx \beta$ is *false* if $\alpha = \beta$ and true if α, β are two distinct symbols in $\mathcal{A}$. More formally, a substitution σ is said to be a *solution* of a node constraint ϕ iff for any $(\alpha \not\approx \beta)$ occurring in ϕ we have $\alpha\sigma \neq \beta\sigma$ and $\alpha\sigma, \beta\sigma \in \mathcal{A}$. We denote by $sol(\phi)$ the set of solutions of ϕ. A substitution σ is said to be a *counter-solution* of a node constraint ϕ iff there exists $(\alpha \not\approx \beta)$ in ϕ s.t. we have $\alpha\sigma = \beta\sigma$. We denote by $csol(\phi)$ the set of counter-solutions of ϕ.

Clearly, if $\sigma \in csol(\phi)$ then $\sigma\theta \in csol(\phi)$ for any substitution θ, and $\sigma \notin sol(\phi)$. Similarly, if $\sigma \in sol(\phi)$ then $\sigma\theta \in sol(\phi)$ for any substitution θ, and $\sigma \notin csol(\phi)$. If σ is a ground substitution and $dom(\sigma)$ contains all the variables occurring in ϕ, then we have either $\sigma \in sol(\phi)$ or $\sigma \in csol(\phi)$.

A *constrained term-graph* is a pair $[\![G \mid \phi]\!]$ where G a term-graph and ϕ a node constraint. For the sake of clarity, $[\![G \mid \top]\!]$ is denoted by G. We are now in position to introduce our notion of term-graph rewrite rule.

Definition 2. *A term-graph rewrite rule is an expression of the form* $[\![L \mid \phi]\!] \rightarrow_\alpha R$ *where:*

1. $[\![L \mid \phi]\!]$ *is a constrained term-graph (the left-hand side of the rule).*
2. *R is a sequence of actions, s.t. if R contains an action of the form $\beta.i \gg \gamma$ then $\beta \in dom(L)^2$.*
3. $\alpha \in dom(L)$ *(α is the root of the rule).*

Example 1. The following rules insert an element α before a cell β in a doubly linked list δ. A doubly linked list cell is denoted by a term-graph $dll(\alpha, \beta, \delta)$ where α denotes the previous cell, β the value of the cell and δ the next cell (tail).

---

[2] This ensures that the action is always applicable on L.

- $\lambda{:}insert(\alpha,\beta,\delta{:}nil) \to_\lambda \lambda \gg \delta$. If δ is nil then the result is δ^3
- $[\![\lambda{:}insert(\alpha,\beta,\delta{:}dll(\delta_1,\delta_2,\delta_3)) \mid \beta \not\approx \delta]\!] \to_\lambda \lambda.3 \gg \delta_3$. If β is distinct from δ then α must be inserted into the tail of δ.
- $\lambda{:}insert(\alpha,\beta,\beta{:}dll(\beta_1,\beta_2,\beta_3)) \to_\lambda \lambda{:}dll(\beta_1,\alpha,\beta), \beta_1.3 \gg \lambda, \beta.1 \gg \lambda$. Otherwise, we create a new cell $\lambda{:}dll(\beta_1,\alpha,\beta)$, we redirect the first argument of β to λ and the last argument of the cell before β to λ.

Note we may have $\beta = \beta_1 = \beta_3$ in the last rule (circular list of length 1). Thanks to the flexibility of our language, we do not need to give any specific rule for this particular case (this is essential from a practical point of view).

Definition 3. *(Rewriting Step) Let G be a ground term-graph. Let $\rho = [\![L \mid \phi]\!] \to_\alpha R$ be a rewrite rule. We write $G \to_\rho H$ iff the following holds:*

- *There exists a ground substitution σ of the variables occurring in L s.t. $L\sigma \subseteq G$ and $\sigma \in sol(\phi)$.*
- *$H = G_{[R\sigma]}$.*

If $\mathcal{R}$ is a set of rewrite rules, then we write $G \to_{\mathcal{R}} H$ if $G \to_\rho H$ for some $\rho \in \mathcal{R}$. As usual $\to_{\mathcal{R}}$ denotes the reflexive and transitive closure of $\to_{\mathcal{R}}$.

Remark 1. The substitution σ can be easily computed by using standard unification: for any $\alpha \in dom(L)$, one has to find a node $\beta \in dom(G)$ s.t. $\alpha\sigma = \beta$ and $ref_L(\alpha)\sigma = ref_G(\beta)$. Of course, there may be several solutions (as in the usual case: a term may contain several distinct subterms matched by the left-hand side of a given rule). An important difference with the usual case is that even if we fix the value of the root node α in L, there may be still several solutions, except if all the nodes in L are accessible from α.

Example 2. Let ρ be the following rule: $\alpha{:}f(\beta,\delta{:}g(\gamma)) \to_\alpha \alpha{:}h(\beta,\delta), \delta.1 \gg \alpha$. This rule transforms a term-graph $\alpha{:}f(\beta,\delta{:}g(\gamma))$ into $\alpha{:}h(\beta,g(\alpha))$. Let $G = \lambda_1{:}f(\lambda_2{:}g(\lambda_4), \lambda_3{:}g(\lambda_4{:}a))$. We apply the rule ρ on G. We denote by L the left-hand side of ρ. We try to find a substitution σ s.t. $L\sigma \subseteq G$. Since $head_L(\alpha) = f$, we must have $head_G(\alpha\sigma) = f$, thus $\alpha\sigma = \lambda_1$. Since we must have $ref_G(\alpha\sigma) = ref_{L\sigma}(\alpha\sigma)$ we have $\beta\sigma = \lambda_2$ and $\delta\sigma = \lambda_3$. Then since $ref_G(\lambda_3) = L\sigma(\lambda_3)$, we have $g(\lambda_4) = L(\delta)\sigma = g(\gamma)\sigma$, thus $\gamma\sigma = \lambda_4$.

We obtain the term-graph:

$$\lambda_1{:}f(\lambda_2{:}g(\lambda_4), \lambda_3{:}g(\lambda_4{:}a))_{[\lambda_1{:}h(\lambda_2,\lambda_3), \lambda_3.1 \gg \lambda_1]}$$

$$= \lambda_1{:}h(\lambda_2{:}g(\lambda_4), \lambda_3{:}g(\lambda_4{:}a))_{[\lambda_3.1 \gg \lambda_1]} = \lambda_1{:}h(\lambda_2{:}g(\lambda_4), \lambda_3{:}g(\lambda_1)).$$

Example 3. We consider the rule ξ defined as follows: $\alpha{:}f(\beta{:}a,\delta{:}a) \to_\alpha \beta{:}b, \delta{:}c$. Let G be the term-graph $\lambda_1{:}f(\lambda_2{:}a, \lambda_2)$. We apply ξ on G. The only possible substitution is: $\sigma = \{\alpha \mapsto \lambda_1, \beta \mapsto \lambda_2, \delta \mapsto \lambda_2\}$. We obtain the term-graph: $\lambda_1{:}f(\lambda_2{:}a, \lambda_2)_{[\lambda_2{:}b,\lambda_2{:}c]} = \lambda_1{:}f(\lambda_2{:}b, \lambda_2)_{[\lambda_2{:}c]} = \lambda_1{:}f(\lambda_2{:}c, \lambda_2)$.

---

[3] For the sake of clarity we write the constrained term-graph $[\![\lambda{:}insert(\alpha,\beta,\delta{:}nil) \mid \top]\!]$ without brackets since its constraint part is $\top$.

5   Narrowing

5.1   Term-Graph Substitutions

We need to introduce some further notations. Two term-graphs G and H are said to be *disjoint* (written $G \parallel H$) if $dom(G) \cap dom(H) = \emptyset$. Two term-graphs G and H are said to be *compatible* (written $G \bowtie H$) iff for any $\alpha \in \mathcal{N}$ s.t. $ref_G(\alpha)$ and $ref_H(\alpha)$ are defined, we have $ref_G(\alpha) = ref_H(\alpha)$ (i.e. G, H coincide on the intersection of their domains). Obviously, if $G \parallel H$ then $G \bowtie H$.

If G, H are compatible then $G \cup H$ denotes the minimal term-graph G' s.t. $G \subseteq G'$ and $H \subseteq G'$ (it is clear that G' always exists). If $G \bowtie H$ then $G \setminus H$ denotes the (minimal) term-graph I s.t. $I \parallel H$ and $H \cup I = G$.

The notion of g-substitution is the analogue of the notion of substitution for terms. When instantiating a term-graph, one has not only to specify the value of the variables occurring in it, but also to define the references corresponding to the nodes that are introduced by the substitution. Clearly, these nodes should be distinct from the ones already occurring in the considered term-graph.

Definition 4. *(g-substitution)* A g-substitution *is a pair* $\varsigma = (\sigma, G)$ *where* σ *is a substitution and* G *a term-graph s.t. if* $x \in dom(\sigma)$ *then* $x\sigma \in \mathcal{N}(G)$. *A g-substitution of a term-graph* H *is a g-substitution* $\varsigma = (\sigma, G)$ *s.t.* σ *is compatible with* H *and* $G \bowtie H\sigma$. *In this case,* $H\varsigma$ *denotes the term-graph:* $H\sigma \cup G$.

For instance, if $H = \alpha{:}f(\beta, \delta)$ and $\varsigma = (\{\beta \to \delta\}, \delta{:}g(\delta, \lambda{:}a))$, then $H\varsigma = \alpha{:}f(\delta{:}g(\delta, \lambda{:}a), \delta)$. Note that δ cannot have a reference in H distinct from the one in ς, according to the previous definition, since it is defined in $\delta{:}g(\delta, \lambda{:}a)$.

(σ, G) is said to be *ground* if σ, G are ground. If $\varsigma = (\sigma, G)$ then σ_ς denotes the substitution σ and Gr_ς denotes the term-graph G. If ϑ is a g-substitution of Gr_ς, then $\varsigma\vartheta$ denotes the g-substitution $(\sigma_\varsigma\sigma_\vartheta, \mathrm{Gr}_\varsigma\theta \cup \mathrm{Gr}_\vartheta)$ (composition of ς and ϑ). ς is said to be *more general* than ϑ if there is a substitution ϱ s.t. $\varsigma\varrho = \vartheta$.

By a slight abuse of notation, we write $\varsigma \in sol(\phi)$ (resp. $\varsigma \in csol(\phi)$) iff $\sigma_\varsigma \in sol(\phi)$ (resp. $\sigma_\varsigma \in csol(\phi)$). Similarly, if ϕ is a node constraint (resp. a node or a sequence of actions) then $\phi\varsigma$ denotes the expression $\phi\sigma_\varsigma$ (note that if G is a term-graph, then $G\sigma_\varsigma$ is not equal to $G\varsigma$ if Gr_ς is not empty).

5.2   Symbolic Handling of Actions

An *s-graph* is either a term-graph or an expression of the form $apply(G, \tau)$ where G is a term-graph and τ is a sequence of actions. A *g-term* is a triple $[\![G \mid \phi]\!]^\tau$ where G is an s-graph, ϕ a node constraint and τ a sequence of actions. τ denotes in some sense the "history" of G, i.e. the set of actions applied for getting G. ϕ imposes additional constraint on the variables occurring in G.

The use of g-terms allows us to handle actions in a symbolic way (i.e. without performing them explicitly). $G_{[\tau]}$ and $apply(G, \tau)$ have very different meanings: $G_{[\tau]}$ denotes the term-graph obtained by applying τ on G, whereas $apply(G, \tau)$ is merely a syntactic object. We relate these two notions by associating semantics to g-terms. If $\mathcal{G}$ is a g-term, then $value(\mathcal{G})$ is a term-graph defined as follows:

$value(\llbracket G \mid \phi \rrbracket^\tau) \overset{\text{def}}{=} G$ if G is term-graph, and $value(\llbracket apply(G, \chi) \mid \phi \rrbracket^\tau) \overset{\text{def}}{=} G_{[\chi]}$ otherwise (note that $value(\llbracket G \mid \phi \rrbracket^\tau)$ does not depend on ϕ and τ).

5.3   Narrowing Steps

Obviously, an action can be applied on a node α only if every other node β occurring in the term-graph is known to be distinct from α. Otherwise, we do not know whether β is to be redirected or not, hence we cannot apply the action. The next definition formalizes this notion.

Let a be an action s.t. $r(a) = \{\alpha\}$. A node β is said to be a-isolated in a constrained term-graph $\llbracket G \mid \phi \rrbracket$ iff either β is syntactically equal to α or if the application of the action a cannot affect the occurrences of β in $\llbracket G \mid \phi \rrbracket$, i.e. iff one of the following conditions holds: either $\beta = \alpha$, or $\beta \notin dom(G)$ and a is not a global redirection, or $\beta \not\approx \alpha$ occurs in ϕ, or $\alpha, \beta \in \mathcal{A}$. A constrained term-graph $\llbracket G \mid \phi \rrbracket$ is said to be *ready* for an action a if any node in $\llbracket G \mid \phi \rrbracket$ is a-isolated. Roughly speaking a node β is a-isolated if one has enough information to decide whether β is affected by a or not. This ensures that a behaves in a similar way *for all possible values of β.*

Our narrowing algorithm uses several rules. The first one corresponds to the usual narrowing step and is defined as follows.

$$\llbracket G \mid \psi \rrbracket^\tau \quad_{\rho,\varsigma} \quad \llbracket H \mid \psi' \rrbracket^{\tau\sigma}$$

If:

- G is a term-graph, $\rho = \llbracket L \mid \phi \rrbracket \to_\alpha R$ is a rewrite rule.
- σ is a most general substitution compatible with L and G s.t.:
  - $L\sigma \bowtie G\sigma$ ($L\sigma$ and $G\sigma$ must be compatible),
  - $\alpha\sigma \in dom(G\sigma)$ (the root of the rule occurs in the considered term-graph).
- $H = apply(G\sigma \cup L\sigma, R\sigma)$, $G' \overset{\text{def}}{=} L\sigma \setminus G\sigma$, $\varsigma \overset{\text{def}}{=} (\sigma, G')$.
- $\psi' = \psi\sigma \wedge \phi\sigma \wedge \bigwedge_{\beta \in r(\tau\sigma), \delta \in dom(G')} \beta \not\approx \delta$.

ψ' inherits from the constraints in ψ and in ϕ. The additional disequations $\beta \not\approx \delta$ express the fact that every synthesized node δ (i.e. every node occurring in $dom(L\sigma)$ but not in $dom(G\sigma)$) should be distinct from the nodes β that have been previously redirected. This property is essential for soundness. The g-substitution (σ, G') plays a role similar to the one of unifiers in term narrowing. In contrast to the rewrite step, this first narrowing rule does not explicitly apply the sequence of actions R on the term-graph $G\sigma \cup L\sigma$ but only *encodes* them into the g-term. The reason is that the actions are not necessarily applicable at this point, since the term-graph may not be ready for them.

Example 4. Let ρ be the following rule: $\alpha{:}f(\beta, \delta{:}g(\gamma{:}a, \zeta)) \to_\alpha \alpha{:}h(\beta, \delta), \delta.1 \gg \alpha$. Let $G = \llbracket \lambda_1{:}f(\lambda_2, \lambda_3) \mid \top \rrbracket^\tau$ (where $\lambda_1, \lambda_2, \lambda_3$ are variables). We assume that $\tau = \lambda_1{:}f(\lambda_2, \lambda_3)$ (hence λ_1 has been redirected). We apply the narrowing rule on G, using the rule ρ. We denote by L the left-hand side of ρ. We try to find a substitution σ, compatible with G and L, s.t. $\alpha\sigma$ occurs in $dom(G)$ and $G\sigma \bowtie L\sigma$. Since $head_L(\alpha) = f$, we must have $head_G(\alpha\sigma) = f$, thus $\alpha\sigma = \lambda_1$. Since we must have $ref_{G\sigma}(\alpha\sigma) = ref_{L\sigma}(\alpha\sigma)$ we have $\beta\sigma = \lambda_2$ and $\delta\sigma = \lambda_3$.

Obviously, the obtained substitution satisfies the desired conditions. The term-graph G' in the definition of the narrowing rule is $\lambda_3{:}g(\gamma{:}a,\zeta)$ (γ,ζ are variables). We obtain:

$$[\![apply(\ (\lambda_1{:}f(\lambda_2,\lambda_3),\lambda_3{:}g(\gamma{:}a,\zeta))\ ,\ \lambda_1{:}h(\lambda_2,\lambda_3),\lambda_3.1 \gg \lambda_1)\ |\ \lambda_1 \not\approx \lambda_3 \wedge \lambda_1 \not\approx \gamma]\!]^\tau.$$

Note that the disequations $\lambda_1 \not\approx \lambda_3 \wedge \lambda_1 \not\approx \gamma$ have been added in the constraint part of the g-term. This is due to the fact that since λ_1 has already been redirected, the nodes synthesized during the application of the narrowing rule should be distinct from λ_1. The actions $\lambda_1{:}h(\lambda_2,\lambda_3)$ and $\lambda_3.1 \gg \lambda_1$ are not performed at this point but only stored into the g-term.

Additional narrowing rules are required to handle g-terms of the form $[\![apply(G,\tau)\ |\ \phi]\!]^\xi$, i.e. to explicitly apply the actions introduced by the previous rule. The first one – denoted by T – is trivial: it simply transforms a g-term into a term-graph in case of empty sequences of actions.

$$[\![apply(G,\epsilon)\ |\ \phi]\!]^\tau \quad T,_\emptyset\ [\![G\ |\ \phi]\!]^\tau.$$

The second and third rules, respectively denoted by A and A^+, apply the first action in the sequence and then proceed to the next ones (assuming that the considered term-graph is ready for this action). A^+ handles node creations, whereas A handles all other actions.

$$[\![apply(G,a.\tau)\ |\ \phi]\!]^\xi \quad A,_\emptyset\ [\![apply(G_{[a]},\tau)\ |\ \phi]\!]^{\xi.a}.$$

If a is not a node creation, $[\![G\ |\ \phi]\!]$ is ready for a.

$$[\![apply(G,\alpha^+.\tau)\ |\ \phi]\!]^\xi \quad A^+,_\sigma\ [\![apply(G_{[\alpha^+]},\tau\sigma)\ |\ \phi \wedge \bigwedge_{\beta \in V \cap \mathcal{N}(G)} \beta \not\approx \gamma]\!]^{\xi.\gamma^+}$$

If $\sigma = \{\alpha \to \gamma\}$ where $\gamma = NewNode(G)$.

The added disequations ensure that the variables already present in the term-graph are distinct from the newly created node γ (this is essential to prevent these variables to be unified with γ afterwards).

The above rules are clearly not sufficient to ensure completeness. Consider for instance the following rule: $\lambda{:}f(\alpha{:}a,\beta) \to_\lambda \alpha{:}b$. Assume we want to apply the narrowing rule on the g-term $f(\delta{:}a,\delta'{:}a)$, where δ,δ' denote variables. Then the narrowing rule cannot apply, because we do not know at this point whether $\delta = \delta'$ or not. If $\delta = \delta'$ then the term-graph should be reduced to: $f(\delta{:}b,\delta)$. Otherwise, we should get: $f(\delta{:}b, \delta'{:}a)$. In other words, the considered term-graph is not ready for the action $\delta{:}b$ because δ' is not isolated. Since both options are possible (namely $\delta = \delta'$ or $\delta \neq \delta'$) we need to consider the two possibilities separately, i.e. in two distinct branches. This is done by the two following branching rules, denoted by $B^=$ and $B^{\neq}$ respectively.

$$[\![apply(G, a.\tau) \mid \phi]\!]^{\varsigma} \quad B^=_{,\sigma} \quad [\![apply(G\sigma, (a.\tau)\sigma) \mid \phi\sigma]\!]^{\varsigma\sigma}$$

If $\alpha \in r(a)$, β is a non a-isolated node in G

and σ is a most general substitution compatible with G s.t. $\alpha\sigma = \beta\sigma$.

$$[\![apply(G, a.\tau) \mid \phi]\!]^{\varsigma} \quad B^{\neq}_{,\emptyset} \quad [\![apply(G, a.\tau) \mid \phi \wedge \alpha \not\approx \beta]\!]^{\varsigma}$$

If $\alpha \in r(a)$, β is a non a-isolated node in G.

In both cases, β becomes a-isolated after application of the rule (either because it is instantiated by α or because the disequation $\alpha \not\approx \beta$ is added in the constraints). Applying these rules on the term-graph $[\![apply(f(\delta{:}a, \delta'{:}a), \delta{:}b) \mid \top]\!]$ yields the two following g-terms: $[\![apply(f(\delta{:}a, \delta), \delta{:}b) \mid \top]\!] \qquad [\![f(\delta{:}b, \delta) \mid \top]\!]^{\delta{:}b}$ and $[\![apply(f(\delta{:}a, \delta'{:}a), \delta{:}b) \mid \delta \not\approx \delta']\!] \qquad [\![f(\delta{:}b, \delta'{:}a) \mid \delta \not\approx \delta']\!]^{\delta{:}b}$

Definition 5. *If $\mathcal{R}$ is a set of rewrite rules, then we write $G \; {}_{\mathcal{R},\varsigma} \; H$ if $G \; {}_{\rho,\varsigma} \; H$ for some $\rho \in \mathcal{R} \cup \{T, A, A^+, B^{\neq}, B^=\}$. ${}_{\mathcal{R},\varsigma}$ is inductively defined as follows: $G \; {}_{\mathcal{R},\varsigma} \; H$ iff either $\varsigma = \emptyset$ and $G = H$ or $G \; {}_{\mathcal{R},\varrho} \; G', G' \; {}_{\mathcal{R},\vartheta} \; H$, ϱ is a g-substitution of $G_{\mathcal{R}\vartheta}$ and $\varsigma = \varrho\vartheta$.*

We provide a detailed example of application. We define the following functions $\#$ and *equal* computing respectively the length $\#$ of a circular list (rules ρ_1, ρ_2, ρ_3) and the equality on natural numbers (ξ_1, ξ_2):

$$\alpha{:}\#(\beta) \to_\alpha \alpha{:}\#'(\beta, \beta) \; (\rho_1) \qquad \alpha{:}\#'(\beta_1{:}cons(\beta_2, \beta_3), \beta_3) \to_\alpha \alpha'^+, \alpha{:}s(\alpha'), \alpha'{:}0 \; (\rho_2)$$
$$[\![\beta_1{:}\#'(\beta_2{:}cons(\beta_3, \beta_4), \beta_5) \mid \beta_4 \not\approx \beta_5]\!] \to_{\beta_1} \alpha'^+, \beta_1{:}s(\alpha'), \alpha'{:}\#'(\beta_4, \beta_5) \; (\rho_3)$$
$$\alpha{:}equal(0,0) \to_\alpha \alpha{:}true \; (\xi_1) \qquad \alpha{:}equal(s(\beta_1), s(\beta_2)) \to_\alpha \alpha{:}equal(\beta_1, \beta_2) \; (\xi_2)$$

Assume we want to solve the goal[4]: $\gamma{:}equal(\gamma'{:}\#(\gamma''), s(s(0))) \to \quad true$ (i.e. to find the circular lists γ'' of length 2). We assume that γ'' is a variable and γ, γ' are names. The corresponding narrowing derivation is depicted below. We get the (unique) solution: $\gamma''{:}cons(\beta_3, \beta_4{:}cons(\beta'_2, \gamma''))$ where $\gamma'' \neq \beta_4$ (the other disequations are irrelevant since α' and α'' do not occur in the term-graph). In order to improve the readability we do not specify the sequence of actions occurring in the g-terms, since they can be easily recovered from the previous steps. Moreover, we only give the derivation yielding *true* (the reader can check that all the other derivations fail).

---

[4] There are many ways to define goals in the literature (equations, booleans, expressions,...). In this paper we focus on the basic narrowing steps.

$$\gamma\!:\!equal(\gamma'\!:\!\#(\gamma''), s(s(0)))$$

ρ_1 $\quad apply(\gamma\!:\!equal(\gamma'\!:\!\#(\gamma''), s(s(0))), \gamma'\!:\!\#'(\gamma'', \gamma''))$

A $\quad apply(\gamma\!:\!equal(\gamma'\!:\!\#'(\gamma'', \gamma''), s(s(0))), \epsilon)$

T $\quad \gamma\!:\!equal(\gamma'\!:\!\#'(\gamma'', \gamma''), s(s(0)))$

ρ_3 $\quad \llbracket apply(\gamma\!:\!equal(\gamma'\!:\!\#'(\gamma'', \gamma''\!:\!cons(\beta_3, \beta_4)), s(s(0))),$
$\qquad \alpha'^+, \gamma'\!:\!s(\alpha'), \alpha'\!:\!\#'(\beta_4, \gamma''))$
$\qquad | \ \beta_4 \neq \gamma'' \rrbracket$

A^+ $\quad \llbracket apply((\gamma\!:\!equal(\gamma'\!:\!\#'(\gamma''\!:\!cons(\beta_3, \beta_4), \gamma''), s(s(0))), \alpha'),$
$\qquad \gamma'\!:\!s(\alpha'), \alpha'\!:\!\#'(\beta_4, \gamma''))$
$\qquad | \ \beta_4 \not\approx \gamma'', \beta_3 \not\approx \alpha', \beta_4 \not\approx \alpha', \gamma'' \not\approx \alpha' \rrbracket$

A $\quad \llbracket apply((\gamma\!:\!equal(\gamma'\!:\!s(\alpha'), s(s(0))), \gamma''\!:\!cons(\beta_3, \beta_4), \alpha'),$
$\qquad \alpha'\!:\!\#'(\beta_4, \gamma''))$
$\qquad | \ \beta_4 \not\approx \gamma'', \beta_3 \not\approx \alpha', \beta_4 \not\approx \alpha', \gamma'' \not\approx \alpha' \rrbracket$

A $\quad \llbracket apply(\gamma\!:\!equal(\gamma'\!:\!s(\alpha'\!:\!\#'(\beta_4, \gamma''\!:\!cons(\beta_3, \beta_4))), s(s(0))),$
$\qquad \epsilon)$
$\qquad | \ \beta_4 \not\approx \gamma'', \beta_3 \not\approx \alpha', \beta_4 \not\approx \alpha', \gamma'' \not\approx \alpha' \rrbracket$

T $\quad \llbracket \gamma\!:\!equal(\gamma'\!:\!s(\alpha'\!:\!\#'(\beta_4, \gamma''\!:\!cons(\beta_3, \beta_4))), s(s(0)))$
$\qquad | \ \beta_4 \not\approx \gamma'', \beta_3 \not\approx \alpha', \beta_4 \not\approx \alpha', \gamma'' \not\approx \alpha' \rrbracket$

ξ_1, A $\quad \llbracket \gamma\!:\!equal(\alpha'\!:\!\#'(\beta_4, \gamma''\!:\!cons(\beta_3, \beta_4)), s(0))$
$\qquad | \ \beta_4 \not\approx \gamma'', \beta_3 \not\approx \alpha', \beta_4 \not\approx \alpha', \gamma'' \not\approx \alpha' \rrbracket$

ρ_2 $\quad \llbracket apply(\gamma\!:\!equal(\alpha'\!:\!\#'(\beta_4, \gamma''\!:\!cons(\beta_3, \beta_4\!:\!cons(\beta'_2, \gamma''))), s(0)),$
$\qquad \alpha''^+, \alpha'\!:\!s(\alpha''), \alpha''\!:\!0))$
$\qquad | \ \beta_4 \not\approx \gamma'', \beta_3 \not\approx \alpha', \beta_4 \not\approx \alpha', \gamma'' \not\approx \alpha' \rrbracket$

A^+ $\quad \llbracket apply((\gamma\!:\!equal(\alpha'\!:\!\#'(\beta_4, \gamma''\!:\!cons(\beta_3, \beta_4\!:\!cons(\beta'_2, \gamma''))), s(0)), \alpha''),$
$\qquad \alpha'\!:\!s(\alpha''), \alpha''\!:\!0)$
$\qquad | \ \beta_4 \not\approx \gamma'', \beta_3 \not\approx \alpha', \beta_4 \not\approx \alpha', \gamma'' \not\approx \alpha', \gamma'' \not\approx \alpha'', \beta_3 \not\approx \alpha'', \beta_4 \not\approx \alpha'', \beta'_2 \not\approx \alpha'' \rrbracket$

A, A, T $\quad \llbracket \gamma\!:\!equal(\alpha'\!:\!s(\alpha''\!:\!0), s(0)), \gamma''\!:\!cons(\beta_3, \beta_4\!:\!cons(\beta'_2, \gamma''))$
$\qquad | \ \beta_4 \not\approx \gamma'', \beta_3 \not\approx \alpha', \beta_4 \not\approx \alpha', \gamma'' \not\approx \alpha', \gamma'' \not\approx \alpha'', \beta_3 \not\approx \alpha'', \beta_4 \not\approx \alpha'', \beta'_2 \not\approx \alpha'' \rrbracket$

ξ_2, A, T $\quad \llbracket \gamma\!:\!equal(\alpha''\!:\!0, 0), \gamma''\!:\!cons(\beta_3, \beta_4\!:\!cons(\beta'_2, \gamma''))$
$\qquad | \ \beta_4 \not\approx \gamma'', \beta_3 \not\approx \alpha', \beta_4 \not\approx \alpha', \gamma'' \not\approx \alpha', \gamma'' \not\approx \alpha'', \beta_3 \not\approx \alpha'', \beta_4 \not\approx \alpha'', \beta'_2 \not\approx \alpha'' \rrbracket$

ξ_1, A, T $\quad \llbracket \gamma\!:\!true, \gamma''\!:\!cons(\beta_3, \beta_4), \beta_4\!:\!cons(\beta'_2, \gamma'')$
$\qquad | \ \beta_4 \not\approx \gamma'', \beta_3 \not\approx \alpha', \beta_4 \not\approx \alpha', \gamma'' \not\approx \alpha', \gamma'' \not\approx \alpha'', \beta_3 \not\approx \alpha'', \beta_4 \not\approx \alpha'', \beta'_2 \not\approx \alpha'' \rrbracket$

6   The Properties of the Narrowing Relation

Soundness and Completeness are defined w.r.t. the ground rewriting rule introduced in Definition 3. **Soundness** ensures that every narrowing derivation can be related to a sequence of rewriting steps, operating at the ground level. More precisely, if we have $\llbracket G \mid \top \rrbracket \quad {}_{\mathcal{R}, \varsigma} \ \llbracket H \mid \phi \rrbracket^\tau$ then for any ground instance ϑ of H solution of ϕ, we should have $G\varsigma\vartheta \rightarrow_{\mathcal{R}} H\vartheta$. Unfortunately, this property does not hold for all substitutions ϑ. Indeed, if ϑ contains a node α on which a global redirection is performed during the narrowing derivation, then the term-graph obtained by rewriting from $G\varsigma\vartheta$ is different from $H\vartheta$, since any instance of α in ϑ should be redirected during the rewriting process. Thus we will assume that ϑ contains no such node (this property is always satisfied if we restrict ourself to irreducible derivations, see Definition 6). Similarly, ϑ should not contain any created node.

Theorem 1. *(Soundness) Let $\mathcal{R}$ be a set of rewrite rules. Let $[\![G \mid \psi]\!]^\tau$ and $[\![H \mid \psi']\!]^{\tau'}$ be two g-terms s.t. $[\![G \mid \psi]\!]^\tau \xrightarrow{}_{\mathcal{R},\varsigma} [\![H \mid \psi']\!]^{\tau'}$. Let ϑ be a ground g-substitution of H s.t. $\vartheta \in sol(\psi')$ and Gr_ϑ contains no node α s.t. $\alpha \in \mathcal{C}$ or $\alpha \gg \beta \in \tau'$, for some $\beta \in \mathcal{N}$. $\varsigma\vartheta$ is a g-substitution of G. Moreover value($[\![G \mid \psi]\!]^\tau \varsigma\vartheta) \to_{\mathcal{R}}$ value($[\![H \mid \psi]\!]^{\tau'}\vartheta$).*

In particular, this implies that if G, H are two term-graphs s.t. $[\![G \mid \top]\!]^\epsilon \xrightarrow{}_{\mathcal{R},\varsigma} [\![H \mid \phi]\!]^\tau$ and ϑ is a g-substitution of H satisfying the above conditions, then $G\varsigma\vartheta \to_{\mathcal{R}} H\vartheta$.

Completeness expresses the fact that every rewriting derivation from a ground instance of a considered term-graph G can be subsumed by narrowing from $[\![G \mid \top]\!]$. More precisely, if ς is a g-substitution of G s.t. $G\varsigma \to H$ then there should exist a narrowing derivation $[\![G \mid \top]\!] \xrightarrow{}_{\mathcal{R},\varrho} [\![G' \mid \phi]\!]^\tau$ and a g-substitution $\vartheta \in sol(\phi)$ s.t. $G'\vartheta = H$ and $\varsigma = \varrho\vartheta$. This property does not hold for every substitution ς, but only for those that are irreducible w.r.t. the considered set of rewrite rules. The definition of an irreducible substitution is more complicated than in the usual case (i.e. for standard terms), since we have to take into account global redirections. Roughly speaking a term-graph G will be considered as reducible iff a rule can be applied on a node in $dom(G)$ (1) *or* if a rule can globally redirect a node in G (2) *or* if it contains a created node (3). More formally:

Definition 6. *A g-substitution $\varsigma = (\sigma, G)$ is said to be $\mathcal{R}$-irreducible iff the following conditions hold:*

1. *There is no rule $L \to_\lambda R \in \mathcal{R}$ s.t. there exists a substitution θ of the variables in L s.t. $L\theta \bowtie G$ and $\lambda\theta \in dom(G)$.*
2. *There is no rule $L \to_\lambda R \in \mathcal{R}$ s.t. there exists a substitution θ of the variables in L and a node $\alpha \in \mathcal{N}(L)$ s.t. $L\theta \bowtie G$, $\alpha\sigma \in \mathcal{N}(G)$, R contains an action of the form $\alpha \gg \beta$ for some $\beta \in \mathcal{N}$.*
3. *$\mathcal{N}(G) \cap \mathcal{C} = \emptyset$.*

If a constructor based signature is used, then Condition 1 simply states that ς contains no defined function. Similarly, Condition 2 can be easily guaranteed if only nodes labeled by defined functions can be globally redirected (this is a rather natural restriction). Condition 3 simply expresses the fact that ς should not contain any created node.

Theorem 2. *(Completeness) Let $\mathcal{R}$ be a set of rewrite rules. Let G, H be two term-graphs and let ς be an $\mathcal{R}$-irreducible ground substitution of the variables in G s.t. $G\varsigma \to_{\mathcal{R}} H$. For any ϕ s.t. $\varsigma \in sol(\phi)$ and for any sequence of actions τ s.t. $r(\tau) \subseteq dom(G)$, there exists ϱ s.t. $[\![G \mid \phi]\!]^\tau \xrightarrow{}_{\mathcal{R},\varrho} [\![G' \mid \psi]\!]^{\tau'}$ and $\vartheta' \in sol(\psi)$ s.t. $\varsigma = \varrho\vartheta'$, G' is a term-graph, and value($[\![G' \mid \psi]\!]\vartheta') = H$.*

In particular, if $G\varsigma \to_{\mathcal{R}} H$ and ς is $\mathcal{R}$-irreducible, then there exist ϱ s.t. $[\![G \mid \top]\!]^\epsilon \xrightarrow{}_{\mathcal{R},\varrho} [\![G' \mid \psi]\!]^{\tau'}$ and $\vartheta \in sol(\psi)$ s.t. $\varsigma = \varrho\vartheta$, G' is a term-graph, and $G'\vartheta = H$.

7   Conclusion

We have shown that narrowing could be extended to a large class of term-graph rewrite systems. The considered rewrite rules allow one to fully handle data-structures with pointers thanks to the actions like pointer redirections, node redefinition and creation. These results are the first ones concerning the narrowing relation in such a wide class of term-graph rewrite systems. It is also the first narrowing-based algorithm able to synthesize cyclic data-structures as answers in a context where bisimilar graphs are considered as equal only if they are identical. In this paper we were rather interested in the basic definition of narrowing, its soundness and completeness. Optimal term-graph narrowing strategies as studied in [7,8,1] are out of the scope of this paper, but a matter of future work. The considered term-graph rewrite systems are not always confluent. We proposed in [6] the use of term-graphs with priority in order to recover the confluence property within orthogonal systems. The future narrowing strategies should certainly integrate the priority over the nodes of a term-graph in addition to neededness properties.

References

1. S. Antoy, D. W. Brown, and S.-H. Chiang.   Lazy context cloning for non-deterministic graph rewriting. In *Third International Workshop on Term Graph Rewriting, TERMGRAPH*, pages 61–70, 2006.
2. S. Antoy, R. Echahed, and M. Hanus. A needed narrowing strategy. *Journal of the ACM*, 47(4):776–822, July 2000.
3. A. Bakewell, D. Plump, and C. Runciman. Checking the shape safety of pointer manipulations. In *RelMiCS*, pages 48–61, 2003.
4. A. Bakewell, D. Plump, and C. Runciman. Specifying pointer structures by graph reduction. In *AGTIVE*, pages 30–44, 2003.
5. H. Barendregt, M. van Eekelen, J. Glauert, R. Kenneway, M. J. Plasmeijer, and M. Sleep. Term Graph Rewriting. In *PARLE'87*, pages 141–158. Springer, LNCS 259, 1987.
6. R. Caferra, R. Echahed, and N. Peltier. Rewriting term-graphs with priority. In *Proceedings of PPDP (Principle and Practice of Declarative Programming)*. ACM, 2006.
7. R. Echahed and J.-C. Janodet.   Admissible graph rewriting and narrowing. In *IJCSLP*, pages 325–342, 1998.
8. R. Echahed and J. C. Janodet. Completeness of admissible graph collapsing narrowing. In *Proc. of Joint APPLIGRAPH/GETGRATS Workshop on Graph Transformation Systems (GRATRA 2000)*, March 2000.
9. A. Habel and D. Plump.   Term graph narrowing. *Mathematical Structures in Computer Science*, 6:649–676, 1996.
10. M. Hanus. The integration of functions into logic programming : from theory to practice. *Journal of Logic Programming*, 19&20:583–628, 1994.
11. M. K. Rao. Completeness results for basic narrowing in non-copying implementations. In *Proc. of Joint Int. Conference and Symposium on Logic Programming*, pages 393–407. MIT press, 1996.
12. H. Yamanaka. Graph narrowing and its simulation by graph reduction. Research report IIAS-RR-93-10E, Institute for Social Information Science, Fujitsu Laboratories LDT, June 1993.

Molecular Analysis of Metabolic Pathway with Graph Transformation

Karsten Ehrig[*], Reiko Heckel, and Georgios Lajios[**]

Department of Computer Science, University of Leicester, United Kingdom
{karsten, reiko, gl51}@mcs.le.ac.uk

Abstract. Metabolic pathway analysis is one of the tools used in biology and medicine in order to understand reaction cycles in living cells. A shortcoming of the approach, however, is that reactions are analysed only at a level corresponding to what is known as the 'collective token view' in Petri nets, i.e., summarising the number of atoms of certain types in a compound, but not keeping track of their identity.

In this paper we propose a refinement of pathway analysis based on hypergraph grammars, modelling reactions at a molecular level. We consider as an example the citric acid cycle, a classical, but non-trivial reaction for energy utilisation in living cells. Our approach allows the molecular analysis of the cycle, tracing the flow of individual carbon atoms based on a simulation using the graph transformation tool AGG.

1 Introduction

From the beginning biology has been one of the main application areas of graph transformations [3]. In recent years this line of research has been renewed by proposals for modelling recombination of DNA sequences in cells [7] and other biochemical reactions [10,12].

One major argument in favour of graph transformation-based models for biological systems and chemical reactions is their inherent concurrency, allowing reactions to take place simultaneously as long as they involve different resources and to keep track of causal dependencies and conflicts between them. So far, little of the concurrency concepts available for graph transformation systems have actually been applied in this area. This paper can be seen as a first attempt to identify interesting questions and possible solutions based on a well-known, but non-trivial case study.

In particular we are interested in the analysis of causal dependencies between biochemical reactions. Given a metabolic pathway (a sequence of reactions) we would like to be able to trace the history of particular atoms or molecules. This is relevant, for example, when trying to anticipate the outcome of experiments

---

[*] Work was partially supported through the IST-2005-16004 Integrated Project SENSORIA: Software Engineering for Service-Oriented Overlay Computers.

[**] Research was partially funded by European Community's Human Potential Programme under contract HPRN-CT-2002-00275, [SegraVis].

A. Corradini et al. (Eds.): ICGT 2006, LNCS 4178, pp. 107–121, 2006.

using radioactive isotopes of such atoms. Such questions have been crucial to the detailed understanding of the nature of reactions like the citric acid cycle.

To be able to answer them we propose a new hypergraph model for chemical compounds which refines the classical representation in terms of structural formulae in two different ways.

– Our representation keeps track of the identity of atoms or molecular components by means of the identities of hyperedges. In contrast, when writing down chemical reactions with structural formulae, the identities of the reacting atoms are not explicitly represented in the notation. In situations where several atoms of the same element are involved, this lack of information leads to ambiguity as to where a new atom is placed in the resulting molecule. Our graph transformation-based model allows to track atom identities by graph homomorphisms between the graphs representing the compounds before and after the reaction.

This refinement is comparable to the relation of graph transformation systems and place-transition Petri nets in the collective token view. While the latter record and change the numbers of tokens on certain places (resulting in the rewriting of a multiset of places), graph transformation allows the rewriting of sets of nodes (and edges), each with their individual identity.
– Modelling atoms as hyperedges, each connected to an ordered sequence of nodes, the relative spatial orientation of different molecular components is recorded through the ordering of the nodes connected to a hyperedge.

Using this model we are able to trace the dependencies between different steps in the reaction based on individual atoms and their spatial arrangement. The approach is illustrated by simulating the (at that time surprising) outcome of a classical experiment that led to a deeper understanding of the citric acid cycle. We also provide an encoding of the model in terms of attributed bipartite graphs that can be implemented in the AGG system for simulation and analysis.

The paper is organised as follows. In the next section, we introduce our running example, the *citric acid cycle*, and explain why the exact configuration of atoms in 3-dimensional space is important for understanding biochemical reactions properly. Our formal model, based on hypergraph transformation systems, is presented in Section 3. In Section 4, we show that the carbon flow in the citric acid cycle can be analysed with the graph transformation tool AGG. A conclusion closes this paper.

2   Molecular Analysis

2.1   Basic Reactions

Metabolic pathway analysis is one of the tools in biology and medicine in order to understand reaction cycles in living cells. A shortcoming of the approach, however, is that reactions are analysed at the level of structural formulae, summarising the number of atoms of certain types in a compound without keeping track of their identity. *Molecular analysis* aims at understanding chemical reactions at the level of individual atoms or component molecules.

As an example consider the *citric acid cycle* (also known as the *tricarboxylic acid cycle*, the *TCA cycle*, or the *Krebs cycle*) [13]. This cycle is a series of chemical reactions of central importance in all living cells that utilise oxygen as part of cellular respiration. Starting with *acetyl-CoA*, one of the resulting products of the chemical conversion of carbohydrates, fats and proteins, the citric acid cycle produces fast usable energy in the form of *NADH*, *GTP*, and *FADH$_2$* which are precursors of the well known *adenosine-tri-phosphate (ATP)*. The citric acid cycle is shown in Fig. 1.

Fig. 2 shows reaction 1 of the citric acid cycle in detail, focussing on the molecular interactions between *oxaloacetate* and *acetyl-CoA*. *Acetyl-CoA* derived from *pyruvate* is the output of the metabolism of glucose known as glycolysis and therefore of major importance for the energy metabolism in living cells. *CoA* known as coenzyme A is just a transport agent for the acetyl group as input for the critic acid cycle.

But how is the metabolism of the acetyl group in the citric acid cycle? The pathway analysis in Fig. 1 shows the acetyl group as input of reaction 1 (see Fig. 2). To analyse the cycle more precisely the first important question is: How is the acetyl group metabolised in the cycle, i.e. in which output agents can we find the two C-atoms of the input acetyl group? Taking a first look would provide the answer: The acetyl group is metabolised into two CO_2 molecules. But our further analysis will show that after one cycle the C-atoms of the acetyl group are still in the agents of the cycle.

To analyse this problem more precisely, we have to go one step further in the molecular constitution of the agents. Before we introduce the spatial constitutions of molecules in general in Section 2.2, we would like to introduce this issue with respect to our example.

The output agent of reaction 1, *citrate*, has two CH_2COO groups, one on the top and one on the bottom (see Fig. 2). To fit into the enzyme *aconitase* catalysing reaction 2 (see Fig. 3), only the CH_2COO group marked with 3 is able to fit into the enzyme due to 3-dimensional spatial relations (see Fig. 4). This utilised CH_2COO group will provide the resulting agent CO_2 in reaction 4. But the CH_2COO group provided from the Acetyl-CoA in the beginning of the cycle is not utilised in this reaction. So the main remaining question for our molecular analysis is: Where is the acetyl group metabolised in the citric acid cycle?

This question is important for a deeper analysis of the citric acid cycle and can not be directly answered with the current pathway analysis techniques. Biological pathways are like a black box: We can measure the input and output agents but we know nothing about the reactions in between. To analyse them we have to go down to the molecular level to observe how input agents are utilised during the reaction process. As we have seen in reaction 2, reactions in living cells are more complicated than classical chemical reactions because most of the biological reactions are catalysed by enzymes accepting only special molecular structures.

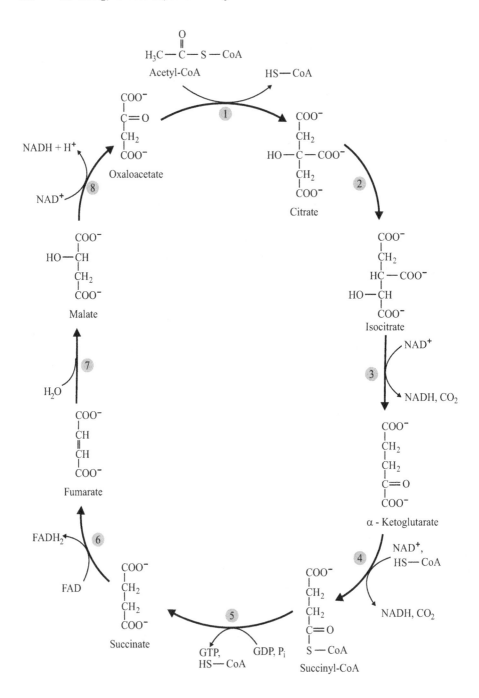

Fig. 1. Citric Acid Cycle

Apart from the structural formula also the 3-dimensional configuration of the molecules has to be taken into account. To model the configuration we need suitable techniques that can be provided by graph transformation as shown in Section 3.

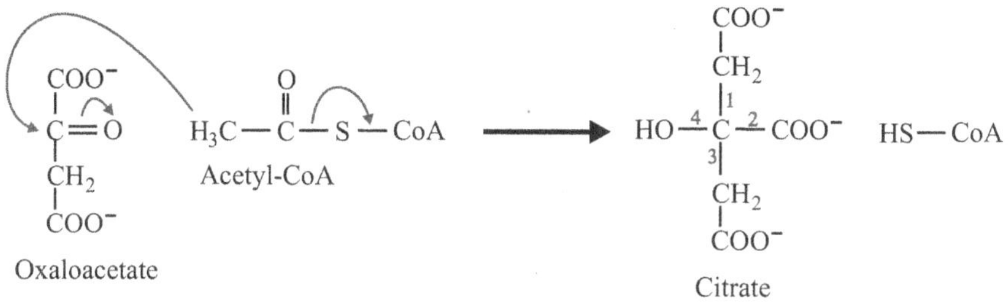

Fig. 2. Reaction 1 of the citric acid cycle

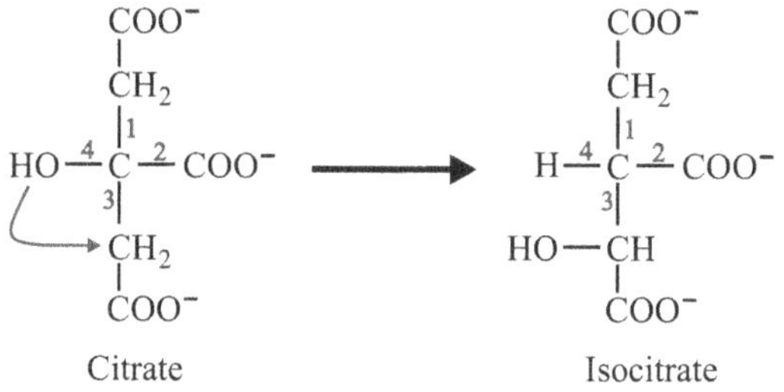

Fig. 3. Reaction 2 of the citric acid cycle

2.2   Spatial Configuration

Spatial configuration plays a key role in many chemical reactions. The arrangement of atoms of a molecular entity in space distinguishes *enantiomers* which have different chemical properties. They often smell and taste differently, and the difference with respect to their pharmacological effect can be serious, as receptors in the human body interact only with drug molecules having the proper absolute configuration. *Chiral* molecules are mirror images of each other, but can not be superimposed by translation and rotation. Figure 5 shows a sample chiral molecule: *glyceraldehyde*. The entantiomers *L-glyceraldehyde* and *D-glyceraldehyde* act like left and right hands, which are equal except for their arrangement in 3-dimensional space.

Chirality plays a role in several of the molecules and enzymes involved in the citric acid cycle. Isocitrate and malate are chiral molecules, citrate is *prochiral*, i.e. would become chiral if one of two identical ligands (attached atoms or

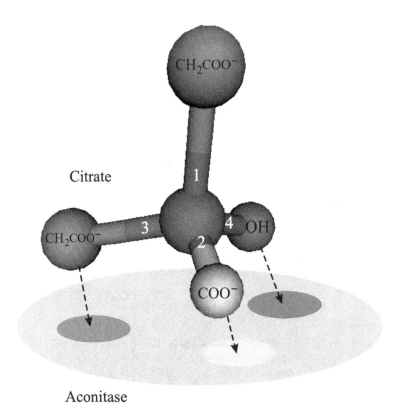

Fig. 4. Binding of the Citrate at the Aconitase Enzyme Surface

groups) is replaced by a new substituent. This also results in asymmetric phenomena when citrate is isomerised to isocitrate with the enzyme *aconitase*, which
have been explored using radiocarbon to mark individual atoms [13, Chap. 13].
Isocitrate can only be processed as a D-isomer by *isocitrate dehydrogenase*, while
the *L-isomer* does not react, or can even stop the entire cycle by bounding to
the enzyme, which happens in the absence of magnesium metal compound, as
was discovered recently [8]. These examples show that a formal representation
of metabolic pathways should always cope with the stereochemical aspects.

In chemistry, there are several naming conventions for the distinction of enantiomers. Notably, there is the classification according to optical activity, (+) or
(-), the D/L- and the R/S-classification scheme. All are used in special areas
of chemistry for either historical or practical reasons, but they are not directly
convertible into each other. So, for instance a molecule with positive optical activity (+) can be either D or L, and there is no general rule to determine this.
The D/L-convention is based on relating the molecule to glyceraldehyde, which
is one of the smallest commonly-used chiral molecules.

When writing down structural formulas, lines depict bonds approximately in
the plane of the drawing; bonds to atoms above the plane are shown with a wedge

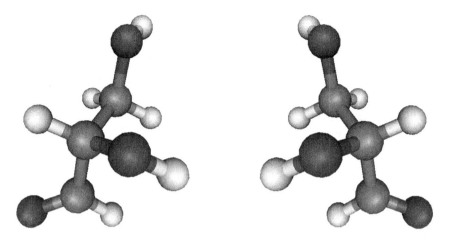

Fig. 5. 3-dimensional view of L-glyceraldehyde and D-glyceraldehyde (from left to right), created with MolSurf `www2.chemie.uni-erlangen.de/services/molsurf/`.

(starting from an atom in the plane of the drawing at the narrow end of the wedge); and bonds to atoms below the plane are shown with dashed lines (see Fig. 6).

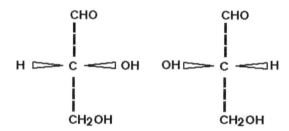

Fig. 6. Structural formulas with stereochemical information. On the left hand side, D-glyceraldehyde is shown, on the right hand side, L-glyceraldehyde.

We will establish in the next section a hypergraph approach to model molecular reactions which considers stereochemical configuration by comparing the molecules to the structure of D-glyceraldehyde.

3   Molecular Reaction Modelling with Graph Transformation

3.1   Hypergraph Approach

Given a ranked set of labels $\mathcal{A} = (\mathcal{A}_n)_n$, an $\mathcal{A}$-labelled hypergraph (V, E, s, l) consists of a set V of vertices, a set E of edges, a function $s : E \to V$ assigning each edge a sequence of vertices in V, and an edge-labelling function $l : E \to \mathcal{A}$

such that, if $length(s(e)) = n$ then $l(e) = A$ for $A \in \mathcal{A}_n$, i.e., the rank of the labels determines the number of nodes the edge is attached to. A morphism of hypergraphs is a pair of functions $\phi_V : V_1 \to V_2$ and $\phi_E : E_1 \to E_2$ that preserve labels and assignments of nodes, that is, $l_2 \circ \phi_E = l_1$ and $\phi_V \circ s_1 = s_2 \circ \phi_E$. A morphism thus has to respect the atom represented by an edge and also its chemical valence (number of bonds).

Labelled hypergraphs can be considered as hierarchical graph structures. As shown by Löwe [6], pushouts can be computed elementwisely for all hierarchical graph structures and therefore the standard graph transformation approaches can be applied. A *graph transformation rule* is a span of injective hypergraph morphisms $p = (L \xleftarrow{l} K \xrightarrow{r} R)$, called a *rule span*. The left-hand side L contains the items that must be present for an application of the rule, the right-hand side R those that are present afterwards, and the gluing graph K specifies the "gluing items", i.e., the objects which are read during application, but are not consumed.

A *direct transformation* $G \xRightarrow{p(o)} H$ is given by a *double-pushout (DPO) diagram* $o = \langle o_L, o_K, o_R \rangle$ as shown below, where (1), (2) are pushouts and top and bottom are rule spans. We assume that the match o_L is an *injective* graph homomorphism.

$$
\begin{array}{ccccc}
L & \xleftarrow{\;l\;} & K & \xrightarrow{\;r\;} & R \\
{\scriptstyle o_L}\big\downarrow & (1) & \big\downarrow{\scriptstyle o_K}\;(2) & & \big\downarrow{\scriptstyle o_R} \\
G & \xleftarrow{\;g\;} & D & \xrightarrow{\;h\;} & H
\end{array}
$$

If we are not interested in the rule and diagram of the transformation we will write $G \xRightarrow{t} H$ or just $G \Longrightarrow H$.

3.2 Structural Modeling with Hypergraphs

We use hypergraphs (V, E, s, l) to model molecules and their reactions, interpreting the hyperedges as atoms and the nodes as bonds between them. The string $s(e)$ of vertices incident to an edge $e \in E$ gives the specific order of the bonds to other atoms, coding also their spatial configuration, as we will see. As ranked set of labels, we use

$$
\begin{aligned}
\mathcal{A}_1 &= \{\text{H}, \text{CH}_3, \text{OH}, \ldots\} \\
\mathcal{A}_2 &= \{\text{O}, \text{CH}_2, \text{S}, \ldots\} \\
\mathcal{A}_3 &= \{\text{CH}, \text{N}, \ldots\} \\
\mathcal{A}_4 &= \{\text{C}, \text{S}, \ldots\} \\
&\;\;\vdots
\end{aligned}
$$

to denote elements of the periodic system or entire chemical groups. The rank of a label models the valence of an atom. For instance, a carbon atom with $l(e) = \text{C}$

always has $s(e) = v_1v_2v_3v_4$, a word of length 4. Hence, we define C as a label of rank 4. For elements with more than one possible valence (e.g. sulphur), the corresponding label can belong to several of the sets $\mathcal{A}_n$.

Given an organic molecule, we represent the 3-dimensional configuration of the ligands of a C atom as a hypergraph by relating it to D-glyceraldehyde, one of the simplest chiral organic compounds. We impose a numbering on the ligands of a carbon atom such that a substitution of ligand 1 by OH, ligand 2 by CHO, ligand 3 by CH$_2$OH, and ligand 4 by H would result in D-glyceraldehyde.

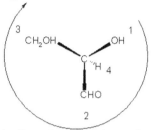

This convention defines the spatial arrangement of the ligands unambiguously. Substitution of ligands may change the angles between the ligands, and they often differ from the regular tetrahedral angle of 109 28′, but the so called *angle strain* [5] does not affect the uniqueness of the molecule represented by our notation. We will not consider angle deviations in the rest of this paper and always assume that the tetrahedron surrounding a carbon atom is regular.

As an example, we give the representation of the prochiral molecule *citrate* as a hypergraph (see Fig 7):

$$V = \{v_1, v_2, \ldots, v_6\}, E = \{e_1, e_2, \ldots, e_7\},$$

$$s(e_1) = v_1, s(e_2) = v_1v_2, s(e_3) = v_3, s(e_4) = v_2v_3v_4v_5, s(e_5) = v_4,$$

$$s(e_6) = v_5v_6, s(e_7) = v_6$$

$$l(e_1) = \text{COO}^-, l(e_2) = \text{CH}_2, s(e_3) = \text{OH}, s(e_4) = \text{C}, s(e_5) = \text{COO}^-,$$

$$s(e_6) = \text{CH}_2, s(e_7) = \text{COO}^-$$

The above representation allows different vertex-labellings for the same molecule. For instance, changing $s(e_4) = v_5v_4v_5v_3$ in the above example, it would still represent citrate, as the arrangements only differ with respect to a 120 rotation around the axis determined by the OH-group.

In order to determine equivalent resprentations, we investigate the symmetries of the involved geometric structure. It is known that the symmetry group T_d of the regular polyhedron is isomorphic to the symmetric group S_4 which consists of all permutations of a set of cardinality 4. Via this isomorphism, we can relate different representations to isomorphic polyhedra. The concept of chirality can be translated into symmetry groups as follows: The normal subgroup T of T_d of order 12 consisting of identity, 4 rotations by 120 clockwise (seen from a vertex), 4 rotations by 120 anti-clockwise and 3 rotations by 180 (axis through the centres of two opposite edges) preserves the stereochemistry of a molecule, whereas the group elements not contained in this subgroup, i.e. 6 reflections and 6 rotoreflections, transform a molecule into its chiral counterpart. The isomorphism $T_d \rightarrow S_4$ maps T to the alternating group A_4 of all even permutations. We build the factor group $T_d/T \cong S_4/T_4$, yielding $\mathbb{Z}_2$, which represents the two

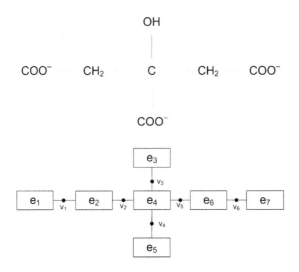

Fig. 7. Structural formula and hypergraph representation of citrate

variants of configuration. So two 4-letter words are equivalent iff they lie on the same orbit. Thus, from the 24 different permutations of the four different letters forming the word $s(e)$ for $e \in E$, any two can be regarded equivalent if they are related by an even permutation [4].

This leads to the concept of symmetry rules. The group of orientation preserving transformations A_4 is generated by $a = (123)$ and $b = (12)(34)$. It thus suffices to give two symmetry rules as shown in Fig. 8. By applying these rules repeatedly, every permutation of the ligands which preserves chirality can be achieved. Normal forms of representations could be introduced following the standards of applied chemistry where priorities are defined for the most common organic residue groups.

A common principle in chemistry is to use shortcut notation of standard groups such as OH, COOH, CH_2 as well as of whole residues of a complex molecule in order to focus on the main structures of interest in a specific reaction. The chemist regards these shortcuts as equivalent with the expanded notation, and so should our formalization do. We therefore introduce rules which expand and collapse shortcut notations so that molecules can be processed by the reaction rules regardless of the chosen representation. An example is given in Fig. 9.

A graph grammar for molecular reaction modelling thus consists of three classes of rules: symmetry rules, expansion and collapsing rules, and reaction rules. Examples for the latter will be given in the next section.

As discussed above, prochirality plays a role in the isomerisation of *citrate* to *isocitrate*. Thus, even though *citrate* is not chiral, the binding of *citrate* to *aconitase* fixes its spatial configuration and selects between the two COO⁻ groups (see Fig. 4). This is important when investigating the way of marked atoms in molecular reactions, for instance the carbon flow in the citrate cycle. Do the

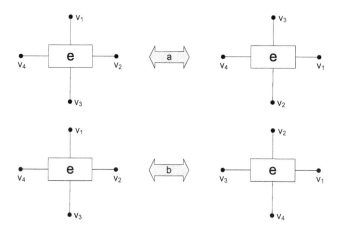

Fig. 8. Symmetry Rules

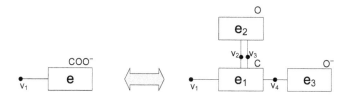

Fig. 9. Expansion and collapsing rule

carbon atoms coming from *acetyl-CoA* in step 1 remain in the cycle or are they metabolised to another substance? Or, even, does that happen with a certain probability within one passage through the whole cycle? By marking the atom with the radioactive isotope ^{14}C, this question can be answered experimentally. A careful investigation of the reaction and its stereochemical aspects also reveals the answer, and we will show how this can be done in an automated way. A manual tracking is error-prone and not feasible in more complicated situations.

4   Tool Support

For tool support we use the graph transformation tool environment AGG [11,1]. To model the hypergraph representation presented in Section 3, we have to define a mapping between this approach and a typed attributed graph grammar system as represented by AGG.

Fig. 10 shows the basic representation of a C atom in AGG: The hyperedge labelled with C is represented by the C node in AGG and the nodes are represented by the four square nodes around the C node. The order of the C atom bonds is modelled by the edge attribute *o:Int*. Starting with the bond on the top with $o = 1$, the bonds are numbered with increasing numbers with

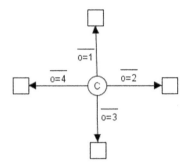

Fig. 10. Modelling of a C atom in AGG

$o_{i+1} = (o_i + 1) \bmod n$ and n number of bindings. The number of bindings is usually a unique number depending on the atom. The C atom has 4 bindings, the O atom 2 bindings and the H atom 1 binding.

Fig. 11 shows reaction 2 of the citric acid cycle (see Fig. 3 in Section 2) modelled in AGG. As shown in Fig. 4, the enzyme *aconitase* accepts only the source agent *citrate* with the indicated o edge attribute order of the *1:C* atom in the left-hand side of Fig. 11. In this reaction the OH group of the *1:C* atom is exchanged with the OH group of the *3:C* atom. This leads to the new agent *isocitrate*.

Note that unchanged nodes and edges during the reaction have to be mapped with a unique number from the left-hand side to the right-hand side. Unmapped nodes and edges of the left-hand side are deleted and unmapped nodes and edges of the right-hand side are created during the graph transformation step. The mapping preserves the value of the edge attribute o. Newly created edges on the right-hand side of the rule have to be assigned with the edge attribute o in the way described above.

Molecular Analysis in AGG: Modelling the 8 reactions of the citric acid cycle and a start graph containing the source agents, the whole cycle can be simulated in AGG. For further analysis it may be also suitable to analyse a specific part of the cycle by changing the start graph accordingly.

One important question is the metabolism of the *acetyl* group as source agent of reaction 1. We would expect that the C atoms of the acetyl group are contained in the CO_2 target agents of reactions 3 and 4. But marking these C atoms with additional attributes in AGG (which may correspond to radioactive marking in practice) shows a surprising result: the C atoms are still in the cycle, we can find them in the *oxaloacetate* at the end of the cycle. What happened? Further analysis shows that we can find the marked C atoms in the target CO_2 agents after a second turn of the cycle. This is the result of the special *prochiral* behavior of the aconitase enzyme shown in Fig. 4. The C atoms of the acetyl group fit into the enzyme only in this special three dimensional configuration, modelled by the o edge attribute in AGG. This results in an exactly defined configuration of the target agents.

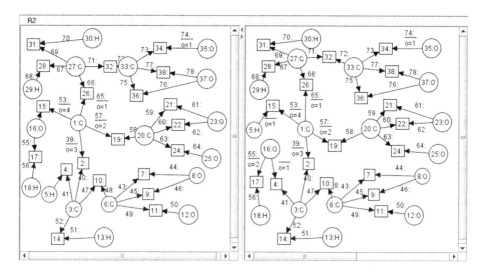

Fig. 11. Reaction 2 of the citric acid cycle in AGG

5   Related Work

The use of Graph Transformation for biological systems has a long history, as pointed out in the overview paper [9], but early applications were mostly devoted to the field of morphogenesis. Our approach focuses on biochemistry, a field which gained much importance in the last decades because of the growth of biotechnology. Providing automated assistance for analyzing biochemical reactions can help in understanding the principles which govern the processes in living cells.

Several formal approaches to chemical and biological systems have been proposed. In [7], graph replacement systems are used to describe DNA processing. Nucleotides or polynucleotides form the vertices of graphs from which bigger structures are composed. Reactions are modelled by transformation rules. In contrast, our hypergraph approach represents the atoms as edges, allowing a more detailed description of the spatial configuration of the molecules, which is quite important when tracing atoms in reaction cycles, as we have seen. Another difference is that we incorporated symmetry and expansion rules into the graph grammar, thereby retaining more flexibility for different but equivalent representations.

Process calculus was also proposed as a modelling technique for biological systems [2], but aiming a higher level of abstraction than our molecular description. Issues of concurrency, which also play a role in living cells, were proposed to be modeled with process calculi. For molecular analysis, concurrent actions are implicitly modeled by specifying the reaction for singular molecules, assuming that there are e.g. 10^{23} molecules involved. Classical concurrency questions concerning sharing of resources, deadlocks, etc. are not reasonably investigated on this scale of abstraction. Our graph-based model allows for similar extensions with respect to time or stochastic aspects as other approaches like process calculi and petri nets, but provides a more direct visual specification of the molecular configuration, making it more feasible for the chemical expert than other computer science modelling techniques.

6   Conclusion

In this paper we have shown how molecular analysis of metabolic pathway reactions could be performed using hypergraph transformation, supported by the attributed graph grammar tool environment AGG through a representation as bipartite graphs. With the citric acid cycle we have applied this approach to a well-known, but non-trivial case study.

It has been shown that metabolic pathway analysis is not just a trivial reaction system. To understand the reactions it is very important to be able to trace the history of particular atoms or molecules. In the citric acid cycle we have shown that at least two turns of the reaction cycle have to be considered to understand the metabolism of the agents, an insight which can not be obtained at the more abstract level.

Our approach may also be helpful for further analysis of unknown (or not well-known) reaction pathways in living cells. Understanding these pathways is important for the treatment of diseases or the analysis of drug metabolism in certain situations. Since the genomic code differs slightly in each living cell the reaction pathways may differ, too.

The traceability of radioactive metabolites is very important in the treatment of cancer. Since the reaction pathways in cancer cells differ from the normal behaviour, the traceability of radioactive metabolites is a significant marker for finding metastases.

References

1. Attributed Graph Grammar (AGG) Tool Environment, http://tfs.cs. tu-berlin. de/agg, 2006.
2. L. Cardelli, S. Pradalier, *Where Membranes Meet Complexes*. Proceedings of BioConcur 2005.
3. V. Claus, H. Ehrig, G. Rozenberg (eds.), *Graph-Grammars and Their Application to Computer Science and Biology*. International Workshop, Bad Honnef, LNCS 73, pages 389–400, Springer 1979.
4. F. A. Cotton, *Chemical Applications of Group Theory*. Wiley 1990.
5. *IUPAC Basic Terminology of Stereochemistry*. Pure & Applied Chemistry, Vol. 68, No. 12, pp. 2193-2222, 1996.
6. M. Löwe, *Algebraic Approach to Single-Pushout Graph Transformation*. Theoretical Computer Science Vol. 109(1&2), pages 181–224, 1993.
7. J. S. McCaskill, Ulrich Niemann, *Graph Replacement Chemistry for DNA Processing*. Lecture Notes in Computer Science, Volume 2054, pages 103–116, Springer 2001.

8. A.D. Mesecar, D.E. Koshland Jr., *A new model for protein sterospecificity*. Nature Vol. 403, pages 614–615, 2000.
9. F. Rosselló, G. Valiente, *Graph Transformation in Molecular Biology*. In H.-J. Kreowski, U. Montanari, F. Orejas, G. Rozenberg, G. Taentzer (Ed.), *Formal Methods in Software and System Modeling*, Lecture Notes in Computer Science 3393, pages 116-133, Springer 2005.
10. F. Rosselló, G. Valiente, *Analysis of Metabolic Pathways by Graph Transformation*. Proceedings 2nd International Conference of Graph Transformation (ICGT'04), Springer LNCS 3256, pages 70–82, 2004.
11. G. Taentzer, *AGG: A Graph Transformation Environment for Modeling and Validation of Software*, Proc. Application of Graph Transformations with Industrial Relevance (AGTIVE'03), Pfaltz, J. and Nagl, M., pages 446–453, Charlottesville/Virgina, USA, 2003, http://tfs.cs.tu-berlin.de/agg.
12. Maneesh K. Yadav, Brian P. Kelley, and Steven M. Silverman, *The Potential of a Chemical Graph Transformation System*. Proceedings 2nd International Conference of Graph Transformation (ICGT'04), Springer LNCS 3256, pages 83–95, 2004.
13. G.L. Zubay, W. W. Parson, D. E. Vance, Principles of Biochemisty, Vol. 2, 1995.

Matrix Approach to Graph Transformation: Matching and Sequences

Pedro Pablo Pérez Velasco and Juan de Lara

Escuela Politécnica Superior
Universidad Autónoma de Madrid
{pedro.perez, juan.delara}@uam.es

Abstract. In this work we present our approach to (simple di-)graph transformation based on an algebra of boolean matrices. Rules are represented as boolean matrices for nodes and edges and derivations can be efficiently characterized with boolean operations only. Our objective is to analyze properties inherent to rules themselves (without considering an initial graph), so this information can be calculated at specification time. We present basic results concerning well-formedness of rules and derivations (compatibility), as well as concatenation of rules, the conditions under which they are applicable (coherence) and permutations. We introduce the *match*, which permits the identification of a grammar rule left hand side inside a graph. We follow a similar approach to the single pushout approach (SPO), where dangling edges are deleted, but we first adapt the rule in order to take into account any deleted edge. To this end, a notation borrowed from functional analysis is used. We study the conditions under which the calculated data at specification time can be used when the match is considered.

1 Introduction

Graph Transformation [11] is becoming increasingly popular in computer science as it provides a formal basis for graph manipulation. Transformations of this data structure are central to many application areas, such as visual languages, visual simulation, picture processing and model transformation (see [5] and [11] vol.2 for some applications).

The classical algebraic approach to graph transformation is based on *category theory* [3], and provides a rich body of theoretical results(see [11] vol.1). Thus, graph transformations expressed as graph rewriting become not only graphical and intuitive but also formal, declarative and high-level models, subject themselves to analysis [11] [5] [6]. Nonetheless, methods to increase efficiency and new analysis techniques that can be efficiently implemented in tools are needed for real industrial applications.

In contrast to the categorical-algebraic approach, we propose an algebraic characterization based on boolean matrix algebra. In this way, simple digraphs can be represented as boolean matrices and productions as matrices for edge and node deletion and addition, together with a graph L (also represented with matrices) that must be present in the host graph in order for the rule to be applicable. Therefore, the effects of a production $p : L \rightarrow R$ can be modelled using boolean matrix operations only. This purely algebraic approach constitutes a different perspective from algebraic-categorical approaches, as it provides an operational characterization of most concepts (closer to implementation) and has the potential for efficient implementation and parallelization.

A. Corradini et al. (Eds.): ICGT 2006, LNCS 4178, pp. 122–137, 2006.

In our work [10], most analysis is made independently of the host graph. The advantages of this approach are twofold. First, all properties under study are *inherent* to the graph transformation system and second, it has the practical advantage that the analysis can be performed by a tool in the phase of specification of the grammar, independently of any host graph. We present concepts such as coherence (potential applicability of a sequence), minimal initial digraph (smallest graph with enough elements to execute a sequence), rule permutation coherence and G-congruence (potential sequential independence). These concepts provide a rich amount of information about productions and how they are related to each other, including limitation in their application, dependencies and dynamical behaviour. To the best of our knowledge, some of these results are new, for example we have studied conditions for coherence of rule advancement and delay an arbitrary number of positions in a sequence. For space limitations, some proofs are omitted, but can be found in [10].

In addition, we introduce the match as an *operator* modifying the rule by *including* the context in which it is applied. We use a similar approach to SPO [4], where the dangling edges are deleted. Thus, the rule is adapted to include the edges that would become dangling and explicitly delete them. Our goal is to use the information calculated about the grammar at specification time once the initial host graph is considered. In this work, we study how this information is modified when a host graph is taken into account. We also introduce a bra-ket operational notation for rules similar to that of functional analysis for operators (also known as *Dirac Notation*) [1]. Thus, productions can be depicted as $R = \langle L, p \rangle$, splitting the static part (initial state, L) from the dynamics (element addition and deletion, p).

The paper is organized as follows. Section 2 presents the characterization of graphs and productions in our approach, together with rule sequences, minimal initial digraph, permutation and G-congruence. Section 3 presents our approach to handle the match. Section 4 revisits the properties calculated for rules in section 2, and study how they are affected by the match. Section 5 presents the conclusions and future work.

2   Characterization and Basic Properties

This section presents an informal introduction to the basic concepts in our approach. In subsection 2.1, we start defining simple digraphs, which can be represented as boolean matrices, introduce basic operations on these matrices and show a characterization of graph transformation rules using them. We formulate the conditions for a production to be *compatible* (i.e. it defines a simple digraph) and the concept of *completion*, where matrices representing graphs are modified – *arranged* – to permit operations between them. In subsection 2.2, we present production concatenation together with the concept of *coherence*. We present the minimal initial digraph, the conditions for sequence permutations to be coherent and the concept of potential sequential independence.

2.1   Simple Digraphs and Productions

A graph $G = (V, E)$ consists of two sets, one of nodes $V = \{V_i \mid i \in I\}$ and one of edges $E = \{(V_i, V_j) \in V \times V\}$. In this paper we are concerned with *simple digraphs*, "simple" meaning that only two arrows are allowed between two nodes (one in each direction), and "di-" because arrows have a direction. A simple digraph G is uniquely determined by its *adjacency matrix* A_G, whose element a_{ij} is one if $(i,j) \in E$, and zero otherwise. As we will delete and add edges and nodes, a *nodes vector* V_G is also associated to our digraph G, with its elements equal to one if the corresponding node is present in G and zero otherwise.

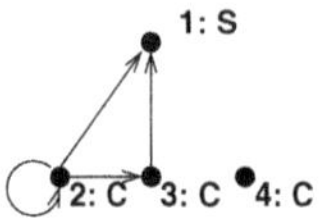

$$M^E_{C-s} = \begin{bmatrix} 0 & 0 & 0 & 0 & 1 \\ 1 & 1 & 1 & 0 & 2 \\ 1 & 0 & 0 & 0 & 3 \\ 0 & 0 & 0 & 0 & 4 \end{bmatrix} \qquad M^N_{C-s} = \begin{bmatrix} 1 & 1 \\ 1 & 2 \\ 1 & 3 \\ 1 & 4 \end{bmatrix}$$

(a)   (b)

Fig. 1. (a) A Simple Digraph Representing a Client-Server System (b) Matrix Representation

Fig. 1(a) shows a digraph representing a client-server system. Links between the clients and the server represent that the client is connected to a server. Links between clients represent a directed communication channel, while a loop link represents a message. The matrix representation of the previous graph is shown in Fig.1(b).

The boolean product between two adjacency matrices $M_G = (g_{ij})_{i,j\ 1,\dots,n}$ and $M_H = (h_{ij})_{i,j\ 1,\dots,n}$ is defined as $(M_G \odot M_H)_{ij} = \bigvee_{k=1}^{n} (g_{ik} \wedge h_{kj})$.

Next, we are interested in formulating the properties (that we call *compatibility*) that should be fulfilled by a boolean matrix and a vector of nodes to define a simple digraph. We want to forbid edges incident to nodes that do not belong to the digraph. We first define the norm $\|\cdot\|_1$ of a vector $N = (v_1, \dots, v_n)$ as $\|N\|_1 = \bigvee_{i=1}^{n} v_i$.

Proposition 1. *A pair* (M, N), *where* M *is an adjacency matrix and* N *a vector of nodes, is compatible if and only if they verify* $\left\| (M \vee M^t) \odot \overline{N} \right\|_1 = 0.$ [1]

Now we consider productions and their characterization. We define a production as a morphism – in the sense of category theory – which transforms a simple digraph into another one, $p : L \to R$. We can describe a production p with two matrices for edges and two vectors for nodes. Therefore a production can be specified as functions between boolean matrices and vectors.

Definition 1 (Production). *A production* p *is a morphism between two simple digraphs* L *and* R, *and can be specified by the tuple* $p = \left(L^E, R^E; L^N, R^N \right)$ *where* E *stands for edge and* N *for node.* L *is the* left hand side *(LHS) and* R *is the* right hand side *(RHS).*

A production models deletion and addition of edges and nodes, carried out in the order just mentioned, i.e., first deletion and then addition. These actions can be represented with two matrices for edges (e^E, r^E) and two vectors for nodes (e^N, r^N), which can be calculated as:[2] $e = L\,\overline{(L\,R)} = L\,\overline{R}$ and $r = R\,\overline{(L\,R)} = R\,\overline{L}$.

---

[1] Where t denotes transposition.

[2] Superindices E and N shall be omitted if, for example, the formula applies to both cases or if it is clear from context which we refer to. Moreover, the **and** operator ($\wedge$) will also be omitted.

Fig. 2 shows a rule that creates a communication channel between two clients connected to the same server. The deletion matrix e^E (and vector e^N) is zero, while the addition matrix r^E has a unique non-zero element at position $(2, 3)$ and the addition vector for nodes is zero. From previous definitions, a number of conditions are immedi-

$$L^E_{CC} = \begin{bmatrix} 0 & 0 & 0 & 1 \\ 1 & 0 & 0 & 2 \\ 1 & 0 & 0 & 3 \end{bmatrix} \qquad R^E_{CC} = \begin{bmatrix} 0 & 0 & 0 & 1 \\ 1 & 0 & 1 & 2 \\ 1 & 0 & 0 & 3 \end{bmatrix}$$

(a)   (b)

Fig. 2. (a) Create Channel Rule (b) Matrix Representation of Rule (only for edges)

ate (see next proposition). The first two state that elements cannot be rewritten (erased and created or vice versa) by a rule application. This is a consequence of the way in which matrices e and r are calculated.[3] The last two conditions say that if an element is in the RHS, then it is not deleted, and that if the element is in the LHS, it is not created.

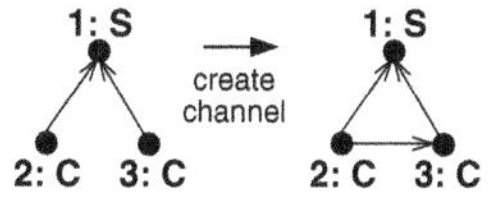

Proposition 2. *Let* $p : L \to R$ *be a production, the following identities hold for both edges and nodes:* $r\,\bar{e} = r,\ e\,\bar{r} = e,\ R\,\bar{e} = R,\ L\,\bar{r} = L.$

Finally we are ready to characterize a production $p : L \to R$ using deletion and addition matrices, starting from its LHS: $R = r \vee \bar{e}\,L$ (for both edges and nodes). It could be the case that the production erases a node but leaves some incident edges (*dangling edges*). Some conditions have to be imposed on matrices and vectors of nodes and edges to keep compatibility when a rule is applied (i.e., to avoid dangling edges):

1. An incoming edge cannot be added to a node that is going to be deleted or, using the norm, $\left\| r^E \odot e^N \right\|_1 = 0$. Similarly, for outgoing edges: $\left\| \left(r^E \right)^t \odot e^N \right\|_1 = 0$. Note how, vector e^N has a 1 in position i, if the node has to be deleted. Row i in matrix r^E depicts the outgoing edges for node i, and has a 1 in column j if edge (i, j) has to be added. Therefore vector $r^E \odot e^N$ contains elements $(\vee_{j=1}^{n} r^E_{ij} \wedge e^N_j)_{i\ 1,\dots,n}$ with a 1 in position i, if there is some newly added edge from node i to some node j which is deleted by the production. The transposition of r^E checks for new edges starting from deleted nodes.
2. Deleting a node with some incoming edge is forbidden, if the edge is not deleted as well: $\left\| \overline{e^E}\,L^E \odot e^N \right\|_1 = 0$. For outgoing edges: $\left\| \left(\overline{e^E}\,L^E \right)^t \odot e^N \right\|_1 = 0$. Matrix $\overline{e^E}\,L^E$ contains the edges in the rule's LHS that are not deleted, therefore $\overline{e^E}\,L^E \odot e^N$ results in a vector with a one in position i if some node j is deleted and has an incident edge coming from i (and the edge is not deleted). The transposition of $\overline{e^E}\,L^E$ checks for outgoing edges from deleted nodes.

---

[3] This contrasts with the DPO approach, in which edges and nodes can be rewritten in a single rule. This can be useful to forbid the rule application if the dangling condition is violated. Section 3 explains how to deal with dangling edges in this approach.

3. It is not possible to add an incoming edge to a node which is neither present in the LHS nor added by the production: $\left\| r^E \odot \left(\overline{r^N \ L^N} \right) \right\|_1 = 0$. Similarly, for edges starting in a given node: $\left\| \left(r^E \right)^t \odot \left(\overline{r^N \ L^N} \right) \right\|_1 = 0$. In this case, $\overline{r^N \ L^N}$ is a vector containing a 1 in position i if node i does not belong to the LHS and is not going to be added.

4. It is not possible for an edge to reach a node which does not belong to the LHS and which is not going to be added: $\left\| \left(\overline{e^E} L^E \right) \odot \left(\overline{r^N \ L^N} \right) \right\|_1 = 0$. For outgoing edges: $\left\| \left(\overline{e^E} L^E \right)^t \odot \left(\overline{r^N \ L^N} \right) \right\|_1 = 0$. In this case, $\overline{e^E} L^E$ is a matrix with a 1 in the edges that are in the LHS and not deleted.

Thus we arrive naturally at the next proposition:

Proposition 3. *Let $p : L \to R$ be a production, if previous conditions in items 1-4 are fulfilled then $R^E = r^E \vee \left(\overline{e^E} \ L^E \right)$ and $R^N = r^N \vee \left(\overline{e^N} \ L^N \right)$ are compatible.*

which is easily proved, as we have to check that $\left\| (M \vee M^t) \odot \overline{N} \right\|_1 = 0$, with $M = r^E \vee \overline{e^E} L^E$ and $\overline{N} = \overline{r^N} \left(e^N \vee \overline{L^N} \right)$. Therefore,

$$
\left(M \vee M^t \right) \odot \overline{N} = \left[\left(r^E \vee \overline{e^E} L^E \right) \vee \left(r^E \vee \overline{e^E} L^E \right)^t \right] \odot \left[\overline{r^N} \left(e^N \vee \overline{L^N} \right) \right] =
$$

$$
= \left[r^E \vee \overline{e^E} L^E \vee \left(r^E \right)^t \vee \left(\overline{e^E} L^E \right)^t \right] \odot \left(e^N \vee \overline{r^N \ L^N} \right) \qquad (1)
$$

Conditions in items 1-4 are taken from this identity.

For the rule in Fig. 2, it is easy to check that (R^E, R^N) are compatible, as vector $\overline{N}$ has all elements equal to zero (because e^N and $\overline{L^N}$ are zero).

Up to now we have assumed that when operating with matrices and vectors these had the same size, but in general matrices and vectors represent graphs with different sets of nodes or edges, although probably with some common subsets. Moreover, the elements in both matrices can appear in a different order. An operation called *completion* modifies matrices (and vectors) to allow some specified operation. Suppose we want to operate with two matrices representing the edges of two graphs (a similar operation can be defined for vectors of nodes). In this way, first a common subset C of elements are identified, and it is moved up in the matrices, maintaining the order. Then, the common subset is sorted in the second matrix to obtain the same order as in the first one. Then, the elements present in the first matrix but not in the second one are added to the second one (i.e. rows and columns of zeros), sorted like in the first one. Similarly, the elements present in the second matrix but not in the first one are added to the first one (i.e. rows and columns of zeros), sorted like in the second one.

For example, if we have to operate the graph in Fig. 1 with the LHS of rule in Fig. 2, then the matrix of edges and the vector of nodes of the rule have to be enlarged. If we identify nodes and edges with the same label, we get the following result:

$$L_{CC}'^E = \begin{bmatrix} 0 & 0 & 0 & 0 & | & 1 \\ 1 & 0 & 0 & 0 & | & 2 \\ 1 & 0 & 0 & 0 & | & 3 \\ 0 & 0 & 0 & 0 & | & 4 \end{bmatrix} ; \quad L_{CC}'^N = \begin{bmatrix} 1 & | & 1 \\ 1 & | & 2 \\ 1 & | & 3 \\ 0 & | & 4 \end{bmatrix}$$

where an additional column and row has been added to the edge matrix and an additional element has been added to the nodes vector. In this case, the matrices for the graph in Fig. 1 remain the same. Note how, if we had assumed other identification of nodes in the different graphs, the completion procedure would have produced a different result. Once the matrices and vectors of the two graphs are completed, we can define any graph transformation (i.e. any morphism on simple digraphs) as two boolean functions (for the edges matrix and for the nodes vector, which we have modelled with e and r). These functions may change arbitrarily 0's and 1's in the matrix of edges and vector of nodes (and thus we have to check compatibilty after their application).

2.2 Concatenation, Permutations and Minimal Initial Digraph

It is possible to define sequences of rules and the order in which they are to be applied.

Definition 2 (Concatenation). *Given a set of productions* $\{p_1, \ldots, p_n\}$*, the notation* $s_n = p_n; p_{n-1}; \ldots; p_1$ *defines a sequence of productions establishing an order in their application, starting with* p_1 *and ending with* p_n*.*

A concatenation is said to be *coherent* if actions carried out by one production do not prevent[4] the application of those coming afterwards. Fig. 3 shows more rules for the example. Messages are depicted as self-loops, which can be sent through channels. For example sequence *remove_channel; send; message_ready; create_channel* is coherent, as link $(2, 3)$ is created by the first rule (*create_channel*), used by rule *send* and then deleted by the last rule. We assume an identification of nodes in the different rules having the same numbers, but other combinations could be studied as well.[5]

The conditions for coherence of a concatenation of two rules $s_2 = p_2; p_1$ are:

1. The first production – p_1 – does not delete any edge used by p_2: $e_1^E L_2^E = 0$.
2. p_2 does not add any edge used, but not deleted, by p_1: $r_2^E L_1^E \overline{e_1^E} = 0$.
3. No common edges are added by both productions: $r_1^E r_2^E = 0$.

The first condition is needed because if p_1 deletes one edge used by p_2, then p_2 is not applicable. The last two conditions are needed in order to obtain a simple digraph (with at most one edge in each direction between two nodes). Applying the first two identities in proposition 2, the three previous equalities can be transformed into $R_1^E \overline{e_2^E} r_2^E \vee L_2^E e_1^E \overline{r_1^E} = 0$ and similar for nodes.

---

[4] Potentially, because no actual application of productions to a host graph is considered.
[5] Hence, completion is not unique – there may exist several ways to identify nodes across productions – depending on how rules are defined or the operation to be performed.

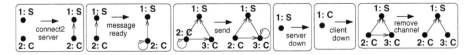

Fig. 3. Additional Rules for the Client-Server Example

Our objective is to obtain a closed formula to represent these conditions for the case with n productions. For this purpose, we introduce a graphical notation for boolean equations: a single arrow means $\wedge$, while a fork (more than one arrow starting in the same node) stands for $\vee$. These diagrams are useful to understand how the formulas change depending on the number of productions. As an example, the representation of coherence equations for two productions (for edges) is shown in Fig. 4(left). The figure also shows the equations for three and five productions.

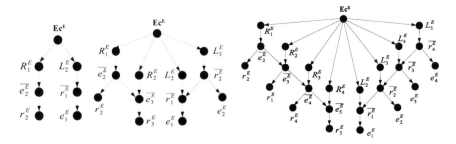

Fig. 4. Graph for Sequence of Length 2 (left), 3(middle) and 5(right)

Analysing the graphs for sequences of increasing size, we arrive at the following theorem concerning sequences of arbitrary size. The proof is not included here, it can be found at [10].

Theorem 1 (Sequence Coherence). *The concatenation* $s_n = p_n; \ldots; p_1$ *is coherent if*

$$\bigvee_{i=1}^{n} \left(R_i \, \nabla_{i+1}^{n} \left(\overline{e_x}\, r_y \right) \vee L_i \, \triangle_1^{i-1} \left(e_y\, \overline{r_x} \right) \right) = 0 \tag{2}$$

where

$$\triangle_{t_0}^{t_1}\left(F(x,y) \right) = \bigvee_{y=t_0}^{t_1} \left(\bigwedge_{x=y}^{t_1} \left(F(x,y) \right) \right) ; \nabla_{t_0}^{t_1}\left(G(x,y) \right) = \bigvee_{y=t_0}^{t_1} \left(\bigwedge_{x=t_0}^{y} \left(G(x,y) \right) \right)$$

E.g., sequence $s_1 = remove\_channel; send; message\_ready; create\_channel$ is coherent but $send; message\_ready; remove\_channel$ is not, because the first production ($remove\_channel$) deletes edge $(2,3)$ needed by $send$ one step afterwards. The resulting matrix of the coherence formula has a one in such position and zeros elsewhere. In this way, the resulting matrix of the formula is useful to indicate where the potential coherence problems are. On the other hand, sequence $s_2 = remove\_channel; send;$

create_channel is coherent, but it is worth stressing that edge $(2, 2)$ needs to be supplied by the host graph, because rule *send* needs a self loop representing a message and we know that such element is not added by any rule before *send*. Altogether, coherence allows the grammar designer to check dependencies between rules, and to realize possible conflicts, some of which can be solved if the initial graph provides enough edges and nodes. This is related to the notion of *minimal initial digraph*, which is a graph containing the necessary nodes and edges for a rule (or sequence) to be applicable.

Theorem 2 (Minimal Initial Digraph). *Given a coherent concatenation of productions* $s_n = p_n; \ldots; p_1$, *its minimal initial digraph is defined by:* $M_n = \bigtriangledown_1^n \left(\overline{r_x} L_y \right)$.

One graph is easily obtained which contains enough nodes and edges to execute a coherent sequence: $\bigvee_{i=1}^n L_i$. However, this graph can be made smaller, so for example, for production p_1 we only include in M_n elements which are in the LHS, but not added. In a similar way, for p_2 we include elements in its LHS if they are not added by p_2 nor p_1. Therefore, we have $M_n = \left(\overline{r}_1 L_1 \right) \vee \left(\overline{r}_1 L_2 \right) \left(\overline{r}_2 L_2 \right) \vee \cdots \vee \left(\overline{r}_1 L_n \right) \cdots \left(\overline{r}_n L_n \right)$, which is the expanded form of $\bigtriangledown_1^n \left(\overline{r_x} L_y \right)$. Note how, we assume a given identification of nodes and edges in the different productions of the sequence, that is, a certain way of completing each matrix. The calculation of the minimal initial digaph for sequence $s_2 = remove\_channel; send; create\_channel$ is shown in Fig.5 as an example.

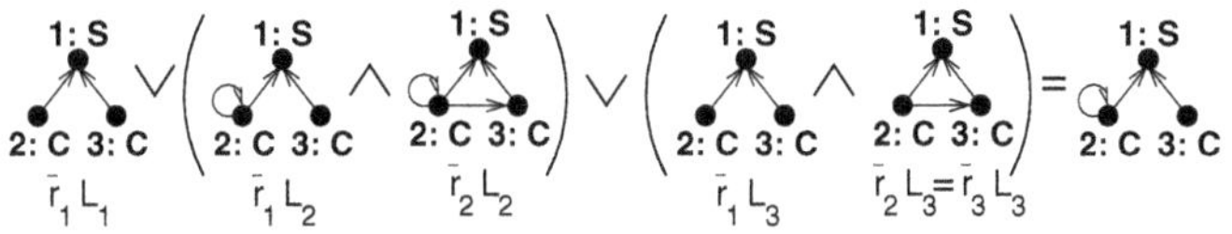

Fig. 5. Minimal Digraph for Sequence s_2

The **image of a concatenation** $s_n = p_n; \ldots; p_1$ (please, refer to [10]) almost can be seen as a production $s_n = (r_s, e_s)$, where $r_s = \triangle_1^n \left(\overline{e_x} r_y \right)$ and $e_s = \bigvee_{i=1}^n e_i$, i.e.,

$$s_n \left(M_n \right) = \bigwedge_{i=1}^n \left(\overline{e_i} M_n \right) \vee \triangle_1^n \left(\overline{e_x} r_y \right) = r_s \vee \overline{e_s} M_n \qquad (3)$$

However, in this case, it is not true that $r_s \overline{e_s} = r_s$, which in particular implies that it is important to *delete* elements (apply e_s) before *addition* takes place (r_s application).

The following result states conditions to keep coherence in case of permuting one production inside a sequence [10].

Theorem 3 (Production Permutations). *Consider coherent productions* $t_n = p_\alpha; p_n; p_{n\,1}; \ldots; p_1$ *and* $s_n = p_n; p_{n\,1}; \ldots; p_1; p_\beta$ *and permutations* ϕ *and* δ.

1. $\phi \left(t_n \right)$ *is coherent if:* $e_\alpha^E \bigtriangledown_1^n \left(\overline{r_x^E} L_y^E \right) \vee R_\alpha^E \bigtriangledown_1^n \left(\overline{e_x^E} r_y^E \right) = 0$.

2. $\delta \left(s_n \right)$ *is coherent if:* $L_\beta^E \triangle_1^n \left(\overline{r_x^E} e_y^E \right) \vee r_\beta^E \triangle_1^n \left(\overline{e_x^E} R_y^E \right) = 0$.

where ϕ advances the last production to the front, that is, moves the left-most rule to the right $n-1$ positions in a sequence of n rules. Thus, ϕ has associated permutation $\phi = [1 \quad n \quad n-1 \ldots 3 \quad 2]$. In a similar way, δ delays the first production $n-1$ positions in a sequence of n rules, moving it to the last position. Thus, $\delta = [1 \quad 2 \ldots n-1 \quad n]$. For sequence $t_2 = send; create\_channel; remove\_channel$, $\phi(t_2) = create\_channel; remove\_channel; send$ is coherent.

G-congruence guarantees that two coherent and compatible concatenations have the same output starting with G as minimal initial digraph. The conditions to be fulfilled are known as *Congruence Conditions* (CC). A coherent and compatible concatenation s_n and a coherent and compatible permutation of it, $\sigma(s_n)$, which besides have the same minimal initial digraph G (*G-congruent*) are *potentially sequential independent*. For advancement and delaying of productions, the congruence conditions are (see [10]):

$$CC(\phi, s_n) = L_n \nabla_1^{n-1}(\overline{e_x} r_y) \vee r_n \nabla_1^{n-1}(\overline{r_x} L_y) = 0 \tag{4}$$

$$CC(\delta, s_n) = L_1 \nabla_2^n(\overline{e_x} r_y) \vee r_1 \nabla_2^n(\overline{r_x} L_y) = 0 \tag{5}$$

For sequence $s = send; create\_channel; remove\_channel$, $CC(\phi, s) = 0$, therefore we obtain the same result by advancing $send$ twice. As s and $\phi(s)$ have the same initial digraph (the one in Fig. 5, plus edge $(2, 3)$), they are potential sequential independent. Symbol $\perp$ denotes potential sequence independence, thus we can write $send \perp (create\_channel; remove\_channel)$ in previous example. Note that it is possible to check sequential independence between a rule and a sequence, in contrast with results in the algebraic-categorical approach.

3   Match, Extended Match and Production Transformation

Matching is the operation of identifying the LHS of a rule inside a host graph. This identification is not necessarily unique, thus becoming a source of non determinism.

Definition 3 (Match). *Given a production $p : L \to R$ and a simple digraph G, any $m : L \to G$ total injective morphism is known as a match (for p in G).*

Recalling the notion of *completion*, a match can be interpreted as one of the possible ways to *complete* L in G. We do not explicitly care about types or labels in our matrices ("S" and "C" in the examples), but this can be thought as restrictions for the *completion* procedure, which cannot identify elements with different types.

Fig.6(a) displays a production p and a match m for p in G. It is possible to close the diagram, making it commutative ($m \circ p = p \circ m$), using the pushout construction [5] on category **Pfn(Graph)** of simple digraphs and partial functions (see [9]). This categorical construction for relational graph rewriting is carried out in [9] in their Theorem 3.2 and Corollary 3.3. Proposition 3.5 in [9] gives a sufficient condition to decide if a given rewriting square like the one in Fig.6(a) can be closed.

Definition 4 (Direct Derivation). *Given $p : L \to R$ and $m : L \to G$ as in Fig.6(a), $d = (p, m)$ is called a direct derivation with result $H = p(G)$.*

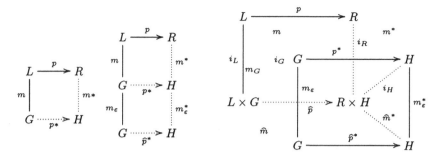

Fig. 6. (a) Production plus Match. (b) Neighbourhood. (c) Extended Match and Production.

If a concatenation $s_n = p_n; \ldots; p_1$ is considered together with the set of matchings $m_n = \{m_1, \ldots, m_n\}$, then $d_n = (s_n, m_n)$ is a **derivation**.

When applying a rule to a host graph, the main problem to concentrate on is that of so-called *dangling edges*, which is differently addressed in SPO and DPO. In DPO, if an edge comes to be dangling then the rule is not applicable (for that match), while SPO allows the production to be applied, deleting any dangling edge. In this paper we propose an SPO-like behaviour. Fig.6(b) shows our strategy to handle dangling edges:

1. Morphism m shall identify rule's left hand side in the host graph.
2. A neighbourhood of $m(L) \subseteq G$ covering all relevant extra elements is selected (performed by m_ε[6]), taking into account all dangling edges not considered by match m with their corresponding source and target nodes.
3. Finally, p is enlarged (through operator T_ε, see definition below) erasing any otherwise dangling edge.

Definition 5 (Extended Match). *Given a production $p : L \to R$, a host graph G and a match $m : L \to G$, the extended match $\widehat{m} : L \times G \to G$ is a morphism whose image is $m(L) \bigcup \varepsilon$, where ε is the set of dangling edges and their source and target nodes.*

Coproduct (see Fig.6(c)) is used for coupling L and G, being the first embedded into the second by morphism m. We use the notation $\underline{L} \stackrel{def}{=} m_G(L) \stackrel{def}{=} (m_\varepsilon \circ m)(L)$ i.e., extended digraphs are underlined and defined by composing m and m_ε.

Example. Consider the digraph L, the host graph G and the morphism match depicted on the left side of Fig. 7. On the top right side in the same figure, $m(L)$ is drawn, and $m_G(L)$ on the bottom right side. Nodes 2 and 3 and edges $(2, 1)$, $(2, 3)$ and $(2, 2)$ have been added to $m_G(L)$. The edges would become dangling in the image "graph" of G by p, $p(G)$. Note how this composition is possible, as m and m_ε are functions between boolean matrices which have been completed.

Once we are able to complete the rule's LHS, we have to do the same for the rest of the rule. To this end we define an operator $T_\varepsilon : \mathfrak{G} \to \mathfrak{G}'$, where $\mathfrak{G}$ is the original grammar and $\mathfrak{G}'$ is the grammar transformed once T_ε has modified the production. The

---

[6] Recall that morphisms are functions on boolean matrices and vectors.

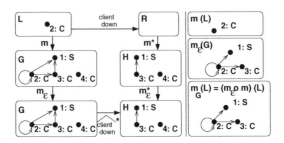

Fig. 7. Matching and Extended Match

notation that we use from now on is borrowed from functional analysis [1]. Bringing this notation to graph grammar rules, a rule is written as $R = \langle L, p \rangle$ (separating the static and dynamic parts of the production) while the grammar rule transformation including matchings is: $\underline{R} = \langle m_G(L), T_\varepsilon p \rangle$.

Proposition 4. *With notation as above, production p can be extended to consider any dangling edge, $\underline{R} = \langle m_G(L), T_\varepsilon p \rangle$.*

Proof

What we do is to split the identity operator in such a way that any problematic element is taken into account (erased) by the production. In some sense, we first add elements to p's LHS and afterwards enlarge p to erase them. Otherwise stated, $m_G = T_\varepsilon^{-1}$ and $T_\varepsilon = m_G^{-1}$, so in fact we have $R = \langle L, p \rangle = \langle L, (T_\varepsilon^{-1} \circ T_\varepsilon) p \rangle = \langle m_G(L), T_\varepsilon(p) \rangle = \underline{R}$. The equality $\underline{R} = R$ is valid strictly for edges.

The effect of considering a match can be interpreted as a new production concatenated to the original production. Let $p_\varepsilon \overset{def}{=} T_\varepsilon$,

$$\underline{R} = \langle m_G(L), T_\varepsilon(p) \rangle = \langle T_\varepsilon(m_G(L)), (p) \rangle = \tag{6}$$
$$= p(T_\varepsilon(m_G(L))) = p \, ; p_\varepsilon \, ; m_G(L) = p \, ; p_\varepsilon(\underline{L})$$

Considering the match can be interpreted as a temporary modification of the grammar, so it can be said that the grammar modifies the host graph and – temporarily – the host graph interacts with the grammar.

If we think of m_G and T_ε as productions respectively applied to L and $m_G(L)$, it is necessary to specify their erasing and addition matrices. To this end, we introduce matrix ε, with elements in row i and column i equal to one if node i is to be erased by p, and zero otherwise (see definition 5). This matrix considers any potential dangling edge.

For m_G we have that $\underline{e}^N = \underline{e}^E = 0$, and $\underline{r} = L\overline{L}$ (for both nodes and edges), as the production has to add the elements in $\underline{L}$ that are not present in L. Let $p_\varepsilon = \left(e_{T_\varepsilon}^E, r_{T_\varepsilon}^E ; e_{T_\varepsilon}^N, r_{T_\varepsilon}^N \right)$, then $e_{T_\varepsilon}^N = r_{T_\varepsilon}^E = r_{T_\varepsilon}^N = 0$ and $e_{T_\varepsilon}^E = \varepsilon \wedge \underline{L}^E$.

Example. Consider rules depicted in Fig. 8, in which *server_down* is applied to model a server failure. We have

$$e^E = r^E = L^E = \begin{bmatrix} 0|1 \end{bmatrix} e^N = \begin{bmatrix} 1|1 \end{bmatrix} ; \; r^N = \begin{bmatrix} 0|1 \end{bmatrix} ; \; L^N = \begin{bmatrix} 1|1 \end{bmatrix} ; \; R^E = R^N = \emptyset$$

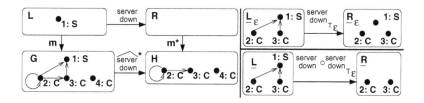

Fig. 8. Full Production and Application

Once m_G and operator T_ε have been applied, the resulting matrices are

$$r^E = \begin{bmatrix} 0\,0\,0 & 1 \\ 1\,0\,0 & 2 \\ 1\,0\,0 & 3 \end{bmatrix} \; ; \; \underline{L}^E = \begin{bmatrix} 0\,0\,0 & 1 \\ 1\,0\,0 & 2 \\ 1\,0\,0 & 3 \end{bmatrix} \; ; \; \underline{R}^E = \begin{bmatrix} 0\,0 & 2 \\ 0\,0 & 3 \end{bmatrix} \; ; \; e^E_{T_\varepsilon} = \begin{bmatrix} 0\,0\,0 & 1 \\ 1\,0\,0 & 2 \\ 1\,0\,0 & 3 \end{bmatrix}$$

Matrix r^E, besides edges added by the production, specifies those to be added by m_G to the LHS in order to consider any potential dangling edge (in this case $(2,1)$ and $(3,1)$). As neither m_G nor production *server_down* delete any element, $\underline{e}^E = 0$. Finally, p_ε removes all potential dangling edges (check out matrix $e^E_{T_\varepsilon}$) but it does not add any, so $r^E_{T_\varepsilon} = 0$. Vectors for nodes have been omitted.

Let $T_\varepsilon = \left(T_\varepsilon{}^N, T_\varepsilon{}^E \right)$ be the adjoint operator of T_ε. Define e^E_ε and r^E_ε respectively as the erasing and addition matrices of $T_\varepsilon(p)$. It is clear that $r^E_\varepsilon = \underline{r}^E = r^E$ and $e^E_\varepsilon = e^E \vee \varepsilon \underline{L}^E$, so

$$\underline{R}^E = \left\langle \underline{L}^E, T_\varepsilon(p) \right\rangle = r^E_\varepsilon \vee \overline{e^E_\varepsilon}\, \underline{L}^E = r^E \vee \overline{\left(e^E \vee \varepsilon \underline{L}^E \right)}\, \underline{L}^E =$$
$$= r^E \vee \overline{\left(\overline{\varepsilon} \vee \underline{L}^E \right)}\, \overline{e^E} \underline{L}^E = r^E \vee \overline{e^E}\, \overline{\varepsilon}\, \underline{L}^E$$

The previous identities show that $\underline{R}^E = \left\langle \underline{L}^E, T^E_\varepsilon(p^E) \right\rangle = \left\langle \overline{\varepsilon}\, \underline{L}^E, p^E \right\rangle$, which proves that $T_\varepsilon = \left(T_\varepsilon{}^N, T_\varepsilon{}^E \right) = (id, \overline{\varepsilon})$.

Summarizing, when a given match m is considered for a production p, the production itself is first modified in order to consider all potential dangling edges. m is automatically transformed into a match which is free from any dangling element and, in a second step, a pre-production p_ε is appended to form the concatenation $\widehat{p} = p \; ; p_\varepsilon$

4   Revision and Extension of Basic Concepts

In this section we brush over all concepts and theorems introduced in section 2, completing them by considering matchings.

Let $s_n = p_n; \ldots; p_1$ be a concatenation. As there is a match for every production in the sequence, it is eventually transformed into $s_n = p_n; p_{\varepsilon,n}; \ldots; p_1; p_{\varepsilon,1}$. Fig.9 displays the corresponding derivation. For **compatibility**, the main difference when considering matchings is that the sequence is increased in the number of productions so it shall be necessary to check more conditions.

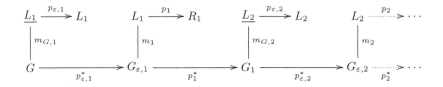

Fig. 9. Productions and ε-productions in a Concatenation

4.1 Initial Digraph Set

Concerning the *minimal initial digraph*, one may have different ways of completing the rule matrices, depending on the matches. Therefore, we no longer have a unique initial digraph, but a set.

Definition 6 (Initial Digraph Set). *Given s_n a sequence, its associated initial digraph set $\mathfrak{M}(s_n)$ is the set of simple digraphs M_i such that*

1. *M_i has enough nodes and edges for every production of the concatenation to be applied in the specified order, and*
2. *M_i has no proper subgraph with previous property*

$\forall M_i \in \mathfrak{M}(s_n)$. *Every element $M_i \in \mathfrak{M}(s_n)$ is said to be an initial digraph for s_n.*

It is easy to see that $\mathfrak{M}(s_n) \neq \emptyset, \forall s_n$ finite sequence of productions. The initial digraph set contains all graphs that can potentially be identified by matches in concrete host graphs. In section 2.1, coherence was used in an absolute way but now, due to matching, coherence is a property depending on the given initial digraph. Hence, we now say that s_n is coherent with respect to initial digraph M_i.

For the initial digraph set, we can define the *maximal initial digraph* as the element $M_n \in \mathfrak{M}(s_n)$ which considers all nodes in p_i to be different. This element is unique up to isomorphism, and corresponds to considering the parallel application of every production in the sequence. In a similar way, $M_i \in \mathfrak{M}(s_n)$ in which all possible identifications are performed are known as *minimal initial digraphs*, which in general are not unique. As an example, left of Fig. 10 shows the minimal digraph set for sequence $s_2 = remove\_channel; remove\_channel$, which is not coherent, as the link between two clients is deleted twice. In this way, the initial digraphs should provide two links. It is possible to provide some structure $\mathfrak{T}(s_n)$ to set $\mathfrak{M}(s_n)$ (see the right of Fig. 10). Every node in $\mathfrak{T}$ represents an element of $\mathfrak{M}$, and a directed edge from one node to another stands for one operation of identification between corresponding nodes in LHS and RHS of productions of the sequence s_n. Node M_7 is the maximal initial digraph, as it only has outgoing edges. The structure $\mathfrak{T}$ is known as graph-structured stack, in our case with single root node.

4.2 Coherence

Coherence formulas do not change, except that now there are conditions for all ε-productions. When considering the match, coherence is similar to conflict detection in

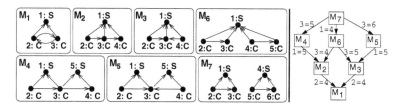

Fig. 10. Initial Digraph Set for $s_2 = remove\_channel; remove\_channel$

critical pairs [5] [6], where an important issue is efficiency [8]. We believe our approach is a contribution in improving the efficiency in finding this kind of conflicts.

The functional notation introduced so far can be used to re-enunciate Theorem 1 for coherence, deriving conditions which resemble those of perpendicular vectors and kernel of a function. Let $q_{L_i} = \triangle_1^{n-1}(\overline{r_x}\,e_y)$ and $q_{R_i} = \nabla_{i+1}^n(\overline{e_x}\,r_y)$, then $s_n = p_n; \ldots; p_1$ is coherent if $\langle L_i, q_{L_i} \rangle = \langle R_i, q_{R_i} \rangle = 0$.

In addition, when the host graph is not considered, if nodes are identified across rules, it can be the case that some dangling edge appears in the concatenation. For example, given $p_2; p_1$, suppose that rule p_1 uses but does not delete edge $(4, 1)$, that rule p_2 specifies the deletion of node 1 and that we have identified both nodes 1. It is mandatory to add one ε-production $p_{\varepsilon,2}$ to the grammar, which conceptually is of a different nature than those previously discussed. The latter dangling edges appear in the context where the rule is applied, but not in other rules. We have an unavoidable problem of coherence between p_1 and $p_{\varepsilon,2}$ if we wanted to advance the application of $p_{\varepsilon,2}$ to p_1. Hence, we split the set of edges deleted by ε-productions into two disjoint classes:

- **External.** Any edge not appearing explicitly in the grammar rules, i.e., edges of the host graph "in the surroundings" of the actual initial digraph. Examples are edges $(2, 1)$ and $(3, 1)$ in Fig.8.
- **Internal.** Any edge used or appended by a previous production in the concatenation. One example is the previously mentioned edge $(4, 1)$.

ε-productions can be classified accordingly in **internal ε-productions** if any of its edges is internal and **external ε-production** otherwise. External ε-productions cannot be considered during rule specification which, in turn, may spoil coherence, compatibility, etc. One way to handle this problem is to check the conditions under which all ε-productions can be advanced to the front of the sequence. Given a host graph G in which s_n – coherent and compatible – is to be applied, and assuming a match which identifies s_n's actual initial digraph (M_n) in G, we check whether for some $\widehat{m}$ and $\widehat{T_\varepsilon}$, which respectively represent all changes to be done to M_n and all modifications to s_n, it is correct to write $H_n = \left\langle \widehat{m}(M_n), \widehat{T_\varepsilon}(s_n) \right\rangle$, where H_n would be the piece of the final state graph H corresponding to the image of M_n.

Example. Let $s_2 = p_2; p_1$ be a coherent and compatible concatenation. Using operators we can write $H = \langle m_{G,2}(\langle m_{G,1}(M_2), T_{\varepsilon,1}(p_1) \rangle), T_{\varepsilon,2}(p_2) \rangle$, which is equivalent to $H = p_2; p_{\varepsilon,2}; p_1; p_{\varepsilon,1}(M_2)$, with actual initial digraph twice modified $M_2 = m_{G,2}(m_{G,1}(M_2)) = (m_{G,2} \circ m_{G,1})(M_2)$.

Definition 7 (Exact Derivation). *Let $d_n = (s_n, m_n)$ be a derivation with actual initial digraph M_n, concatenation $s_n = p_n; \dots; p_1$, matches $m_n = \{m_{G,1}, \dots, m_{G,n}\}$ and ε-productions $\{p_{\varepsilon,1}, \dots, p_{\varepsilon,n}\}$. It is an* exact *derivation if there exist $\widehat{m}$ and $\widehat{T_\varepsilon}$ such that $H_n = d_n(M_n) = \left\langle \widehat{m}(M_n), \widehat{T_\varepsilon}(s_n) \right\rangle$.*

Previous equation might be satisfied if once all matches are calculated, the following identity holds: $p_n; p_{\varepsilon,n}; \dots; p_1; p_{\varepsilon,1} = p_n; \dots; p_1; p_{\varepsilon,n}; \dots; p_{\varepsilon,1}$. Equation (3) allows us to consider a concatenation almost as a production, justifying operators $\widehat{T_\varepsilon}$ and $\widehat{m}$ and our abuse of the notation (recall that brakets apply to productions and not to sequences).

Proposition 5. *With notation as before, if $p_{\varepsilon,j} \perp (p_{j-1}; \dots; p_1)$, $\forall j$, then d_n is exact.*

Proof

Operator $\widehat{T_\varepsilon}$ modifies the sequence adding a unique ε-production, the composition[7] of all ε-productions $p_{\varepsilon,i}$. To see this, if one edge is to dangle, it should be eliminated by the corresponding ε-production, so no other ε-production deletes it unless it is added by a subsequent production. But by hypothesis there is sequential independence of every $p_{\varepsilon,j}$ with respect to all preceeding productions and hence $p_{\varepsilon,j}$ does not delete any edge used by $p_{j-1}, \dots, p_1$. In particular no edge added by any of these productions is erased.

In definition 7, $\widehat{m}$ is the extension of the match m which identifies the actual initial digraph in the host graph, so it adds to $m(M_n)$ all nodes and edges to distance one to nodes that are going to be erased. A symmetrical reasoning to that of $\widehat{T_\varepsilon}$ shows that $\widehat{m}$ is the composition of all $m_{G,i}$.

With definition 7 and proposition 5 it is feasible to get a concatenation where all ε-productions are applied first, and all grammar rules afterwards, recovering the original concatenation. Despite some obvious advantages, all dangling edges are deleted at the beginning, which may be counterintuitive or even undesired. For example, if the deletion of a particular edge is used for synchronization purposes. The following corollary states that exactness can only be ruined by internal ε-productions. Let s_n be a sequence to be applied to a host graph G and $M_k \in \mathfrak{M}(s_n)$.

Corollary 1. *With notation as above, assume there exists at least one match in G for M_k that does not add any internal ε-production. Then, d_n is exact.*

Proof (sketch)

All potential dangling elements are edges surrounding the actual initial digraph. It is thus possible to adapt the part of the host graph modified by the sequence at the beginning, so applying proposition 5 we get exactness.

5   Conclusions and Future Work

In this paper we have presented a new approach to simple digraph transformation based on an algebra of boolean matrices. We have shown some results (coherence, minimal initial digraphs, permutation, G-congruence) that can be calculated on the graph

---

[7] Given a sequence of productions, their composition is one production which performs the same operations, see [10] for the formal definition.

transformation system, independent of the host graph. We have introduced the match, and how to handle dangling edges by generating ε-productions which are applied previous to the original rule in order to delete dangling edges.

We believe that the main difference of our approach with respect to others is that we use boolean operators to represent graph manipulations. Other approaches such as DPO and SPO use a categorical representation of the operations, which, on the one hand makes the approach more general, but on the other, makes bigger the gap between specification and implementation on tools. In addition, we believe that concepts like initial digraph, coherence, arbitrary sequences of finite length are easier to express and study in our framework than using category theory. Concerning additional related work, the relational approach of [9] uses also exclusively a categorical approach for operations. Other approaches such as logic-based [12], algebraic-logic [2], relation-algebraic [7] are more distant from ours.

With respect to future work, we are working on application conditions, studying the structure of $\mathfrak{M}(s_n)$, bringing to our framework techinques from Petri nets, considering more general types of graphs and implementing the current concepts in a tool.

Acknowledgements. This work has been sponsored by the Spanish Ministry of Science and Education, project TSI2005-08225-C07-06. The authors would like to thank the referees for their useful comments.

References

1. Braket notation intro: http://en.wikipedia.org/wiki/Bra-ket_notation.
2. Courcelle, B. 1990. *Graph Rewriting: An Algebraic and Logic Approach* Handbook of Theoretical Computer Science, Vol. B. pp.: 193-242.
3. Ehrig, H. 1979. *Introduction to the Algebraic Theory of Graph Grammars.* In V. Claus, H. Ehrig, and G. Rozenberg (eds.), 1st Graph Grammar Workshop, pages 1-69. LNCS 73.
4. Ehrig, H., Heckel, R., Korff, M., Löwe, M., Ribeiro, L., Wagner, A., Corradini, A. 1999. *Algebraic Approaches to Graph Transformation - Part II: Single Pushout Approach and Comparison with Double Pushout Approach.* In [11] Vol.1, pp.: 247-312.
5. Ehrig, H., Ehrig, K., Prange, U., Taentzer, G. 2006. *Fundamentals of Algebraic Graph Transformation.* Springer.
6. Heckel, R., Küster, J. M., Taentzer, G. 2002. *Confluence of Typed Attributed Graph Transformation Systems.* Proc. ICGT'2002. LNCS 2505, pp.: 161-176. Springer.
7. Kahl, W. 2002. *A Relational Algebraic Approach to Graph Structure Transformation* Tech.Rep. 2002-03. Universität der Bundeswehr München.
8. Lambers, L., Ehrig, H., Orejas, F. 2006. *Efficient Conflict Detection in Graph Transformation Systems by Essential Critical Pairs.* Proc. GT-VMT'06, to appear in ENTCS (Elsevier).
9. Mizoguchi, Y., Kuwahara, Y. 1995. Relational Graph Rewritings. Theoretical Computer Science, Vol 141, pp. 311-328.
10. Pérez Velasco, P. P., de Lara, J. 2006. *Towards a New Algebraic Approach to Graph Transformation: Long Version.* Tech. Rep. of the School of Comp. Sci., Univ. Autónoma Madrid. http://www.ii.uam.es/~jlara/investigacion/techrep_03_06.pdf.
11. Rozenberg, G. (managing ed.) 1999. *Handbook of Graph Grammars and Computing by Graph Transformation. Vol.1 (Foundations), Vol.2(Applications, Languages and Tools), Vol.3., (Concurrency, Parallelism and Distribution).* World Scientific.
12. Schürr, A. *Programmed Graph Replacement Systems.* In [11], Vol.1, pp.: 479 - 546.

String Generating Hypergraph Grammars with Word Order Restrictions

Martin Riedl[1], Sebastian Seifert[1], and Ingrid Fischer[2,*]

[1] Computer Science Institute, University of Erlangen–Nuremberg, Germany
Martin.Riedl@informatik.stud.uni-erlangen.de,
Sebastian.T.Seifert@stud.informatik.uni-erlangen.de
[2] ALTANA Chair for Bioinformatics and Information Mining, University of
Konstanz, Germany
Ingrid.Fischer@inf.uni-konstanz.de

Abstract. Discontinuous constituents and free word order pose constant problems in natural language parsing. String generating hypergraph grammars have been proven useful for handling discontinuous constituents. In this paper we describe a new notation for hypergraph productions that allows on-the-fly interconnection of graph parts with regard to user-defined constraints. These constraints handle the order of nodes within the string hypergraph. The HyperEarley parser for string generating hypergraph grammars [1] is adapted to the new formalism. A German example is used for the explanation of the new notation and algorithms.

1 Free Word Order in Natural Languages

A prominent difference between natural languages lies the order of words or constituents (groups of words that belong together on a syntactic level) in the various types of sentences. E.g. declarative sentences in English have a fixed word order with the subject first, followed by the verb, the objects and finally prepositional phrases. There are also languages with nearly no order restrictions at all: in Hungarian the excessive use of word endings ensures that sentences can be understood despite the completely free word order. In German, there is a mixture. Declarative sentences have the finite verb in the second position and the infinite verb in the last but one position. Other sentence parts, especially between the finite and infinite part of the verb, can take variable positions. This is a combination of fixed and free word order.

A German example sentence is given in (1). The first line is in German, the second a word-by-word translation into English and the third an idiomatically correct translation:

(1) Die Nachricht wurde durch einen Boten   von Marathon nach Athen gebracht der dann starb
    The news   was   by   a   messenger from Marathon to   Athens brought who then died
    The news was brought from Marathon to Athens by a messenger, who then died.

---

* This research was done while Ingrid Fischer was employed at the Chair of Computer Science 2, University of Erlangen-Nuremberg.

A. Corradini et al. (Eds.): ICGT 2006, LNCS 4178, pp. 138–152, 2006.

Note that the finite verb "wurde" ("was") is in the second position and the infinite verb "gebracht" ("brought") in the last but one position. In between these positions, the source "von Marathon" ("from Marathon"), the target "nach Athen" ("to Athens") and the agent "durch einen Boten" ("by a messenger") of the action are listed. Their positions are not fixed as shown in (2). All of these variations have the same meaning.

(2) Die Nachricht **wurde** durch einen Boten   von Marathon nach Athen **gebracht** der dann starb
     The news   **was**   by   a   messenger from Marathon to   Athens **brought** who then died

     Die Nachricht **wurde** von Marathon durch einen Boten   nach Athen **gebracht** der dann starb
     The news   **was**   from Marathon by   a   messenger to   Athens **brought** who then died

     Die Nachricht **wurde** von Marathon nach Athen durch einen Boten   **gebracht** der dann starb
     The news   **was**   from Marathon to   Athens by   a   messenger **brought** who then died

Several other combinations of agent, source and target are invalid. The source "von Marathon" ("from Marathon") must appear before the target "nach Athen" ("to Athens"). Otherwise, the meaning of the sentence changes: the messenger seems to be born in Marathon and it is not clear that the message's transport starts there (3).

(3) Die Nachricht wurde nach Athen durch einen Boten   von Marathon gebracht der dann starb
     The news   was   to   Athens by   a   messenger from Marathon brought who then died.

In addition to word order, the running example (1) demonstrates another problem of German syntax: "Bote" ("messenger") is described more closely through the relative clause "der dann starb" ("who then died"). This relative clause does not follow the noun "Bote" ("messenger") directly, but is moved to the last position of the sentence (after the infinite verb). Together, the noun and the relative clause form a discontinuous constituent. Discontinuous constituents are separated by one or more other constituents but still belong together on a semantic or syntactic level. This connection between the two parts cannot be expressed with general context–free Chomsky grammars [2].

The desired phrase structure tree for example 1 is shown in Fig. 1.[1] To shorten the representation, triangles are used to indicate that several (terminal) words are generated from one nonterminal symbol. The most important part of the tree is the derivation of the verb phrase (*VP*) into an auxiliary verb (*Aux*), the prepositional phrases denoting the agent (*PP*), the source (*PPS*) and the target (*PPT*) of the action and the infinite verb part, the second participle (*Part2*). The prepositional phrase (*PP*) for the agent contains the discontinuous constituent and is transformed into a preposition (*Prep*), a noun phrase (*NP*) and the relative clause (*RelCl*) in the last position of the example sentence. The root of the tree *S* starts a sentence that is split into a noun phrase (*NP*) and a verb phrase (*VP*).

---

[1] It is possible to construct a weakly equivalent context–free Chomsky grammar to parse such a sentence, using some workaround for the discontinuous constituent like attaching one of its parts in another production than the other.

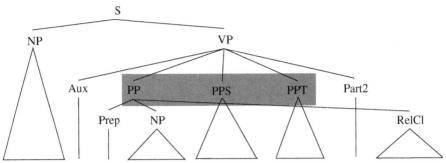

Fig. 1. The phrase structure tree for "Die Nachricht wurde durch einen Boten von Marathon nach Athen gebracht der dann starb." ("The news was brought from Marathon to Athens by a messenger who then died.")

In [3] string generating hypergraph grammars were used to construct such trees based on the the context–free substitution of hyperedges with hypergraphs. To parse sentences with these grammars, an Earley based algorithm, called HyperEarley, was presented in [4]. The Earley algorithm [5] is a well–known $\mathcal{O}(n^3)$ parsing algorithm for context–free grammars that is particularly suited for natural language processing.

Free word order is possible in example (1) between the three prepositional phrases denoting agent, source and target as shown in example 2. These phrases are shaded in Fig. 1. Every variation between the order of *PP*, *PPS*, *PPT* is possible as long as the source *PPS* is mentioned before the target *PPT*.

In this paper, string generating hypergraph grammars and their HyperEarley parser are extended to allow on-the-fly interconnection of graph parts with constraints for free word order and word order variations. In the next section a short introduction into string generating hypergraph grammars is given. Based on the example in (1), word order constraints and their notation within the grammar are introduced in Section 3. The extension of the HyperEarley algorithm is described in Section 4. The paper ends with a conclusion and an outlook.

2 String Generating Hypergraph Grammars

Hyperedge replacement grammars have been studied extensively in the last decades. Introductions and applications can be found in [6,7]. A subset of hypergraph grammars, context–free string generating hypergraph grammars, are described in detail in [6,8,9]. The definitions used in this paper are briefly summarized:

A labeled **hypergraph** (E, V, s, t, l, b, f) consists of finite sets of **hyperedges** E and of nodes V, a source function s and a target function $t : E \rightarrow V$ which assign a sequence of source respectively target nodes to each hyperedge, a labeling function $l : E \rightarrow \Sigma$ where Σ is a finite alphabet, and sequences of external source nodes b and external target nodes f. A hypergraph's **type** is the pair $(|b|, |f|)$. A hyperedge's type is likewise defined as $(|s(e)|, |t(e)|)$ where $e \in E$. A **string** (hyper-)graph consists solely of hyperedges of type $(1, 1)$ connected via nodes being source to one edge and target to one edge. Only the single external source node and the single external target node re connected to one and not two hyperedges. A string hypergraph represents a linear sequence $\langle s_1 s_2 \ldots s_n \rangle \in \Sigma^n$ of edge labels.

A context–free hyperedge replacement grammar $G = (N, T, P, S)$ consists of the finite sets T of **terminal** edge labels, N of **non-terminal** edge labels, P of **productions** and a start graph S called the **axiom**. A context–free production or rule $p = (L, R)$, commonly written $L \rightarrow R$, is a pair of hypergraphs with left-hand-side (L) and right-hand-side graph (R). L is a **singleton** hypergraph, i.e. a graph consisting of a single hyperedge e with external nodes b and f such that $s(e) = b$ and $t(e) = f$. The types of both L and R need to be the same. Each production describes a replacement of a single hyperedge as described by L with the graph R. R's external nodes are merged with L's nodes, respecting order, so that are R is added disjointly (except for the external nodes) to the host graph the production is applied to. A context–free hyperedge replacement grammar is a **string generating hypergraph grammar (SGHG)** if the language generated by the grammar consists only of string hypergraphs. For natural language processing, the hyperedges of a string hypergraph are labeled with the words of the sentence they represent. Σ is the union of words (terminals T) and names of syntactic units (nonterminals N). A string graph's hyperedge labels, read in order from the external source to the external target node, form the underlying sentence.

Please note that for the rest of this paper we assume that the string generating hypergraph grammars are reduced, cycle–free and ϵ–free [2]. If a hyperedge replacement grammar generates a string language, the start graph must be of type $(1, 1)$. A prominent property of such SGHG is that each node is source for at most one hyperedge and target for at most one hyperedge; otherwise no string language will be generated [4].

3   Word Order Constraints in SGHGs

With the help of SGHG, phenomena of discontinuity in natural language can be easily modeled by a context–free grammar. However, modeling the syntactic structure of sentences in which some parts may be reordered freely becomes a tedious task, since the number of productions representing reorderings of n free parts grows with $n!$. Though the ID/LP approach [10] developed for alleviating this problem in classical (flat) context–free grammars cannot easily be transferred to hypergraph based linguistic modeling, it has inspired a somewhat similar notation. The main idea of ID/LP is to distinguish immediate dominance (ID) constraints from linear precedence (LP) constraints. The left hand side of a phrase structure rule (i.e. context–free Chomsky rule) dominates the symbols on its right hand side. The order of the symbols of the right hand side is the linear

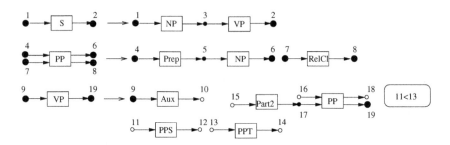

Fig. 2. String generating hyperedge replacement rules with word order constraints for example (1)

precedence of the rule. Traditional phrase structure rules incorporate immediate dominance and precedence into a single rule. In contrast ID/LP maintains separate rule sets. In our approach, we also separate immediate dominance and linear precedence, but only within one rule.

This method is described with the help of example (1). The main rules to construct this sentence and its variations are shown in Fig. 2.[2] The first rule replaces *S* (sentence) with *NP* (noun phrase) and *VP* (verb phrase). This rule is identical to the well known $S \rightarrow NP\ VP$ in phrase structure grammars. For this rule, no hyperedges or word order constraints are necessary. Numerical node labels indicate the sequence of source and target nodes of the left and right hand side and map external nodes onto each other. External nodes are drawn larger than internal nodes.

The second rule handles the discontinuous constituent "durch einen Boten ... der dann starb" ("by a messenger who then died"). The prepositional phrase (*PP*) has type $(2,2)$. The rule's right hand side contains the preposition (*Prep*) and the noun phrase (*NP*) leading to "durch einen Boten" ("by a messenger"). These parts have to follow each other in this order.[3] Since this is not the case for the relative clause generated from the symbol *RelCl*, there is no connection between the relative clause and the preposition with the noun phrase on the right hand side. Between both parts other constituents can be inserted.

The third rule has to deal with varying word order. The verb phrase (*VP*) is split into the auxiliary (*Aux*), the second participle (*Part2*) and three prepositional phrases. The prepositional phrases denoting the grammatical source and target of the verb (*PPS, PPT*) as well as *Aux* and *Part2* have one source and one target node. The prepositional phrase *PP* for the agent of the verb has two source and two target nodes. This rule is not an SGHG rule if applied literally, since there are internal nodes that are either not a source or not a target. These nodes are drawn in white. Such an "illegal" rule represents a set of legal rules

---

[2] In this paper, hyperedges are drawn as rectangles with their label inside. Nodes are drawn as circles, the connection between nodes and edges is marked with arrows.

[3] Of course in a real world grammar, a rule must be included that splits "einen Boten" ("a messenger") into a determiner and a noun. This rule is omitted here, since it does not offer any new insights.

with different orderings of their parts based on the white nodes. The additional information "11 < 13" indicates a **constraint** on linear precedence of nodes in the final string graph. To write down constraints on freely attachable nodes, the node numbers are reused. For the sake of simplicity in this paper, all nodes have unique numbers even between different rules.

The white nodes are called **open nodes**, the black nodes are **closed nodes**. It is shown in [4] that in SGHG each internal node is source for one hyperedge and target for one hyperedge and each external node is either source to one hyperedge or target for one hyperedge. Otherwise no string language is generated. These nodes are the so–called closed nodes in the remainder of the paper.

For open nodes on the right hand side of productions, these requirements are not fulfilled immediately but inbound or outbound hyperedges are chosen during the derivation. **Open internal nodes** are either source or target node to exactly one hyperedge. This means that open internal nodes lack an outbound or an inbound hyperedge. In Fig. 2 in the third rule $10, 11, 12, 13, 14, 15, 16, 18$ are open internal nodes. $10, 12, 14, 18$ have only an inbound hyperedge whereas $11, 13, 15, 16$ have only outbound hyperedges. **Open external nodes** are not connected to any hyperedge. External nodes define the type of a graph (the right hand side and the left hand side of a rule must have the same type). There are no open external nodes in Fig. 2. All external nodes are closed.

The idea of an open node is to have the possibility to choose between different hyperedges of the right hand side that might connect to that node by another open node. When combining open nodes during the derivation, the various string graph fragments are combined to form a single string graph, i.e. no open nodes are left and no cycles produced. Different word orders can thus be produced. For our running example, this means that, if we regard the finally derived string as a "path" through the intermediate graphs, that the right hand side of the *VP*–production is entered through node 9 and the *Aux* hyperedge. The outbound node 10 is an open node, so we can choose freely from any open node without an inbound hyperedge as a possible successor. Here, we can continue with the nodes $11, 13, 15, 16$. This means *PPS*, *PPT*, *PP* or *Part2* might follow *Aux*.

Not all possible word order variations are desirable. In our example, we must make sure that the source of the action is mentioned before the target of the action. This is achieved through constraints over node labels. The constraint $11 < 13$ in Fig. 2 states that the open node 11 must be entered before the open node 13. The constraint $12 < 13$ would have had the same effect. In the generative model, constraints are accumulated during derivation and restrain free combination of string graphs at the end.

In general C is a set of **order relations** (constraints): Let O be the set of all open nodes. An order relation between two open nodes $o_1 < o_2$ with $o_1, o_2 \in O$ means that o_1 must be before o_2 in the final string sequence. The set of order relations applied to an open string graph can be used to create a directed acyclic graph (DAG). The nodes of the DAG are the open nodes. If the set of order relations does not fulfill the criteria of a DAG (e.g. there are

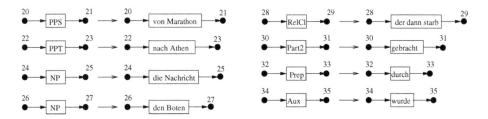

Fig. 3. Missing rules for the analysis of example (1) leading to the derivation tree in Fig. 1

cycles inside the graph), the constraints are not well defined. The DAG can be checked for implicit cyclic dependencies that occur through statically connected hyperedges by merging nodes that are statically connected in the open graph [11].

A **hypergraph grammar rule with restriction in word–order** consists of a triple (L, R, C), L and R being the actual rule consisting of a left hand side and a right hand side that may have open nodes, and C being a set of constraints which restrict reordering of the free parts.

For the real–world application of the rules in Fig. 2 to example (1), several productions are missing. Fig. 3 contains the complementing rules to build the example sentence (1). The combination of rules from Fig. 2 and Fig. 3 produces the derivation tree given in Fig. 4 and together they form the example grammar. All sentences given in example (2) are also produced by the grammar.[4]

This section concludes with the definition of the example graph grammar (N, T, P, S) consisting of nonterminal symbols N, in our example all linguistic acronyms, terminal symbols T, here the words of example (1), the productions P given in Fig. 2 and Fig. 3 and the start symbol S, the whole *sentence* in our application context.

4   Parsing SGHG with Restriction in Word Order

In [4] an Earley based parser [5] called **HyperEarley** for string generating hypergraph grammars was introduced. Its main data structure is called a **chart**. For the chart, positions at the beginning of the string, between the words and at the end of the string of words to be parsed are numbered. This numbering scheme is easily transferred onto string hypergraphs. The nodes in the hypergraph are numbered from 0 to 8 as done at the bottom of Fig. 4.[5] When parsing a sentence $s_0 s_1 \ldots s_{n-1}$ consisting of n words, the chart is a $(n+1) \times (n+1)$ table. In our

---

[4] Please note that it is not possible to generate the following sentence without the discontinuous constituent. Different rules are necessary.

Die Nachricht	**wurde**	von	Marathon	nach	Athen	durch	einen	Boten	der	dann	starb	**gebracht**
The news	**was**	from	Marathon	to	Athens	by	a	messanger	who	then	died.	**brought**

[5] Please do not confuse the numbers in Fig. 4 (the chart positions) with the numbers in Fig. 2 (the node labels for constraints).

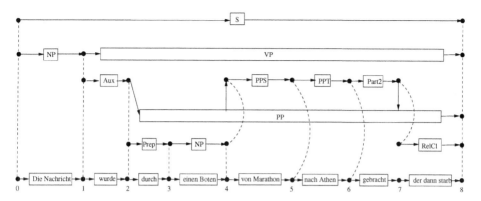

Fig. 4. The derivation tree for the application of the rules given in Fig. 2 to example (1)

running example we have a 9×9 table. In this table, sets of **chart entries** are stored. A chart entry at position (i, j) contains information about the partial derivation trees constructed for the substring $s_i \ldots s_{j-1}$.

Entries are never removed from the chart and are immutable after creation. They consist of information about the currently used grammar production and the progress made in completing the subtree given by this production. This is visualized by so–called **dotted rules**, where the dot marks the parsing progress. The dot is one special node on the right hand side of the rule that marks which parts of the right hand side of the rule have already been found. It is also called **current node**. If the dot is at an external target node of the rule's right hand side, the chart entry is **inactive**. Otherwise it is **active**, i.e. ready to accept a terminal or an inactive chart entry. Note that (different to the classical Earley algorithm) an inactive chart entry must not necessarily be finished; there might still be hyperedges to be processed via another external source node.

The HyperEarley algorithm, like the Earley algorithm, consists of three steps that alternate until the possibilities to apply one of them are exhausted. These steps are **shift** handling terminal hyperedges, **predict** inserting new active chart entries and **complete** combining active with inactive chart entries to generate new entries. When applied to a chart entry e and a hyperedge h that is part of the rule's right hand side e, the method **parts(e,h)** returns either a previously created chart entry describing a derivation of h or null.

The parsing of grammars with open nodes and word order constraints is inspired by [12,13,14] handling ID/LP grammars with Earley–like algorithms.[6] The main idea is that possible connections of the right hand side's parts are delayed as long as possible. If the connection of an open node cannot be delayed further, all possible connections are handled in parallel.

---

[6] While it is possible to translate SGHG productions with open nodes and word order constraints into productions without open nodes and constraints, this leads to an explosion of rules within the grammar and of chart entries, slowing down the parsing process considerably [12,13,14].

Algorithm 4.1 `predict`

Parameters:
e: active chart entry
grammar: the parser's current grammar

```
 1: entry–list← {}
 2: node–list ← list of nodes that can be connected to the current node or external
    source node of start symbol's rules.
 3: for all  n in node–list do
 4:    h = hyperedge following n, h labeled with nonterminal
 5:    if parts(e,h) is defined then
 6:       add continuation(parts(e,h), e) to entry–list
 7:    else
 8:       for all rules r where label(lefthandside(r))=label(h) do
 9:          entry–list ← generate–prediction(r,n)
10:       end for
11:    end if
12: end for
13: for all chart entries c in entry–list do
14:    if c is not in chart entry (to(e), to(e)) then
15:       add c to chart entry (to(e), to(e))
16:       predict(c)
17:    end if
18: end for
```

4.1 Insertion of New Chart Entries: `predict`

`predict` (see Alg. 4.1) is called with an active chart entry e and is applied whenever new active entries are inserted into the chart.

A parse starts with the prediction of the start symbol S. In this case e is `null`. `node-list` is filled with the external source nodes of the start symbol's rules as there is no current node yet (line 2). In line 4, h equals the S-hyperedge. `parts`(e, h) is not defined, so for all productions with left hand side labeled S, new predictions are generated in line 9. These predictions are inserted in the chart as new entries in lines 13–15. `to`(e) is defined as 0 as e is `null`. Finally in line 16 `predict` is called recursively with the newly generated chart entry. Recursion stops when terminal rules are reached. In our case, rules for NP have to be predicted.

This is an example for the second case for `predict`'s application. A new, active chart entry is inserted if another active chart entry expects a nonterminal symbol. In this case, the first inserted chart entry for S has its current node set to node 1 so that chart entries for the hyperedge labeled with the non–terminal NP must be predicted. In Fig. 3 two rules for NP are given. In line 2 `node-list` is set to 24, 26. Two new entries are inserted into the chart starting at $(0, 0)$ with current nodes 24 and 26, see Fig 3.

The second case is especially interesting when free word order is possible within a rule. In Fig. 2 this is the case for the third rule. Let's assume the situation given in Fig. 5. Here the current node is at node 10; the auxiliary

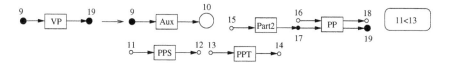

Fig. 5. The large node is the current node in this rule during the parsing process. It remains open for which hyperedge it will be predicted next.

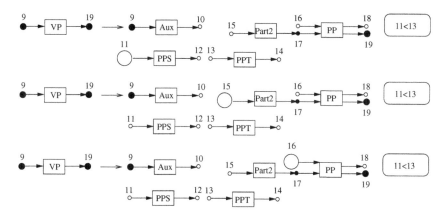

Fig. 6. The large node is the current node in this rule during the parsing process. These chart entries are predicted after Fig. 5.

verb has already been processed. The question is, which new entry is predicted next. As 10 is an open target node, it may connect to every other open source node on the right hand side. Open source nodes are $11, 13, 15, 16$. But the open node 13 cannot be used, because otherwise the constraint stating $11 < 13$ is not fulfilled. In Fig. 6 the three rules for the new chart entries are shown. In Alg. 4.1 in line 2, the possible new current nodes are calculated. In our example, **node-list** contains the nodes $11, 15, 16$. These new chart entries are inserted in the lines 13–15.

In the third case, **predict** inserts **continuation entries**, active entries that restart the parsing of an inactive entry through another external source node. In this case **parts(e, h)** is not **null** (line 5) but returns another chart entry, signifying that the current node has already been moved over this hyperedge once. In our running example, the second and third rule of Fig. 2 lead to this situation. The second rule for PP can be predicted twice depending on the current node in the third rule. For the first prediction it is at node 16, for the second prediction at node 17.

4.2   Handling Edges Labeled with Terminal Symbols: shift

The task of **shift** as shown in Alg. 4.2 is to process the ith hyperedge of the input string graph. In line 2 the main loop over all active chart entries e ending at position $i - 1$ starts. In **node-list** (line 3), all source nodes of the rule's

---

Algorithm 4.2 shift
___
Parameters:
t, ith-hyperedge in input string graph labeled with terminal
 1: entry–list ← {}
 2: **for all** active entries e where to$(e)= i - 1$ **do**
 3:     node–list ← list of source nodes that can be identified with the current node.
 4:     **for all** n in node–list, n not external target node **do**
 5:         h = hyperedge following n, h labeled with terminal
 6:         **if** label(h) = label of hyperedge t **then**
 7:             entry-list ← generate–new–chart–entries(e,t,n)
 8:         **end if**
 9:     **end for**
10:     **for all** chart entries c in entry–list **do**
11:         insert c into chart$[$from$(e),$to$(t)]$
12:         **if** c is inactive **then**
13:             complete(e)
14:         **else**
15:             predict(e)
16:         **end if**
17:     **end for**
18: **end for**

---

right hand side in chart entry e that can be identified with the current node are collected. If the current node is a closed node, it is only the current node itself. For an open node, it might be several different (open) nodes. As for predict, these nodes are calculated based on the DAG generated from open nodes and word order constraints. The loop over the collected nodes (line 4) determines for each node n the hyperedge h following n. If h matches the terminal t in label and type, e can be extended using t. In line 7 the new chart entry is generated. Finally, all newly generated chart entries collected in entry-list must be inserted in the chart (lines 10–17). A chart entry is inserted at the beginning of the active entry from(e) to the end of the terminal entry to(t). If the newly generated entry is inactive complete is called, otherwise predict.

In Fig. 7 shifting over "von Marathon" ("from Marathon") is shown.[7]

4.3   Combination of Active and Inactive Chart Entries: complete

complete handles inactive chart entries ia. For an inactive chart entry, the dot is at an external target node. The inactive entry can be used by an active chart entry to advance its own current node. In line 2 the main loop over all active chart entries e with to(e) = from(ia) is given. If the current node of the active entry is an open node, then (line 3) the nodes that are not external target

---

[7] Please note that in our rules given in Figs. 2, 3 terminal and nonterminal symbols are not mixed on the right hand side of a rule. There is only one terminal symbol on the right hand side in Fig. 3. This is often the case in natural language applications but not necessarily so.

Fig. 7. Shifting in the rule handling "von Marathon" ("from Marathon")

Algorithm 4.3 `complete`

Parameters:
ia, an inactive chart entry

```
 1: entry–list ← {}
 2: for all active entries e where to(e)=from(ia) do
 3:     node–list ← list of nodes in e that can connect to the current node
 4:     for all n in node–list, n not external target node do
 5:         h = hyperedge following n
 6:         if expects(e, ia) then
 7:             entry–list ← generate–new–chart–entries(e, ia, n)
 8:         end if
 9:     end for
10: end for
11: for all chart entries e in entry–list do
12:     insert e into chart[from(ia), to(ia)]
13:     if e is inactive then
14:         complete(e)
15:     else
16:         predict(e)
17:     end if
18: end for
```

---

nodes and can be connected to the current node must be calculated. For all these nodes the following hyperedge h is determined in line 5. The function `expects(e, ia)` in line 6 is extended compared to the original Earley algorithm. `expects(e, ia)` determines if a given inactive edge *ia* is accepted for completion of e. Please note that parsing of an inactive edge is not necessarily finished; an edge is inactive if the current node, the dot, has reached a target node of the rule. There might be several external target nodes. If the label or type of the left hand side of the inactive chart entry's rule differs from e's expected nonterminal edge label or type, `expects(e, ia)` is false. If the node used to enter *ia* does not correspond to the current node in e, `expects(e,ia)` is false. And if *ia* is a continuation chart entry, but the inactive entry that has been continued does not match `parts(e, h)`, *ia* represents a different derivation of the hyperedge than the one assumed the last time it was traversed; therefore, `expects(e, ia)` is false. It is true otherwise. If `expects` returns true, a new chart entry is generated (line 7). As usual in lines 10–16, the newly generated entries are inserted into the chart. If the new entry is inactive, `complete` is called, otherwise `predict`. Deciding whether a chart entry is active or inactive

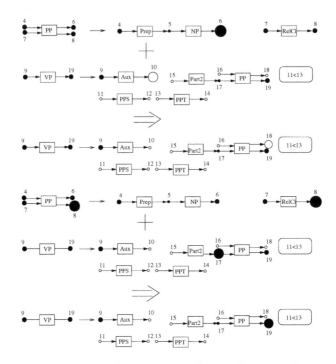

Fig. 8. Completion steps in the running example

is more tricky than for the usual Earley algorithm. If, after completion, an open internal target node is current, it might be possible to merge it with either an internal source node or with an external target node (depending on fulfillment of constraints and similar consistency considerations). In the first case, the entry would have to be counted as active, in the latter as inactive. Instances of both variants are therefore generated, if necessary. Furthermore, care must be taken not to generate a finished entry, i.e. an inactive entry with no unused external nodes remaining, as long as there are open parts left.

In Fig. 8 two applications of `complete` are shown. In both situations the *PP*–rule is inactive and combined with the active *VP*–rule. They differ in the external target nodes used. For Fig. 8, top, it must be first determined which of the open nodes can be used after node 10. Node 16 is possible. `expects` then checks whether the label of the inactive chart entry's left hand side and the label of the hyperedge following the current node match. Both are labeled *PP*. Additionally it checks whether the rank of both hyperedges match and whether the same entry node (external source node) is used. In our case it is the first entry node. After `complete` as in Fig. 8, top, is used, the *PP-rule* must be repredicted. Then the completion as in Fig. 8, bottom, can take place after several steps.

Finally a main procedure `parse` is needed taking sentences to be parsed as input, and returning all derivation trees. This procedure is similar to the `parse` procedure in [1]. It starts with `predict` to insert chart entries for the start symbol *S*. `shift` is called each time `predict` and `complete` do not lead to new entries.

5   Conclusion and Future Work

Extending string generating hypergraph grammars with free nodes and word order constraints has been proven useful to model free word order languages as German or Hungarian. For phrases with free word order consisting of n words, $n!$ phrase structure rules are necessary compared to one rule in the new formalism. Depending on the amount of free word order within a natural language this can lead to huge savings in grammar size.

The extension of the Earley parsing algorithm was based on the the ID/LP Earley parser of S. Shieber [12]. Parsing is in the worst case exponentially in grammar size. The argument in [13,14] for Shieber's ID/LP parser can easily be transferred to HyperEarley. Nevertheless using formalisms like ID/LP or Hyper-Earley is useful. In [15] approaches for Head Driven Phrase Structure Grammar based on discontinuous constituents and free word order versus continuous constituents are compared. It is shown, that the former generates significantly less chart entries than the latter approaches. The analysis is based on two large grammars for German.

Several extensions to the parser described are possible. Especially for German it is necessary to address positions in a sentence directly. In a German declarative clause the finite verb is always in the second position and the infinite verb in the last or last but one position. The second position cannot be addressed yet in opposite to the last position. All other parts of the declarative sentence can be moved freely around these two fixed positions. Also there might be one or no element in the last position after the infinite verb. To express this kind of constraints order lists as used in dependency grammars are necessary [16]. Also the ideas of *partially ordered multiset context–free grammars* [17] can be transferred onto SGHG to allow for more descriptive power.

References

1. Seifert, S., Fischer, I.: Parsing String Generating Hypergraph Grammars. In Ehrig, H., Engels, G., Parisi-Presicce, F., Rozenberg, G., eds.: 2nd International Conference on Graph Transformations (ICGT04). Number 3256 in Lecture Notes On Computer Science, Rome, Italy, Springer-Verlag (2004) 352 – 267
2. Jurafsky, D., Martin, J.H.: Speech and Language Processing: An Introduction to Natural Language Processing, Computational Linguistics, and Speech Recognition. Prentice Hall (2000)
3. Fischer, I.: Modelling Discontinuous Constituents with Hypergraph Grammars. In Pfaltz, J.L., Nagl, M., Böhlen, B., eds.: 2nd International Workshop on Applications of Graph Transformation with Industrial Relevance (AGTIVE'03). Number 3062 in Lecture Notes in Computer Science, Charlottesville, USA, Springer Verlag (2005)
4. Sebastian Seifert: Ein Earley-Parser für Zeichenketten generierende Hypergraphgrammatiken. Studienarbeit, Lehrstuhl für Informatik 2, Universität Erlangen-Nürnberg (2004)
5. Earley, J.: An Efficient Context–Free Parsing Algorithm. Communications of the ACM **13**(2) (1970) 94–102

6. Habel, A.: Hyperedge Replacement: Grammars and Languages. Number 643 in Lecture Notes in Computer Science. Springer-Verlag, Berlin (1992)
7. Drewes, F., Habel, A., Kreowski, H.J.: Hyperedge Replacement Graph Grammars. In Rozenberg, G., ed.: Handbook of Graph Grammars and Computing by Graph Transformation. Vol. I: Foundations. World Scientific (1997) 95–162
8. Engelfriet, J., Heyker, L.: The string generating power of context-free hypergraph grammars. Journal of Computer and System Sciences **43** (1991) 328–360
9. Engelfriet, J., Heyker, L.: Context–free hypergraph grammars have the same term-generating power as attribute grammars. Acta Informatica **29** (1992) 161–210
10. Gazdar, G., Klein, E., Pullum, G., Sag, I.: Generalized Phrase Structure Grammar. Harvard University Press (1985)
11. Martin Riedl: Wortstellungsrestriktionen für Zeichenketten generierende Hypergraphgrammatiken. Studienarbeit, Lehrstuhl für Informatik 2, Universität Erlangen-Nürnberg (2005)
12. Shieber, S.M.: Direct Parsing of ID/LP Grammars. Linguistics and Philosophy **7**(2) (1984) 135–154
13. Barton, G.E.: On the Complexity of ID/LP Parsing. Computational Linguistics **11** (1985) 205–218
14. Barton, G.E.: The Computational Difficulty of ID/LP Parsing. In: Proceedings of the 23rd conference on Association for Computational Linguistics July 08-12. (1985) 76–81
15. Müller, S.: Continuous or discontinuous constituents? a comparison between syntactic analyses for constituent order and their processing systems. Research on Language and Computation, Special Issue on Linguistic Theory and Grammar Implementation **2**(2) (2004) 209–257 http://www.cl.uni-bremen.de/~stefan/Pub/discont.html.
16. Barta, C., Dormeyer, R., Spiegelhauer, T., Fischer, I.: Word Order and Discontinuities in a Dependency Grammar for Hungarian. In Zoltan, A., Csendes, D., eds.: Proceedings of the 2nd Conf. on Hungarian Computational Linguistics (MSZNY), Szeged Hungary, Juhasz Nyomda (2004) 19–27
17. Mark-Jan Nederhof, G.S., Shieber, S.: Partially Ordered Multiset Context-Free Grammars And Free-Word-Order Parsing. In: 8th International Workshop of Parsing Technologies (IWPT), Nancy, France (2003)

Composition and Decomposition of DPO Transformations with Borrowed Context[*]

Paolo Baldan[1], Hartmut Ehrig[2], and Barbara König[3]

[1] Dipartimento di Informatica, Università Ca' Foscari di Venezia, Italy
[2] Institut für Softwaretechnik und Theoretische Informatik,
Technische Universität Berlin, Germany
[3] Institut für Informatik und interaktive Systeme, Universität Duisburg-Essen,
Germany

Abstract. Double-pushout (DPO) transformations with borrowed context extend the standard DPO approach by allowing part of the graph needed in a transformation to be borrowed from the environment. The bisimilarity based on the observation of borrowed contexts is a congruence, thus facilitating system analysis. In this paper, focusing on the situation in which the states of a global system are built out of local components, we show that DPO transformations with borrowed context defined on a global system state can be decomposed into corresponding transformations on the local states and vice versa. Such composition and decomposition theorems, developed in the framework of adhesive categories, can be seen as a first step towards an inductive definition, in sos style, of the labelled transition system associated to a graph transformation system. As a special case we show how an ordinary DPO transformation on a global system state can be decomposed into local DPO transformations with borrowed context using the same production.

1 Introduction

Graph transformations [7] have been applied successfully to several areas of software and system engineering, including syntax and semantics of visual languages, visual modelling of behaviour and programming, metamodelling and model transformation, refactoring of models and programs. Almost invariably the underlying idea is the same: the states of a system are modelled by suitable graphs and state changes are represented by graph transformations. Consequently, the behaviour of the system is expressed by a transition system, where states are reachable graphs and transitions are induced by graph transformations. The transition system can be the basis for defining various notions of abstract behavioural equivalences, e.g., trace, failures and bisimulation equivalence. These, in turn, can be used to provide a solid theoretical justification for various constructions and techniques in the above mentioned areas of system

---

[*] Research partially supported by the EC RTN 2-2001-00346 Project SegraVis, the MIUR Project ART, the DFG project SANDS and CRUI/DAAD Vigoni "Models based on Graph Transformation Systems: Analysis and Verification".

A. Corradini et al. (Eds.): ICGT 2006, LNCS 4178, pp. 153–167, 2006.

engineering, e.g., for the formalisation of behavioural refinement, or to show semantical correctness of refactoring and model transformation.

The applicability of these techniques generally requires the considered behavioural equivalence to be a congruence: two systems—seen as equivalent from the point of view of an external observer—must be equivalent also in all possible contexts or environments.

Unfortunately, behavioural equivalences defined over unlabelled transition systems naively generated by using transformation rules often fail to be congruences. The same problem arises for several other computational formalisms which can be naturally endowed with an operational semantics based on unlabelled reductions, such as the λ-calculus [2] or many process calculi with mobility or name passing, e.g., the π-calculus [11] or the ambient calculus [4].

In order to overcome this problem recently there has been a lot of interest in the automatic derivation of labelled transition systems where bisimilarity is a congruence for reactive systems endowed with an (unlabelled) reduction semantics (see, e.g., [10,8,6,12]). In particular, in the case of double-pushout (DPO) graph rewriting this has led to an extension of the approach, called DPO approach with borrowed contexts [6]. Intuitively a label C of a transition represents the (minimal) context that must be "added" to the current state in order to allow the transformation or reduction step to be performed.

In this paper, we focus on the situation in which the states of a global system are built out of local states of the components of the systems. Then we show that DPO transformations with borrowed context defined on a global system state can be decomposed into corresponding transformations on the local states. Vice versa we study the conditions under which local transformations can be composed to yield global ones. The main results of this paper are composition and decomposition theorems for DPO transformations with borrowed context in the framework of rewriting systems over adhesive categories [9]. As a special case we show how an ordinary DPO transformation on a global system state can be decomposed into local DPO transformations with borrowed context using the same production.

These composition and decomposition results can be seen as a first step towards a structural operational semantics for adhesive rewriting systems, i.e., towards a framework where the transition system associated to a graph transformation system can be defined inductively, in SOS style. Compare for instance the inductive CCS rule stating that from $P \xrightarrow{a} P'$ and $Q \xrightarrow{\bar{a}} Q'$ (where a is an action and $\bar{a}$ its corresponding coaction) one can derive $P \mid Q \xrightarrow{\tau} P' \mid Q'$ (where the label τ stands for a silent transition). Intuitively $P \xrightarrow{a} P'$ means that P can move to Q' if the environment performs an output on channel a and, similarly, Q can move if the environment performs an input on a. The two local moves can be combined leading to a transition for $P \mid Q$ where nothing is "borrowed" from the environment (as expressed by the τ-label).

Having an inductive way of specifying the behaviour of a graph can lead to a new understanding of system semantics and new proof techniques. E.g., inductive definitions can be quite useful when comparing the semantics of two calculi, as in [3].

The rest of the paper is structured as follows. In Section 2 we introduce the basics of adhesive categories and of the DPO approach with borrowed contexts. In Section 3 we introduce a category of transformations with borrowed contexts, which is the basis for the formalisation of the composition and decomposition theorems for transformations given in Sections 4 and 5, respectively. Finally, in Section 6 we conclude and outline directions of future research. Proofs of all theorems, propositions and lemmas can be found in [1].

2   DPO Transformation with Borrowed Contexts

Adhesive categories have been introduced in [9], as categories where pushouts along monomorphisms are so-called Van-Kampen squares (see Condition 3 in the definition below). We will only briefly sketch the theory of adhesive categories.

Definition 1 (Adhesive category). *A category* **C** *is called* adhesive *if*

1. **C** *has pushouts along monos;*
2. **C** *has pullbacks;*
3. *Given a cube diagram as shown on the right with: (i) $A \to C$ mono, (ii) the bottom square a pushout and (iii) the left and back squares pullbacks, we have that the top square is a pushout iff the front and right squares are pullbacks.*

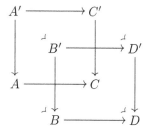

The category **Set** of sets and functions is adhesive. Adhesive categories enjoy closure properties, for instance if **C** is adhesive then so is any functor category $\mathbf{C}^{\mathbf{X}}$, any slice category $\mathbf{C}{\downarrow}C$ and any co-slice category $C{\downarrow}\mathbf{C}$. Therefore, since the category of graphs and graph morphisms is a functor category **Graph** $\cong$ **Set** $^{\leftleftarrows}$, it is adhesive.

A subobject of a given object T is an isomorphism class of monomorphisms to T. Binary intersections of subobjects exist in any category with pullbacks. In adhesive categories also binary unions of subobjects exist and can be obtained by taking the pushout over their intersection. Moreover, the lattice of subobjects is distributive.

Theorem 2 ([9]). *For an object T of an adhesive category* **C***, the partially ordered set $Sub(T)$ of subobjects of T is a distributive lattice. Given two subobjects $A, B \in Sub(T)$, the meet $A \cap B$ is (the isomorphism class of) their pullback, while the join $A \cup B$ is (the isomorphism class of) their pushout in* **C** *over their intersection.*

$$A \cap B \quad PB \quad T \qquad\qquad A \cap B \quad PO \quad A \cup B \cdots\cdots T$$

The following lemma will be useful in the future where we have to show that certain squares in adhesive categories are pullbacks or pushouts. It follows directly from Theorem 2.

Lemma 3. *Consider the following diagram where all arrows are mono. The square below is a pullback if and only if $A = B \cap C$ (all objects are seen as subobjects of E). Furthermore the square is a pushout if and only if $A = B \cap C$ and $D = B \cup C$.*

$$
\begin{array}{ccc}
A & \longrightarrow & B \\
\downarrow & & \downarrow \\
C & \longrightarrow & D \longrightarrow E
\end{array}
$$

We next define rewriting with borrowed contexts on objects (e.g., over graphs) with interfaces, as introduced in [6]. Intuitively, the borrowed context is the smallest extra context which must be added to the object being rewritten in order to obtain an occurrence of the left-hand side. The extra context can be added only using the interface.

Definition 4 (Borrowed contexts, transformations). *Let* **C** *be a fixed adhesive category and let $r = (L \leftarrow I \rightarrow R)$ be a rewriting rule. A DPO transformation with borrowed context—short transformation—t (of r) is a diagram in* **C** *of the following form, where all arrows are mono:*

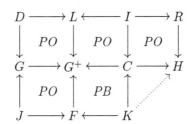

In this case we write $(J \rightarrow G) \overset{r,m}{\Longrightarrow} (K \rightarrow H)$ where $m = G \leftarrow D \rightarrow L$ is the partial match. *If instead we want to focus on the interaction with the environment we say that $J \rightarrow G$ makes a transition with borrowed context $J \rightarrow F \leftarrow K$ and becomes $K \rightarrow H$ (written: $(J \rightarrow G) \overset{J \quad F \quad K}{\longrightarrow} (K \rightarrow H)$).*

For a given transformation t_i we will denote the objects occurring in the corresponding diagram by D_i, G_i, J_i, G_i^+, C_i, H_i, F_i, K_i.

The squares in the diagram above have the following meaning: the upper left-hand square merges the left-hand side L and the object G to be rewritten according to a partial match $G \leftarrow D \rightarrow L$ of the left-hand side in G. The resulting object G^+ contains a total match of L and can be rewritten as in the standard DPO approach, which produces the two remaining squares in the upper row. The pushout in the lower row gives us the borrowed (or minimal) context F which is missing in order to obtain a total match of L, along with a morphism $J \rightarrow F$ indicating how F should be attached to G. Finally, the interface for the

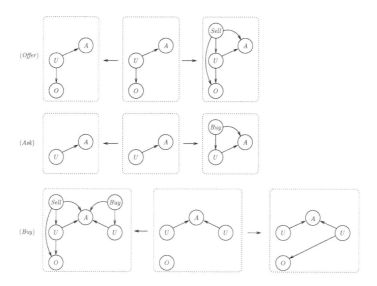

Fig. 1. Rewriting system Market

resulting object H is obtained by "intersecting" the borrowed context F and the object C via a pullback. Roughly, the new interface includes what is preserved of the old interface and of the context borrowed from the environment. The two pushout complements that are constructed in Definition 4 may not exist. In this case no rewriting step is possible.

It has been shown in [6] that bisimilarity on the transition system labelled with borrowed contexts is a congruence with respect to cospan composition.

Example. Consider the category **Graph** of labelled graphs and label-preserving morphisms. Take the rewriting system Market in **Graph** depicted in Fig. 1, which can be interpreted as a very high-level description of the interactions between users of an electronic market place. Graph nodes are represented as circles, with their label inside. Edges are directed and unlabelled. Users, represented as U-labelled nodes, can possess objects, denoted by O-labelled nodes, and they can be connected to one (or more) market places, represented by A-labelled nodes.

A user possessing some objects can autonomously decide to offer one of them to other users, on a market place, expressed by rule *(Offer)*. A user can also ask for something to buy on a market he is connected to, expressed by rule *(Ask)*. A request and an offer, after some negotiation which is not modelled, can meet, the object is sold and moved from the seller to the buyer, modelled by rule *(Buy)*.

An example of a transformation with borrowed context using production *(Buy)* can be found in Fig. 3. It is applied to the graph with interface $J_1 \rightarrow G_1$ in Fig. 2. The graph G_1 includes a market place A, with a user U, possessing two objects and trying to sell one of them. Note that the borrowed context consists of an additional user playing the role of a buyer. In other words, the existence of

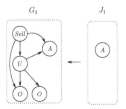

Fig. 2. The graph with interface $J_1 \to G_1$

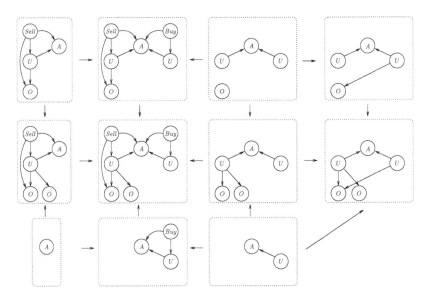

Fig. 3. A transformation with borrowed context t_1 over $J_1 \to G_1$, using rule *(Buy)*

the transformation expresses the fact that rule *(Buy)*, can be applied assuming that the context provides a user which buys the object sold by the user in G_1.

Remark: Note that we obtain the well-known case of DPO transformations if we consider total matches $L \to G$ instead of partial matches $G \leftarrow D \to L$, which implies $G = G^+$. In this case we can take any interface object J, for instance the initial object—if it exists in the category—which implies that F and K are also initial objects.

3   Transformation Morphisms

A first step towards the composition of transformations is the formalisation of the intuitive idea of embedding of a transformation into another. This is done by introducing a suitable notion of transformation morphism.

Definition 5 (Transformation morphisms). *Let t_1, t_2 be two transformations for a fixed rewriting rule $L \leftarrow I \rightarrow R$.* A transformation morphism $\theta \colon t_1 \rightarrow t_2$ *consists of arrows $D_1 \rightarrow D_2$, $G_1 \rightarrow G_2$, $G_1^+ \rightarrow G_2^+$, $C_1 \rightarrow C_2$, $H_1 \rightarrow H_2$, $J_2 \rightarrow J_1$, $F_2 \rightarrow F_1$ and $K_2 \rightarrow K_1$ such that the diagram below commutes. (The arrows $L \rightarrow L$, $I \rightarrow I$, $R \rightarrow R$ in the diagram are the identities.)*

A transformation morphism is called componentwise mono *if it is composed of monos only.*

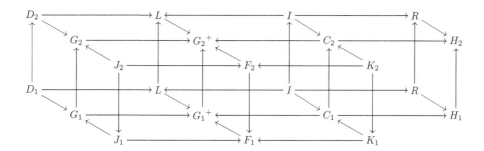

The intuition—at least if all arrows are mono—is that a morphism "embeds" transformation t_1 into t_2. Thus, G_1 (the object being rewritten) is mapped into G_2 and the same holds for D_1 (the partial match), G_1^+, C_1 and H_1. Furthermore, since G_1 is contained in G_2, it might be necessary to borrow more context from the environment. Hence F_1 can be larger than F_2 and the same holds for the inner and outer interfaces of F_1 (denoted by J_1 and K_1). For instance J_1 might have to be larger than J_2 since more context has to be attached. Hence the "squares" J_2, J_1, G_1, G_2 and F_2, F_1, G_1^+, G_2^+ and K_2, K_1, C_1, C_2 are not real squares, but will be called *horseshoes* in the following.

The complexity of our proofs stems from the fact that these horseshoes have to be taken into account. Otherwise it would be possible to simply work in a functor category.

Definition 6 (Category of transformations). *The category having as objects transformations and as arrows transformation morphisms is denoted by* **Trafo**. *Composition of transformation morphisms is defined componentwise.*

Example. Consider the graph with interface $J_3 \rightarrow G_3$ in Fig. 4. The graph G_3 includes a market place with two users. The first one possesses two objects and is trying to sell one of them. The second user is looking for an object to buy. A transformation for $J_3 \rightarrow G_3$, using rule *(Buy)*, can be found in Fig. 5. Observe that in this case the given graph already includes all what is needed for applying rule *(Buy)* and thus nothing is actually borrowed from the context. Thus only the interface is exposed in the label, i.e., the graph $F_3 = J_3$. It is not difficult to see that there is an obvious transformation morphism $\theta_1 : t_1 \rightarrow t_3$, where t_1 is the transformation in Fig. 3.

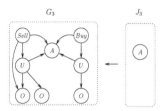

Fig. 4. The graph with interface $J_3 \to G_3$.

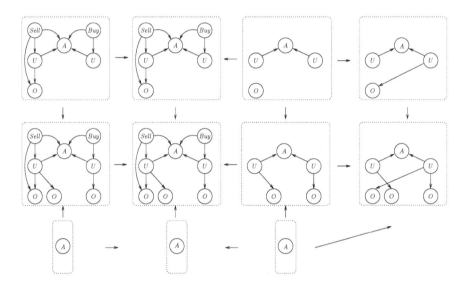

Fig. 5. A transformation with borrowed context t_3 over $J_3 \to G_3$, using rule *(Buy)*

Although the definition of transformation morphisms does not impose any condition on the vertical squares or horseshoes, we can infer some properties by taking into account that all horizontal squares are either pullbacks or pushouts (along monos, and thus also pullbacks).

Lemma 7 (Properties of transformation morphisms). *For a transformation morphism θ as defined in Definition 5 it holds that:*

- *The squares I, I, L, L and I, I, R, R and C_1, C_2, G_1^+, G_2^+ and C_1, C_2, H_1, H_2 are pushouts.*
- *If the arrows $G_1^+ \to G_2^+$, $C_1 \to C_2$ and $H_1 \to H_2$ are mono, the squares L, L, G_1^+, G_2^+ and I, I, C_1, C_2 and R, R, H_1, H_2 and K_2, K_1, F_2, F_1 and D_1, D_2, G_1, G_2 are pullbacks.*

4   Composition of Transformations

In this section we study a composition mechanism for transformations. More precisely we show that given two transformations t_1, t_2, using the same production, with a common subtransformation t_0, the two transformations can be combined via a pushout. We will give sufficient conditions for the existence of this pushout and show how it can be constructed.

We first consider a simpler category where objects are pushouts and we show how to construct pushouts in this setting.

Lemma 8 (Pushouts in the category of pushouts). *Let* **C** *be a fixed adhesive category. Consider the category of pushouts in* **C***, where objects are pushouts p_i of the form*

$$
\begin{array}{ccc}
A_0^i & \longrightarrow & A_2^i \\
\downarrow & & \downarrow \\
A_1^i & \longrightarrow & A_3^i
\end{array}
$$

and an arrow $\varphi\colon p_1 \to p_2$ consists of four arrows $(\varphi_0, \varphi_1, \varphi_2, \varphi_3)$ (with $\varphi_i\colon A_i^1 \to A_i^2$) which connect the corners of the squares such that the full diagram (which is a cube) commutes.

Given two arrows $\varphi^1\colon p_0 \to p_1$, $\varphi^2\colon p_0 \to p_2$ in this category, a pushout $\psi^1\colon p_1 \to p_3$, $\psi^2\colon p_2 \to p_3$ can be computed by constructing four pushouts of the arrows φ_i, ψ_i, provided that these pushouts exist. Then the resulting (pushout) square is composed of the four mediating arrows.

Note that even if the pushouts p_0, p_1, p_2 consist only of monos, the resulting pushout square p_3 does not necessarily consist of monos. Hence in our case this property has to be shown by different means.

We next introduce a property ensuring the composability of transformations.

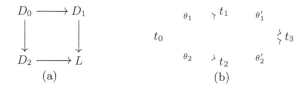

(a)   (b)

Fig. 6. Composition of transformations

Definition 9 (Composable transformation morphisms). *Let $\theta_i\colon t_0 \to t_i$ with $i \in \{1, 2\}$ be transformation morphisms. We say that θ_1 and θ_2 are composable if*

1. θ_1, θ_2 are componentwise mono and
2. the square in the underlying category $\mathbf{C}$ in Fig. 6(a) (where the top and right arrows appear in θ_1 and the left and bottom arrows appear in θ_2) is a pullback.

Intuitively the second condition in the definition above requires that the partial match for t_0 is the intersection of the partial matches for t_1, t_2.

Theorem 10 (Composition of transformations). *Let $\theta_i \colon t_0 \to t_i$ with $i \in \{1,2\}$ be two composable transformation morphisms. Then the pushout of θ_1, θ_2 exists (see Fig. 6(b)) and can be obtained in the following way:*

- *Construct $D_3, G_3, G_3^+, C_3, H_3$ by taking pushouts and J_3, F_3, K_3 by taking pullbacks. For instance D_3 is constructed by taking the pushout of $D_0 \to D_1$, $D_0 \to D_2$, where these two arrows are taken from θ_1 respectively θ_2. This produces the transformation morphisms $\theta_i' \colon t_i \to t_3$.*
- *In order to construct the arrows in t_3 we proceed as follows:*
  - *Most arrows can be immediately obtained as mediating arrows. This is the case for $D_3 \to G_3, D_3 \to L, G_3 \to G_3^+, L \to G_3^+, C_3 \to G_3^+, C_3 \to H_3, J_3 \to F_3, K_3 \to F_3, I \to C_3, R \to H_3$.*
  - *Furthermore construct $J_3 \to G_3$ by composing $J_3 \to J_1 \to G_1 \to G_3$. Similarly for $F_3 \to G_3^+$ and $K_3 \to C_3$.*

5   Decomposition of Transformations

In the previous section we have shown how to compose larger transformations out of smaller ones. Here we are going into the opposite direction and show under which conditions transformations can be split into smaller ones. That is, given a transformation of $J \to G$ and a decomposition of G into subobjects G_1, G_2, is it possible to find transformations for these subobjects, such that the composition of these transformations yields the original transformation?

5.1   Projecting Transformations

In order to be able to formulate the decomposition of transformations, we will first show how to project a transformation to a subobject of G, i.e., to a subobject of the object to be rewritten. We identify some conditions which ensure that a transformation can be projected over a subobject of the rewritten object. Roughly, the interface of the subobject must be sufficiently large to guarantee that the needed context can be actually borrowed.

Definition 11 (Extensibility). *Let t_2 be a transformation and and let $J_2 \to J_1 \to G_1 \to G_2$ be a factorisation of the arrow $J_2 \to G_2$. Then the transformation is called* extensible *with respect to this factorisation, whenever there exists a subobject F_1 of U_2 (the pushout of $G_2^+ \leftarrow C_2 \to H_2$) such that*

$$G_1 \cup L = G_1 \cup F_1 \qquad G_1 \cap F_1 = J_1.$$

The definition above basically requires (in lattice-theoretic terms) that the pushout complement F_1 of $J_1 \to G_1 \to G_1^+$ exists, where $G_1^+ = G_1 \cup L$. Note that in adhesive categories the pushout complement of monos is unique (if it exists).

The extensibility condition given in Definition 11 can be difficult to work with. Below we give an alternative handier condition, sufficient for extensibility.

Lemma 12 (Sufficient condition for extensibility). *Let t_3 be a transformation and let $J_3 \to J_1 \to G_1 \to G_3$ be a factorisation of the arrow $J_3 \to G_3$. Then t_3 is extensible with respect to this factorisation if the pushout complement X_{13} of $J_1 \to G_1 \to G_3$ exists, i.e., there exists an object X_{13} and morphisms such that the square below is a pushout.*

$$
\begin{array}{ccc}
J_1 & \longrightarrow & G_1 \\
\downarrow & & \downarrow \\
X_{13} & \longrightarrow & G_3
\end{array}
$$

In this case set $F_1 = (X_{13} \cup F_3) \cap (G_1 \cup L)$.

Essentially, the sufficient condition requires that the interface of the smaller object G_1 is sufficiently large to allow to get the larger object G_3 by extending G_1 along its interface.

Now let t_i be a transformation over an object with interface $J_i \to G_i$ ($i \in \{1,2\}$) and let $J_2 \to J_1 \to G_1 \to G_2$ be a factorisation of $J_2 \to G_2$. We say that a transformation morphism $\theta : t_1 \to t_2$ is *consistent* with the factorisation if it has the arrows $J_2 \to J_1$ and $G_1 \to G_2$ as components.

Proposition 13 (Projection of transformations). *Let t_2 be a transformation and let $J_2 \to J_1 \to G_1 \to G_2$ be a (mono) factorisation of the morphism $J_2 \to G_2$ such that t_2 is extensible with respect to this factorisation. Then there exists a unique transformation t_1 of $J_1 \to G_1$, with a componentwise mono transformation morphism $\theta : t_1 \to t_2$, consistent with the factorisation.*

The objects of this transformation can be constructed as follows:

1. *Construct U_2 as the pushout of $C_2 \to G_2^+$ and $C_2 \to H_2$. Now all objects can be considered as subobjects of U_2.*
2. *The object F_1 is given by the extensibility property above, which requires that $G_1 \cup L = G_1 \cup F_1$ and $G_1 \cap F_1 = J_1$. Set $D_1 = G_1 \cap D_2$, $G_1^+ = G_1 \cup L$, $C_1 = G_1^+ \cap C_2$, $H_1 = C_1 \cup R$, $K_1 = F_1 \cap C_1$.*

5.2   Decomposing Transformations

As a first step towards the decomposition of a transformation, we introduce a suitable decomposition for an object with interface.

Definition 14 (Proper decomposition). *Let $J_3 \to G_3$ be an object with interface. Then a proper decomposition of $J_3 \to G_3$ is a cube as shown below* *where all arrows are mono, the square G_0, G_1, G_2, G_3 is a pushout and and the square J_0, J_1, J_2, J_3 is a pullback. (Note that the four remaining "squares" are horseshoes.)*

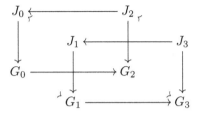

Theorem 15 (Decomposition of transformations). *Let t_3 be a transformation of an object with interface $J_3 \to G_3$. Consider a proper decomposition of $J_3 \to G_3$ as in Definition 14 and assume that the transformation t_3 is extensible with respect to the factorisations $J_3 \to J_1 \to G_1 \to G_3$ and $J_3 \to J_2 \to G_2 \to G_3$.*

Then there are transformations t_i for $J_i \to G_i$ (where $i \in \{0, 1, 2\}$) with componentwise mono transformation morphisms $\theta_j : t_0 \to t_j$, $\theta'_j : t_j \to t_3$ (where $j \in \{1, 2\}$) forming a pushout in the category of transformations (see the diagram in Theorem 10). These transformation morphisms can be obtained via projections as described in Proposition 13.

Observe that, if in the cube in Theorem 15 above we have the special (but very typical) case where $J_0 = J_1 = J_2 = J_3 = G_0$ (and all arrows between these objects are the identities), the sufficient extensibility condition of Lemma 12 is satisfied: in the terminology of this lemma $X_{13} = G_2$ and $X_{23} = G_1$.

In a sense, composition and decomposition are inverse to each other up to isomorphism. The fact that composition is the inverse of decomposition has been shown directly in Theorem 15. On the other hand, since projections are unique (by Proposition 13), there is—up to isomorphism—only one way to decompose a transformation according to a proper decomposition of the rewritten object (see Definition 14). Hence, also decomposition is the inverse of composition.

Next we discuss the special case where a DPO rewriting step with trivial borrowed context is decomposed, leading to transformations with possibly non-empty borrowed contexts. Assume that $G = G_3$ can be split into G_0, G_1, G_2 as in the pushout diagram below on the left and consider a DPO rewriting step for G_3. Then this step can be extended to a transformation with borrowed context for G_3 (with interface G_0) with a total match of the left-hand side.

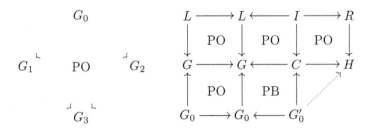

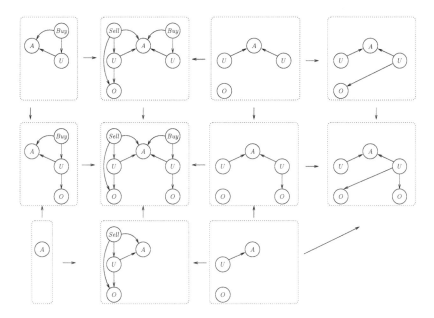

Fig. 7. A transformation with borrowed context t_2 over $J_2 \to G_2$, using rule *(Buy)*

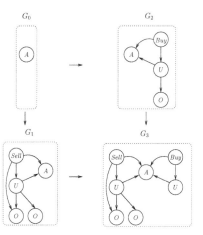

Fig. 8. Decomposition of transformations

In this case we can set $J_0 = J_1 = J_2 = J_3 = G_0$ and obtain a proper decomposition of $J_3 \to G_3$ as in Definition 14 (the top square is trivially pullback and the bottom square is a pushout by assumption). Then, decomposing transformation t_3 as described in Proposition 15 leads to three transformations t_0, t_1, t_2, with—in general—partial matches D_0, D_1, D_2.

Example. Consider the graph with interface $J_3 \to G_3$ in Fig. 4. Note that there is an obvious factorisation $J_3 \to J_1 \to G_1 \to G_3$ of $J_3 \to G_3$. Furthermore, the transformation t_3 in Fig. 5, which uses rule *(Buy)*, is extensible along such factorisation. In fact, the sufficient condition given by Lemma 12 is satisfied.

Therefore we can project the transformation t_3 in Fig. 5 along such factorisation thus obtaining the transformation t_1 over $J_1 \to G_1$ depicted in Fig. 3. As already noted, as an effect of projecting the transformation over a smaller graph, the borrowed context becomes non-trivial (larger than the interface): the rule can be applied assuming that the context provides a user which buys the object sold by the user in G_1.

More generally, consider the diagram in Fig. 8, where morphisms are the inclusions suggested by the shapes of the graphs. This is a pushout in **Graph**. Moreover, we can imagine all graphs G_i to have an interface given by $J_i = G_0$. Then the conditions of Proposition 15 are satisfied: we can project the transformation t_3 in Fig. 5 to transformations over $J_i \to G_i$ ($i \in \{0, 1, 2\}$). The projection over $J_1 \to G_1$ leads to the transformation t_1 in Fig. 3, while the projection over $J_2 \to G_2$ leads to the transformation t_2 in Fig. 7. Both t_1 and t_2 project to the same derivation t_0 over $J_0 \to G_0$. The pushout of the obtained transformations can be computed, according to Theorem 10 to obtain t_3 again.

6   Conclusion and Comparison to Related Work

In this paper, focusing on a setting in which a system is built out of smaller components, we discussed how derivations with borrowed context over the global state can be decomposed into transformations over the local state of each single component using the same rule. Vice versa, we showed that, under suitable consistency conditions, local transformations can be composed to give rise to a transformation over the global system state.

We remark that the form of composition described in this paper is quite different from amalgamation as described for instance in [5]. There two transformations for different rules are amalgamated producing a transformation for the amalgamated rule. In our case, instead, the rule is fixed and the transformations differ with respect to the context that has to be borrowed from the environment. By composing objects and hence transformations we obtain additional structure which might reduce the borrowed context.

The composition and decomposition results can be seen as a basic step towards the possibility of defining transformations only for "atomic objects" and assemble all possible transformations out of these atomic transformations, and thus, towards an inductive definition, in SOS style, of the transition system of a graph transformation system (more generally an adhesive rewriting system). In addition to the composition result we will also need the possibility to compose an evolving object with a passive context and to have rules for handling restrictions of the interface. This would correspond to the communication, parallel composition and restriction rules for process calculi. Additionally, composition would be even more natural and closer to process calculi if performed over so-called rewriting

steps, hiding the internal details, rather than on full transformations. That is—
in the terminology of Definition 4—we would like to observe only the object with
interface $J \to G$, the resulting object $K \to H$ and the label or borrowed context $J \to F \leftarrow K$, but not the objects D, G^+, C which are only auxiliary or
intermediate objects. We plan to extend our approach to this setting.

References

1. P. Baldan, H. Ehrig, and B. König. Composition and decomposition of DPO
   transformations with borrowed contexts. Technical report, Universität Duisburg-
   Essen, 2006.
2. H. P. Barendregt. *The Lambda Calculus—its Syntax and Semantics*, volume 103
   of *Studies in Logic and Foundations of Mathematics*. North-Holland, 1984.
3. F. Bonchi, F. Gadducci, and B. König. Process bisimulation via a graphical encoding. In *Proc. of ICGT '06*. Springer, 2006. LNCS, to appear.
4. L. Cardelli and A. D. Gordon. Mobile ambients. In *Proc. of FoSSaCS '98*, pages
   140–155. Springer-Verlag, 1998. LNCS 1378.
5. A. Corradini, U. Montanari, F. Rossi, H. Ehrig, R. Heckel, and M. Löwe. Algebraic
   approaches to graph transformation—part I: Basic concepts and double pushout
   approach. In G. Rozenberg, editor, *Handbook of Graph Grammars and Computing
   by Graph Transformation, Vol. 1: Foundations*, chapter 3. World Scientific, 1997.
6. H. Ehrig and B. König. Deriving bisimulation congruences in the DPO approach to
   graph rewriting. In *Proc. of FOSSACS '04*, pages 151–166. Springer, 2004. LNCS
   2987.
7. H. Ehrig, H.-J. Kreowski, U. Montanari, and G. Rozenberg, editors. *Handbook of
   Graph Grammars and Computing by Graph Transformation, Vol.3: Concurrency,
   Parallellism, and Distribution*. World Scientific, 1999.
8. O. H. Jensen and R. Milner. Bigraphs and transitions. In *Proc. of POPL 2003*,
   pages 38–49. ACM, 2003.
9. S. Lack and P. Sobociński. Adhesive and quasiadhesive categories. *RAIRO –
   Theoretical Informatics and Applications*, 39(3), 2005.
10. J. J. Leifer and R. Milner. Deriving bisimulation congruences for reactive systems.
    In *Proc. of CONCUR 2000*, 2000. LNCS 1877.
11. R. Milner. The polyadic π-calculus: a tutorial. In *Logic and Algebra of Specification*.
    Springer-Verlag, Heidelberg, 1993.
12. V. Sassone and P. Sobociński. Reactive systems over cospans. In *Proc. of LICS
    '05*, pages 311–320. IEEE, 2005.

Process Bisimulation
Via a Graphical Encoding*

Filippo Bonchi[1], Fabio Gadducci[1], and Barbara König[2]

[1] Dipartimento di Informatica, Università di Pisa
[2] Institut für Informatik und Interaktive Systeme, Universität Duisburg-Essen

Abstract. The paper presents a case study on the synthesis of labelled transition systems (LTSs) for process calculi, choosing as testbed Milner's Calculus of Communicating System (CCS). The proposal is based on a graphical encoding: each CCS process is mapped into a graph equipped with suitable *interfaces*, such that the denotation is fully abstract with respect to the usual structural congruence.

Graphs with interfaces are amenable to the synthesis mechanism based on *borrowed contexts* (BCs), proposed by Ehrig and König (which are an instance of *relative pushouts*, originally introduced by Milner and Leifer). The BC mechanism allows the effective construction of an LTS that has graphs with interfaces as both states and labels, and such that the associated bisimilarity is automatically a congruence.

Our paper focuses on the analysis of the LTS distilled by exploiting the encoding of CCS processes: besides offering some technical contributions towards the simplification of the BC mechanism, the key result of our work is the proof that the bisimilarity on processes obtained via BCs coincides with the standard strong bisimilarity for CCS.

1 Introduction

The dynamics of a computational device is often defined by a *reduction system* (RS): a set, representing the space of possible states of the device; and a relation among these states, representing the possible evolutions of the device. This is e.g. the case of the paradigmatic functional language, the λ-calculus: the *β-reduction rule* $(\lambda x.M)N \Rightarrow M[N/x]$ models the application of a functional process $\lambda x.M$ to the actual argument N, and the reduction relation is then obtained by freely instantiating and contextualising the rule.

While RSs have the advantage of conveying the semantics with relatively few compact rules, their main drawback is poor compositionality, in the sense that the dynamic behaviour of arbitrary standalone terms can be interpreted only by inserting them in the appropriate context, where a reduction may take place. In fact, simply using the reduction relation for defining equivalences between components fails to obtain a compositional framework, and in order to recover a suitable congruence it is often necessary to verify the behaviour of single components

---

* Research partially supported by the DFG project SANDS, the EU RTN 2-2001-00346 SEGRAVIS and IST 2004-16004 SENSORIA, and the MIUR PRIN 2005015824 ART.

A. Corradini et al. (Eds.): ICGT 2006, LNCS 4178, pp. 168–183, 2006.

under any viable execution context. This is the road leading from contextual equivalences for the λ-calculus to barbed and dynamic equivalences for the π-calculus. In these approaches, though, proofs of equivalence are often tedious and involuted, and they are left to the ingenuity of the researcher.

A standard way out of the impasse, reducing the complexity of such analyses, is to express the behaviour of a computational device by a *labelled transition system* (LTS). Should the label associated to a component evolution faithfully express how that component might interact with the whole of the system, it would be possible to analyse *in vitro* the behaviour of a single component, without considering all contexts. Thus, a "well-behaved" LTS represents a fundamental step towards a compositional semantics of the computational device. It is not always straightforward, though, to identify the right "label" that should be distilled, starting from a previously defined RS. Indeed, after Milner's proposal of an alternative semantics for the π-calculus [17] based on reactive rules modulo a structural congruence on processes, inspired by the CHAM paradigm [4], an ongoing stream of research has been investigating the relationship between the LTS semantics for process calculi and their more abstract RS semantics.

Early attempts by Sewell [22] devised a strategy for obtaining an LTS from an RS by adding contexts as labels on transitions. The technique was refined by Leifer and Milner [15] who introduced *relative pushouts* (RPOs) in order to capture the notion of *minimal context* activating a reduction. The generality of this proposal (and its bicategorical formulation due to Sassone and Sobocinski [20]) allows it to be applied to a large class of formalisms. More importantly, such attempts share the basic property of synthesising a congruent bisimulation equivalence, thus ensuring that the resulting LTS semantics is compositional. However, for the time being there are few case studies which either involve rich calculi, or succeed in making comparisons with standard behavioural equivalences. To tackle a fully-fledged case study is the main aim of this paper.

Our starting point for the synthesis of an LTS are the graphical techniques proposed for modelling the reduction semantics of nominal calculi in [10,12]: processes are encoded in *graphs with interfaces*, an instance of *cospan categories* [11], and process reduction is simulated by *double-pushout* (DPO) rewriting [1]. Since the category of cospans over graphs admits RPOs [21], its choice as the domain of the encoding for nominal calculi ensures that the synthesis of an LTS can be performed, and that a *compositional* observational equivalence is obtained.

The key technical point is the use of the *borrowed context* (BC) technique [8] as a tool to equip graph transformation in the DPO style with an LTS semantics. Graphs with interfaces are amenable to the synthesis mechanism based on BCs (which are in turn an instance of RPOs): this allows the construction of an LTS that has graphs with interfaces as both states and labels, and such that the associated bisimilarity is automatically a congruence. Exploiting the BC technique, also large case studies may be taken into account: until now the difficulties in the presentation of the LTSs obtained via the use of RPOs forced to restrict the analysis to simple case studies, relying either on standard (ground) term rewriting [15], or on extremely simplified variants of process calculi [20]:

more elaborated proposals using bigraphs [18,14] result in infinitely branching LTSs, banning recursive processes or failing to capture standard bisimilarity.

Summing up, the aim of our work is straightforward: to present a fully-fledged case study on the synthesis of LTSs for process calculi, choosing as testbed Milner's Calculus of Communicating System (CCS). More precisely, the paper focuses on the analysis of the LTS obtained by exploiting the BC technique and the encoding of CCS (recursive) processes into unstructured graphs, along the lines of the methodology sketched above. Besides offering some technical contributions towards the simplification of the BC synthesis mechanism, the key result is the proof that the bisimilarity on (recursive) processes obtained via BCs coincides with the standard strong bisimilarity for CCS. We believe that our work may offer novel insights on the synthesis of LTSs, as well as offering further evidence of the adequacy of graph-based formalisms for system design and verification.

The extended version of the paper [5] contains additional examples, categorical notations and detailed proofs.

2   Two Operational Semantics for CCS

This section introduces CCS [16] and two alternative operational semantics: the classical LTS semantics and the reduction semantics.

Definition 1 (processes). *Let $\mathcal{N}$ be a set of* names, *ranged over by $a, b, c, \ldots$; $\tau \notin \mathcal{N}$ an invisible* name; *$\Delta = \{a, \bar{a} \mid a \in \mathcal{N}\} \uplus \{\tau\}$ a set of prefixes, ranged over by δ; and finally, X a set of* agent variables, *ranged over by $x, y, \ldots$. An* open process *P is a term generated by the (mutually recursive) syntax*

$$P ::= M, \ (\nu a)P, \ P_1 \mid P_2, \ rec_x.P \quad M ::= \mathbf{0}, \ \delta.P, \ M_1 + M_2, \ \delta.x$$

A process *is a term such that each occurrence of an agent variable x is in the scope of a rec_x-operator. We let $P, Q, R, \ldots$ range over the set $\mathcal{P}$ of processes, and $M, N, O \ldots$ range over the set $\mathcal{S}$ of* summations.

The standard definition for the set of free names of a process P, denoted by $\mathbf{fn}(P)$, is assumed. Similarly for α-conversion with respect to the *restriction* operators $(\nu a)P$: the name a is bound in P, and it can be freely α-converted.

The classical observational semantics, *bisimilarity*, is given over an inductively defined *labelled transition system* (LTS). We spell out the LTS, and denote by $\sim_{CCS}$ the standard strong bisimilarity, without formally introducing it.

Definition 2 (labelled transition system). *The* transition relation *for processes is the relation $L_{CCS} \subseteq \mathcal{P} \times \Delta \times \mathcal{P}$ inductively generated by the set of axioms and inference rules below (where $P \xrightarrow{\delta} Q$ means that $\langle P, \delta, Q \rangle \in L_{CCS}$).*

$$\frac{}{\delta.P \xrightarrow{\delta} P} \qquad \frac{P \xrightarrow{a} Q, \ R \xrightarrow{\bar{a}} S}{P \mid R \xrightarrow{\tau} Q \mid S} \qquad \frac{P \xrightarrow{\delta} Q}{(\nu a)P \xrightarrow{\delta} (\nu a)Q} \ a \notin \mathbf{fn}(\delta)$$

$$\frac{P \xrightarrow{\delta} Q}{P \mid R \xrightarrow{\delta} Q \mid R} \qquad \frac{P \xrightarrow{\delta} Q}{P + R \xrightarrow{\delta} Q} \qquad \frac{P\left[^{rec_x.P}/_x\right] \xrightarrow{\delta} Q}{rec_x.P \xrightarrow{\delta} Q}$$

As usual, we avoided presenting the symmetric counterparts of those three inference rules involving the parallel and sum operators; moreover, the substitution operator is supposed not to capture any name, possibly through α-conversion.

The behavior of a process P can also be described as a relation over *abstract processes*, obtained by closing a set of basic rules under structural congruence.

Definition 3 (structural congruence). *The* structural congruence *for processes is the relation* $\equiv \subseteq \mathcal{P} \times \mathcal{P}$, *closed under process construction and α-conversion, inductively generated by the set of axioms below.*

$$P \mid Q = Q \mid P \qquad P \mid (Q \mid R) = (P \mid Q) \mid R \qquad P \mid 0 = P$$

$$M + N = N + M \qquad M + (N + O) = (M + N) + O \qquad M + 0 = M$$

$$(\nu a)(\nu b)P = (\nu b)(\nu a)P \qquad (\nu a)(P \mid Q) = P \mid (\nu a)Q \ \ for \ a \notin \mathbf{fn}(P) \qquad (\nu a)0 = 0$$

$$(\nu a)(M + \delta.P) = M + \delta.(\nu a)P \ \ for \ a \notin \mathbf{fn}(M + \delta.0) \qquad rec_x.P = P[{}^{rec_x.P}/_x]$$

Definition 4 (reduction semantics). *The* reduction relation *for processes is the relation* $R_{CCS} \subseteq \mathcal{P} \times \mathcal{P}$, *closed under the structural congruence $\equiv$, inductively generated by the set of axioms and inference rules below (where $P \to Q$ means that $\langle P, Q \rangle \in R_{CCS}$).*

$$\frac{}{a.P + M \mid \bar{a}.Q + N \to P \mid Q} \qquad \frac{}{\tau.P + M \to P}$$

$$\frac{P \to Q}{(\nu a)P \to (\nu a)Q} \qquad \frac{P \to Q}{P \mid R \to Q \mid R}$$

There is a main difference with respect to the standard reduction semantics for CCS, namely, the axiom schema concerning the distributivity of the restriction operators with respect to the prefix operators, even if they have been already considered in the literature, see e.g. [9]. These equalities do not change substantially the reduction semantics, and they indeed hold in all the observational equivalences we are aware of. In particular, two congruent processes are also strongly bisimilar. Most importantly, they allow a simplified presentation of the graphical encoding: we refer the reader to [12] for a more articulate analysis.

The LTS semantics specifies how a system, seen as a single component, may interact with the environment, and it allows the definition of an observational equivalence by means of bisimilarity. On the other hand, the RS semantics specifies how a system, seen as the whole, evolves. The latter is usually more natural, but it does not take in account the interactions, and consequently, does not provide any "good" notion of behavioral equivalence. The main aim of the theory of reactive systems proposed by Milner in [15] is to systematically derive an LTS from an RS semantics. In this paper, exploiting a graphical encoding of processes, we derive an LTS from a graph rewriting semantics. More precisely, in the next sections we introduce a graphical encoding of CCS processes which preserves the reduction semantics. The encoding is then used to distill an LTS with pairs of graph morphisms as labels: the main result of the paper states that the resulting bisimilarity coincides with the standard strong bisimilarity.

3   Graphs and Their Extension with Interfaces

We recall a few definitions concerning (typed hyper-)graphs, and their extension with *interfaces*, referring to [6] for a more detailed introduction.

Definition 5 (graphs). *A* (hyper-)graph *is a four-tuple* $\langle V, E, s, t \rangle$ *where V is the set of nodes, E is the set of edges and $s, t : E \rightarrow V$ are the source and target functions. An* (hyper-)graph *morphism is a pair of functions* $\langle f_V, f_E \rangle$ *preserving the source and target functions.*

The corresponding category is denoted by **Graph**. However, we often consider *typed graphs* [7], i.e., graphs labelled over a structure that is itself a graph.

Definition 6 (typed graphs). *Let T be a graph. A* typed graph *G over T is a graph $|G|$, together with a graph morphism $t_G : |G| \rightarrow T$. A morphism between T-typed graphs $f : G_1 \rightarrow G_2$ is a graph morphism $f : |G_1| \rightarrow |G_2|$ consistent with the typing, i.e., such that $t_{G_1} = t_{G_2} \circ f$.*

The category of graphs typed over T is denoted T-**Graph**: it coincides with the slice category **Graph** $\downarrow T$. In the following, a chosen type graph T is assumed.

In order to inductively define the encoding for processes, we need to provide operations over typed graphs. The first step is to equip them with suitable "handles" for interacting with an environment.

Definition 7 (graphs with interfaces). *Let J, K be typed graphs. A* graph with input interface J and output interface K *is a triple $= \langle j, G, k \rangle$, for G a typed graph and $j : J \rightarrow G$, $k : K \rightarrow G$ the input and output* morphisms.

Let   and   be graphs with the same interfaces. An interface graph morphism *$f :   \Rightarrow $ is a typed graph morphism $f : G \rightarrow H$ between the underlying graphs that preserves the input and output interface morphisms.*

We let $J \xrightarrow{j} G \xleftarrow{k} K$ denote a graph with interfaces J and K.[1] If the interfaces J, K are *discrete*, i.e., they contain only nodes, we simply represent them by sets. Moreover, if K is the empty set, we often denote a graph with interfaces simply as a graph morphism $J \rightarrow G$. In order to define our encoding processes, we introduce two binary operators on graphs with discrete interfaces.

Definition 8 (two composition operators). *Let $= I \xrightarrow{j} G \xleftarrow{k} K$ and $' = K \xrightarrow{j'} G' \xleftarrow{k'} J$ be graphs with discrete interfaces. Then, their* sequential *composition is the graph with discrete interfaces $\circ   ' = I \xrightarrow{j''} G'' \xleftarrow{k''} J$, for G'' the disjoint union $G \uplus G'$, modulo the equivalence on nodes induced by $k(x) = j'(x)$ for all $x \in N_{G'}$, and j'', k'' the uniquely induced arrows.*

---

[1] With an abuse of notation, we sometimes refer to the image of the input and output morphisms as inputs and outputs, respectively. More importantly, in the following we often refer implicitly to a graph with interfaces as the representative of its isomorphism class, still using the same symbols to denote it and its components.

Let $= J \xrightarrow{j} G \xleftarrow{k} K$ *and* $= J' \xrightarrow{j'} H \xleftarrow{k'} K'$ *be graphs with discrete interfaces. Then, their* parallel composition *is the graph with discrete interfaces* $\otimes$ $= (J \cup J') \xrightarrow{j''} V \xleftarrow{k''} (K \cup K')$, *for V the disjoint union $G \uplus H$, modulo the equivalence on nodes induced by $j(x) = j'(x)$ for all $x \in N_J \cap N_{J'}$ and $k(y) = k'(y)$ for all $y \in N_K \cap N_{K'}$, and j'', k'' the uniquely induced arrows.*

Intuitively, the sequential composition $\circ$ ' is obtained by taking the disjoint union of the graphs underlying   and   ', and gluing the outputs of   with the corresponding inputs of   '. Similarly, the parallel composition $\otimes$ is obtained by taking the disjoint union of the graphs underlying   and   , and gluing the inputs (outputs) of   with the corresponding inputs (outputs) of   . Note that the two operations are defined on "concrete" graphs, even if the result is independent of the choice of the representatives, up-to isomorphism.

A *graph expression* is a term over the syntax containing all graphs with discrete interfaces as constants, and parallel and sequential composition as binary operators. An expression is *well-formed* if all the occurrences of those operators are defined for the interfaces of their arguments, according to Definition 8; its interfaces are computed inductively from the interfaces of the graphs occurring in it, and its *value* is the graph obtained by evaluating all operators in it.

4   From Processes to Graphs with Interfaces

This section presents our graphical encoding for CCS processes. After presenting a suitable type graph, shown in Fig. 1, the composition operators previously defined are exploited. This corresponds to a variant of the usual construction of the tree for a term of an algebra: names are interpreted as variables, so that they are mapped to leaves of the graph and can be safely shared.

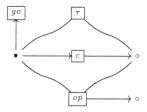

Fig. 1. The type graph T_{CCS} (for $op \in \{rcv, snd\}$)

Intuitively, a graph having as root a node of type • (⋄) corresponds to a process (to a summation, respectively), while each node of type ∘ basically represents a name. Note that the edge *op* stands for a concise representation of two operators, namely *snd* and *rcv*, simulating the two prefixes. There is no operator for simulating either parallel composition or non-deterministic choice. Instead, the operator c is a syntactical device for "coercing" the occurrence of a summation inside a process context (a standard device from algebraic specifications). Finally, the operator *go* is another syntactical device for detecting the "entry" point of the computation, thus avoiding to perform any reduction below the outermost prefix operators: it is later needed for modeling the RS semantics.

The second step is the characterization of a class of graphs, such that all processes can be encoded into an expression containing only those graphs as constants, and parallel and sequential composition as binary operators. Let $p, s \notin \mathcal{N}$: our choice of graphs as constants is depicted in Fig. 2, for all $a \in \mathcal{N}$.

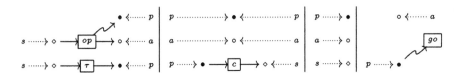

Fig. 2. Graphs op_a (for $op \in \{rcv, snd\}$) and τ; id_p, id_a, and c; 0_p, 0_a, and 0_s; ν_a and *go* (from left to right and top to bottom)

Finally, let us denote id_Γ and 0_Γ as a shorthand for $\bigotimes_{a \ \Gamma} id_a$ and $\bigotimes_{a \ \Gamma} 0_a$, respectively, for a finite set of names $\Gamma \subseteq \mathcal{N}$ (since the ordering is immaterial). The encoding of processes into graphs with interfaces, mapping each finite process into a graph expression, is presented below.

Definition 9 (encoding for finite processes). *Let P be a finite process, and let Γ be a set of names, such that $\mathbf{fn}(P) \subseteq \Gamma$. The (mutually recursive) encodings $P\ {}^p_\Gamma$ and $M\ {}^s_\Gamma$, mapping a process P into a graph with interfaces, are defined by structural induction according to the rules below.*

$$[\![M]\!]^p_\Gamma = \begin{cases} 0_p \otimes 0_\Gamma & \text{if } \mathbf{fn}(M) = \emptyset \\ (c \otimes id_\Gamma) \circ [\![M]\!]^s_\Gamma & \text{otherwise} \end{cases}$$

$$[\![(\nu a)P]\!]^p_\Gamma = \begin{cases} [\![P]\!]^p_\Gamma & \text{if } a \notin \mathbf{fn}(P) \\ (id_p \otimes \nu_b \otimes id_\Gamma) \circ [\![P\{{}^b/a\}]\!]^p_{\{b\} \uplus \Gamma} & \text{for } b \notin \Gamma \ \text{otherwise} \end{cases}$$

$$[\![P \mid Q]\!]^p_\Gamma = [\![P]\!]^p_\Gamma \otimes [\![Q]\!]^p_\Gamma \qquad\qquad [\![M + N]\!]^s_\Gamma = [\![M]\!]^s_\Gamma \otimes [\![N]\!]^s_\Gamma$$

$$[\![0]\!]^s_\Gamma = 0_s \otimes 0_\Gamma \qquad\qquad\qquad\quad [\![\tau.P]\!]^s_\Gamma = (\tau \otimes id_\Gamma) \circ [\![P]\!]^p_\Gamma$$

$$[\![a.P]\!]^s_\Gamma = (rcv_a \otimes id_\Gamma) \circ [\![P]\!]^p_\Gamma \qquad\quad [\![\bar{a}.P]\!]^s_\Gamma = (snd_a \otimes id_\Gamma) \circ [\![P]\!]^p_\Gamma$$

Note the conditional rule for the mapping of $M\ {}^p_\Gamma$. This is required by the use of 0 as the neutral element for both the parallel and the non-deterministic operator: in fact, the syntactical requirement $\mathbf{fn}(M) = \emptyset$ coincides with the semantical constraint $M \equiv 0$.

The mapping is well-defined, since the resulting graph expression is well-formed; moreover, the encoding $P\ {}^p_\Gamma$ is a graph with interfaces $(\{p\} \cup \Gamma, \emptyset)$. Our encoding is sound and complete (even if not surjective), as stated by the proposition below (adapted from [10]).

Proposition 1. *Let P, Q be finite processes, and let Γ be a set of names, such that $\mathtt{fn}(P) \cup \mathtt{fn}(Q) \subseteq \Gamma$. Then, $P \equiv Q$ if and only if $P \,{}^P_\Gamma = Q \,{}^P_\Gamma$.*

Note in particular how the lack of restriction operators is dealt with simply by manipulating the interfaces, even if the price to pay is the presence of "floating" axioms for prefixes, as shown by Fig. 3.

Fig. 3. Encoding for both $[\![(\nu b)a.\bar{b}.0]\!]^P_{\{a\}}$ and $[\![a.(\nu b)\bar{b}.0]\!]^P_{\{a\}}$

4.1 Tackling Recursive Processes

In order to show how recursive processes can be encoded as suitable infinite graphs, the first step is to consider a complete partial order on graphs.

Definition 10 (graph order). *Let   ,   be graphs with interfaces (J, K). Then, $\sqsubseteq_{J,K}$ if there exists a mono $f:$ $\Rightarrow$.*

Thus, we consider the standard subgraph relationship, partitioned over interfaces. These partial orders are complete with respect to ω-chains, and it is noteworthy that the encoding $0 \,{}^P_\Gamma$ is the bottom of the order for those graphs with interfaces $(\{p\} \cup \Gamma, \emptyset)$.

Definition 11. *Let $P[x]$ be an open process, such that the single agent variable x may occur free in P. Let $\mathcal{C} = \{ P_i \,{}^P_\Gamma \mid i \in \mathbb{N} \}$ be a chain where $P_0 = P[^0/x]$ and $P_{i+1} = P[^{P_i}/x]$. Then, $rec_x.P \,{}^P_\Gamma$ denotes the least upper bound of $\mathcal{C}$.*

In other terms, each open process $P[x]$ defines an ω-chain on the graphs with interfaces $(\{p\} \cup \Gamma, \emptyset)$, and $rec_x.P \,{}^P_\Gamma$ is the least upper bound of this chain, computed as the least fixed point starting from the bottom element, i.e., $0 \,{}^P_\Gamma$.

Of course, two recursive expressions may be mapped to isomorphic graphs with interfaces, even if they are not structurally congruent, nor can be unfolded to the same expression. Nevertheless, the extended encoding is clearly still sound.

5   On Graphs with Interfaces and Borrowed Contexts

This section introduces the *double-pushout* (DPO) approach to the rewriting of graphs with interfaces and its extension with *borrowed contexts* (BCs).

Definition 12 (graph production). *A T-typed graph production is a span $L \xleftarrow{l} I \xrightarrow{r} R$ with l mono in T-**Graph**. A typed graph transformation system (GTS) $\mathcal{G}$ is a tuple $\langle T, P, \pi \rangle$ where T is the type graph, P is a set of production names and π is a function mapping each name to a T-typed production.*

Definition 13 (derivation of graphs with interfaces)
Let $J \to G$ and $J \to H$ be two graphs with interfaces. Given
a production $p : L \xleftarrow{l} I \xrightarrow{r} R$, a match of p in G is a
morphism $m : L \to G$. A direct derivation from $J \to G$ to
$J \to H$ via p and m is a diagram as depicted in the right,
where (1) and (2) are pushouts and the bottom triangles com-
mute. In this case we write $J \to G \Longrightarrow J \to H$.

$$
\begin{array}{ccccc}
L & \xleftarrow{l} & I & \xrightarrow{r} & R \\
m\downarrow & (1) & \downarrow & (2) & \downarrow \\
G & \longleftarrow & C & \longrightarrow & H \\
 & \nwarrow_{k} \uparrow & & & \\
 & J & & &
\end{array}
$$

The morphism $k : J \to C$ such that the left triangle commutes is unique, whenever it exists. If such a morphism does not exists, then the rewriting step is not
feasible. Moreover, note that the canonical DPO derivations can be seen as a
special instance of these, obtained considering as interface J the empty graph.

In these derivations, the left-hand side L of a production must occur completely in G. On the contrary, in a *borrowed context* (BC) derivation the graph L
might occur partially in G, since the latter may interact with the environment
through J in order to exactly match L. Those BCs are the "smallest" extra contexts needed to obtain the image of L in G. The mechanism was introduced in
[8] in order to derive an LTS from direct derivations, using BCs as labels. The
following definition is lifted from [2], extending the original one by including also
morphisms that are not necessarily mono. Note that the labels derived in this
way correspond to the labels derived via relative pushouts in a suitable category.

Definition 14 (rewriting with borrowed contexts). *Given a production*
$p : L \xleftarrow{l} I \xrightarrow{r} R$, *a graph with interfaces $J \to G$ and a mono $d : D \rightarrowtail L$, we*
say that $J \to G$ reduces to $K \to H$ with transition label $J \rightarrowtail F \leftarrow K$ via p and
d if there are graphs G^+, C and additional morphisms such that the diagram
below commutes and the squares are either pushouts (PO) or pullbacks (PB). In
this case we write $J \to G \xrightarrow{J \rightarrowtail F \leftarrow K} K \to H$, which is also called rewriting step
with borrowed context.

$$
\begin{array}{ccccccc}
D & \rightarrowtail & L & \leftarrow & I & \rightarrow & R \\
\downarrow & PO & \downarrow & PO & \downarrow & PO & \downarrow \\
G & \rightarrowtail & G^+ & \leftarrow & C & \rightarrow & H \\
\uparrow & PO & \uparrow & PB & \uparrow & \nearrow & \\
J & \rightarrowtail & F & \leftarrow & K & &
\end{array}
$$

Consider the diagram above. The upper left-hand square merges the left-hand
side L and the graph G to be rewritten according to a partial match $G \hookleftarrow D \rightarrowtail$
L. The resulting graph G^+ contains a total match of L and can be rewritten
as in the standard DPO approach, producing the two remaining squares in the
upper row. The pushout in the lower row gives us the borrowed (or minimal)
context F which is missing in order to obtain a total match of L, along with a
morphism $J \rightarrowtail F$ indicating how F should be pasted to G. Finally, we need an
interface for the resulting graph H, which can be obtained by "intersecting" the
borrowed context F and the graph C via a pullback.

Note that two pushout complements that are needed in Definition 14, namely
C and F, may not exist. In this case, the rewriting step is not feasible.

6   From Process Reductions to Graph Rewrites

Following [10], this section introduces the rewriting system $\mathcal{R}_{CCS}$, showing how it simulates the reduction semantics for processes: it is quite simple, since it contains just two rules, depicted in the upper right corner of Fig. 4 (the remaining graphs in this figure are explained later). The first rule models a τ-transition, whereas the second models synchronisation. Note that, in order to disable reduction inside prefixes, we enrich our encoding, attaching an edge go on the root of each process. So, let $P\,{}^g_\Gamma = P\,{}^p_\Gamma \otimes go$. Moreover, for any graph   with interfaces $(\{p\} \cup \Gamma, \emptyset)$, let $reach(\ )$ be the graph with the same interfaces reachable from the image of the roots $\{p\} \cup \Gamma$.

It seems noteworthy that two rules suffice for recasting the reduction semantics of the calculus. First of all, the structural rules are taken care of by the fact that graph morphisms allow for embedding a graph into a larger one, thus simulating the closure of reduction by context. Second, no distinct instance of the rules is needed, since graph isomorphism takes care of the closure with respect to structural congruence, as well as of the renaming of the free name.

Proposition 2 (reductions vs. rewrites). *Let P be a processes, and let Γ be a set of actions such that $\mathtt{fn}(P) \subseteq \Gamma$. If $P \to Q$, then $\mathcal{R}_{CCS}$ entails a direct derivation $P\,{}^g_\Gamma \Longrightarrow G$ via an injective match, such that $reach(G) = Q\,{}^g_\Gamma$. Viceversa, if $\mathcal{R}_{CCS}$ entails a direct derivation $P\,{}^g_\Gamma \Longrightarrow G$ via an injective match, then there exists a process Q such that $P \to Q$ and $reach(G) = Q\,{}^g_\Gamma$.*

The correspondence holds since the go operator forces the match to be applied only on top, thus forbidding the occurrence of a reduction inside the outermost prefixes. The condition on reachability is needed since, during the reduction, some process components may be discarded, in correspondence of the solving of non-deterministic choices. The restriction to injective matches is necessary in order to ensure that the two edges labeled by c can never be merged together. Intuitively, allowing their coalescing would correspond to the synchronization of two summations, i.e., as allowing a reduction $a.P + \bar{a}.Q \to P \mid Q$.

7   The Synthesised Transition System

This section contains the main results of our paper. Its aim is to apply the BC synthesis mechanism to $\mathcal{R}_{CCS}$, and then to analyse the resulting LTS. Proving along the way a few general results on the technique, we show that the LTS is finitely branching (when quotiented up to isomorphism) and equivalent to a succinct $\to_C$ whose transitions have a direct interpretation as process transitions. The main theorem of the section states that $\to_C$ induces on (the encoding of) processes the standard strong bisimilarity.

7.1   Reducing the Borrowing

In order to know all the possible transitions originating from a graph with interfaces $J \to G$, all the subgraphs D's of L_s and L_τ and all the mono mappings

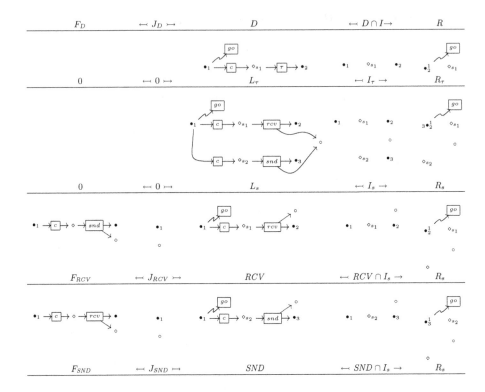

Fig. 4. Productions *synch*: $L_s \overset{l_s}{\leftarrow} I_s \rightarrow R_s$ and τ: $L_\tau \overset{}{\leftarrow} I_\tau \rightarrow R_\tau$ (upper right area). The table illustrates the derivation rules for the concise LTS (0 is the empty graph).

into G should be analysed. To shorten this long and tedious procedure, we show here two pruning techniques for restricting the space of possible D's.

First, note that those items of a left-hand side L that are not in D have to be pasted to G through J. Thus, consider a node n of D corresponding to n' in L such that n' is the source or the target of some edge e that does not occur in D. Since the edge e is in L but not in D, it must be added to G through J, and thus n must be also in J. A node such as n is called a *boundary node*.

Let us now consider *SND*—as shown in Fig. 4— as a subgraph of L_s. Its root is a boundary node since it has an ingoing edge that occurs in L_s but not in *SND*. Also the name in *SND* is a boundary node, since in L_s there is an ingoing edge that does not occur in *SND*. Hence this node must be mapped to a node occurring in the interface J of G.

The notion of boundary nodes is formally captured by the categorical notion of *initial pushout* (as recalled in [5]). Since our category has initial pushouts, the previous discussion is formalized by the proposition below.

Proposition 3. *Let $p : L \overset{l}{\leftarrow} I \overset{r}{\rightarrow} R$ be a production and $d : D \rightarrowtail L$ a mono such that diagram (i) in Fig. 5 is the initial pushout of d. If a graph $J \rightarrow G$ can* *perform a* BC *rewriting step* via *p and d then there exist a mono D ↣ G and a morphism $J_D \to J$ such that diagram (ii) in Fig. 5 commutes.*

This proposition allows to heavily prune the space of all possible D's. As far as our case study is concerned, we can exclude all those D's having among boundary nodes a summation node (depicted by ◇) since these never appear in the interface J of a graph resulting from the encoding of some process. For the same reason, we can exclude all those D's having among their boundary nodes a continuation process node (any of those two nodes depicted by • that are not the root) observing that the only process node in the interface J is the root node.

A further pruning—partially based on proof techniques presented in [8]—is performed by excluding all those D's which generate a BC transition that is not relevant for the bisimilarity. In general terms, we may always exclude all the D's that contain only nodes, since those D's can be embedded in every graph (with the same interface) generating the same transitions. Concerning our case study, those transitions generated by a D having the root node without the edge labeled *go* are also not relevant. In fact, a graph can perform a BC transition using such a D if and only if it can perform a transition using the same D with a *go* edge outgoing from the root. Note indeed that the resulting states of these two transitions only differ for the number of *go* edges attached to the root: the state resulting after the first transition has two *go*'s, the state resulting after the second transition only one. These states are bisimilar, since the number of *go*'s does not change the behavior.

The previous remarks are summed up by the following lemma.

Lemma 1. *Bisimilarity on the* LTS *synthesized by* BCs *coincides with bisimilarity on the* LTS *obtained by considering as partial matches D the graphs L_s, SND and RCV (shown in Fig. 4) as subgraphs of L_s, and only the graph L_τ as subgraph of L_τ.*

7.2 Strong Bisimilarity vs. BC Bisimilarity

Exploiting the remarks of the previous section, we first introduce a concise LTS containing only those BC transitions that are needed to establish the borrowed bisimilarity. Then, we use this concise LTS to prove our main theorem on the correspondence between the borrowed and the CCS bisimilarity.

Proposition 4. *Let $p : L \hookleftarrow I \to R$ be a production of $\mathcal{R}_{CCS}$; $d : D \hookrightarrow L$ a mono such that in Fig. 5, diagram (i) is the initial pushout of d and diagram (iii) is a pullback; and $J \hookleftarrow G$ a graph with interfaces. Then $J \hookleftarrow G \xrightarrow{J \hookleftarrow F \hookrightarrow K} K \to H$ via p and d if and only if there exists a mono $D \hookrightarrow G$, a graph V and a morphism $J_D \to J$ such that the central square of diagram (iv) in Fig. 5 commutes and F and H are constructed as illustrated there.*

The proposition above is a key step in the definition of a concise LTS. In fact, it tells us how to construct the label F and the resulting state H, just starting from a set of minimal rules of the form $R \leftarrow D \cap I \hookrightarrow D \hookleftarrow J_D \hookrightarrow F_D$. Given

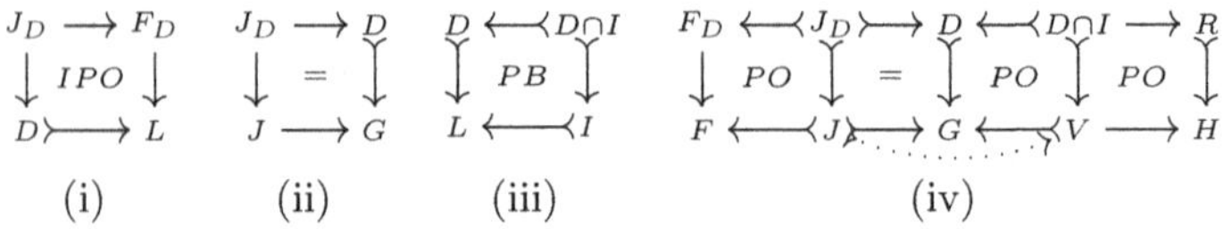

$$
\begin{array}{cccc}
\begin{array}{ccc}
J_D & \longrightarrow & F_D \\
\downarrow\!\! \scriptstyle{IPO} & & \downarrow \\
D & \rightarrowtail & L
\end{array}
&
\begin{array}{ccc}
J_D & \longrightarrow & D \\
\downarrow & = & \downarrow \\
J & \longrightarrow & G
\end{array}
&
\begin{array}{ccc}
D & \longleftarrow\!\!\!< & D \cap I \\
\downarrow & \scriptstyle{PB} & \downarrow \\
L & \longleftarrow\!\!\!< & I
\end{array}
&
\begin{array}{ccccccccc}
F_D & \longleftarrow\!\!\!< & J_D & \rightarrowtail & D & \longleftarrow\!\!\!< & D \cap I & \longrightarrow & R \\
\downarrow\!\scriptstyle{PO} & & \downarrow & = & \downarrow\!\scriptstyle{PO} & & \downarrow\!\scriptstyle{PO} & & \downarrow \\
F & \longleftarrow\!\!\!< & J & \rightarrowtail & G & \longleftarrow\!\!\!< & V & \longrightarrow & H
\end{array}
\\[2mm]
(i) & (ii) & (iii) & (iv)
\end{array}
$$

Fig. 5. Diagrams used in the propositions of Section 7

a mono $D \rightarrowtail G$, the resulting state H can be computed in a DPO step, i.e., all the items of G matched by D and not in $D \cap I$ are removed and replaced by R. This transition is possible only if there exists a morphism $J_D \rightarrow J$ such that the central diagram commutes. In this case, the resulting label F is computed as the pushout between the minimal label $J_D \rightarrow F_D$ and $J_D \rightarrow J$.

We thus now define a concise transition system, starting from the set of rules, of the form $R \leftarrow D \cap I \rightarrow D \leftarrow J_D \rightarrow F_D$, that are depicted in Fig. 4. The main difference with respect to the standard transition system is that the interface J of a graph is never enlarged by a transition, but always remains the same.

Definition 15 (concise transition system). *Let the graph D be either SND, RCV, L_s or L_τ; and let J_D, F_D, $D \cap I$ and R be the graphs defined according to Fig. 4. Then, $J \rightarrowtail G \xrightarrow{\ J \ F \ J\ }_C J \rightarrow H$ if and only if a diagram as the one illustrated in Fig. 5 (iv) can be constructed, where the morphism $J \rightarrow H$ is uniquely induced by $H \leftarrow V \leftarrow G \leftarrow J$.*

Note that the pushout complement of $D \cap I \rightarrow D \rightarrow G$ always exists because for each D as in Fig. 4 all the nodes of $D \cap I$ are in D, and thus we have a transition for each $D \rightarrow G$ and for each $J_D \rightarrow J$ such that the central diagram commutes. Moreover, the morphism $J \rightarrow V$ always exists (since J is discrete and V contains all nodes of G) and it is unique (since $V \rightarrow G$ is mono).

More precisely, consider either SND or RCV as D: the existence of a morphism $J_D \rightarrow J$ means that the name used in the synchronisation must occur in the interface. Whenever either L_s or L_τ is D, J_D is the empty graph 0 and thus a morphism always exists. In these two latter cases the label of the transition is always the span of identities on J and the resulting state is exactly the state obtained from a DPO direct derivation.

The difference between $\rightarrow$ and $\rightarrow_C$ can be explained via an analogy to the CCS-like transition $a.0 + b.0 \xrightarrow{\ \bar{b}.P+M\ } P$. The concise LTS forgets about P and M, and the transition represented in $\rightarrow_C$ is $a.0 + b.0 \xrightarrow{\ \bar{b}.0\ }_R 0$. This operation is performed without changing the resulting bisimilarity, as stated below.

Proposition 5. *Let $\sim$ be the BC bisimilarity, and let $\sim_C$ be the bisimilarity defined on $\rightarrow_C$. Then $\sim_C$ and $\sim$ coincide for all those graphs with discrete interfaces belonging to the image of our encoding.*

The previous proposition allows a simpler proof of the correspondence between strong bisimilarity for CCS and the one resulting from the BC construction.

Theorem 1. *Let P, Q be processes, and let Γ be a set of names, such that $\mathtt{fn}(P) \cup \mathtt{fn}(Q) \subseteq \Gamma$. Then $P \overset{g}{\underset{\Gamma}{\ }} \sim Q \overset{g}{\underset{\Gamma}{\ }}$ if and only if $P \sim_{CCS} Q$.*

Proof. Here we give just a brief sketch of the proof. First of all, note that the set of inference rules below define the same LTS of Definition 2, for $A \subseteq \mathcal{N}$ a finite set of names, Q, R and S processes, and M and N summations.

$$\frac{P \equiv (\nu A)((\tau.Q + M) \mid R)}{P \overset{\tau}{\to} (\nu A)(Q \mid R)} \qquad \frac{P \equiv (\nu A)((\bar{a}.Q + M) \mid (a.R + N) \mid S)}{P \overset{\tau}{\to} (\nu A)(Q \mid R \mid S)}$$

$$\frac{P \equiv (\nu A)((a.Q + M) \mid R) \quad a \notin A}{P \overset{a}{\to} (\nu A)(Q \mid R)} \qquad \frac{P \equiv (\nu A)((\bar{a}.Q + M) \mid R) \quad a \notin A}{P \overset{\bar{a}}{\to} (\nu A)(Q \mid R)}$$

The correspondence between the concise LTS $\to_C$ and the standard LTS of CCS seems then evident, since each of those inference rules above exactly corresponds to a rule $R \leftarrow D \cap I \quad D \quad J_D \quad F_D$ in Fig. 4.

For instance the third rule above corresponds to the third row $D = RCV$ in Fig. 4. Indeed, $P \equiv (\nu A)((a.Q + M) \mid R)$ if and only if RCV can be embedded in G where $J \quad G$ is $P \overset{g}{\underset{\Gamma}{\ }}$. The condition $a \notin A$ is satisfied if and only if a occurs in the interface J, i.e., if and only if there exists a morphism $J_{RCV} \to J$ such that everything commutes. If such a condition is satisfied a transition in $\to_C$ is performed with label $J \quad F \quad J$ where $J \quad F$ is (part of) the pushout between $J_{RCV} \quad J$ and $J_{RCV} \quad F_{RCV}$. Since the latter morphism is fixed, $J \quad F$ depends only $J_{RCV} \quad J$, i.e., it depends only on the name of J corresponding to the unique name of J_{RCV}, that here we have called a. Then, for each graph with interface J such that RCV occurs inside, and such that the unique name of RCV occurs in J with name a, a transition is performed with a label depending only on a. Roughly, this label can be thought of as a context corresponding to $- \mid \bar{a}.0 \overset{g}{\underset{\Gamma}{\ }}$ with $J = \{p\} \cup \Gamma$. The resulting state $(\nu A)(Q \mid R)$ does not exactly correspond to the state resulting from $\to_C$, since the latter contains those graphs that represent discarded choices. However, these summations are not connected anymore to the reachable graph, and thus they do not influence in any way the behavior of the resulting graph.

The second rule corresponds to the second row $D = L_s$. In fact, $P \equiv (\nu A)((\bar{a}.Q + M) \mid (a.R + N) \mid S)$ if and only if L_s can be embedded into G where $J \quad G$ is $P \overset{g}{\underset{\Gamma}{\ }}$. There are no other conditions on this rule and this is exactly expressed by the fact that J_{L_s} is 0. The τ-label exactly corresponds to the label of $\to_C$ given by the span of identities on J.

8   Conclusions and Further Work

Our paper presents a case study in the synthesis of LTSs for process calculi. A sound and complete graphical encoding for processes is exploited in order to apply the BC mechanism for automatically deriving an LTS: states are graphs with interfaces, labels are cospans of graph morphisms, and two (encodings of) processes are strongly bisimilar in the distilled LTS if and only if they are also strongly bisimilar according to the standard LTS.

We consider our case study to be relevant for the reasons outlined below.

Technically, its importance lies in the pruning techniques that have been developed in order to cut to a manageable size the borrowed LTS: they exploit abstract categorical definitions, such as initial pushouts, yet resulting in a simplified LTS with the same bisimulation relation (see Proposition 3).

Methodologically, its relevance is due to its focussing on a fully-fledged case study, including also possibly recursive processes: most examples in the literature restrain themselves to the finite fragment of a calculus, as it happens for the encoding of CCS processes into bigraphs by Milner in [18].

In order to further illustrate the advantages (and the possibilities for future developments) of our approach, let us consider the latter proposal, similar in aim to our work. It is noteworthy that the encoding into graphs with interfaces allows the use of two rewriting rules only: intuitively, these rules are *non-ground* since they can be both contextualized *and* instantiated. This feature results in synthetising a finitely branching (also for possibly recursive processes) LTS: this seems one of the key advantages of the borrowed context technique with respect to the bigraphical approach, where reaction rules must be ground, hence infinite in number and inducing an infinitely branching LTS already for finite processes.

This non-groundness supports our hope to use the BC mechanism for distilling a set of inference rules, instead of characterizing directly the set of possible labelled transitions. This should be obtained by extending Proposition 4 and offering an explicit construction of the interface K for the target state of a transition: its construction was irrelevant for our purposes here, since the reuse of the interface J of the starting state does not change the bisimilarity. A related composition result is already presented in [3].

Finally, we consider promising the combined use of a graphical encoding (into graphs with interfaces) and of the BC techniques, and we plan to test its expressiveness by capturing also nominal calculi. We feel confident that our approach could be safely extended to those calculi whose distinct feature is name fusion [19], while it might fail for calculi where a more flexible notion of name scoping is needed, as suggested by preliminary results on the π-calculus in [13].

References

1. P. Baldan, A. Corradini, H. Ehrig, M. Löwe, U. Montanari, and F. Rossi. Concurrent semantics of algebraic graph transformation. In H. Ehrig, H.-J. Kreowski, U. Montanari, and G. Rozenberg, editors, *Handbook of Graph Grammars and Computing by Graph Transformation*, volume 3, pages 107–187. World Scientific, 1999.
2. P. Baldan, A. Corradini, T. Heindel, B. König, and P. Sobociński. Processes for adhesive rewriting systems. In W. Aceto and A. Ingólfsdóttir, editors, *Foundations of Software Science and Computation Structures*, volume 3921 of *Lect. Notes in Comp. Sci.* Springer, 2006.
3. P. Baldan, H. Ehrig, and B. König. Composition and decomposition of DPO transformations with borrowed contexts. This volume.
4. G. Berry and G. Boudol. The chemical abstract machine. *Theor. Comp. Sci.*, 96: 217–248, 1992.

5. F. Bonchi, F. Gadducci, and B. König. Process bisimulation via a graphical encoding. Technical Report TR-06-07, Dipartimento di Informatica, Università di Pisa, 2006.

6. A. Corradini and F. Gadducci. An algebraic presentation of term graphs, via gs-monoidal categories. *Applied Categorical Structures*, 7:299–331, 1999.

7. A. Corradini, U. Montanari, and F. Rossi. Graph processes. *Fundamenta Informaticae*, 26:241–265, 1996.

8. H. Ehrig and B. König. Deriving bisimulation congruences in the DPO approach to graph rewriting. In I. Walukiewicz, editor, *Foundations of Software Science and Computation Structures*, volume 2987 of *Lect. Notes in Comp. Sci.*, pages 151–166. Springer, 2004.

9. J. Engelfriet and T. Gelsema. Multisets and structural congruence of the π-calculus with replication. *Theor. Comp. Sci.*, 211:311–337, 1999.

10. F. Gadducci. Term graph rewriting and the π-calculus. In A. Ohori, editor, *Programming Languages and Semantics*, volume 2895 of *Lect. Notes in Comp. Sci.*, pages 37–54. Springer, 2003.

11. F. Gadducci and R. Heckel. An inductive view of graph transformation. In F. Parisi-Presicce, editor, *Recent Trends in Algebraic Development Techniques*, volume 1376 of *Lect. Notes in Comp. Sci.*, pages 219–233. Springer, 1997.

12. F. Gadducci and U. Montanari. A concurrent graph semantics for mobile ambients. In S. Brookes and M. Mislove, editors, *Mathematical Foundations of Programming Semantics*, volume 45 of *Electr. Notes in Theor. Comp. Sci.* Elsevier Science, 2001.

13. F. Gadducci and U. Montanari. Observing reductions in nominal calculi *via* a graphical encoding of processes. In A. Middeldorp *et al.*, editor, *Processes, terms and cycles (Klop Festschrift)*, volume 3838 of *Lect. Notes in Comp. Sci.*, pages 106–126. Springer, 2005.

14. O. H. Jensen and R. Milner. Bigraphs and transitions. In G. Morriset, editor, *Principles of Programming Languages*, pages 38–49. ACM Press, 2003.

15. J. Leifer and R. Milner. Deriving bisimulation congruences for reactive systems. In C. Palamidessi, editor, *Concurrency Theory*, volume 1877 of *Lect. Notes in Comp. Sci.*, pages 243–258. Springer, 2000.

16. R. Milner. *Communication and Concurrency*. Prentice Hall, 1989.

17. R. Milner. The polyadic π-calculus: A tutorial. In F.L. Bauer, W. Brauer, and H. Schwichtenberg, editors, *Logic and Algebra of Specification*, volume 94 of *Nato ASI Series F*, pages 203–246. Springer, 1993.

18. R. Milner. Pure bigraphs: Structure and dynamics. *Information and Computation*, 204:60–122, 2006.

19. J. Parrow and B. Victor. The fusion calculus: Expressiveness and simmetry in mobile processes. In V. Pratt, editor, *Logic in Computer Science*, pages 176–185. IEEE Computer Society Press, 1998.

20. V. Sassone and P. Sobociński. Deriving bisimulation congruences using 2-categories. *Nordic Journal of Computing*, 10:163–183, 2003.

21. V. Sassone and P. Sobociński. Reactive systems over cospans. In *Logic in Computer Science*, pages 311–320. IEEE Computer Society Press, 2005.

22. P. Sewell. From rewrite rules to bisimulation congruences. *Theor. Comp. Sci.*, 274:183–230, 2004.

Toposes Are Adhesive

Stephen Lack[1] and Paweł Sobociński[2,*]

[1] School of Computing and Mathematics, University of Western Sydney, Australia
[2] Computer Laboratory, University of Cambridge, United Kingdom

Abstract. Adhesive categories have recently been proposed as a categorical foundation for facets of the theory of graph transformation, and have also been used to study techniques from process algebra for reasoning about concurrency. Here we continue our study of adhesive categories by showing that toposes are adhesive. The proof relies on exploiting the relationship between adhesive categories, Brown and Janelidze's work on generalised van Kampen theorems as well as Grothendieck's theory of descent.

Introduction

Adhesive categories [11, 12] and their generalisations, quasiadhesive categories [11] and adhesive HLR categories [6], have recently begun to be used as a natural and relatively simple general foundation for aspects of the theory of graph transformation, following on from previous work in this direction [5]. By covering several "graph-like" categories, they serve as a useful framework in which to prove structural properties. They have also served as a bridge allowing the introduction of techniques from process algebra to the field of graph transformation [13, 7].

From a categorical point of view, the work follows in the footsteps of distributive and extensive categories [4] in the sense that they study a particular relationship between certain finite limits and finite colimits. Indeed, whereas distributive categories are concerned with the distributivity of products over coproducts and extensive categories with the relationship between coproducts and pullbacks, the various flavours of adhesive categories consider the relationship between certain pushouts and pullbacks.

Adhesive categories are defined to be the categories with pullbacks where pushouts along monomorphisms are van Kampen [11] and as a consequence such pushouts can be considered as being well-behaved with respect to pullbacks. As we shall explain, related work includes Grothendieck's theory of descent (see [9] for an overview) and generalised approaches to the van Kampen theorem [1].

As shown in [11], adhesive categories are closed under several categorical constructions, thus if $\mathbf{C}$ and $\mathbf{D}$ are adhesive then so is their product $\mathbf{C} \times \mathbf{D}$, choosing any object $C \in \mathbf{C}$, the slice category $\mathbf{C}/C$ and coslice category $C/\mathbf{C}$ are adhesive and, for any category $\mathbf{X}$, the functor category $[\mathbf{X}, \mathbf{C}]$ is adhesive. It is also

* Research partially supported by EPSRC grant EP/D066565/1. The first author gratefully acknowledges the support of the Australian Research Council.

known that the category of sets and functions **Set** is adhesive, in particular this means that any presheaf topos [**X**, **Set**] is adhesive. These constructions are useful because adhesive categories satisfy many of the so-called HLR-axioms, and as a consequence, several important theorems about the rewriting theory of double-pushout transformations can be proved at the level of adhesive categories. Indeed, it is perhaps surprising that so many of the axioms, which were not known previously to be related, hold automatically in any adhesive category. One of the original contributions of this paper is a proof that adhesive categories satisfy the *special pullback-pushout property*, one of the aforementioned axioms.

The central part of the paper is devoted to studying the relationship between toposes and adhesive categories. Topos theory has many different facets and applications within mathematics and computer science. One of the interesting properties of toposes is that they have finite limits and colimits, and these behave somewhat as they do in **Set**. Indeed, while toposes enjoy much more structure than adhesive categories, adhesive categories themselves have certain finite colimits (pushouts along monomorphisms) and limits (pullbacks) which are well-behaved with respect to each other. The question of whether toposes are adhesive is thus a very natural one.

As we have shown in [11, 12], there are adhesive categories which are not toposes. Here we prove that the converse is not true – indeed, the main contribution of the paper is Theorem 26, the conclusion of which states that toposes are adhesive. From a computer science perspective, this means that the theory developed for adhesive categories can be applied to any topos, not just a presheaf. This is a significant development, since topos theory is a well-established mathematical discipline with many important results and wide-reaching applications.

One interesting example of a category which is a topos and not a presheaf is the *Schanuel topos*. It has been used (see [8], for example) to model languages with name binding, such as the Pi-calculus. Our theorem thus allows us to apply the rewriting theory developed for adhesive categories to such a setting. While we do not study this example in detail within the present paper, we plan to study such systems as part of future work. The Schanuel topos actually arises as a full subcategory of a presheaf category with objects the (atomic) *sheaves*. It is a general fact that any such category of sheaves is a topos.

The proof of our theorem relies on exploiting the connections between adhesive categories (or more generally, van Kampen squares), Brown and Janelidze's generalised van Kampen theorems and Grothedieck's theory of descent. Indeed, in order to prove that toposes are adhesive, we must show that pushouts along monomorphisms are van Kampen. To do so, we split such a pushout into two: one pushout with all morphisms mono, and one with two monomorphisms and two epimorphisms. The former is also a pullback in any topos, and a theorem of Brown and Janelidze's (here recalled as Theorem 23) guarantees that it satisfies the van Kampen theorem – which implies that the original pushout is a van Kampen square. Here we also prove that pushouts of the latter kind are van

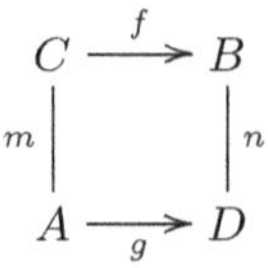

$$
\begin{array}{ccc}
C & \xrightarrow{\ f\ } & B \\
{\scriptstyle m}\big\downarrow & & \big\downarrow{\scriptstyle n} \\
A & \xrightarrow[\ g\ ]{} & D
\end{array}
$$

Fig. 1. Pushout diagram

Kampen in a topos, it is the most difficult and technical result of the paper. This completes the proof because, as we show in Lemma 2, van Kampen squares compose in categories with pullbacks and pushouts.

Structure of the paper. In Section 1 we recall two equivalent ways of defining van Kampen squares and show that van Kampen squares compose in categories with pushouts and pullbacks. We recall the definition of adhesive categories and prove that adhesive categories enjoy the special pullback-pushout property. Section 2 recalls the fragments of descent theory and topos theory necessary for our main result. In Section 3, we recall the theorem of Brown and Janelidze, which forms one of the ingredients of the proof of our main Theorem 26. Theorem 25 is the other main ingredient, and its proof relies on the background introduced in Section 2. We conclude in Section 4 with several directions for future work. The paper is relatively self-contained, although we omit the proofs of well-known results and instead provide references to standard sources.

1   Van Kampen Squares and Adhesive Categories

Adhesive and quasiadhesive categories are defined using certain pushout diagrams which are called van Kampen squares. We refer the reader to [11,12] for an introduction to, the enumeration of the basic properties of, and the applications of adhesive and quasiadhesive categories. Here we shall concentrate on the definitions of van Kampen squares and derive the properties which will be needed for the proof of our main result. We shall also prove that adhesive categories satisfy the so-called special pullback-pushout property, which is one of the many HLR axioms, the majority of which have already been shown in [11,12] to hold in any adhesive category.

We shall need both the axiomatic and the "equivalence of categories" versions of the definitions of van Kampen squares [12]. In particular, using the former, we shall show that van Kampen squares compose in categories with pushouts and pullbacks, and that adhesive categories satisfy the special pullback-pushout property. The fact that van Kampen squares compose, together with the latter, equivalent, way of defining van Kampen squares will be used in the proof of our main Theorem 26. The latter definition of van Kampen squares also makes it possible to establish a relationship between van Kampen squares and Brown and Janelidze's generalised van Kampen theorems, as we shall explain in §3.

Definition 1. A van Kampen square is a pushout diagram as in Fig 1 which satisfies the following condition:

- for any commutative cube, as illustrated, of which Fig 1 forms the bottom face and the back faces are pullbacks: the front faces are pullbacks iff the top face is a pushout.

The following lemma shows that, in categories with pushouts and pullbacks, van Kampen squares paste together to give van Kampen squares.

Lemma 2. *Consider the illustrated commutative diagram in a category with pushouts and pullbacks. If (1) and (2) are van Kampen then so is (1)+(2).*

Proof. Straightforward; in order to show that the combined pushout is stable under pullback it suffices to break up a cube into two cubes, using the existence of pullbacks. Conversely, a cube with its top face a pushout, can be split into two using the existence of pullbacks and pushouts. □

We shall now recall an equivalent definition of van Kampen squares which will be useful for the purposes of this paper. The reader is referred to [12] for the proof that the definitions are equivalent. The alternative definition is stated by saying that a certain functor, induced by the diagram in Fig 1, is required to be an equivalence of categories. We begin by defining the codomain category of the functor.

Definition 3. Let $\mathbf{C}/A \times_{\mathbf{C}/C} \mathbf{C}/B$ denote the category with objects commutative diagrams of pullbacks, as illustrated, and arrows the obvious morphisms between such diagrams.

For a morphism $u: U \to V$ we shall write $u : \mathbf{C}/V \to \mathbf{C}/U$ for the functor given by pulling back along u. Referring to Fig 1, the functors n and g induce a functor

$$\mathsf{Pb}: \mathbf{C}/D \to \mathbf{C}/A \times_{\mathbf{C}/C} \mathbf{C}/B.$$

Using the functor Pb, we can define the property of square (1) being van Kampen as follows:

Definition 4. The pushout diagram of Fig 1 is said to be van Kampen whenever one of the following equivalent conditions holds:

(i) Pb is an equivalence of categories;
(ii) the pushout is stable under pullback, and the functor Pb is essentially surjective on objects.[1]

---

[1] A functor $F: \mathbf{C} \to \mathbf{D}$ is said to be *essentially surjective on objects* when, for every object $D \in \mathbf{D}$, there exists an object $C \in \mathbf{C}$ such that $FC \cong D$.

Definition 5 (Adhesive categories). A category with pullbacks and pushouts along monomorphisms is said to be adhesive if any pushout square as in Fig 1, in which m is a monomorphism, is van Kampen.

Examples of adhesive categories include **Set** (see [11]) and the category of graphs **Graph**. The fact that the latter is adhesive follows from the fact that **Set** is adhesive and the fact that the functor category $[\mathbf{X}, \mathbf{C}]$ is adhesive whenever $\mathbf{C}$ is. Thus, in particular, any presheaf topos is adhesive. In §3 we shall show that any topos is adhesive, thus providing several new examples of adhesive categories – for instance, the Schanuel topos [8].

Adhesive categories have found an application as a foundation for parts of the theory of graph transformation. Indeed, it has been shown in [11, 12] that adhesive and quasiadhesive categories satisfy many of the previously proposed HLR-axioms [5]. Here we shall extend this thesis by showing that another of the aforementioned axioms holds in adhesive categories, the so called special pullback-pushout property. Actually, we are able to prove a more general property by requiring less assumptions about the arrows in the diagram below.

Lemma 6 (Special pullback-pushout property). *Suppose that the illustrated commutative diagram in an adhesive category has m, n and l mono. Suppose that (1) is a pushout and (1)+(2) is a pullback, then (2) is a pullback.*

$$\begin{array}{ccccc} A & \xrightarrow{f} & C & \xrightarrow{p} & E \\ {\scriptstyle m}\downarrow & (1) & {\scriptstyle n}\uparrow & (2) & \downarrow{\scriptstyle l} \\ B & \xrightarrow{g} & D & \xrightarrow{q} & F \end{array}$$

Proof. Suppose we have an object X and morphisms $\alpha\colon X \to D$ and $\beta\colon X \to E$ such that $q\alpha = l\beta$. We shall show that there exists $k\colon X \to C$ such that $nk = \alpha$ and $pk = \beta$. Notice that it suffices to show the existence of such a morphism, uniqueness follows since n is mono.

Construct the illustrated cube by taking pullbacks. Now $qg\alpha_1 = q\alpha g' = l\beta g'$, and we use the fact that $(1)+(2)$ is a pullback to derive the existence of a unique morphism $h\colon X_1 \to A$ such that $mh = \alpha_1$ and $pfh = \beta g'$.

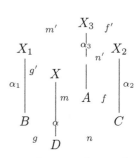

Also note that $m\alpha_3 = \alpha_1 m' = mhm'$, and using the fact that m is mono, $\alpha_3 = hm'$ (†). Also, $lp\alpha_2 = qn\alpha_2 = q\alpha n' = l\beta n'$. Since l is mono, we have that $p\alpha_2 = \beta n'$ (‡).

We shall use the fact that the top face of the cube is a pushout to derive the existence of the required morphism. Indeed, we have $\alpha_2 f' = fa_3 = fhm'$ where we used (†) to derive the last equality. Thus we get a unique $k\colon X \to C$ such that $kg' = fh$ and $kn' = \alpha_2$.

It remains to show that k satisfies the necessary properties, that is, $nk = \alpha$ and $pk = \beta$. Indeed, we have $nkg' = nfh = gmh = g\alpha_1 = \alpha g'$ and $nkn' = n\alpha_2 = \alpha n'$. Using the fact that the top face of the cube is a pushout, and in particular, the uniqueness of the mediating morphism, we have $nk = \alpha$.

Similarly, $pkg' = pfh = \beta g'$ and $pkn' = p\alpha_2 = \beta n'$, where we used (‡) to derive the last equality. This implies that $pk = \beta$ and we are finished.   □

2   Toposes and Descent

In order to prove that toposes are adhesive, we shall first recall the necessary background in this section: basic aspects of descent theory as well as the definition and several well-known properties of toposes.

We start in §2.1 by recalling a little of Grothendieck's theory of descent. We refer the reader to [9], for example, for a more detailed account. Historically, the theory arose from algebraic geometry and has had an impact on many disciplines within mathematics and computer science, including algebraic topology, logic and type theory. Descent theory, as we shall show, is also closely related to van Kampen squares.

In §2.2 we recall some elementary facts about toposes. See [10] for an overview of topos theory. Toposes have been widely used by mathematicians and computer scientists interested in logic, topology, geometry or category theory.

We relate these two topics by recalling that epimorphisms in toposes are effective for descent. This, in fact, is a consequence of a more general fact that regular epimorphisms in locally-cartesian closed categories are effective descent morphisms.

2.1   Descent

Recall that a morphism $p\colon X \to D$ is said to be a regular epimorphism if it is the coequaliser of some morphisms $w_1, w_2\colon W \to X$. We recall below a well-known lemma which relates regular epimorphisms and pushout diagrams.

Lemma 7. *If $p\colon X \to D$ has a kernel pair $p_1, p_2\colon P \to X$ (the diagram is a pullback square) then p is a regular epimorphism iff the pullback is also a pushout.*

$$\begin{array}{ccc} P & \xrightarrow{p_2} & X \\ {\scriptstyle p_1}\downarrow & & \downarrow{\scriptstyle p} \\ X & \xrightarrow{p} & D \end{array}$$

Proof. ($\Leftarrow$) If the diagram is a pushout square then it follows immediately that p is a coequaliser of p_1 and p_2;

($\Rightarrow$) If p is regular epi then it is the coequaliser of $w_1, w_2\colon W \to X$. Since the diagram above is a pullback, there exists a unique morphism $w\colon W \to P$ such that $p_1 w = w_1$ and $p_2 w = w_2$. From this it follows that p is the coequaliser of p_1 and p_2. Again using the fact that the square is a pullback, there exists an arrow $h\colon X \to P$ such that $p_1 h = \mathrm{id}_X$ and $p_2 h = \mathrm{id}_X$. It follows that for any $\alpha\colon X \to Y$ and $\beta\colon X \to Y$ if $\alpha p_1 = \beta p_2$ then $\alpha = \beta$. The universal property of pushouts thus follows from the universal property of coequalisers.   □

The conclusion of the following lemma ensures that the evident induced morphism to the vertex of any pushout diagram is a regular epimorphism.

Lemma 8. *Given a pushout diagram as in Fig 1 in a category with coproducts, $[g, n]\colon A + B \to D$ is a regular epimorphism.*

Proof. It is the coequaliser of the pair $i_1 m, i_2 f\colon C \to A + B$.   □

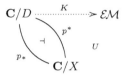

Fig. 2. The adjunction $p_* \dashv p^*$ and functor to the Eilenberg-Moore category $\mathcal{EM}$, with the obvious forgetful functor U.

We shall now recall some basic facts about descent. Let $\mathbf{C}$ be a category with pullbacks and $p\colon X \to D$ a morphism in $\mathbf{C}$. The pullback functor $p\colon \mathbf{C}/D \to \mathbf{C}/X$ always has a left adjoint, given by composition with p. Recall that the right adjoint $p\colon \mathbf{C}/D \to \mathbf{C}/X$ is said to be *monadic* if the unique comparison functor K from $\mathbf{C}/D$ to the Eilenberg-Moore category $\mathcal{EM}$ generated by the monad arising from the adjunction is an equivalence of categories (see Fig 2).

The following definition states when a morphism is said to be *effective for descent*. The intuitive idea is that, given an effective descent morphism $p\colon X \to D$, one can reason about the structure of a category $\mathbf{C}/D$, which may be difficult, by reasoning about certain algebras over $\mathbf{C}/X$ – thus in a sense "descending" along p. Indeed, referring to the diagram of Fig 2, to say that p is effective for descent is to say that the comparison functor $K\colon \mathbf{C}/D \to \mathcal{EM}$, where $\mathcal{EM}$ is the Eilenberg-Moore category (category of algebras) of the monad with endofunctor $p\,p\colon \mathbf{C}/X \to \mathbf{C}/X$, is an equivalence of categories; ie p is monadic.

Definition 9. If p is monadic, we shall say either that p is *effective for descent* or that it is an *effective descent* morphism.

In the particular case of the monad $p \dashv p$, the Eilenberg-Moore category can be characterised as a category of certain pullback diagrams in $\mathbf{C}$. This characterisation will prove useful in the proof of our main result.

Lemma 10. *The Eilenberg-Moore category $\mathcal{EM}$ of the monad induced by the adjunction $p \dashv p$ is the category whose:*

- *objects are diagrams of pullbacks, as illustrated in the left diagram of Fig 3, where $p_1, p_2\colon P \to X$ is the kernel pair of p;*
- *arrows are pairs α, β which combine into a commutative diagram, as illustrated in the other diagram of Fig 3.*

The comparison functor $K\colon \mathbf{C}/D \to \mathcal{EM}$ (see Fig 2) takes an object $d\colon D' \to D$ of $\mathbf{C}/D$ to the rear faces of the cube of pullbacks formed from d and the pullback diagram of Lemma 7.

Proof. See [9, §3.4]. Note that the category discussed there is the category of descent data: its objects are pairs $\langle x\colon X' \to X, \, \xi\colon p_1 x \to p_2 x \rangle$ where ξ is an iso in $\mathbf{C}/P$. This category is clearly isomorphic to the category described above.   □

Remark 11. The category $\mathcal{EM}$ is *not* the same as $\mathbf{C}/X \times_{\mathbf{C}/P} \mathbf{C}/X$ of Definition 3. Indeed, if p is a regular epimorphism and we start with the pushout diagram of

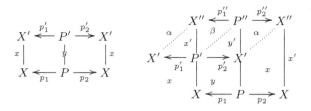

Fig. 3. Objects and morphisms of $\mathcal{EM}$

Lemma 7, then it makes sense to compare the two categories. While the objects of both are pairs of pullback diagrams, an object of the latter category can potentially involve two maps $X' \to X$ rather than one.

The next lemma relates the effective descent morphisms and regular epimorphisms. The two classes coincide in any locally cartesian closed category – a category $\mathbf{C}$ in which every slice $\mathbf{C}/C$ is cartesian closed.

Lemma 12. *In a locally cartesian closed category, a morphism is effective for descent (cf Definition 9) iff it is a regular epimorphism.*

Proof. See [9], for example.   □

We have the following useful fact as a consequence of Lemmas 12 and 8. It states that, in a category with coproducts, the evident morphism induced by an arbitrary pushout is effective for descent.

Corollary 13. *Given a pushout diagram of Fig 1 in a locally cartesian closed category with coproducts, the induced arrow $[g, n] \colon A + B \to D$ is an effective descent morphism.*   □

Now letting $p = [g, n]$ and using Corollary 13, Lemma 10, the comparison functor $K \colon \mathbf{C}/D \to \mathcal{EM}$ which takes an object $d \colon D' \to D$ to the back faces of the cube of pullbacks, as illustrated, is an equivalence of categories. This fact, and in particular the fact that K is essentially surjective on objects, will form an important part of the proof of Theorem 25, the hardest step of the proof of our main Theorem 26.

2.2   Toposes

Here we list the properties of toposes which we shall use to prove our main theorem. We refer the reader to [10] for a more thorough account of topos theory. We give a standard definition of toposes below; note that the actual statement of the definition is not important for the purposes of this paper, instead we shall list the precise properties of toposes we require in the remainder of this section.

Definition 14. A topos is a category **C** which:

   (i) is cartesian closed and has equalisers (and consequently, all finite limits);
  (ii) has a subobject classifier.

It follows from the axioms above that toposes have finite colimits [10, A2.2.9] and are locally cartesian closed [10, A2.3.4]. In particular, the latter implies that:

Proposition 15. *The pullback functors* $u : \mathbf{C}/V \to \mathbf{C}/U$ *have right adjoints, and so preserve all colimits.*   □

The proof of Theorem 25, the hardest step in the proof of our main result, relies on the fact that toposes are *extensive* [4]. Extensive categories can be said to have well-behaved coproducts, in a similar sense to how pushouts along monomorphisms can be said to be well-behaved in adhesive categories. Here we give the (axiomatic) definition and a well-known characterisation.

Definition 16 (Extensive categories). A category **C** is said to be *extensive* if it has finite coproducts, pull-backs along coproduct injections, and satisfies the following property:

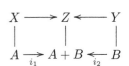

  – given a commutative diagram, as illustrated, the top row is a coproduct diagram iff the two squares are pullbacks.

The following result states two properties of coproducts in extensive categories which, if they hold in arbitrary categories, are enough to show extensivity. Interestingly, it is unknown whether a similar characterisation can be given for adhesive categories.

Proposition 17. *A category with finite coproducts and pullbacks along coproduct injections is extensive iff (1) it has coproducts which are stable under pullback and (2) pulling back one coproduct injection along the other results in the initial object.*[2]

Proof. See [4, Proposition 2.14].   □

We shall now recall some other well-known properties of toposes, one of which is that toposes are extensive:

Lemma 18. *If* **C** *is a topos then:*

   (i) *epimorphisms in* **C** *are regular and are stable under pullback;*
  (ii) *monomorphisms in* **C** *are regular and are stable under pushout;*
 (iii) *pushouts along monomorphisms in* **C** *are pullbacks;*

---

[2] Coproducts are said to be *disjoint* if they satisfy property (2) and coproduct injections are monomorphisms. When coproducts are stable under pullback, the fact that injections are monomorphisms is derivable from (2) (cf [4, Lemma 2.13]).

(iv) **C** *is extensive;*

(v) all arrows $f: A \to B$ in **C** *can be factorised into an epimorphism $e: A \to C$ followed by a monomorphism $m: C \to B$ with $me = f$; moreover, the factorisation is unique up to isomorphism in the obvious way.*

Proof. For the first parts of *(i)* and *(ii)* see [10, A1.4.9]. The fact that epimorphisms are stable under pullback follows from Proposition 15. For the second part of *(ii)* and for *(iii)* see [10, A2.4.3]. For part *(iv)*, we know from Proposition 17 that it is enough to check for stability of coproducts under pullback and disjointness. For disjointness see [10, A2.4.4], stability follows from Proposition 15. For part *(iv)* see [10, A2.3.5].   □

Using Lemma 12 and part *(i)* of Lemma 18, we obtain the following well-known result:

Corollary 19. *In a topos, the classes of epimorphisms and effective descent morphisms coincide.*   □

We shall require one technical lemma, which holds in toposes. It concerns certain diagrams of pullbacks – indeed, it is well known that given a diagram (as illustrated below) where the right square is a pullback then the left square is a pullback if and only if the exterior of the diagram is a pullback. The lemma below gives a sufficient condition for the right square being a pullback when the exterior of the diagram and the left square are pullbacks.

Lemma 20. *Consider in a topos a diagram, as illustrated, with g an epimorphism. If the left square and the exterior are pullbacks then so is the right square.*

$$
\begin{array}{ccccc}
A & \xrightarrow{f} & C & \xrightarrow{p} & E \\
{\scriptstyle s}\downarrow & & {\scriptstyle t}\downarrow & & {\scriptstyle u}\downarrow \\
B & \twoheadrightarrow{g} & D & \xrightarrow{q} & F
\end{array}
$$

Proof. This follows immediately from [2, Lemma 4.6] and Corollary 19. Indeed, as shown in [2], it suffices to require that g is an effective descent morphism.   □

We conclude this section by recalling the notion of an equivalence relation in a cartesian category, a generalisation of the usual notion of equivalence relation, and recalling that equivalence relations in toposes are *effective*.

Definition 21 (Equivalence relation). Suppose that **C** is a category with finite products. By a *relation*, we mean a monomorphism $\langle a, b \rangle : R \to A \times A$. A relation is said to be an *equivalence relation* if all three of the following hold:

- it is *reflexive*: there exists a morphism $r: A \to R$ such that $ar = br = \mathrm{id}_A$;
- it is *symmetric*: there exists a morphism $s: R \to R$ such that $as = b$ and $bs = a$;
- it is *transitive*: referring to the illustrated pullback diagram, there exists a morphism $t: P \to R$ such that $at = ap$ and $bt = bq$.

$$
\begin{array}{ccc}
P & \xrightarrow{q} & R \\
{\scriptstyle p}\downarrow & & {\scriptstyle a}\downarrow \\
R & \xrightarrow{b} & A
\end{array}
$$

Proposition 22. *In a topos, equivalence relations are* effective, *that is, they are the kernel pairs of their coequaliser.*

Proof. See [10, A2.4.1]. □

3 Toposes are Adhesive

Having recalled the necessary background theory, in this section we shall prove that toposes are adhesive (Theorem 26) which is the main technical contribution of the paper. Recall from [12] that the converse does not hold – indeed there are adhesive categories which are not toposes (for instance the category of pointed sets **Set**).

The proof itself relies on the fact that, in a topos, a pushout along a monomorphism can be broken up into two pushouts – (1) one with all arrows monomorphisms and (2) one with two monomorphisms and two epimorphisms.

Using the fact that van Kampen squares compose (cf Lemma 2), it suffices to show that, in toposes, pushouts of kinds (1) and (2) are van Kampen squares. The fact that pushouts of kind (2) are van Kampen (Theorem 25) is the most difficult and technical part of our proof. The fact that pushouts of kind (1) are van Kampen squares follows immediately from a well-known theorem of Brown and Janelidze [1]:

Theorem 23 (Brown and Janelidze). *Suppose that* **C** *is an extensive category with finite limits. Given a pullback diagram, as illustrated, with all morphisms mono, the induced (cf paragraph following Definition 3) functor* $\mathsf{Pb}\colon \mathbf{C}/D \to \mathbf{C}/A \times_{\mathbf{C}/C} \mathbf{C}/B$ *is an equivalence of categories if and only if the map* $[g,n]\colon A + B \to D$ *induced by* g *and* n *is an effective descent morphism.*

$$
\begin{array}{ccc}
C & \xrightarrow{f} & B \\
m\downarrow & & \downarrow n \\
A & \xrightarrow{g} & D
\end{array}
$$

Proof. See [1, Proposition 3.2]. □

As an immediate application of the above theorem, we are able to show that pushouts in toposes with all arrows monomorphisms are van Kampen.

Corollary 24. *A pushout as in Fig 1 in a topos, with all arrows monomorphic, is van Kampen.*

Proof. First note that by Lemma 18(*iii*), such a pushout is also a pullback, and by Corollary 13 we know that the arrow $[g,n]$ induced by the pushout is a (regular) epimorphism. Toposes have finite limits and are extensive (cf Lemma 18(*iv*)), thus we can apply Theorem 23 to obtain that Pb is an equivalence of categories – in other words, the pushout is van Kampen (cf Definition 4). □

The second class of pushouts we shall consider are pushouts where two of the morphisms are epimorphisms and two are monomorphisms. The following fact is the most technical part of our main result:

Theorem 25. *A pushout as in Fig 1 in a topos, with* f *(and so* g*) epimorphic and* m *and* n *monomorphic, is van Kampen.*

Proof. Using the second part of Definition 4 and the stability of pushouts under pullback (cf Proposition 15), it will suffice to show that the functor $\mathsf{Pb}\colon \mathbf{C}/D \to \mathbf{C}/A \times_{\mathbf{C}/C} \mathbf{C}/B$ induced by such a pushout is essentially surjective on objects. In other words, given a diagram as in Definition 3 with both squares pullbacks, we must find a map $d\colon D' \to D$ whose pullbacks along g and n are, respectively, a and b. By extensivity (cf Lemma 18(*iv*)), this amounts to finding d whose pullback along $p = [g, n]\colon A + B \to D$ is $a + b\colon A' + B' \to A + B$.

But, by Corollary 13, p is an effective descent map. Using the fact that $K\colon \mathbf{C}/D \to \mathcal{EM}$ is essentially surjective on objects (cf paragraph following Corollary 13), it suffices to show that the pullback of $a + b$ along p_1 coincides with its pullback along p_2,

$$
\begin{array}{ccccc}
A' + B' & \xleftarrow{\;p'_1\;} & P' & \xrightarrow{\;p'_2\;} & A' + B' \\
{\scriptstyle a+b}\big\downarrow & & \big\downarrow & & \big\downarrow{\scriptstyle a+b} \\
A + B & \xleftarrow[\;p_1\;]{} & P & \xrightarrow[\;p_2\;]{} & A + B
\end{array}
$$

where $p_1, p_2\colon P \to A + B$ are the projections of the kernel pair of p – thus showing that the diagram is an object of the Eilenberg-Moore category $\mathcal{EM}$. By extensivity, P is given by $A_2 + C + C + B$, where $g_1, g_2\colon A_2 \to A$ is the kernel pair of $g\colon A \to D$. It follows that the projections themselves are:

$$
p_1 = [g_1, m] + [f, \mathrm{id}_B]\colon A_2 + C + C + B \to A + B
$$

$$
p_2 = [g_2 + f,\; m + \mathrm{id}_B]\colon A_2 + C + C + B \to A + B.
$$

Using extensivity once more, to show that the pullbacks of $a + b$ along p_1 and p_2 agree, it suffices to show that the pullbacks along each of the components of p_1 and p_2 agree. And since m a and f b agree, all that remains is to check that $g_1 a$ and $g_2 a$ agree. To do so, we form the pullback in diagram (*i*) below and then show that the squares of diagram (*ii*) are pullbacks.

$$
\begin{array}{ccc}
A'_2 \xrightarrow{\;\langle g'_1, g'_2 \rangle\;} A' \times A' & \qquad A'_2 \xrightarrow{\;g'_i\;} A' & \qquad A' \xrightarrow{\;g'_i d'\;} A' \\
{\scriptstyle a_2}\big\downarrow \qquad\qquad \big\downarrow{\scriptstyle a\ \ a} & \qquad {\scriptstyle a_2}\big\downarrow \qquad \big\downarrow{\scriptstyle a} & \qquad {\scriptstyle a}\big\downarrow \qquad\quad \big\downarrow{\scriptstyle a} \\
A_2 \xrightarrow[\;g_1, g_2\;]{} A \times A & \qquad A_2 \xrightarrow[\;g_i\;]{} A & \qquad A \xrightarrow[\;g_i d\;]{} A \\
(i) & \qquad (ii) & \qquad (iii)
\end{array}
$$

Let $d\colon A \to A_2$ be the unique map satisfying $g_1 d = g_2 d = \mathrm{id}_A$, and similarly let $d'\colon A' \to A'_2$ be the unique map satisfying $g'_1 d' = g'_2 d' = \mathrm{id}_{A'}$ and $a_2 d' = da$. Then the squares of diagram (*iii*) are clearly pullbacks.

Let $f_1, f_2\colon C_2 \to C$ be the kernel pair of f. Then there are pullback squares (*iv*), and so pullback squares (*v*). Let $m_2\colon C_2 \to A_2$ be the unique map satisfying $g_i m_2 = m f_i$ for $i = 1$ and 2. Similarly, let $m'_2\colon C'_2 \to A'_2$ be the unique map satisfying $g'_i m'_2 = m' f'_i$ for $i = 1$ and 2 as well as $a_2 m'_2 = m_2 c_2$. Thus we get the pullback squares (*vi*).

$$C_2' \xrightarrow{f_i'} C' \xrightarrow{f'} B' \qquad C_2' \xrightarrow{f_i'} C'' \xrightarrow{m'} A' \qquad C_2' \xrightarrow{g_i'm_2'} A' \qquad A' + C_2' \xrightarrow{[d',m_2']} A_2' \xrightarrow{g_i'} A'$$

(with vertical maps c_2, c, b / c_2, c, a / c_2, a / $a+c_2, a_2, a$)

$$C_2 \xrightarrow[f_i]{} C \xrightarrow[f]{} B \qquad C_2 \xrightarrow[f_i]{} C \xrightarrow[m]{} A \qquad C_2 \xrightarrow[g_i m_2]{} A \qquad A + C_2 \xrightarrow[[d,m_2]]{} A_2 \xrightarrow[g_i]{} A$$

$$(iv) \qquad\qquad (v) \qquad\qquad (vi) \qquad\qquad (vii)$$

Using extensivity and the fact that diagrams (iii) and (vi) are pullbacks, the exteriors and the left hand squares of diagram (vii) are pullbacks, so that the right hand squares will be pullbacks, and the proof complete, provided that $[d, m_2]\colon A + C_2 \to A_2$ is an epimorphism (cf Lemma 20).

To see that $[d, m_2]$ is an epimorphism, consider the map $[\Delta, \langle g_1 m_2, g_2 m_2\rangle]\colon$ $A + C_2 \to A \times A$ induced by the diagonal $\Delta\colon A \to A \times A$ and $\langle g_1 m_2, g_2 m_2\rangle\colon C_2 \to A \times A$, and factorise it as an epimorphism $[h_1, h_2]\colon A + C_2 \to R$ followed by a monomorphism $\langle r_1, r_2\rangle\colon R \to A \times A$. We shall show that R is A_2, with $h_1 = d$ and $h_2 = m_2$, so that $[d, m_2]$ is an epimorphism, as required.

If we regard R as a relation on A, it is clearly reflexive, since by construction it contains the diagonal; it is symmetric, since the relations A and C_2 are so. The pullback $(A + C_2) \times_A (A + C_2)$ is given by $A + C_2 + C_2 + C_3$, where $C_3 = C \times_B C \times_B C$, and the "composition" map $C_3 \to C_2$ sending a triple (c_1, c_2, c_3) of generalized elements of C to (c_1, c_3), induces an evident map $A + C_2 + C_2 + C_3 \to A + C_2$, which in turn induces a map $R \circ R \to R$ showing that the relation R is transitive and so an equivalence relation.

In a topos, an equivalence relation is the kernel pair of its coequaliser (cf Proposition 22), but the coequaliser of $r_1, r_2\colon R \to A$ is the coequaliser of the maps $r_1[h_1, h_2], r_2[h_1, h_2]\colon A + C_2 \to A$, since $[h_1, h_2]$ is epi. This in turn is the coequaliser of the maps $g_1 m_2$ and $g_2 m_2$ and so, using the definition of the g_i, it is the coequaliser of $m f_1, m f_2\colon C_2 \to A$.

$$\begin{array}{ccc} C_2 & \xrightarrow{f_2} & C \\ {\scriptstyle f_1}\downarrow & & \downarrow{\scriptstyle f} \\ C & \xrightarrow{f} & B \end{array}$$

Using the fact that f is epi, the diagram to the right is a pushout (cf Lemma 7). Thus a map $w\colon A \to W$ satisfying $wmf_1 = wmf_2$ induces a unique map $v\colon B \to W$ satisfying $vf = wm$; and so a unique map $u\colon D \to W$ satisfying $ug = w$ and $un = v$. Clearly $g\colon A \to D$ coequalises $m f_1, m f_2$; using the universal property of coequalisers we obtain that u is an isomorphism. This proves that the coequaliser of the projections of R is $g\colon A \to D$, and so that R is the kernel pair of g; but the kernel pair of g is A_2, and this now proves that $[d, m_2]\colon A + C_2 \to A_2$ is an epimorphism, as claimed. □

We are now able to combine these results in order to deduce our main contribution:

Theorem 26. *Toposes are adhesive.*

Proof. Consider the pushout of Fig 1 in a topos $\mathbf{C}$, with m a monomorphism. We shall show that it is a van Kampen square. As a consequence of Proposition 15, all pushouts are stable under pullback.

By parts (*iii*) and (*ii*) of Lemma 18, such a pushout is also a pullback and the map n is also a monomorphism. Factorise $g\colon A \to D$ as a epimorphism $q\colon A \to F$ followed by a monomorphism $k\colon F \to D$, and form the pullback squares as illustrated. It follows immediately that j is a monomorphism. Using Lemma 18(*i*), r is an epimorphism.

$$
\begin{array}{ccc}
C & \xrightarrow{r} E & \xrightarrow{j} B \\
m\downarrow & (\ ) \quad l\downarrow\ (\ ) & \downarrow n \\
A & \xrightarrow{q} F & \xrightarrow{k} D
\end{array}
$$

The exterior of the diagram above is a pushout by assumption. We know by Proposition 15 that pushouts are stable under pullback – the stability of this pushout under pullback along k implies that square (†) is also a pushout, and so square (‡) is also a pushout by the usual cancellation properties of pushouts.

If each of these squares is van Kampen then the conclusion of Lemma 2 implies that so is the exterior; thus it will suffice to consider separately square (†) with r and q epimorphisms and m and l monomorphism, and square (‡) with l, n, j and k all monomorphisms. The fact that the latter is van Kampen follows from Corollary 24, while the fact that the former is van Kampen follows from Theorem 25. □

Remark 27. Recall from [11] that the converse of Theorem 26 does not hold. Indeed, adhesive categories are closed under the coslice construction and thus in general are not even extensive.

4 Conclusion

Throughout the paper we have concentrated on the class of adhesive categories which has many examples of interest to computer scientists, in particular those interested in the theory of graph transformation. We have shown that adhesive categories satisfy the special pullback-pushout lemma, which was previously taken as one of the HLR axioms.

Our main result is that toposes are adhesive; the proof relies on exploiting the relationship between van Kampen squares, descent theory [9] and Brown and Janelidze's work [1] on generalised van Kampen theorems. More concretely, we prove that pushouts along monomorphisms in toposes are van Kampen by splitting them into two pushouts and proving that each is van Kampen – the fact that one is van Kampen follows from Brown and Janelidze's well-known theorem and the proof of the other relies on the fact that epimorphisms in toposes are effective for descent.

In future work, we plan to study the ramifications of the fact that toposes are adhesive by using the rewriting theory developed for adhesive categories to study languages with name-passing which are modelled using the Schanuel topos. We also plan to extend our main theorem to show that certain classes of quasitoposes [14] are quasiadhesive. Such a result would not only prove to be of theoretical interest, but would also allow us simple proofs of the quasiadhesivity of many categories of interest to the graph transformation community. This is because it is possible to show that they arise via so called *Artin gluing* [3].

References

1. R. Brown and G. Janelidze. Van Kampen theorems for categories of covering morphisms in lextensive categories. *Journal of Pure and Applied Algebra*, 119:255–263, 1997.

2. A. Carboni, G. Janelidze, G. M. Kelly, and R. Paré. Localization and stabilization for factorization systems. *Applied Categorical Structures*, 5(1):1–58, 1997.

3. A. Carboni and P. T. Johnstone. Connected limits, familial representability and Artin glueing. *Mathematical Structures in Computer Science*, 5:441–449, 1995.

4. A. Carboni, S. Lack, and R. F. C. Walters. Introduction to extensive and distributive categories. *Journal of Pure and Applied Algebra*, 84(2):145–158, February 1993.

5. H. Ehrig, A. Habel, H.-J. Kreowski, and F. Parisi-Presicce. Parallelism and concurrency in high-level replacement systems. *Mathematical Structures in Computer Science*, 1, 1991.

6. H. Ehrig, A. Habel, J. Padberg, and U. Prange. Adhesive high-level replacement categories and systems. In *Proceedings of the 2nd International Conference on Graph Transformation ICGT '04*, 2004.

7. H. Ehrig and B. König. Deriving bisimulation congruences in the DPO approach to graph rewriting. In *Foundations of Software Science and Computation Structures FoSSaCS '04*, volume 2987 of *Lecture Notes in Computer Science*, pages 151–166. Springer, 2004.

8. M. Fiore and S. Staton. Comparing operational models of name-passing process calculi. *Information and Computation*, 2005. To appear.

9. G. Janelidze and W. Tholen. Facets of descent, I. *Applied Categorical Structures*, 2:245–281, 1994.

10. P. T. Johnstone. *Sketches of an Elephant: A topos theory compendium, vol 1.* Clarendon Press, 2002.

11. S. Lack and P. Sobociński. Adhesive categories. In *Foundations of Software Science and Computation Structures, FoSSaCS '04*, volume 2987 of *Lecture Notes in Computer Science*, pages 273–288. Springer, 2004.

12. S. Lack and P. Sobociński. Adhesive and quasiadhesive categories. *Theoretical Informatics and Applications*, 39(3):511–546, 2005.

13. V. Sassone and P. Sobociński. Reactive systems over cospans. In *Logic in Computer Science, LiCS '05*, pages 311–320. IEEE Press, 2005.

14. O. Wyler. *Lecture Notes on Topoi and Quasitopoi.* World Scientific, 1991.

Graph Transactions as Processes[*]

Paolo Baldan[1], Andrea Corradini[2], Luciana Foss[2,3,**], and Fabio Gadducci[2]

[1] Dipartimento di Informatica, Università Ca' Foscari di Venezia, Italy
[2] Dipartimento di Informatica, Università di Pisa, Italy
[3] Instituto de Informática, Universidade Federal do Rio Grande do Sul, Brasil

Abstract. *Transactional graph transformation systems* (T-GTSs) have been recently proposed as a mild extension of the standard DPO approach to graph transformation, equipping it with a suitable notion of *atomic* execution for computations. A typing mechanism induces a distinction between stable and unstable items, and a *transaction* is defined as a shift-equivalence class of computations such that the starting and ending states are stable and all the intermediate states are unstable.

The paper introduces an equivalent, yet more manageable definition of transaction based on graph processes. This presentation is used to provide a universal characterisation for the class of transactions of a given T-GTS. More specifically, we show that the functor mapping a T-GTS to a graph transformation system having as productions exactly the transactions of the original T-GTS is the right adjoint to an inclusion functor.

Keywords: Graph processes, refinement, transactions, zero-safe nets.

1 Introduction

Graph transformation systems (GTSs) are a flexible formalism for the specification of complex systems, that may take into account aspects such as object-orientation, concurrency, mobility and distribution [9,10]. In fact, graphs can be naturally used to provide a structured representation of the states of a system, which highlights its subcomponents and their logical or physical interconnections. Then, the events occurring in the system, which are responsible for the evolution from one state into another, are modelled as the application of suitable transformation rules. Such a representation is precise enough to allow the formal analysis of the system under scrutiny, as well as amenable of an intuitive, visual representation, which can be easily understood also by a non-expert audience.

Along the years several enrichments of the original framework have been introduced, extending GTSs with structuring concepts that are needed to master the complexity of large specifications. Several modularity and refinement notions have been proposed, providing basic mechanisms for encapsulation, abstraction and information hiding (see, e.g., [11,14,13]).

---

[*] Supported by the CNPq-CNR IQ-MOBILE II, the EC RTN 2-2001-00346 SEGRAVIS, the EU IST-2004-16004 SENSORIA and the MIUR PRIN 2005015824 ART.
[**] Supported by CAPES and CNPq.

A. Corradini et al. (Eds.): ICGT 2006, LNCS 4178, pp. 199–214, 2006.

In a *top-down* approach to the specification of a complex system, one can start describing each operation of the system as a single "abstract" rule. Then, each abstract rule is refined to a computation, describing in a more concrete way the activity performed and possibly the use of temporary resources. In order to guarantee that the behaviour of the refined system is correct with respect to the abstract specification, each computation corresponding to an abstract rule has to be executed "atomically", i.e., either it completes successfully, or the effects of a partial execution should not be visible at the abstract level: in one word, the computation refining an abstract rule must be a *transaction*.

The notion of transaction has been originally defined and studied in the realm of database management systems, and only later it has been considered in programming and specification formalisms, like process calculi, programming languages and Petri nets. A transaction represents a unit of interaction with the management system, that is treated in a coherent and reliable way, independently of other transactions, and that must be either entirely completed or aborted. Ideally, the following *ACID properties* should be guaranteed for each transaction

- *Atomicity*: either all of the tasks of a transaction are performed (and the transaction is *committed*) or none of them are;
- *Consistency*: the database is in a legal state when the transaction begins and when it ends;
- *Isolation*: no operation outside the transaction can see the data in an intermediate state;
- *Durability*: the effects of a committed transaction are persistent.

The above properties are also meaningful for characterising transactions in specification formalisms of concurrent/distributed systems, where the interaction now occurs with the environment: *atomicity*, *consistency* and *isolation* carry on with equal relevance, while only *durability* does not have a clear meaning anymore since no persistent repository of data is modelled.

Transactions can be introduced in different ways in a modelling, specification or programming formalism. In *control-centered formalisms*, like process calculi and programming languages, where the execution of computations is ruled by expressive control mechanisms, typically new control structures are introduced for starting/committing transactions. In *data-centered formalisms*, like rewriting formalisms and (possibly High-Level) Petri nets, where the control structures are typically poor and the emphasis is on the structure of the state that evolves during a computation, transactions are more naturally defined indirectly, by identifying parts of the state which represent temporary (or "unstable") resources, only visible within a transaction. This is the approach that has been taken for *zero-safe nets* [4], which is a reference model for our work on transactional GTSs.

Zero-safe nets are Place/Transition Petri nets equipped with a distinguished subset of *zero places*. The places model resources that are consumed or produced by transitions and the zero places model resources that are invisible to the exterior of a step. A step in a zero-safe net starts at a *stable marking* (i.e., containing no zero places), evolves through *unstable markings* and ends in a stable marking. Stable tokens produced in a step are "frozen" and delivered at the end.

Inspired by the work on zero-safe nets, *transactional graph transformation systems* (T-GTSs), introduced in [1], are a mild extension to the double-pushout (DPO) approach to graph transformation, providing a simple way of expressing transactional activities. The basic tool is a typing mechanism for graphs which induces a distinction between *stable* and *unstable* graph items. Given a typed graph, representing a system state, we can identify a subgraph which represent its "stable" part, i.e., the fragment of the state which is visible from an external observer. Transactions in a T-GTS are thus abstract, "minimal" computations starting from a completely stable graph, evolving through graphs with unstable items and eventually ending up in a new stable state.

In this paper we elaborate further on transactional GTSs. At first we obtain an alternative characterisation of transactions as *graph processes*, by exploiting the results in [2]. Next we show how the internal structure of transactions can be abstracted away, by considering an *abstract* GTS associated to the T-GTS: unstable items disappear and each distinct transaction becomes a single atomic production, which rewrites the starting stable state to the final stable state.

The main result of the paper shows that the operation mapping each T-GTS to its abstract counterpart is characterised as a universal construction in the categorical setting. More specifically, such construction is turned into a functor between the corresponding categories of systems, which is right adjoint to the inclusion functor in the opposite direction. The result is obtained by equipping T-GTSs with a notion of *implementation morphism*, allowing to map a single production to a whole transaction. This provides a solid theoretical justification to the notion of abstract GTS associated to a T-GTS: according to an intuitive interpretation of categorical adjunctions, it states that the constructed abstract GTS is the best approximation of the given T-GTS in the class of ordinary GTSs.

2   Double-Pushout Rewriting

This section briefly summarises the basics of double-pushout (DPO) graph rewriting [8] for directed (multi-)graphs (but definitions and results of the paper generalise easily, for example, to hypergraphs, which are used indeed in the examples). Without loss of generality, as shown in [12], we consider rewriting with *injective matches* only. Graphs are equipped with a *typing morphism* to a fixed *type graph*, which plays an essential role when distinguishing between stable and unstable items in a given graph.

Formally, a *graph* is a tuple $\langle V, E, s, t \rangle$, where V and E are the (disjoint) sets of nodes and edges, and $s, t: E \to V$ are the source and target functions. Sometimes, abusing the notation, G denotes the disjoint union $V_G \uplus E_G$; e.g. writing $x \in G$ means that x is either a node or an edge of the graph G. Given a graph T, a *typed graph* G over T is a graph $|G|$, together with a graph morphism $t_G: |G| \to T$. A *morphism* between T-typed graphs $f: G_1 \to G_2$ is a graph morphism $f: |G_1| \to |G_2|$ respecting the typing, i.e., such that $t_{G_1} = t_{G_2} \circ f$. The category of T-typed graphs and typed graph morphisms is denoted by $T\text{-}\mathbf{Graph}$.

Rewriting rules, called *T-typed productions*, are tuples $q: L_q \xleftarrow{l_q} K_q \xrightarrow{r_q} R_q$, where q is the name of the production, L_q, K_q and R_q are T-typed graphs (called the *left-hand side*, the *interface* and the *right-hand side* of the production, respectively), and l_q, r_q are injective morphisms. Without loss of generality, we always assume that l_q is an inclusion.

A rule q specifies that an occurrence of the left-hand side L_q in a larger graph can be rewritten into the right-hand side R_q, preserving the interface K_q.

Formally, given a typed graph G, a production q, and an injective *match* $g: L_q \to G$, a *direct derivation* δ from G to H using q, g exists, written $\delta: G \xRightarrow{q,g} H$, if the diagram to the right can be constructed, where both squares are pushouts in T-**Graph**.

$$q: L_q \xleftarrow{l_q} K_q \xrightarrow{r_q} R_q$$
$$g \downarrow \qquad \downarrow k \qquad \downarrow h$$
$$G \xleftarrow{b} D \xrightarrow{d} H$$

A graph transformation system is then defined as a collection of rules, over a fixed graph of types.

Definition 1 (graph transformation system). *A T-typed graph transformation system (*GTS*) is a tuple $\mathcal{G} = \langle T, P, \pi \rangle$, where T is a graph, P is a set of production names and π is a function mapping production names in P to T-typed productions.*

A *derivation* in a GTS $\mathcal{G}$ is a sequence of direct derivations via productions of $\mathcal{G}$

$$G_0 \xRightarrow{q_0,g_0} G_1 \xRightarrow{q_1,g_1} \dots\dots \xRightarrow{q_n,g_n} G_{n+1}.$$

A two-steps derivation $G \xRightarrow{q_1,g_1} X \xRightarrow{q_2,g_2} H$ as in the diagram below is called *sequential independent* [8,12] if there are two morphisms $s: L_2 \to D_1$ and $u: R_1 \to D_2$ such that $d_1 \circ s = g_2$ and $b_2 \circ u = h_1$. Intuitively, the images in X of the left-hand side of q_2 and of the right-hand side of q_1 overlap only on items that are preserved by both derivation steps.

$$L_1 \xleftarrow{l_1} K_1 \xrightarrow{r_1} R_1 \qquad\qquad L_2 \xleftarrow{l_2} K_2 \xrightarrow{r_2} R_2$$
$$g_1 \downarrow \quad k_1 \downarrow \quad \searrow^{s} \quad \nearrow \quad \swarrow \quad u \quad \downarrow k_2 \quad \downarrow h_2$$
$$G \xleftarrow{b_1} D_1 \xrightarrow{d_1} X \xleftarrow{b_2} D_2 \xrightarrow{d_2} H$$

In this case, according to the Parallelism Theorem (Theorem 7.8 in [12]), we can apply to G a suitably defined *proper quotient* q of the parallel rule $q_1 + q_2$, obtaining an equivalent direct derivation from G to H via an injective match g. Furthermore, there is an equivalent derivation $G \xRightarrow{q_2,g_2'} X' \xRightarrow{q_1,g_1'} H$ where the two derivation steps are "switched". The equivalence on derivations induced by switchings of sequential independent direct derivations is called *shift-equivalence* [8].

We now equip GTSs with a suitable notion of morphism, allowing us to look at them as objects of a category. This is essential to provide a characterisation of some interesting constructions with universal properties, as shown in Section 5. We shall use a variant of the morphisms in [6,3], where the type graphs are related by a partial morphism rather than by an arbitrary span.

A *partial morphism* $f: G_1 \rightharpoonup G_2$ is a total morphism from a subgraph of G_1, called $dom(f)$, to G_2, and is equivalently depicted as $G_1 \overset{l_f}{\hookleftarrow} dom(f) \overset{r_f}{\rightarrow} G_2$. Given an object A of a category $\mathcal{C}$, the *slice category* $\mathcal{C}{\downarrow}A$ has all $\mathcal{C}$-arrows with target A as objects; an arrow $h: f \rightarrow g$ in $\mathcal{C}{\downarrow}A$ is a $\mathcal{C}$-arrow h such that $g \circ h = f$.[1]

Let $m: A \rightarrow B$ be an arrow in a category $\mathcal{C}$ with pullbacks. Chosen a pullback square as (1) to the right for any $f: D \rightarrow B$, the *pullback functor* along $m: A \rightarrow B$, denoted $m^* : \mathcal{C}{\downarrow}B \rightarrow \mathcal{C}{\downarrow}A$, maps an object $(f: D \rightarrow B) \in \mathcal{C}{\downarrow}B$ to $(m^*(f): m^*(D) \rightarrow A) \in \mathcal{C}{\downarrow}A$. Given arrows $m: A \rightarrow B$ and $f: D \rightarrow B$ of $\mathcal{C}$, we write $g \cong m^*(f)$ if there exists an arrow $C \rightarrow D$ such that square (2) to the right is a pullback.

$$
\begin{array}{ccc}
m^*(D) & \longrightarrow & D \\
{\scriptstyle m^*(f)}\big\downarrow & (1) & \big\downarrow{\scriptstyle f} \\
A & \underset{m}{\longrightarrow} & B
\end{array}
$$

$$
\begin{array}{ccc}
C & \longrightarrow & D \\
{\scriptstyle g}\big\downarrow & (2) & \big\downarrow{\scriptstyle f} \\
A & \underset{m}{\longrightarrow} & B
\end{array}
$$

Definition 2 (GTS **morphism**). *Let* $\mathcal{G}_1 = \langle T_1, P_1, \pi_1 \rangle$ *and* $\mathcal{G}_2 = \langle T_2, P_2, \pi_2 \rangle$ *be* GTS*s. A* GTS *morphism* $f: \mathcal{G}_1 \rightarrow \mathcal{G}_2$ *is a pair* $f = \langle f_T, f_P \rangle$, *where*

- $f_T: T_1 \rightharpoonup T_2$ *is a partial graph morphism;*
- $f_P: P_1 \rightarrow P_2 \cup \{\emptyset\}$ *is a total function on production names, where* $\emptyset: (\quad \leftarrow \quad \rightarrow \quad)$ *is the empty production;*

such that productions are preserved, i.e., for all $p \in P_1$, *with* $f_P(p) = q$, *there are morphisms* $f_\iota^L(p)$, $f_\iota^K(p)$ *and* $f_\iota^R(p)$ *such that the diagram to the right commutes, and* $f_\iota^X(p) \cong t_{X_p}(l_{f_T})$ *for* $X \in \{L, K, R\}$.

The category with GTS*s as objects and the corresponding morphisms as arrows is denoted by* **GTS**.

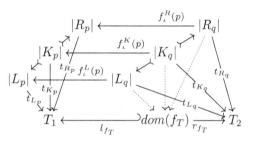

Chosen a pullback functor l_{f_T}, the partial morphism $f_T: T_1 \rightharpoonup T_2$ induces a *re-typing functor* $f_T^* : T_1\text{-}\mathbf{Graph} \rightarrow T_2\text{-}\mathbf{Graph}$, defined on objects as $f_T^*(t_G: |G| \rightarrow T_1) = r_{f_T} \circ l_{f_T}^*(t_G)$. The condition on morphisms involving the pullback squares ensures that all the items in X_p whose type is preserved by f_T occur in $X_{f_P(p)}$. Thus, GTS morphisms are simulations (see e.g. [6,3]), meaning that, for a deriva-tion ρ in $\mathcal{G}_1$, (any choice of) the retyped diagram $f_T^*(\rho)$ is a derivation in $\mathcal{G}_2$.

3   Transactional Graph Transformation Systems

In this section we first recall the basics of *transactional* GTSs [1]. Next we intro-duce the notion of morphism between such systems and we show that morphisms preserve transactions. The basic idea underlying transactional GTSs consists in distinguishing between stable and unstable resources and defining transactions as "minimal" computations which start and end in stable states. The distinction

---

[1] Thus, for example, T-**Graph** = **Graph**$\downarrow T$.

between stable and unstable items in a graph is induced by specifying a subgraph of the type graph, which is intended to represent the stable types.

Definition 3 (transactional GTS). *A transactional* GTS *(T-GTS) is a pair* $\mathcal{Z} = \langle \mathcal{G}, T_s \rangle$, *where* $\mathcal{G}$ *is a* T*-typed* GTS *(the* underlying GTS *of* $\mathcal{Z}$*) and* $i_s: T_s \hookrightarrow T$ *is a subgraph of the type graph of* $\mathcal{G}$*, called the* stable type graph.

We denote by $\mathcal{S}: T\text{-}\mathbf{Graph} \to T_s\text{-}\mathbf{Graph}$ the functor that maps each graph G typed over T to its subgraph consisting of its stably-typed items only, and each morphism to its restriction to stable items: thus $\mathcal{S}$, called the *stabilising functor*, is a concrete choice for the pullback functor i_s.

   The stabilising functor can be applied point-wise to any production of a given T-GTS, thus producing a GTS typed over the stable type graph.

Definition 4 (stabilised GTS). *Given a* T-GTS $\mathcal{Z} = \langle\langle T, P, \pi \rangle, T_s \rangle$, *the* stabilised GTS $\mathcal{S}(\mathcal{Z})$ *is given by* $\langle T_s, P, \pi' \rangle$, *where* $\pi'(q) = \mathcal{S}(\pi(q))$ *for any* $q \in P$.

By construction, there is an obvious GTS morphism from a T-GTS $\mathcal{Z}$ to its stabilised GTS $\mathcal{S}(\mathcal{Z})$, given by the pair $\langle id_{T_s}: T \to T_s, id_P \rangle$. Since GTS morphisms are simulations, the following result trivially holds.

Proposition 1. *Let* $\mathcal{Z}$ *be a* T-GTS *and let* $\rho = G_0 \overset{q_1, g_1}{\Longrightarrow} G_1 \overset{q_2, g_2}{\Longrightarrow} \ldots \overset{q_n, g_n}{\Longrightarrow} G_n$ *be a derivation in* $\mathcal{G}$. *Then* $\mathcal{S}(\rho)$, *defined as below, is a derivation in* $\mathcal{S}(\mathcal{Z})$.

$$\mathcal{S}(\rho) = \mathcal{S}(G_0) \overset{q_1,\ (g_1)}{\Longrightarrow} \mathcal{S}(G_1) \overset{q_2,\ (g_2)}{\Longrightarrow} \ldots \overset{q_n,\ (g_n)}{\Longrightarrow} \mathcal{S}(G_n)$$

Let us come to the definition of transaction in a T-GTS. Inspired by the approach for Petri nets proposed in [4], and extended to nets with read arcs in [5], we introduce *stable steps*, *transactions* and *abstract transactions*. In the following, let $\mathcal{Z} = \langle \mathcal{G}, T_s \rangle$ be an arbitrary but fixed T-GTS.

   Let us first define a graph G as *stable* if it consists only of stable items, i.e., if $|\mathcal{S}(G)| = |G|$, and *unstable* otherwise. A stable step is, intuitively, a computation which starts and ends in stable states. Moreover, stable items which are generated are "frozen", in the sense that they can not be preserved nor consumed by other productions inside the same step; similarly, stable items which are deleted cannot be preserved by other productions. Therefore, the dependencies between productions occurring in a step are induced by unstable items: this implies that at the abstract level, where unstable items are forgotten, all such productions are applicable in parallel.

Definition 5 (stable step and transaction). *A stable step is a derivation* $\rho = G_0 \overset{q_1, g_1}{\Longrightarrow} G_1 \overset{q_2, g_2}{\Longrightarrow} \ldots \overset{q_n, g_n}{\Longrightarrow} G_n$ *which enjoys the following properties*

1. G_0 *and* G_n *are stable graphs;*
2. *the derivation* $\mathcal{S}(\rho)$ *is equivalent in* $\mathcal{S}(\mathcal{G})$ *to a direct derivation via a proper quotient of the rule* $q_1 + \ldots + q_n$ *and a suitable match* g.

A transaction *is a stable step additionally satisfying*

3. *the match g is an isomorphism;*
4. *no intermediate graph G_i (i $\neq$ 0, n) is stable.*

By condition 3, the start graph contains exactly what the transaction needs to reach a successful end. Notice that this condition defines what is a transaction, but then, in a computation, a transaction can be embedded into a larger context. By condition 4 no sub-derivation of ρ is a transaction.

Actually, since we are considering a concurrent model of computations, the fact that all the intermediate graphs are not stable should not be related to the specific order in which productions are applied. Rather, this property should still hold for any shift-equivalent derivation.

When combining shift-equivalence with an equivalence which abstracts also with respect to the concrete identities of items in the involved graphs, i.e., which considers graphs up to isomorphism, we obtain the so-called *abstract truly-concurrent equivalence* [8]. The equivalence class of a derivation ρ with respect to such equivalence will be denoted by $[\rho]_a$ and called an *abstract trace*.

We are now able to introduce the notion of *abstract* transaction.

Definition 6 (abstract transaction). *An* abstract transaction *is an abstract trace* $[\rho]_a$ *such that any derivation* $\rho' \in [\rho]_a$ *is a transaction.*

A simple transactional GTS, presented in [1], tests the equality between integer expressions involving natural numbers, represented as sequences $S(S(\ldots S(0)\ldots))$, and a sum operator. Figure 1 shows some of the productions, whose numbering refers to the original system. The type graph and its stable subgraph, not depicted here, can be inferred from the labeling observing that dashed items (hyper-edges depicted as boxed and nodes depicted as circles) are not stable.

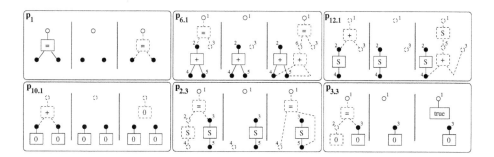

Fig. 1. Some productions of the T-GTS testing equality of natural numbers

Figure 2 shows the sequence of graphs of a derivation starting from the stable graph representing the expression $S(0) + 0 = S(0)$, and using the productions of Figure 1 in the given order. Intuitively, the derivation starts by making unstable the top operator =, and then triggering the evaluation of the sum operator. The evaluation of $+$ does not modify the stable part of the graph: it builds the result using unstable items, which are then consumed by the evaluation of the equality operator. The last graph is stable, and it includes the result of the evalutation on the node to which the original equality operator was attached.

It is not difficult to check that this derivation is a transaction, as it satisfies all conditions of Definition 5; furthermore its equivalence class is an abstract transaction, since all shift-equivalent derivations are transactions as well.

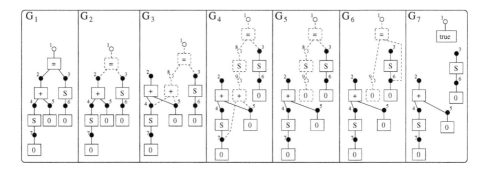

Fig. 2. A sample derivation evaluating $\mathsf{S}(0) + 0 = \mathsf{S}(0)$

We now extend the definition of GTS morphisms to transactional GTSs, explaining how morphisms behave with respect to the stable/unstable items.

Definition 7 (T-GTS morphism). *Let $\mathcal{Z}_1 = \langle \mathcal{G}_1, T_{1s} \rangle$ and $\mathcal{Z}_2 = \langle \mathcal{G}_2, T_{2s} \rangle$ be T-GTSs. A T-GTS morphism $f: \mathcal{Z}_1 \to \mathcal{Z}_2$ is a GTS morphism $f: \mathcal{G}_1 \to \mathcal{G}_2$ between the underlying GTSs, such that*

1. *for all $z \in T_1 \setminus T_{1s}$, we have that $f_T(z)$ is defined and $f_T(z) \in T_2 \setminus T_{2s}$;*
2. *for all $z \in T_{1s}$, if $f_T(z)$ is defined then $f_T(z) \in T_{2s}$.*

The category having T-GTSs as objects and the corresponding morphisms as arrows is denoted by **TGTS**.

Note that we require that the type graph component of a morphism preserves both stable and unstable items. Additionally it must be total on unstable items.

In order to ensure that morphisms are simulations in this more general framework, we prove that T-GTS morphisms preserve abstract transactions.

Proposition 2 (morphisms preserve transactions). *Let $f: \mathcal{Z}_1 \to \mathcal{Z}_2$ be a T-GTS morphism and let $[\rho]_a$ be an abstract transaction in $\mathcal{Z}_1$. Then $[f_T(\rho)]_a$ (see the note after Definition 2) is an abstract transaction in $\mathcal{Z}_2$.*

4   Transactions as Processes

Inspired by the classical *non-sequential processes* for Petri nets, *graph processes* have been proposed in [7,2] as a faithful representation of the derivations of a

GTS up to shift-equivalence. A graph process for a T-GTS $\mathcal{Z}$ is defined as an "occurrence grammar" $\mathcal{O}$, i.e., a grammar satisfying suitable acyclicity constraints, equipped with a T-GTS morphism from $\mathcal{O}$ to $\mathcal{Z}$.

The derivations in $\mathcal{O}$ are mapped through the morphism to derivations in $\mathcal{Z}$, which are shown to be shift-equivalent. Vice versa, from each derivation in $\mathcal{Z}$ a process can be obtained by a simple colimit construction, and shift-equivalent derivations yield isomorphic processes. Since abstract transactions are defined as abstract traces, the corresponding processes provide a compact, handier representation for them, that will be exploited in the next section for the definition of *implementation morphisms* among T-GTSs.

In the present paper a process for a T-GTS $\mathcal{Z}$ is defined by an explicit colimit construction for any derivation in $\mathcal{Z}$. A more abstract characterisation based on structural properties can be provided as well, as in [2], but it is not needed here.

Definition 8 (process from a derivation). *Let* $\mathcal{Z} = \langle\langle T, P, \pi\rangle, T_s\rangle$ *be a T-GTS, and let* $\rho = G_0 \overset{q_1,m_1}{\Longrightarrow} G_1 \overset{q_2,m_2}{\Longrightarrow} \ldots \overset{q_n,m_n}{\Longrightarrow} G_n$ *be a derivation in* $\mathcal{Z}$. *A process* ϕ *associated to* ρ *is a T-GTS morphism* $\phi = \langle\phi_T, \phi_P\rangle: \mathcal{O}_\phi \to \mathcal{Z}$, *where* $\mathcal{O}_\phi = \langle\langle T_\phi, P_\phi, \pi_\phi\rangle, T_{\phi_s}\rangle$ *is obtained as follows*

- $\langle T_\phi, \phi_T\rangle$ *is a colimit object (in* T-**Graph***) of the diagram representing derivation* ρ, *as depicted (for a single derivation step) in the diagram below, where* $c_{X_i}: X_i \to T_\phi$ *is the induced injection for* $X \in \{D, G, L, K, R\}$;
- $T_{\phi_s} \hookrightarrow T_\phi = \phi_T(T_s \hookrightarrow T)$;
- $P_\phi = \{\langle q_i, i\rangle \mid i \in \{1,\ldots,n\}\}$;
- $\pi_\phi(\langle q_i, i\rangle) = (\langle|L_i|, c_{L_i}\rangle \overset{l_i}{\leftarrow} \langle|K_i|, c_{K_i}\rangle \overset{r_i}{\rightarrow} \langle|R_i|, c_{R_i}\rangle)$ *(see the diagram to the right); moreover,* $\phi_P(\langle q_i, i\rangle) = q_i$, *for all* $i \in \{1,\ldots,n\}$.

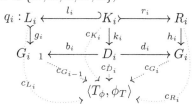

Intuitively, the colimit construction applied to a derivation constructs the graph T_ϕ as a copy of the source graph plus the items created during the rewriting.

As an example, we show the type graph of the process associated to the derivation of Figure 2.

The injections from the graphs of the derivation are implicitly represented by indexing some edges with a *creation index* in the bottom-left corner, and a *deletion index* in the bottom-right one. The creation index is missing in the edges that are not created, i.e., that belong to the start graph, and symmetrically for the deletion index. The image of graph G_i of the derivation, with $i \in \{1,\ldots,7\}$, contains all edges with creation index, if any, smaller than i, and deletion index, if any, larger than or equal to i.

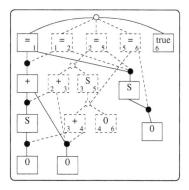

Two processes ϕ and ϕ' for a T-GTS $\mathcal{Z}$ are *isomorphic* if there exists a T-GTS isomorphism $f: \mathcal{O}_\phi \to \mathcal{O}_{\phi'}$ such that $\phi' \circ f = \phi$. An *abstract process* for $\mathcal{Z}$

is an isomorphism class of processes for $\mathcal{Z}$ and it is denoted $[\phi]$ where ϕ is a representative in the class.

Since in a derivation all matches are assumed to be injective, it can be shown that in the associated process all rules are injectively typed in T_ϕ: referring to the diagram after Definition 8, all the morphisms c_{X_i} to T_ϕ are injective for $X \in \{G, D, L, K, R\}$. If $x \in T_\phi$ and $q = \langle q_i, i \rangle$, we say that the production q *consumes* x if x is in the image of c_{L_i} and not in that of c_{K_i}; that q *creates* x if x is in the image of c_{R_i} and not in that of c_{K_i}; and that q *preserves* x if it is in the image of c_{K_i}. This leads to the following net-like notation

$$q = c_{L_i}(|L_i| \setminus l_i(|K_i|)) \qquad q\ = c_{R_i}(|R_i| \setminus r_i(|K_i|)) \qquad \underline{q} = c_{K_i}(|K_i|)$$

We say that q *consumes*, *creates* and *preserves* items in q, q and $\underline{q}$, respectively. Similarly, the sets of productions which consume, create and preserve $x \in T_\phi$ are denoted by x, x and $\underline{x}$, respectively. $Min(\mathcal{O}_\phi)$ denotes the subgraph of T_ϕ consisting of the items x such that $x = \emptyset$, and ϕ the same graph typed over T by the restriction of ϕ_T. The graphs $Max(\mathcal{O}_\phi)$ and ϕ are defined by duality.

Definition 9 (causal relation). *The* causal relation *of a process ϕ is the least transitive and reflexive relation $\leq_\phi$ over $T_\phi \uplus P_\phi$ such that for all $x, y \in T_\phi \uplus P_\phi$ and $q_1, q_2 \in P_\phi$: i) $x \leq_\phi y$ if $x \in\ y$ and ii) $q_1 \leq_\phi q_2$ if $((q_1\ \cap \underline{q_2}) \cup (\underline{q_1} \cap q_2)) \neq \emptyset$.*

It is easy to show that the causal relation is indeed a partial order.

Definition 10 (reachable set). *Let ϕ be a process. For any $\leq_\phi$-left-closed $P' \subseteq P_\phi$, the* reachable set *associated to P' is the set $S_{P'} \subseteq T_\phi$ defined by*

$$x \in S_{P'} \text{ iff } \forall q \in P_\phi \ (x \leq_\phi q \Rightarrow q \notin P') \wedge (q \leq_\phi x \Rightarrow q \in P').$$

We now introduce *transactional processes*, i.e., processes representing abstract transactions. For technical reasons we consider also a wider class of processes, the *unstable transactional processes*, which may start and end in unstable states.

Definition 11 (transactional process). *Let $\mathcal{Z} = \langle \langle T, P, \pi \rangle, T_s \rangle$ be a T-GTS. An* unstable transactional process *is a process ϕ of $\mathcal{Z}$ such that*

1. *for any $x \in T_{\phi_s}$, at most one of the sets x, x , $\underline{x}$ is not empty;*
2. *for any $x \in Min(\mathcal{O}_\phi)$, there exists $q \in P_\phi$ such that either $x \in q$ or $x \in \underline{q}$;*
3. *for any reachable set $S_{P'}$ associated to a non-empty $P' \subset P_\phi$, there exists $x \in S_{P'}$ such that $x \notin Min(\mathcal{O}_\phi) \cup Max(\mathcal{O}_\phi)$.*

If $Min(\mathcal{O}_\phi) \cup Max(\mathcal{O}_\phi) \subseteq T_{\phi_s}$, then ϕ is called transactional process *(t-process). The family of abstract unstable t-processes of $\mathcal{Z}$ is denoted by* **utProc**$(\mathcal{Z})$ *and* **tProc**$(\mathcal{Z}) \subseteq$ **utProc**$(\mathcal{Z})$ *denotes the class of all abstract t-processes of $\mathcal{Z}$.*

Note that if a representative of an abstract process is a(n unstable) transactional one, then all the other members of the equivalence class are so.

Condition 1 implies that each stable item is either in the source or in the target state of the process. Additionally, each stable item that is preserved by at least one production cannot be generated nor consumed in the process itself: this would induce a dependency between productions, violating the defining requirements for transactions (see Definition 5). By condition 2, any item in the source state is used in the computation. Condition 3 ensures that the process is not decomposable into "smaller pieces". It tells that by executing only an initial, non-empty subset P' of the productions of the process, we end up in a graph $S_{P'}$ which is not entirely contained in $Min(\mathcal{O}_\phi) \cup Max(\mathcal{O}_\phi)$, i.e., which contains at least one unstable item. Finally, in a transactional process the source and target states are required to be stable.

For example, the process described after Definition 8 is transactional.

From the theory of graph processes (see [2]) we know that the abstract processes of a T-GTS $\mathcal{Z}$ are in one-to-one correspondence with the abstract traces of $\mathcal{Z}$. More precisely, if $[\rho]_a$ is an abstract trace of $\mathcal{Z}$ and $\rho', \rho'' \in [\rho]_a$ are two derivations, then the processes associated to ρ' and ρ'' are isomorphic. This defines a function TP mapping the abstract traces of $\mathcal{Z}$ to abstract processes for $\mathcal{Z}$. Vice versa, if ϕ is a process for $\mathcal{Z}$, and ρ, ρ' are two derivations of $\mathcal{O}_\phi$, then the retyped derivations $\phi_T(\rho)$ and $\phi_T(\rho')$ of $\mathcal{Z}$ (see the observation after Definition 2) are abstract truly-concurrent equivalent, and thus belong to the same abstract trace. This defines a function PT mapping the abstract processes for $\mathcal{Z}$ to abstract traces of $\mathcal{Z}$. Moreover, it can be proved that functions TP and PT are inverse to each other. By the next proposition they establish an isomorphism between abstract transactions and abstract t-processes: hence, these latter provide an alternative, equivalent characterisation of the former ones.

Proposition 3. *Let $\mathcal{Z}$ be a T-GTS. Then $[\phi]$ is an abstract t-process of $\mathcal{Z}$ iff $PT([\phi])$ is an abstract transaction.*

5   The Abstract System of a Transactional GTS

As mentioned in the introduction, a T-GTS can be seen at two different levels of abstraction. It can be viewed as a standard GTS, where both stable and unstable states, and thus also the internal structure of transactions, are visible. But we can abstract away from the unstable states and observe only complete transactions. Intuitively, this gives rise to another GTS, where abstract transactions of the original T-GTS become productions which rewrite directly the source stable state into the target stable state. This transformation defines a mapping from the objects of the category **TGTS** to those of **GTS**. Interestingly, equipping the category of transactional GTSs with a more general notion of morphism —called *implementation morphism*—, this mapping can be turned into a functor, which is the right adjoint to the inclusion functor in the opposite direction.

We start by introducing the abstract GTS associated to a given T-GTS, where productions are abstract processes of the original T-GTS corresponding to transactions. For technical reasons, it is convenient to define productions as equivalence classes of t-processes which, roughly speaking, are isomorphic when forgetting the stable preserved part. We first define the span induced by a process.

Definition 12 (span underlying a process). *Given a process ϕ for a* T-GTS *$\mathcal{Z}$, the* underlying span *of ϕ is $\Pi(\phi) = \phi \hookleftarrow \phi \cap \phi \hookrightarrow \phi$ (intersection is taken component-wise).*

Given an ut-process ϕ, with $\mathcal{O}_\phi = \langle\langle T_\phi, P_\phi, \pi_\phi\rangle, T_{\phi_s}\rangle$, consider the structure $r(\phi)$, typed over the set of items $T_\phi - Min(\mathcal{O}_\phi) \cap Max(\mathcal{O}_\phi) \cap T_{\phi_s}$, where any component is restricted to such set of types (intuitively, the stable preserved part is forgotten). Then, two ut-processes ϕ_1 and ϕ_2 are *read-equivalent*, written $\phi_1 \simeq_r \phi_2$, if $\Pi(\phi_1) \simeq \Pi(\phi_2)$, i.e., they have the same associated span, and $r(\phi_1) \simeq r(\phi_2)$. A *read ut-process (rut-process)* is defined as an equivalence class of ut-processes with respect to read-equivalence, denoted as $[\phi]_r$ for a representative ϕ. The set of rut-processes of a T-GTS $\mathcal{Z}$ is denoted by **rutProc**$(\mathcal{Z})$. The set of *read t-processes (rt-processes)* **rtProc**$(\mathcal{Z})$ is defined in an analogous way.

In order to associate a concrete span to an abstract process, we need to assume a chosen representative for any equivalence class of processes.

Definition 13 (span underlying abstract process). *Let us assume for each* T-GTS *$\mathcal{Z}$ a* choice function ch *, mapping each rut-process $[\phi]_r$ to a concrete representative* ch *$([\phi]_r) \in [\phi]_r$. The* underlying span *of a rut-process $[\phi]_r$ is defined as $\Pi([\phi]_r) = \Pi(\text{ch} ([\phi]_r))$.*

We are now able to define the abstract system associated with a GTS.

Definition 14 (abstract GTS). *Let $\mathcal{Z} = \langle\mathcal{G}, T_s\rangle$ be a* T-GTS. *The abstract* GTS *associated to $\mathcal{Z}$, denoted by $A(\mathcal{Z})$, is the* GTS *$\langle T_s, \textbf{rtProc}(\mathcal{Z}), \Pi \rangle$ where* **rtProc**$(\mathcal{Z})$ *is the set of rt-processes of $\mathcal{Z}$ and Π is as in Definition 13.*

For instance, in the abstract GTS of the T-GTS recalled in Section 3 the rt-process having the type graph shown after Definition 8 is a production. The corresponding span has graphs G_1 and G_7 as left- and right-hand side, respectively.

An implementation morphism is a T-GTS morphism that maps each given production of the source system to a read unstable *transactional* process of the target system, and also provides a triple of morphisms mapping the production underlying the process to the given production: this additional information is needed to compose implementation morphisms correctly.

Definition 15 (T-GTS implementation morphisms). *For a given* T-GTS *$\mathcal{Z} = \langle\langle T, P, \pi\rangle, T_s\rangle$, let $\widehat{\mathcal{Z}} = \langle\langle T, \textbf{rutProc}(\mathcal{Z}), \Pi \rangle, T_s\rangle$ be the* T-GTS *having all read ut-processes as productions. An implementation pre-morphism $f: \mathcal{Z}_1 \to \mathcal{Z}_2$ is a triple $f = \langle f_T, f_P, f_\iota\rangle$, where $\langle f_T, f_P\rangle: \mathcal{Z}_1 \to \widehat{\mathcal{Z}_2}$ is a* T-GTS *morphism and f_ι is a family $f_\iota = \{f_\iota(p) \mid p \in P_1\}$ such that for each $p \in P_1$, $f_\iota(p) = \langle f_\iota^L(p), f_\iota^K(p), f_\iota^R(p)\rangle$ is a given choice of the three arrows whose existence is required in Definition 2.*

Given two pre-morphisms $\langle f_T, f_P, f_\iota\rangle, \langle f_T, f_P, g_\iota\rangle: \mathcal{Z}_1 \to \mathcal{Z}_2$, let $p \in P_1$ such that $f_P(p) = [\phi]_r$. Then we write $g_\iota(p) \approx f_\iota(p)$ if there are a process automorphism $\alpha: \text{ch} ([\phi]_r) \to \text{ch} ([\phi]_r)$ and a span automorphism $\eta : \Pi ([\phi]_r) \to \Pi ([\phi]_r)$ which restricts to the identity over unstable items, such that $g_\iota(p) = f_\iota(p) \circ \alpha_\Pi \circ \eta$, component-wise ($\alpha_\Pi$ stands for the restriction of α_T to $\Pi ([\phi]_r)$).

An implementation morphism *is an equivalence class of pre-morphisms, where* $\langle f_T, f_P, f_\iota \rangle \approx \langle f_T, f_P, g_\iota \rangle$ *if* $g_\iota(p) \approx f_\iota(p)$ *for all* $p \in P_1$.

Roughly, implementation morphisms are classes of pre-morphisms up to the equivalence induced on the third component by process isomorphisms (note that the type component of an automorphism α : ch $([\phi]_r) \rightarrow$ ch $([\phi]_r)$ restricts to an automorphism over the span Π $([\phi]_r)$). The third component is further quotiented along isomorphisms of the stable subgraph: this is safe because, by the definition of transaction, stable items are not used in composing computations.

In order to provide a correct definition of the category having T-GTSs as objects and implementation morphisms as arrows, we first have to explain how implementation morphisms compose. This is summarised by the next lemma. Given a T-GTS $\mathcal{Z}$ and a production p in $\mathcal{Z}$, below we denote by ϕ_{id_p} the process associated (see Definition 8) to the one-step derivation which applies p to its left-hand side L_p with the identity match.

Lemma 1 (composition and identity for implementation morphisms).
Given a T-GTS $\mathcal{Z}$, *let* $\widehat{\mathcal{Z}}$ *be as in Definition 15. Then, the properties below hold.*

1. *Any* T-GTS *morphism* $f: \mathcal{Z}_1 \rightarrow \widehat{\mathcal{Z}_2}$ *extends to a* T-GTS *morphism* $\widehat{f}: \widehat{\mathcal{Z}_1} \rightarrow \widehat{\mathcal{Z}_2}$.
2. *Given implementation morphisms* $f: \mathcal{Z}_1 \rightarrow \mathcal{Z}_2$ *and* $g: \mathcal{Z}_2 \rightarrow \mathcal{Z}_3$, *let their composition* $g \circ f: \mathcal{Z}_1 \rightarrow \mathcal{Z}_3$ *be the* T-GTS *morphism* $\widehat{g} \circ f: \mathcal{Z}_1 \rightarrow \widehat{\mathcal{Z}_3}$. *Then composition is associative.*
3. *For each* T-GTS $\mathcal{Z}$, *let* $id = \langle id_T, id_P, id_\iota \rangle : \mathcal{Z} \rightarrow \widehat{\mathcal{Z}}$ *be defined as*
   − *the type graph component* id_T *is the identity;*
   − *each production* p *is mapped by* id_P *to the abstract process* $[\phi_{id_p}]_r$;
   − *for each production* p, $id_\iota(p)$ *is a triple of isomorphisms mapping the span* Π $([\phi_{id_p}]_r)$ *to* $L_p \hookleftarrow K_p \hookrightarrow R_p$ *and making the two resulting squares commute.*

   Then id *is well-defined (any choice of* id_ι *determines the same implementation morphism) and it is the identity on* $\mathcal{Z}$.

The proof of the lemma is long and involuted, and we give only some hints. Most interesting is the proof of point *1*. Let $f: \mathcal{Z}_1 \rightarrow \widehat{\mathcal{Z}_2}$ be a T-GTS morphism and ϕ a process for $\mathcal{Z}_1$. Thus ϕ has a set of productions mapped injectively into its type graph T_ϕ. Any such production p is mapped by f_P to a process of $\mathcal{Z}_2$, equipped with morphisms from its minimal and maximal graphs to the left- and right-hand sides of p (given by the component f_ι). Then the process $\widehat{f}_P(\phi)$ is obtained by "gluing" (with a colimit construction) all the processes which are images of productions in ϕ along the intersections in T_ϕ determined by the f_ι component.

The lemma allows to introduce a category with implementation morphims.

Definition 16 (category TGTS^{imp}). *We denote by* **TGTS**^{imp} *the category having transactional* GTSs *as objects and implementation morphisms as arrows.*

Additionally, exploiting point *1* in Lemma 1 we can show that the extension $\widehat{f}: \widehat{\mathcal{Z}_1} \rightarrow \widehat{\mathcal{Z}_2}$ maps *stable* processes to stable processes, i.e., rt-processes of $\mathcal{Z}_1$

are mapped to rt-processes of $\mathcal{Z}_2$. This in turn can be used to prove that the abstraction function for T-GTS can be seen as a functor.

Proposition 4 (abstraction functor). *Function A, mapping a* T-GTS *to its abstract* GTS*, can be extended to a functor* $\mathcal{A}: \mathbf{TGTS}^{imp} \to \mathbf{GTS}$.

Quite obviously, a GTS $\mathcal{G} = \langle T, P, \pi \rangle$ can be seen as a T-GTS $\mathcal{I}(\mathcal{G}) = \langle \langle T, P, \pi \rangle, T \rangle$. This mapping can be extended to an inclusion functor $\mathcal{I}: \mathbf{GTS} \to \mathbf{TGTS}^{imp}$ in the following way: if $f = \langle f_T, f_P \rangle: \mathcal{G}_1 \to \mathcal{G}_2$ is a GTS morphism, then the T-GTS morphism $\mathcal{I}(f) = \langle g_T, g_P, g_\iota \rangle: \mathcal{I}(\mathcal{G}_1) \to \mathcal{I}(\mathcal{G}_2)$ is given as

- $g_T = f_T$;
- for each production $p \in P_1$, its image $g_P(p)$ is the rt-process $[\phi_{f_P(p)}]_r$, where, as above, $\phi_{f_P(p)}$ is the process associated to the one-step derivation obtained by applying $f_P(p)$ to its left-hand side;
- for each production $p \in P_1$, $g_\iota(p)$ is a triple of isomorphisms mapping the span $\Pi_{(2)}([\phi_{f_P(p)}]_r)$ to $L_p \hookleftarrow K_p \hookrightarrow R_p$ and making the two resulting squares commute.

We are now ready to present the main result of the paper.

Theorem 1 (universality of abstraction). *The abstraction functor* $\mathcal{A}: \mathbf{TGTS}^{imp} \to \mathbf{GTS}$ *is right adjoint to the inclusion functor* $\mathcal{I}$.

Proof (Sketch). For each T-GTS $\mathcal{Z}$, we define the component at $\mathcal{Z}$ of the counit $\epsilon : \mathcal{I}(\mathcal{A}(\mathcal{Z})) \to \mathcal{Z}$. This is an implementation morphism, thus a T-GTS morphism $\epsilon : \mathcal{I}(\mathcal{A}(\mathcal{Z})) \to \widehat{\mathcal{Z}}$. Its type graph component is simply the inclusion of the stable type graph into the full type graph, while the component on productions maps each abstract rt-process of $\mathcal{Z}$ to itself. It remains to show that given a GTS $\mathcal{G}$ and a T-GTS $\mathcal{Z}$, for each implementation morphism $f: \mathcal{I}(\mathcal{G}) \to \mathcal{Z}$, there is a unique $h: \mathcal{G} \to \mathcal{A}(\mathcal{Z})$ such that $\epsilon \circ \mathcal{I}(h) = f$.

Now, observe that morphism f maps each production of $\mathcal{G}$ to a rt-process of $\mathcal{Z}$. Since productions in $\mathcal{A}(\mathcal{G})$ are exactly the rt-processes of $\mathcal{G}$, the morphism $h: \mathcal{G} \to \mathcal{A}(\mathcal{G})$ can be defined identically. The proofs of uniqueness and of the fact that $\epsilon \circ \mathcal{I}(h) = f$ are long, but routine. $\square$

6   Conclusions

The present paper carried on the investigation on transactional graph transformation systems, introduced in [1], as a tool for expressing transactional activities in graph transformation. A transaction is defined as a shift-equivalence class of derivations such that the starting and ending states are stable and all the intermediate states are unstable. Thus unstable items are intended to represent temporary resources, only visible within a transaction, and the distinction between stable and unstable items is enforced by a typing mechanism.

The "indirect" definition of transactions based on the dichotomy between stable and unstable items, inspired by the work on zero-safe nets [4], is motivated by our understanding of graph transformation as a data-centered formalism, where the rules of a system are applied non-deterministically, and any form of control on the application of rules has to be encoded in the graphs.

As far as the realm of graph transformation is concerned, though, also more traditional notions of transaction have been considered, most importantly in the design of PROGRES [16]. PROGRES provides a development environment where basic operations, defined by graph transformation rules, can be combined using a rich set of control structures, including traditional programming language constructs, various kinds of non-deterministic choices, as well as transactions. The PROGRES approach is therefore similiar to the way transactions are introduced in programming languages and other control-centered formalism, and as a consequence a direct comparison with our approach is not feasible.

Besides reviewing the basic definitions concerning graph transactions, enriching and streamlining the original proposal, the main result of the present work is the characterisation of the abstract system of a T-GTS, including all transactions as productions, in terms of a universal construction, presented as a right adjoint functor. A key concept introduced in the present paper is that of implementation morphisms among T-GTSs, allowing to map productions to transactional processes. Such morphisms are similar to the refinement morphisms of [11], where productions can be mapped to arbitrary derivations: a deeper analysis of the relationships among the two notions will be a topic of future work.

As mentioned in the introduction, the ACID properties are a canonical way of characterising transactions, even if in our framework only the first three are relevant. Let us discuss informally how such properties are guaranteed by the notion of transaction presented in this paper. *Atomicity* is guaranteed by the fact that computations that do not represent a transaction are forgotten at the abstract level. *Consistency* is guaranteed because the initial and final states of a transaction are all stable, and at the abstract level only stable graphs are considered. *Isolation* is guaranteed by the fact that a transaction, besides starting and ending in stable states, is "minimal", in the sense that all derivations that are shift-equivalent to it are also transactions. Hence, intermediate unstable states are only accessible inside the transaction itself. This implies that, if two transactions can be applied in parallel to a stable graph, then all the direct derivations of either of them are independent of the direct derivations of the other one. Thus, as desired, the transactions can be interleaved in an arbitrary way.

Currently we are working on the definition of a notion of *graph transformation module*, based on the theory presented in this paper. The idea is that a T-GTS can be seen as the implementation of a module, and its abstract GTS as the exported interface. Module composition mechanisms defined as suitable colimits are under investigation, as well as the study of the precise relationship between the new notion of module and existing ones in the literature (as in [11,14,13]).

Acknowledgements. We are mostly indebted to Roberto Bruni and Leila Ribeiro for enlightening discussions about the topic of the paper, as well as to an anonymous referee for pointing out an inconsistency in the submitted version.

References

1. P. Baldan, A. Corradini, F.L. Dotti, L. Foss, F. Gadducci, and L. Ribeiro. Towards a notion of transaction in graph rewriting. In R. Bruni and D. Varró, editors, *Proceedings International Workshop on Graph Transformation and Visual Modeling Techniques*, Electr. Notes in Theor. Comp. Sci. Elsevier, 2006. To appear.

2. P. Baldan, A. Corradini, and U. Montanari. Concatenable graph processes: relating processes and derivation traces. In K.G. Larsen, S. Skyum, and G. Winskel, editors, *Proceedings International Conference on Automata, Languages and Programming*, volume 1443 of *Lect. Notes in Comp. Sci.* Springer, 1998.

3. P. Baldan, A. Corradini, and U. Montanari. Unfolding of double-pushout graph grammars is a coreflection. In H. Ehrig, G. Engels, H.J. Kreowski, and G. Rozenberg, editors, *Proceedings International Workshop on Theory and Application of Graph Transformations*, volume 1764 of *Lect. Notes in Comp. Sci.*, pages 145–163. Springer, 1999.

4. R. Bruni and U. Montanari. Zero-safe nets: Comparing the collective and individual token approaches. *Info. & Comp.*, 156(1-2):46–89, 2000.

5. R. Bruni and U. Montanari. Transactions and zero-safe nets. In H. Ehrig, G. Juhás, J. Padberg, and G. Rozenberg, editors, *Advances in Petri Nets: Unifying Petri Nets*, volume 2128 of *Lect. Notes in Comp. Sci.*, pages 380–426. Springer, 2001.

6. A. Corradini, H. Ehrig, M. Löwe, U. Montanari, and J. Padberg. The category of typed graph grammars and its adjunctions with categories of derivations. In J. Cuny, H. Ehrig, G. Engels, and G. Rozenberg, editors, *Proceedings International Workshop on Graph Grammars and their Application to Computer Science*, volume 1073 of *Lect. Notes in Comp. Sci.* Springer, 1996.

7. A. Corradini, U. Montanari, and F. Rossi. Graph processes. *Fundamenta Informaticae*, 26(3/4):241–265, 1996.

8. A. Corradini, U. Montanari, F. Rossi, H. Ehrig, R. Heckel, and M. Löwe. Algebraic approaches to graph transformation I: Basic concepts and double pushout approach. In Rozenberg [15], chapter 3, pages 163–245.

9. H. Ehrig, G. Engels, H.-J. Kreowski, and G. Rozenberg, editors. *Handbook of Graph Grammars and Computing by Graph Transformation, Vol. 2: Applications, Languages and Tools*. World Scientific, 1999.

10. H. Ehrig, H.-J. Kreowski, U. Montanari, and G. Rozenberg, editors. *Handbook of Graph Grammars and Computing by Graph Transformation. Vol. 3: Concurrency, Parallelism, and Distribution*. World Scientific, 1999.

11. M. Große-Rhode, F. Parisi-Presicce, and M. Simeoni. Formal software specification with refinements and modules of typed graph transformation systems. *Journal of Computer and System Science*, 64(2):171–218, 2002.

12. A. Habel, J. Müller, and D. Plump. Double-pushout graph transformation revisited. *Mathematical Structures in Computer Science*, 11(5):637–688, 2001.

13. R. Heckel, H. Ehrig, G. Engels, and G Tántzer. Classification and comparison of module concepts for graph transformation systems. In Ehrig et al. [9], chapter 17, pages 669–689.

14. H.-J. Kreowski and S. Kuske. Graph transformation units and modules. In Ehrig et al. [9], chapter 15, pages 607–638.

15. G. Rozenberg, editor. *Handbook of Graph Grammars and Computing by Graph Transformation, Vol. 1: Foundations*. World Scientific, 1997.

16. A. Schürr, A. Winter, and A. Zündorf. The PROGRES approach: Language and environment. In Ehrig et al. [9], chapter 13, pages 487–550.

Categorical Foundations
of Distributed Graph Transformation

Hartmut Ehrig[1], Fernando Orejas[2], and Ulrike Prange[1]

[1] Technical University of Berlin, Germany
{ehrig, uprange}@cs.tu-berlin.de
[2] Technical University of Catalonia, Spain
orejas@lsi.upc.edu

Abstract. A distributed graph (N, D) consists of a network graph N and a commutative diagram D over the scheme N which associates local graphs $D(n_i)$ and graph morphisms $D(e) : D(n_1) \rightarrow D(n_2)$ to nodes n_1, n_2 and edges $e : n_1 \rightarrow n_2$ in N.

Although there are several interesting applications of distributed graphs and transformations, even the basic pushout constructions for the double pushout approach of distributed graph transformation could be shown up to now only in very special cases.

In this paper we show that the category of distributed graphs can be considered as a Grothendieck category over a specific indexed category, which assigns to each network N the category of all diagrams D of shape N. In this framework it is possible to give a free construction which allows to construct for each diagram D_1 over N_1 and network morphism $h : N_1 \rightarrow N_2$ a free extension $F_h(D_1)$ over N_2 and to show that the Grothendieck category is complete and cocomplete if the underlying category of local graphs has these properties.

Moreover, an explicit construction for general pushouts of distributed graphs is given. This pushout construction is based on the free construction. The non-trivial proofs for free constructions and pushouts are the main contributions of this paper and they are compared with the special cases known up to now.

1   Introduction

When modelling computation by means of (standard) graph transformation, a graph is supposed to denote the (centralized) state of a given system, and computation steps are modelled as transformations of this graph by means of some productions. To model distributed computation, where the state of the given system is not monolithic, G. Taentzer [1] introduced an extension of graph transformation called distributed graph transformation. The idea is to consider that, on one hand, a graph N (the *network graph*) describes the topology of the given system and, on the other, that the global state is, in some sense, partitioned along that graph. In particular, this is done associating to every node n in N a graph G_n that denotes the local state at this node, and to every edge $e : n \rightarrow n'$ in N a graph morphism $h_e : G_n \rightarrow G_{n'}$. These graph morphisms allow one

A. Corradini et al. (Eds.): ICGT 2006, LNCS 4178, pp. 215–229, 2006.

to describe the shared parts of the local states. Formally, then, a distributed graph is just a functor from the network graph into the category of graphs. In this context, (distributed) graph transformation is defined adapting the double-pushout approach to the (functor) category of distributed graphs.

The practical relevance of distributed graph transformation has been demonstrated in [2, 3], where this approach is used to keep coherence between models of different views. This allows an integrated management of modifications in the code and in the global UML model underlying a software artifact. Using distributed graph transformation we can define in a uniform way different kinds of computation steps. For instance, we can describe not only computations that occur in a single location (i.e. in the graph associated to a given node), but computations that occur simultaneously in several locations that are synchronized through the shared parts of the states involved. Moreover, we can also define transformations on the network, for instance allowing us some forms of refactoring. In some sense, this approach is related to Community (see, e.g. [4]), where the local states are tuples rather than graphs, and Goguens General Systems Theory [5].

Unfortunately, the basic constructions for defining distributed graph transformation as presented in [1] depend on some ad-hoc conditions that, on one hand, limit the power of the approach and, on the other hand, make it difficult to generalize the approach to cases where the states are not modelled as basic graphs, but as attributed graphs or some other kind of arbitrary structures [6, 7]. In particular, even the basic pushout constructions for the double pushout approach of distributed graph transformation could be shown up to now only in very special cases.

In this paper we provide categorical foundations for distributed graph transformation that allow us to provide the basic constructions with full generality. In particular, we generalize distributed graphs to distributed objects, where the local diagrams are not necessarily graphs, but consist of objects and morphisms in a certain category $\mathbf{C}$. Then we show that the category of distributed objects can be considered as a Grothendieck category over a specific indexed category, which assigns to each network N the category of all diagrams D of shape N in $\mathbf{C}$. In this framework it is possible to give a free construction which allows to construct for each diagram D_1 over N_1 and network morphism $h : N_1 \to N_2$ a free extension $F_h(D_1)$ over N_2 and to show that the Grothendieck category is complete and cocomplete if the underlying category of local objects has these properties. Moreover, an explicit construction for general pushouts of distributed objects is given. This pushout construction is based on the free construction. The non-trivial proofs for free constructions and pushouts are the main contributions of this paper and they are compared with the special cases known up to now.

The paper is organized as follows. In section 2 we study the category of distributed objects and present the free diagram extensions. Section 3 is dedicated to the category of distributed objects as a Grothendieck category. In section 4 we show the explicit construction of pushouts of distributed objects and, through an example, how these pushouts are used in distributed graph transformation. In section 5 we introduce persistent morphisms and discuss their role with respect to strongly componentwise pushouts as considered in [1]. Finally, in section 6 we draw some conclusions.

We assume the reader to be familiar with the basic notions of category theory, as presented in, e.g., [4, 8, 9].

2   The Category DisC and Free Diagram Extensions

A distributed graph representing the distributed state of a system can be described, on one hand, by a graph N (the *network graph*) defining the topology of the object and, on the other, associating to every node n in N a graph $D(n)$ that denotes the local state at this node, and to every edge $e : n \rightarrow n'$ in N a graph morphism $D(e) : D(n) \rightarrow D(n')$. In particular, it is assumed that these graph morphisms describe the shared parts of the local states (see example 1).

Formally, in categorical terms, this means that a distributed graph (N, D) consists of the network graph N and a diagram $D : N \rightarrow \mathbf{Graph}$ which associates local graphs $D(n_i)$ and graph morphisms $D(e) : D(n_1) \rightarrow D(n_2)$ to nodes n_1, n_2 and edges $e : n_1 \rightarrow n_2$ in N. However, if we consider that states are not specifically modelled by basic graphs, but by some other kind of structure (as, e.g., typed attributed graphs) then we can easily generalize this definition. In particular, we can consider that a distributed object is a diagram $D : N \rightarrow \mathbf{C}$, where $\mathbf{C}$ is an arbitrary category. Obviously, we may require $\mathbf{C}$ to satisfy some specific properties.

In addition, we require that a diagram $D : N \rightarrow \mathbf{C}$ is commutative. We believe that this should be a consequence of assuming that the morphisms associated to the edges (or to the paths) in N denote the shared parts of the distributed states. In particular, suppose that we have two paths p_1 and p_2 from a node n into n'. According to our intuition, this means that we can consider that for the state at node n, $D(n)$, there is a (not necessarily injective) image $D(p_1)(D(n))$ of $D(n')$, and similarly for $D(p_2)$. Now, if $D(p_1)$ would denote a different morphism from $D(p_2)$, then, it would mean that we could also identify $D(p_1)(D(n))$ and $D(p_2)(D(n))$, which are different parts of the state at n'.

A graph $G = (V, E, s, t)$ consists of a set of nodes (or vertices) V and a set of edges E, with functions $s, t : E \rightarrow V$ assigning a source and target node to each edge, respectively. This concept has been extended to many different kinds of graphs, like hypergraphs, labelled graphs, typed and/or attributed graphs, which we do not define explicitly. Instead, we assume to have some category $\mathbf{C}$ and present the theory of distributed objects on the categorical level, which can be instantiated by various graphs and graph-like structures.

Given a graph $G = (V, E, s, t)$, it can be interpreted as the scheme of a category. This means, the reflexive and transitive closure of G is a category with objects V. Vice versa, a category $\mathbf{C}$ can be seen as a (possibly infinite) graph. In the following, we switch between both concepts as needed in the particular context. As a consequence, we use the terms functor and diagram as synonyms in this context.

Definition 1 (path morphism and commutative functor). *Given a graph N, a functor $D : N \to \mathbf{C}$ (interpreting N as a category) and a path $p : n \overset{e_1}{\to} \dots \overset{e_k}{\to} n'$ in N, we define the path morphism $D(p) : D(n) \to D(n')$ of D along p as $D(p) = D(e_k) \circ \dots \circ D(e_1)$. For the empty path $\epsilon_n : n \overset{0}{\to} n$, $D(\epsilon_n) = id_{D(n)}$.*

A functor $D : N \to \mathbf{C}$ is commutative, if for any two paths $p_1, p_2 : n \to n'$ in N we have $D(p_1) = D(p_2)$.

Remark 1. If D is commutative, we obviously have $D(c) = id_{D(n)}$ for each circle $c : n \to n$ in $\mathbf{C}$. For paths $p : n \to n'$, $p' : n \to n' \overset{f}{\to} n''$ and $p'' : n'' \overset{f'}{\to} n \to n'$ it follows that $D(p') = D(f) \circ D(p)$ and $D(p'') = D(p) \circ D(f')$.

We can now define the category of distributed objects:

Definition 2 (distributed object and distributed morphism). *Given a category $\mathbf{C}$, a distributed object (N, D) over $\mathbf{C}$ (or just a distributed object, if $\mathbf{C}$ is implicit in the given context) consists of a graph N, called network graph, and a commutative functor $D : N \to \mathbf{C}$, called diagram functor.*

A distributed morphism over $\mathbf{C}$ (or just a distributed morphism, if $\mathbf{C}$ is implicit in the given context), $f = (f_N, f_D) : (N_1, D_1) \to (N_2, D_2)$, consists of a graph morphism $f_N : N_1 \to N_2$ and a natural transformation $f_D : D_1 \to D_2 \circ f_N$.

Distributed objects and distributed morphisms over $\mathbf{C}$ form the category $\mathbf{DisC}$.

In particular, we may notice that our previous definition implicitly associates to every network graph N a category consisting of all the commutative functors from N to $\mathbf{C}$. This construction can be extended to a functor.

Definition 3 (functor Diag). *The functor $Diag : \mathbf{Graphs}^{\mathrm{OP}} \to \mathbf{Cat}$ is defined by*

- *for a graph N, $Diag(N) = comFunct[N, \mathbf{C}]$, the category of commutative functors (diagrams) $D : N \to \mathbf{C}$,*
- *for a graph morphism $f : N \to N'$ in $\mathbf{Graphs}$, $Diag(f)(D' : N' \to \mathbf{C}) = D' \circ f : N \to \mathbf{C}$ and $Diag(f)(t : D_1' \to D_2') = t \circ f$.*

In order to construct pushouts and colimits in $\mathbf{DisC}$ in section 3 and 4 we need to show that each commutative diagram $D_1 : N_1 \to \mathbf{C}$ has a free extension $D_2 : N_2 \to \mathbf{C}$ for each network morphism $h : N_1 \to N_2$. In fact, if $\mathbf{C}$ is cocomplete then each network morphism has an associated free construction (extension), leading to a free functor left adjoint to $Diag(h)$. Note, that we only need finite cocompleteness of $\mathbf{C}$ if all network graphs N are finite.

Theorem 1. *If $\mathbf{C}$ is cocomplete then for all network morphisms $h : N_1 \to N_2$ there is a functor $F_h : comFunct[N_1, \mathbf{C}] \to comFunct[N_2, \mathbf{C}]$, that is free with respect to $Diag(h)$.*

Construction. We have to show that there is a free construction $(D_2, u_h^{D_1})$ with $D_2 : N_2 \to \mathbf{C}$ and $u_h^{D_1} : D_1 \to Diag(h)(D_2)$ for each diagram $D_1 : N_1 \to \mathbf{C}$.

For $n_2 \in N_2$ define $N_1(n_2)$ as the full subgraph of N_1 induced by the node set $V(N_1(n_2)) = \{n_1 \in N_1 \mid \exists\ path\ p : h(n_1) \to n_2 \in N_2\}$.

The restriction $D_1|_{N_1(n_2)}\ :\ N_1(n_2)\ \to\ \mathbf{C}$ is a functor. Let $(Col(n_2),\ (i_{n_1}^{n_2})_{n_1\ N_1(n_2)})$ be the colimit of $D_1|_{N_1(n_2)}$, with $i_{n_1}^{n_2} : D_1(n_1) \to Col(n_2)$ and $i_{n_1'}^{n_2} \circ D_1(e_1) = i_{n_1}^{n_2}$ for all $e_1 : n_1 \to n_1' \in N_1(n_2)$.

For an edge $e_2 : n_2 \to n_2'$ we have $N_1(n_2) \subseteq N_1(n_2')$ and therefore $(Col(n_2'),\ (i_{n_1}^{n_2'})_{n_1\ N_1(n_2)})$ is a cocone of $D_1|_{N_1(n_2)}$. This means that there exists a unique morphism $c^{e_2} : Col(n_2) \to Col(n_2')$ with $c^{e_2} \circ i_{n_1}^{n_2} = i_{n_1}^{n_2'}$ for all $n_1 \in N_1(n_2)$.

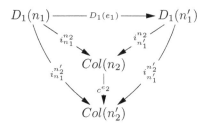

Define $F_h(D_1) = D_2 : N_2 \to \mathbf{C}$ by $D_2(n_2) = Col(n_2)$ and $D_2(e_2) = c^{e_2}$, and $u_h^{D_1} = (i_{n_1}^{\cdot h(n_1)})_{n_1\ N_1}$. □

Proof idea. From the construction it follows that D_2 is a commutative functor and $u_h^{D_1}$ is a well-defined natural transformation.

For a distributed object (N_2, D_2') and a natural transformation $t : D_1 \to D_2' \circ h$ we have to show that there is a unique natural transformation $t\ :\ D_2 \to D_2'$ with $(t\ \circ h) \circ u_h^{D_1} = t$.

For a node $n_2 \in N_2$ and $n_1 \in N_1(n_2)$, by construction there exists a path $p_{n_1}^{n_2} : h(n_1) \to n_2$ in N_2. Since D_2' is commutative, $D_2'(p_{n_1}^{n_2})$ is independent from the chosen path (if there is more than one). Then $(D_2'(n_2), (D_2'(p_{n_1}^{n_2}) \circ t_{n_1})_{n_1\ N_1(n_2)})$ is a cocone of $D_1|_{N_1(n_2)}$ and there exists a unique morphism $t_{n_2} : D_2(n_2) \to D_2'(n_2)$ with $t_{n_2} \circ i_{n_1}^{n_2} = D_2'(p_{n_1}^{n_2}) \circ t_{n_1}$ for all $n_1 \in N_1(n_2)$.

$t\ = (t_{n_2})_{n_2\ N_2}$ is a natural transformation, and the uniqueness follows from the uniqueness of its components.

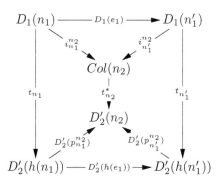

For $n_1 \in N_1$ we have $t_{h(n_1)} \circ i_{n_1}^{h(n_1)} = D_2'(p_{n_1}^{h(n_1)}) \circ t_{n_1} = t_{n_1}$, therefore $(t \circ h) \circ u_h^{D_1} = t$, because $p_{n_1}^{h(n_1)} : h(n_1) \to h(n_1)$ implies $D_2'(p_{n_1}^{h(n_1)}) = id$ (see [10] for more detail).   □

Example 1. Consider the network graphs N_1 and N_2 shown in Fig. 1 on the left hand side, the inclusion $h : N_1 \to N_2$ and the diagram $D_1 : N_1 \to$ **Graphs** shown in Fig. 1 on the right hand side. In the figure on the right, the thick lines represent the network structure, and the diagram morphism is indicated by the small numbers.

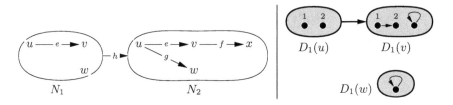

Fig. 1. Two network graphs and a diagram

From the construction we get for each node $n_2 \in N_2$ the corresponding subgraphs $N_1(n_2)$ of N_1, where $N_1(u)$ contains only the node u, $N_1(v)$ contains the nodes u and v and the edge e, $N_1(x) = N_1(v)$ and $N_1(w)$ contains the nodes u and w. The corresponding colimit constructions lead to the following free construction diagram $D_2 : N_2 \to$ **Graphs** over $D_1 : N_1 \to$ **Graphs** with $D_2(u) = Colim(D_1|_{N_1(u)}) = D_1(u)$, and similarly $D_2(v) = D_1(v)$, $D_2(x) = D_1(v)$ and $D_2(w) = D_1(u) \cup D_1(w)$ as shown in Fig. 2.

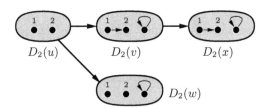

Fig. 2. The corresponding free construction

In section 4 we will use the following decomposition.

Proposition 1. *A distributed morphism* $f = (f_N, f_D) : (N_1, D_1) \to (N_2, D_2)$ *can be decomposed into the following diagram, where* $f_D : F_{f_N}(D_1) \to D_2$ *is the adjunction morphism associated to the morphism* $f_D : D_1 \to Diag(f_N)(D_2)$.

$$(N_1, D_1) \xrightarrow{\ (f_N, f_D)\ } (N_2, D_2)$$

$$(f_N, u_{f_N}^{D_1}) \searrow \qquad \nearrow (id, f_D^*)$$

$$(N_2, F_{f_N}(D_1))$$

Proof. This follows directly from the free construction (Theorem 1).   □

3   DisC as a Grothendieck Category

In this section we show that the category **DisC** can be considered as a Grothendieck category, because there are general categorical results how to construct limits and colimits in Grothedieck categories [4, 11, 12]. We start by defining indexed categories.

Definition 4 (indexed category). *Given a category* **I**, *called index category, an indexed category is a functor* $F : \mathbf{I}^{op} \to \mathbf{CAT}$, *where* **CAT** *denotes the category of all categories.*

Definition 5 (Grothendieck category). *The Grothendieck category* $\mathbf{Gr}(F)$ *of an indexed category* F *has as objects pairs* (i, A) *with* $i \in \mathbf{I}$ *and* $A \in F(i)$. *A morphism* $(i, A) \to (i', A')$ *is a pair* (f, g) *with* $f : i \to i' \in \mathbf{I}$ *and* $g : A \to F(f)(A') \in F(i)$.

Given morphisms $(f, g) : (i, A) \to (i', A')$ *and* $(f', g') : (i', A') \to (i'', A'')$, *the composition is defined by* $(f' \circ f, F(f)(g') \circ g)$. *For an object* (i, A), *the identity* $id_{(i,A)}$ *is given by* (id_i, id_A).

According to [11] we have:

Fact 1. *Let* $F : \mathbf{I}^{op} \to \mathbf{CAT}$ *be an indexed category with Grothendieck category* $\mathbf{Gr}(F)$. *If* **I** *and* $F(i)$ *are complete for all* $i \in \mathbf{I}$, *and* $F(f)$ *is continuous for all* $f : i \to j \in \mathbf{I}$ *then also* $\mathbf{Gr}(F)$ *is complete. If* **I** *and* $F(i)$ *are cocomplete for all* $i \in \mathbf{I}$, *and* $F(f)$ *has a left adjoint for all* $f : i \to j \in \mathbf{I}$ *then also* $\mathbf{Gr}(F)$ *is cocomplete.*

Remark 2. As shown in the proof in [11], limits are constructed componentwise on the index and the functor level. However, this componentwise construction does not work for colimits, where the free construction has to be taken into account.

As a consequence, we can also form the category of distributed objects as the Grothendieck category associated to the indexed category *Diag* as defined in Def. 3.

Theorem 2. *The category* **DisC** *is a Grothendieck category over the indexed category* $Diag : \mathbf{Graphs}^{op} \to \mathbf{Cat}$.

Proof. This is a direct consequence of the definitions of distributed objects and morphisms, the given functor *Diag* and the construction of a Grothendieck category.   □

Theorem 3. *If* **C** *is (co)complete, then also* **DisC** *is (co)complete.*

Proof idea. According to Fact 1, if **C** is (co)complete, **DisC** being (co)complete follows from the facts that **Graphs** is (co)complete, $comFunct[G, \mathbf{C}]$ is (co)complete for all $G \in$ **Graphs** and $Diag(h)$ is continuous (has a left adjoint) for all $h : G \to G' \in$ **Graphs** by Theorem 1 (see [10] for more detail). Note that Theorem 1 shows that $Diag(h)$ has a left adjoint F_h. This means that $Diag(h)$ is a right adjoint and hence continuous. □

4   Graph Transformation in DisC

In this section, we define graph transformations on distributed objects in the double pushout (DPO) approach based on [8]. In particular, we present explicit pushout and pullback constructions in **DisC** and discuss the gluing condition.

Definition 6 (distributed transformation system). *A distributed transformation system $TS = (\mathbf{DisC}, S, P)$ consists of a category **DisC** over some category **C**, a start object S and a set of distributed productions P, where*

1. *a distributed production $p = L \xleftarrow{l} K \xrightarrow{r} R$ consists of distributed objects L, K and R and distributed morphisms $l : K \to L$ and $r : K \to R$,*
2. *a direct distributed transformation $(N, D) \overset{p,m}{\Longrightarrow} (N', D')$ of a distributed object (N, D) via the production p and a match $m : L \to (N, D)$ is given by the following diagram, where (1) and (2) are pushouts in **DisC**,*

$$
\begin{array}{ccccc}
L & \xleftarrow{\ l\ } & K & \xrightarrow{\ r\ } & R \\
\downarrow m & (1) & \downarrow & (2) & \downarrow n \\
(N, D) & \longleftarrow & C & \longrightarrow & (N', D')
\end{array}
$$

3. *a distributed transformation is a sequence $(N_0, D_0) \Rightarrow (N_1, D_1) \Rightarrow \ldots \Rightarrow (N_n, D_n)$ of direct distributed transformations, written $(N_0, D_0) \Rightarrow (N_n, D_n)$,*
4. *the language $L(TS)$ consists of all distributed objects (N, D) in **DisC** derivable from the start object S by a transformation, i.e. $L(TS) = \{(N, D) \mid S \Rightarrow (N, D)\}$.*

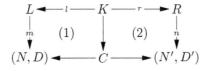

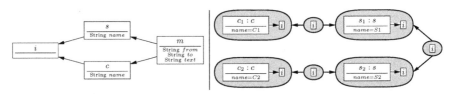

Fig. 3. The type graph and an example of a distributed network

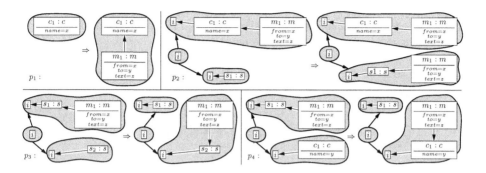

Fig. 4. Example communication productions

Example 2. In the following, we model a small client-server system with asynchronous communication using typed attributed graph transformation. In this case **C** is the category **AGraphs**$_\mathbf{ATG}$ of typed attributed graphs (see [8] for more detail) leading to distributed graphs in **DisC** over typed attributed graphs. The type graph of the local graphs is shown in Fig. 3 on the left hand side. Each client (c) and server (s) has a name and can be connected to an interface connector (i). Messages (m) can be assigned to clients and servers, and they contain the sender ($from$), the receiver (to) and the message itself ($text$).

On the network level, clients and servers can be connected to other servers via the interface connectors. An example of a distributed graph is given in Fig. 3 on the right hand side, where two clients $c1$ and $c2$ are connected to different servers $s1$ and $s2$, which themselves are connected.

The communication between the clients is modeled by graph transformation using communication productions p_1 - p_4 in Fig. 4, that do not change the structure of the underlying network. As usual, only the left- and the right-hand side of the productions are shown - the gluing object is their intersection. First, a client may create a message using the production p_1. Then the message is sent to the server with production p_2. Between different servers, the message can be transmitted using production p_3. If the receiver of the message is connected to the current server where the message is stored, this client can receive the message using production p_4.

Fig. 5 shows some productions for network administration. With production q_1, a new server is added, and q_2 adds a new client.

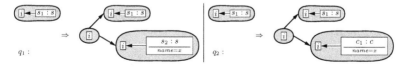

Fig. 5. Example network productions

The result of an application of the production q_2 with c_1 replaced by $x = c_3$ and $s_1 = s_2$ to the distributed graph shown in Fig. 3 is depicted in Fig. 6, where a new client c_3 is added, changing the network graph.

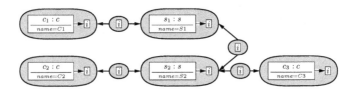

Fig. 6. Application of the distributed production q_2

From Theorem 3 it follows that if **C** is cocomplete the pushout over arbitrary distributed morphisms f and g exists. Since pushouts are the underlying structure of transformations, we want to characterize them more explicitly. The following construction has been introduced as generalized amalgamation in [13, 14].

Theorem 4. *Given distributed morphisms* $f = (f_N, f_D) : (N_0, D_0) \rightarrow (N_1, D_1)$ *and* $g = (g_N, g_D) : (N_0, D_0) \rightarrow (N_2, D_2)$ *in* **DisC**. *According to Proposition 1 these morphisms can be decomposed. Then the diagram in the upper part of Fig. 7 is a pushout over* f *and* g *in* **DisC**, *where* $(1')$ *is a pushout in* **Graphs** *with* $g'_N \circ f_N = h_N = f'_N \circ g_N$ *and* $(4')$ *is a pushout in* $comFunct[N_3, \mathbf{C}]$.

Proof idea. It can be shown that the squares (1), (2), (3) and (4) are pushouts in **DisC**. Then by pushout composition also the complete diagram is a pushout in **DisC** (see [10] for more detail). □

Remark 3. It may be noted that Prop. 1 and the above Theorem are formulated for the category **DisC**, but they hold for any Grothendieck category with free constructions, as shown for a similar general framework in [13, 14].

Example 3. Fig. 8 shows an example pushout construction, as defined above in Fig. 7. The network morphisms can be obtained from the relative positions of the nodes. Square (1) shows the pushout on the network level, and the free extensions of the diagram D_0. Squares (2) and (3) show the corresponding extensions for diagrams (D_2) and (D_1), respectively. Square (4) gives the componentwise pushout on the diagram level.

We also have an explicit construction of pullbacks in **DisC** with complete **C**.

Theorem 5. *Given distributed morphisms* $f = (f_N, f_D) : (N_1, D_1) \rightarrow (N_3, D_3)$ *and* $g = (g_N, g_D) : (N_2, D_2) \rightarrow (N_3, D_3)$ *in* **DisC**. *Then the diagram* (1) *is a pullback over* f *and* g *in* **DisC**, *where* (2) *is a pullback in* **Graphs** *with* $g_N \circ f'_N = h_N = f_N \circ g'_N$ *and* (3) *is a pullback in* $comFunct[N_0, \mathbf{C}]$.

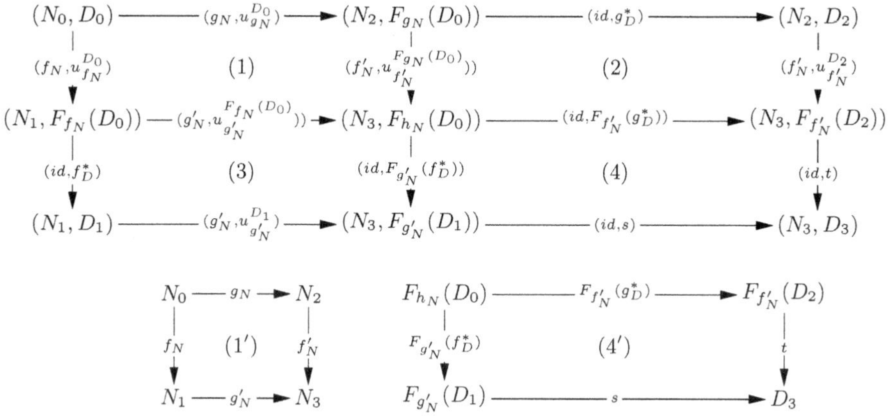

$$(N_0, D_0) \xrightarrow{(g_N, u_{g_N}^{D_0})} (N_2, F_{g_N}(D_0)) \xrightarrow{(id, g_D^*)} (N_2, D_2)$$

$$\downarrow {(f_N, u_{f_N}^{D_0})} \quad (1) \quad \downarrow {(f_N', u_{f_N'}^{F_{g_N}(D_0)})} \quad (2) \quad \downarrow {(f_N', u_{f_N'}^{D_2})}$$

$$(N_1, F_{f_N}(D_0)) \xrightarrow{(g_N', u_{g_N'}^{F_{f_N}(D_0)})} (N_3, F_{h_N}(D_0)) \xrightarrow{(id, F_{f_N'}(g_D^*))} (N_3, F_{f_N'}(D_2))$$

$$\downarrow {(id, f_D^*)} \quad (3) \quad \downarrow {(id, F_{g_N'}(f_D^*))} \quad (4) \quad \downarrow {(id, t)}$$

$$(N_1, D_1) \xrightarrow{(g_N', u_{g_N'}^{D_1})} (N_3, F_{g_N'}(D_1)) \xrightarrow{(id, s)} (N_3, D_3)$$

$$
\begin{array}{ccc}
N_0 \xrightarrow{g_N} N_2 & \qquad F_{h_N}(D_0) \xrightarrow{F_{f_N'}(g_D^*)} F_{f_N'}(D_2) \\
\downarrow f_N \quad (1') \quad \downarrow f_N' & \qquad \downarrow F_{g_N'}(f_D^*) \quad (4') \quad \downarrow t \\
N_1 \xrightarrow{g_N'} N_3 & \qquad F_{g_N'}(D_1) \xrightarrow{\quad s \quad} D_3
\end{array}
$$

Fig. 7. Explicit pushout construction in **DisC**

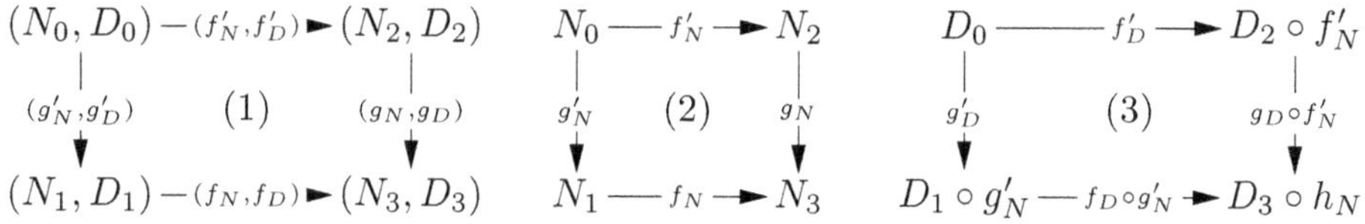

$$
\begin{array}{ccc}
(N_0, D_0) \xrightarrow{(f_N', f_D')} (N_2, D_2) & N_0 \xrightarrow{f_N'} N_2 & D_0 \xrightarrow{f_D'} D_2 \circ f_N' \\
\downarrow {(g_N', g_D')} \quad (1) \quad \downarrow {(g_N, g_D)} & \downarrow {g_N'} \quad (2) \quad \downarrow {g_N} & \downarrow {g_D'} \quad (3) \quad \downarrow {g_D \circ f_N'} \\
(N_1, D_1) \xrightarrow{(f_N, f_D)} (N_3, D_3) & N_1 \xrightarrow{f_N} N_3 & D_1 \circ g_N' \xrightarrow{f_D \circ g_N'} D_3 \circ h_N
\end{array}
$$

Proof. Follows from the proof of Fact 1 in [11] and Remark 2.

5   Persistent Morphisms and Componentwise Pushouts

In [1], the author does not study the construction of general pushouts of distributed graphs. Instead, the paper concentrates on studying when it is possible to build *strongly componentwise pushouts*. Intuitively, if the network morphisms involved are injective, a componentwise pushout of distributed graphs (1) can be seen as the gluing of two distributed graphs (with respect to the common subgraph (N_0, D_0)), where for each node n_3 in N_3, if $n_3 = g_N'(f_N(n_0))$ then the graph at this node, $D_3(n_3)$, is the gluing of the graphs $D_0(n_0)$, $D_1(f_N(n_0))$, and $D_2(g_N(n_0))$ with respect to the corresponding morphisms defined by f_D and g_D. In the following, we call this "componentwise pushouts". But in [1], *strongly componentwise pushouts* are considered with the following additional property: if n_3 is not the image of any node in N_0, but just of a node n_1 in N_1 (respectively n_2 in N_2) then $D_3(n_3)$ is equal to $D_1(n_1)$ (respectively $D_2(n_2)$).

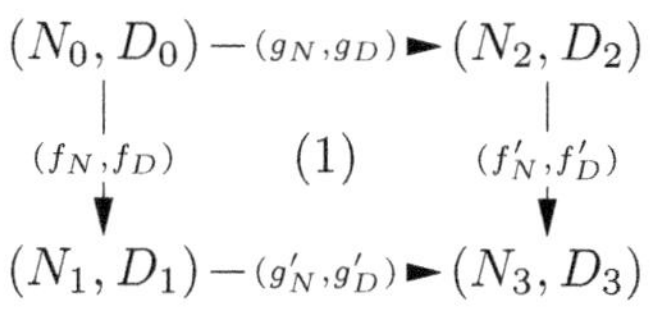

$$
\begin{array}{ccc}
(N_0, D_0) \xrightarrow{(g_N, g_D)} (N_2, D_2) \\
\downarrow {(f_N, f_D)} \quad (1) \quad \downarrow {(f_N', f_D')} \\
(N_1, D_1) \xrightarrow{(g_N', g_D')} (N_3, D_3)
\end{array}
$$

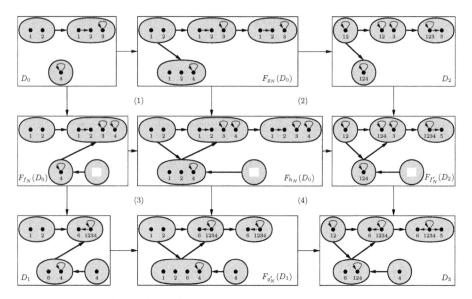

Fig. 8. Example of an explicit pushout construction

This means, in [1], Taentzer provides properties for an if-and-only-if characterization of the existence of strongly componentwise pushouts. Unfortunately, these properties are quite ad-hoc and depend not only on the span of network morphisms, but also on the diagrams, and thus are difficult to generalize to categories of distributed objects over a category **C** different than **Graphs**.

We think that it is important for several applications to have componentwise pushouts, but not necessarily stronlgy componentwise as in [1]. Fig. 8 is an example of a componentwise pushout, which is not strongly componentwise. The upper right node in D_2 has no preimage in D_0, but the local graph is different from the corresponding local graph in D_3. However, in general, arbitrary pushouts of distributed graphs will not be componentwise. The key property to ensure in Proposition 4 componentwise pushouts is that the given network morphisms are *persistent* in the following sense:

Definition 7 (persistent network morphism). *If* **C** *is cocomplete, a morphism* $h : N_1 \to N_2$ *is persistent if for every* D *in* $comFunct[N_1, \mathbf{C}]$ *the unit of the adjunction,* $u_h^D : D \to Diag(h) \circ F_h(D)$, *is an isomorphism.*

For a characterization of persistent morphisms, we need the following property of colimits.

Proposition 2. *Given a commutative functor* $D : N \to \mathbf{C}$ *with colimit object* $Col(D)$ *of* D, *then we have for any* $n \in N$:
If for all $n' \in N$ *there is a path* $p_{n'} : n' \to n$ *in* N *then* $D(n) = Col(D)$.

Proof idea. Since path morphisms are unique, $(D(n), (D(p_{n'}))_{n' \ N})$ is a cocone of D leading to a unique morphism $x : Col(D) \to D(n)$. Using the properties of colimit $Col(D)$ and commutative D it can be shown that x is an isomorphism (see [10] for more detail). □

Taking into account the construction of free functors in Theorem 1, we are now able to characterize persistent network morphisms for cocomplete categories $\mathbf{C}$. Intuitively, a morphism $h : N_1 \to N_2$ is persistent if for a path from $h(n_1)$ to $h(n_2)$ in N_2 there is already a path from n_1 to n_2 in N_1.

Proposition 3. *A morphism $h : N_1 \to N_2$ is persistent if we have for all nodes $n_1, n'_1 \in N_1$ the following property:*
If there exists a path $h(n_1) \to h(n'_1) \in N_2$ then there exists a path $n_1 \to n'_1$ in N_1.

Proof. Given $D : N_1 \to \mathbf{C}$. For $n_1 \in N_1$ we have $Diag(h) \circ F_h(D)(n_1) = F_h(D)(h(n_1)) = Col(D|_{N_1(h(n_1))})$ as defined in the construction of Theorem 1. If $n'_1 \in N_1(h(n_1))$ then there is a path $h(n'_1) \to h(n_1)$ in N_2. The above condition makes sure that there is also a path $n'_1 \to n_1$ in N_1. Applying Proposition 2 with $F = D|_{N_1(h(n_1))}$, this means that $Col(D|_{N_1(h(n_1))}) = D|_{N_1(h(n_1))}(n_1) = D(n_1)$ and hence $Diag(h) \circ F_h(D)(n_1) = D(n_1)$. □

Remark 4. This property is also necessary for persistency for all categories $\mathbf{C}$, where the colimits of an arbitrary $F : \bullet\!\!\rightarrow\!\!\bullet \to \mathbf{C}$ and of its restriction $F|_{\bullet\ \ \bullet} : \bullet\ \ \bullet \to \mathbf{C}$ are in general not isomorphic.

Then, using the construction of pushouts in Theorem 4 we can show that if the associated network morphisms are persistent then the pushouts of the interface graphs are componentwise.

Proposition 4. *If f_N and g_N are persistent, then the pushout in $\mathbf{DisC}$ is a componentwise pushout on the interface network, i.e. $D_3(h_N(n_0))$ is the pushout of $D_1(f_N(n_0)) \overset{f_{D,n_0}}{\longleftarrow} D_0(n_0) \overset{g_{D,n_0}}{\longrightarrow} D_2(g_N(n_0))$ for all $n_0 \in N_0$.*

Proof idea. In **Graphs**, pushouts can be shown to be closed under persistent morphisms. This means that also f'_N, g'_N are persistent and we have $Diag(h_N) \circ F_{h_N}(D_0) = D_0$, $Diag(g'_N) \circ F_{g'_N}(D_1) = D_1$ and $Diag(f'_N) \circ F_{f'_N}(D_2) = D_2$. Since pushouts in functor categories are constructed componentwise, this means that $D_3(h_N(n_0))$ is the pushout of $D_1(f_N(n_0)) \overset{f_{D,n_0}}{\longleftarrow} D_0(n_0) \overset{g_{D,n_0}}{\longrightarrow} D_2(g_N(n_0))$ for all $n_0 \in N_0$ according to the pushout (4) in Fig. 7. □

6   Conclusion

In this paper we have presented categorical foundations for distributed graph transformation that, in our opinion, considerably improve [1]. In particular, we have seen that the category of distributed objects has free constructions and is complete and cocomplete (provided that the underlying category is so). Moreover, we have shown how to explicitly build pushouts using the concept of generalized amalgamation introduced in [13, 14] and we have characterized the class of morphisms (persistent morphisms) that ensure that in a pushout the interfaces will be glued componentwisely and discussed the relationship with [1].

6.1   Towards a Theory of Distributed Graph Transformation

We have provided the basic constructions for defining transformations. However, this is just a first step for fully studying distributed graph transformation in a general setting.

According to Theorem 3, **DisC** is complete and cocomplete provided that **C** is complete and cocomplete. Since the categories **Graphs** of graphs, **Graphs**$_{\mathbf{TG}}$ of typed graphs and **AGraphs**$_{\mathbf{ATG}}$ of typed attributed graphs satisfy both properties [8, Thm. 11.3], **DisGraphs**, **DisGraphs**$_{\mathbf{TG}}$ and **DisAGraphs**$_{\mathbf{ATG}}$ are complete and cocomplete. This means especially that we are able to construct pushouts and pullbacks, which are needed in the DPO approach. The key question is, whether there is a suitable class $\mathcal{M}$ for **DisC** such that $(\mathbf{DisC}, \mathcal{M})$ becomes a (weak) adhesive HLR category. This would allow to instantiate the corresponding theory in [8] to distributed graph transformation over **C**. Unfortunately, for the most obvious choices $\mathcal{M}_1 = Monos \times Monos$, $\mathcal{M}_2 = Persistent$ $Monos \times Monos$ and $\mathcal{M}_3 = Persistent\, Monos \times Mor_{\mathbf{C}}$, $(\mathbf{DisC}, \mathcal{M}_i)$, $i = 1, 2, 3$ is in general not (weak) adhesive HLR, where persistent morphisms are defined in section 5. In order to obtain a (weak) adhesive HLR category, $\mathcal{M}$-morphisms have to be monomorphisms (this rules out choice $\mathcal{M}_3$), pushouts along $\mathcal{M}$-morphisms have to be pullbacks (this rules out choice $\mathcal{M}_1$) and $\mathcal{M}$-morphisms have to be closed under pullbacks (this rules out choice $\mathcal{M}_2$).

But we can show that (injective) persistent network morphisms are closed under pushouts, which implies that at least $\mathcal{M}_3$ is closed under pushouts. This means that we can obtain the Local Church-Rosser Theorem [8, Thm. 5.12] for $(\mathbf{C}, \mathcal{M}_3)$, provided that we require a stronger notion of independence including the $\mathcal{M}_3$ PO-PB decomposition property for the given pair of direct transformations in the corresponding proof.

Moreover we obtain a weaker version of the Embedding Theorem [8, Thm. 6.14], which usually requires an initial pushout (1) over a morphism f.

It is an interesting open question under which conditions initial pushouts over distributed morphisms exist and how they can be constructed for suitable **C**. This would immediately lead to a necessary and sufficient gluing condition [8, Thm. 6.4], which is important for the construction of direct transformations. If we do not have initial pushouts, we can also take some other pushout (1) over f, including the trivial case $B \to C = G \to G'$, and replace the notion of consistency based on pullback-constructions and the boundary object B in (1) by B-consistency depending on the chosen B. In fact, the proof in [8] only uses the pushout property of (1) and not the initiality, which, however, is used for the Extension Theorem [8, Thm. 6.16].

Acknowledgements. This work has been partially supported by the Research Network SEGRAVIS (HPRN-CT-2002-00275) and by the Spanish project GRAMMARS (ref. TIN2004-07925-C03-01).

References

[1] Taentzer, G.: Distributed Graphs and Graph Transformation. Applied Categorical Structures **7**(4) (1999) 431–462

[2] Bottoni, P., Parisi-Presicce, F., Taentzer, G.: Specifying Integrated Refactoring with Distributed Graph Transformations. In Pfaltz, J., Nagl, M., Böhlen, B., eds.: Proc. of AGTIVE 2003. Volume 3062 of LNCS., Springer (2004) 220–235

[3] Bottoni, P., Parisi-Presicce, F., Taentzer, G., Pulcini, S.: Maintaining Coherence between Models with Distributed Rules: From Theory to Eclipse. In Bruni, R., Varró, D., eds.: Proc. of GT-VMT 2006. ENTCS, Elsevier (2006) 81–91

[4] Fiadeiro, J.: Categories for Software Engineering. Springer (2006)

[5] Goguen, J.: Sheaf Semantics for Concurrent Interacting Objects. Mathematical Structures in Computer Science **2**(2) (1992) 159–191

[6] Ehrig, H., Habel, A., Padberg, J., Prange, U.: Adhesive High-Level Replacement Categories and Systems. In Ehrig, H., Engels, G., Parisi-Presicce, F., Rozenberg, G., eds.: Proc. of ICGT 2004. Volume 3256 of LNCS., Springer (2004) 144–160

[7] Ehrig, H., Prange, U., Taentzer, G.: Fundamental Theory for Typed Attributed Graph Transformation. In Ehrig, H., Engels, G., Parisi-Presicce, F., Rozenberg, G., eds.: Proc. of ICGT 2004. Volume 3256 of LNCS., Springer (2004) 161–177

[8] Ehrig, H., Ehrig, K., Prange, U., Taentzer, G.: Fundamentals of Algebraic Graph Transformation. Springer (2006)

[9] Mac Lane, S.: Categories for the Working Mathematician. Volume 5 of Graduate Texts in Mathematics. Springer, New York (1971)

[10] Ehrig, H., Orejas, F., Prange, U.: Categorical Foundations of Distributed Graph Transformation: Long Version. Technical report, TU Berlin (2006)

[11] Tarlecki, A., Burstall, R., Goguen, J.: Some Fundamental Algebraic Tools for the Semantics of Computation: Part 3: Indexed Categories. Theoretical Computer Science **91**(2) (1991) 239–264

[12] Goguen, J.: Information Integration in Institutions. In Moss, L., ed.: Jon Barwise Memorial Volume, Indiana University Press (2006) to appear.

[13] Ehrig, H., Baldamus, M., Orejas, F.: New Concepts for Amalgamation and Extension in the Framework of Specification Logics. Technical Report 91/05, TU Berlin (1991)

[14] Ehrig, H., Baldamus, M., Cornelius, F., Orejas, F.: Theory of Algebraic Module Specification including Behavioural Semantics, Constraints an Aspects of Generalized Morphisms. In Nivat, M., Rattray, C., Rus, T., Scollo, G., eds.: Invited Lecture Proc. of AMAST'91, Springer (1991) 145–172

Dynamic Graph Transformation Systems[*]

Roberto Bruni[1] and Hernán Melgratti[2]

[1] Computer Science Department, University of Pisa, Italy
[2] IMT Lucca Institute for Advanced Studies, Italia
bruni@di.unipi.it, hernan.melgratti@imtlucca.it

Abstract. We introduce an extension of Graph Grammars (GGs), called Dynamic Graph Grammars (DynGGs), where the right-hand side of a production can spawn fresh parts of the type graph and fresh productions operating on it. The features of DynGGs make them suitable for the straightforward modeling of reflexive mobile systems like dynamic nets and the Join calculus. Our main result shows that each DynGG can be modeled as a (finite) GG, so that the dynamically generated structure can be typed statically, still preserving exactly all derivations.

1 Introduction

Graphs can model complex systems at a level of abstraction that is both intuitive and formal. Graph Grammars (GGs) originated in the late 60's as a suitable extension of string grammars: string concatenation is replaced by graph gluing and string rewriting by subgraph replacing. As a model of concurrency, there is also a close analogy between GGs and Petri nets (PNs), as a Petri net can be straightforwardly modeled as a particular GG over discrete graphs. Solid theoretical basis are now available for many different kinds of graph transformation, ranging from the essential node replacement systems [8] and edge replacement systems [6] to the more sophisticated synchronized hyperedge replacement systems [3,11,10] and algebraic approaches to graph rewriting [4,7].

To the best of our knowledge, one extension that has not been deeply investigated in the literature is the use of reflexive productions that can release new rewrite rules. Reflexive systems arise naturally in many areas where graph transformation techniques have been applied with success, like biological and chemical systems and distributed and mobile computing. Moreover, the reflexive extension of many different kinds of rewrite systems have been studied in the literature. In particular, dynamic nets are a mobile extension of PNs, expressive enough to model mobile calculi like π-calculus and Join calculus [1,2]. Dynamic nets are indeed strictly more expressive than PNs.

Exploiting the analogy between PNs and GGs, we propose a reflexive extension of GGs, called *Dynamic Graph Grammars* (DynGGs), whose generality is witnessed by encodings of dynamic nets and Join calculus. However, when posing the question:

"Are Dynamic Graph Grammars more expressive than ordinary ones?"

our main result provides a negative answer: though DynGGs can offer a more convenient abstraction, computationally speaking they are not more expressive than GGs.

---

[*] Research supported by the EU FET-GC2 IST-2004-16004 Integrated Project SENSORIA.

A. Corradini et al. (Eds.): ICGT 2006, LNCS 4178, pp. 230–244, 2006.

From GGs to DynGGs. To complete this informal introduction, we sketch the design choices of DynGGs, introduce some terminology and give a minimal example.

Let T denote a type graph and G_T a graph typed over T. Ordinary T-typed DPO productions are spans of the form $p : (L_T \xleftarrow{l} K_T \xrightarrow{r} R_T)$ such that l and r preserve the typing of items in K_T. (We assume that the reader has some familiarity with GGs, and hence postpone exact formalization to later sections.)

A first simple extension is to consider productions like $p : (L_T \xleftarrow{l} K_T \xrightarrow{r} R_{T'})$ where $T \subseteq T'$, so that fresh types can be generated and used for typing $R_{T'}$. This way the fresh types introduced by p cannot be exploited in the (left-hand side of) productions. The idea is then to spawn also new productions able to operate on items typed in $T' \setminus T$.

A DynGG is thus a triple (T, G_T, P) where T is a type graph, G_T a graph typed over T and P a set of T-typed dynamic productions that take the form $p : (L_T \xleftarrow{l} K_T \xrightarrow{r} \mathcal{G}_{T'})$ where $\mathcal{G}_{T'}$ is again a (T'-typed) DynGG and r relates K_T to the initial graph of $\mathcal{G}_{T'}$. For example, when $P = \emptyset$, then the grammar is roughly a T-typed graph G_T, which is statically fixed and cannot change. Such grammars are called *static*. A production p is *static* if its right-hand side $\mathcal{G}_{T'}$ is a static grammar and $T' = T$. If all productions are static, then the grammar is called *shallow* and is essentially an ordinary GGs: the application of any production can neither change the type graph nor spawn new rules.

Figures 1 and 2 introduce a small ad hoc example whose purpose is to expose a peculiarity of dynamic rewrites. Let T_a be the singleton type graph with just one node a. Let $T_g \supset T_a$ consisting of nodes a and b and two edges $f : a \to b$, and $g : b \to b$.

Take the T_a-typed dynamic grammar $\mathcal{G}_a$ with the dynamic production p in Figure 1. For simplicity, we take the inclusions as legs of the span and draw the typing (dotted lines) only once for each item. The left-hand side (i) of p consists of a T_a-typed graph with just one node n_1, which is preserved by the context (ii), and the right-hand side (iii) spawns a shallow T_g-typed grammar $\mathcal{G}_p$ whose initial graph has, beside n_1, one additional node n_2 and one arc h. The grammar $\mathcal{G}_p$ itself has just one static production $q \in P_p$, illustrated in Figure 2: the left-hand side (i) is a graph with three nodes m_0 and m_1, m_2, which are all preserved (ii), and two arcs h_1, h_2, which are deleted, and the right-hand side (iii) spawns the static T_g-typed grammar $\mathcal{G}_q$ (i.e., a graph) with one additional arc l from m_1 to m_2 and type g.

Assume the initial graph of $\mathcal{G}_a$ is a discrete graph G_0 with one node k typed over a. The application of the production p (with the obvious matching from n_1 to k) spawns a fresh instance $\mathcal{G}_p^1$ of $\mathcal{G}_p$: the type graph becomes T_{g_1} and the underlying graph becomes $G_1 \supset G_0$ with nodes k (typed a) and k_1 (typed b_1) connected by an arc h_1 (typed f_1). Moreover, a production q_1 is now available beside p. A second application of p spawns another fresh instance $\mathcal{G}_p^2$ of $\mathcal{G}_p$: the type graph becomes $T_{g_1} \cup T_{g_2}$ and the underlying graph becomes $G_2 \supset G_1$ with a new node k_2 and an arc $h_2 : k \to k_2$. Again, a production q_2 is now available. Similarly, p can be applied again and again (see Figure 6). However, no suitable matching can ever be found for the application of q_1, q_2, etc. In fact, it is not possible to find two arcs with the same type, say f_i, and the identification condition prevents a non-injective matching of the two arcs in the left-hand side of q_i.

Now compare $\mathcal{G}_a$ with its static, flattened T_g-typed version, where only two productions p and q are available at any time: after two derivation steps with p the underlying graph has two nodes typed over b and two arcs typed over f, and thus q can be applied!

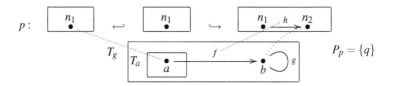

Fig. 1. A dynamic production p

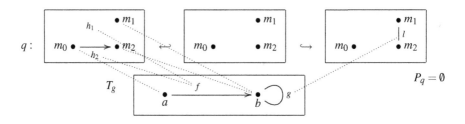

Fig. 2. A static production q

The difference lies in the *separation principle* that DynGGs imposes on items produced as freshly-typed instances by different applications of the same production. We
shall reprise this example to show that incautious encodings of DynGGs intro GGs
would introduce unwanted derivations.

Synopsis. § 2 accounts for some basics of typed GGs [5]. The definition of DynGGs
is original to this contribution and described in § 3, where DynGGs are shown to be a
conservative extension of GGs. § 4 reports some sample encodings of other reflexive
frameworks. The main result of the paper is in § 5, where it is shown that GGs have the
same expressive power as DynGGs. Concluding remarks and future work are in § 6.

2   Typed Graph Grammars

A *(directed) graph* is a tuple $G = \langle N_G, E_G, s_G, t_G \rangle$, where N_G is a set of *nodes*, E_G is
a set of *edges* (or *arcs*), and $s_G, t_G : E_G \to N_G$ are the *source* and *target* functions. We
shall omit subscripts when obvious from the context. A *graph morphism* $f : G \to G'$ is
a couple $f = \langle f_N : N \to N', f_E : E \to E' \rangle$ such that: $s' \circ f_E = f_N \circ s$ and $t' \circ f_E = f_N \circ t$.

Definition 2.1 (Typed Graph). *Given a* graph of types *T, a T-typed graph is a pair*
$\langle |G|, \tau_G \rangle$, *where $|G|$ is the underlying* graph *and $\tau_G : |G| \to T$ is a total morphism.*

In GGs the graph $|G|$ defines the configuration of the system and its items (nodes and
edges) model resources, while τ_G defines the *typing* of the resources. Hence, the un-
derlying graph $|G|$ evolves dynamically, while the type graph T is statically fixed and
cannot change at run-time. For example, when encoding Petri nets in GGs the places
form the discrete graph of types, while markings form the configurations of the system.

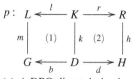

$p: L \xleftarrow{l} K \xrightarrow{r} R$

(a) A DPO direct derivation.

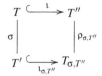

(b) Standard type graph construction.

Fig. 3. DPO and type graph construction

A *morphism* between T-typed graphs $f : G_1 \to G_2$ is a graph morphisms $f : |G_1| \to |G_2|$ such that $\tau_{G_1} = \tau_{G_2} \circ f$. The category of T-typed graphs and their morphisms is denoted by T-**Graph**. Since we work only with typed notions, we will usually omit the qualification "typed", and we will not indicate explicitly the typing morphisms. The following notion of retyping will be used extensively in the context of DynGGs.

Definition 2.2 (Graph Retyping). *Given a T-typed graph $G = \langle |G|, \tau_G \rangle$ and a morphism $\sigma : T \to T'$ we denote by $\sigma \cdot G$ the T'-typed graph $\sigma \cdot G = \langle |G|, \sigma \circ \tau_G \rangle$.*

The key notion to *glue* graphs together is that of a categorical pushout. Roughly, a pushout pastes two graphs by injecting them in a larger graph that is (isomorphic to) their disjoint union modulo the collapsing of some common part. We recall that a *span* is a pair (b,c) of morphisms $b : A \to B$ and $c : A \to C$. A *pushout* of the span (b,c) is then an object D together with two (co-final) morphisms $f : B \to D$ and $g : C \to D$ such that: (i) $f \circ b = g \circ c$ and (ii) for any other choice of $f' : B \to D'$ and $g' : C \to D'$ s.t. $f' \circ b = g' \circ c$ there is a unique $d : D \to D'$ s.t. $f' = d \circ f$ and $g' = d \circ g$. If the pushout is defined, then c and g is called the *pushout complement* of $\langle b, f \rangle$.

A *(T-typed graph) DPO production* $p : (L \xleftarrow{l} K \xrightarrow{r} R)$ is a span of injective graph morphisms $l : K \to L$ and $r : K \to R$. The T-typed graphs L, K, and R are called the *left-hand side*, the *interface*, and the *right-hand side* of the production, respectively.

Definition 2.3 (DPO graph grammar). *A (T-typed) DPO graph grammar $\mathcal{G}$ is a tuple $\langle T, G_{in}, P \rangle$, where G_{in} is the initial (T-typed) graph, P is a set of DPO productions.*

Given a graph G, a production $p : (L \xleftarrow{l} K \xrightarrow{r} R)$, and a *match* $m : L \to G$, a *direct derivation* δ *from G to H using p (based on m)* exists, written $\delta : G \Rightarrow_p H$, if and only if the diagram in Figure 3(a) can be constructed, where both squares are pushouts in T-**Graph**: (1) the rewriting step removes from the graph G the items $m(L - l(K))$, yielding the graph D (with k,b as pushout complement of $\langle m, l \rangle$); (2) then, fresh copies of the items in $R - r(K)$ are added to D yielding H (as a pushout of (k,r)). The interface K specifies both what is preserved and how fresh items must be glued to the existing part. The existence of the pushout complement of $\langle m, l \rangle$ is subject to the satisfaction of the following *gluing conditions* [4]:

- *identification condition:* $\forall x, y \in L$ if $x \neq y$ and $m(x) = m(y)$ then $x, y \in l(K)$;
- *dangling condition:* no arc in $G \setminus m(L)$ should be incident to a node in $m(L \setminus l(K))$.

The identification condition is satisfied by *valid matches*: a match is not valid if it requires a single item to be consumed twice, or to be both consumed and preserved.

A *derivation* is a sequence $\gamma = \{\delta_i : G_{i-1} \Rightarrow_{p_{i-1}} G_i\}_{i \in \{1,\dots,n\}}$ of direct derivations.

3   Dynamic Graph Grammars

As aforementioned, a T-typed dynamic production takes the form: $p : (L_T \xleftarrow{l} K_T \xrightarrow{r} \mathcal{G}_{T'})$ where $\mathcal{G}_{T'}$ is a suitable (T'-typed) dynamic graph grammar. A dynamic graph grammar can contain any number of such productions. Formally:

Definition 3.1 (Dynamic Graph Grammars). *The domain of dynamic graph grammars can be expressed as the least set* DGG *satisfying the equation:*

$$\mathcal{D} = \{(T, G_{in}, P) \mid G_{in} \in \mathbf{Graph}_T \wedge$$
$$\forall p : (L_T \xleftarrow{l} K_T \xrightarrow{r} \mathcal{G}_{T_p}) \in P.(L_T, K_T \in \mathbf{Graph}_T \wedge T \subset T_p \wedge$$
$$\mathcal{G}_{T_p} = (T_p, G_{T_p}, P_p) \in \mathcal{D}) \}$$

where $\mathbf{Graph}_T$ *is the set of* T-*typed graphs and* $r : \iota \cdot K_T \to G_{T_p}$ *is a morphism between* T_p-*typed graphs, where* $\iota : T \hookrightarrow T_p$ *denotes the obvious sub-graph injection.*

Any element $\mathcal{G} = (T, G_{in}, P) \in$ DGG *is called a* Dynamic Graph Grammar. *It is static if* $P = \emptyset$. *It is shallow if* $T = T_p$ *and* $\mathcal{G}_{T_p}$ *is static for all* $p : (L_T \xleftarrow{l} K_T \xrightarrow{r} \mathcal{G}_{T_p}) \in P$.

Note that all Dynamic Graph Grammars are well-founded: since we take DGG as the least set satisfying the recursive domain equation above, the type graphs syntactically appearing in $\mathcal{G} = (T, G_{in}, P) \in$ DGG form a finite tree $\mathcal{T}(\mathcal{G})$ rooted in T, with parent relation given by immediate subsetting (i.e., T_i is parent of T_j iff $T_i \subset T_j$ and no T_k appears in $\mathcal{G}$ such that $T_i \subset T_k \subset T_j$) and where leaves are associated with static grammars. We remark that each type graph $T_p \supset T$ extends T with local declarations $T_p \setminus T$, whose scope is bounded by the specific production p. For simplicity, but without loss of generality, we assume that all additional items introduced by different type graphs inside $\mathcal{G}$ are named differently (i.e., each additional item occurs only in one type graph). We let $\mathbf{T}(\mathcal{G}) = \bigcup_{T_i \in \mathcal{T}(\mathcal{G})} T_i$ denote the *overall flat type graph* of $\mathcal{G}$, and let $\iota_{T_i} : T_i \hookrightarrow \mathbf{T}(\mathcal{G})$ denote the obvious sub-graph inclusion. Note that, by the structuring of $\mathcal{T}(\mathcal{G})$, the type graph $\mathbf{T}(\mathcal{G})$ is just the union of all the leaves of $\mathcal{T}(\mathcal{G})$.

Similarly, all nested productions in $\mathcal{G}$ form the tree $\mathcal{P}(\mathcal{G})$ rooted in P with parent relation given by immediate inclusion (i.e., the set of productions P_i is the parent of P_p iff $p : (L_T \xleftarrow{l} K_T \xrightarrow{r} (T_p, G_{T_p}, P_p)) \in P_i$). Given a T-typed dynamic production $p : (L_T \xleftarrow{l} K_T \xrightarrow{r} \mathcal{G}_{T_p})$ with $\mathcal{G}_{T_p} = (T_p, G_{T_p}, P_p)$ we say that the ordinary $\mathbf{T}(\mathcal{G})$-typed production flat$(p) : (\iota_T \cdot L_T \xleftarrow{l} \iota_T \cdot K_T \xrightarrow{r} \iota_{T_p} \cdot G_{T_p})$ is the *flattening of* p. We let $\mathbf{P}(\mathcal{G}) = \bigcup_{P_i \in \mathcal{P}(\mathcal{G})} \{\text{flat}(p) \mid p \in P_i\}$ denote the *overall set of flat productions* of $\mathcal{G}$. The $\mathbf{T}(\mathcal{G})$-typed shallow grammar $\mathbf{F}(\mathcal{G}) = (\mathbf{T}(\mathcal{G}), \iota_T \cdot G_{in}, \mathbf{P}(\mathcal{G}))$ is called the *flattening of* $\mathcal{G}$.

To define the dynamics of DynGGs we need a more advanced notion of retyping, which can be used to generate fresh items in the type graph. In the following, when considering type graph constructions, we assume that a standard choice of pushout objects satisfying the following requirements is available: Let $T \subset T''$ and $\sigma : T \to T'$

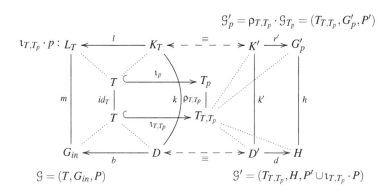

Fig. 4. A direct dynamic derivation

injective, then we denote by $T_{\sigma,T''}$ the pushout object of the inclusion $\iota : T \hookrightarrow T''$ and σ such that T' embeds in $T_{\sigma,T''}$ via set-theoretical inclusion $\iota_{\sigma,T''}$, while T'' embeds via an injection $\rho_{\sigma,T''}$ (see Figure 3(b)) that renames items in $T'' \setminus T$ with fresh names. When $T \subseteq T'$ and σ is the inclusion we replace the subscripting $(-)_{\sigma,T''}$ with $(-)_{T',T''}$.

Definition 3.2 (Fresh Graph Retyping). *Let $T \subset T''$. Given a T''-typed graph $G = \langle |G|, \tau_G \rangle$ and an injection $\sigma : T \to T'$ we let $\sigma \cdot G = \langle |G|, \rho_{\sigma,T''} \circ \tau_G \rangle$.*

Definition 3.3 (Dynamic Retyping). *Given a T-typed Dynamic Graph Grammar $\mathcal{G} = (T, G_{in}, P)$ and an injective morphism $\sigma : T \to T'$ we denote by $\sigma \cdot \mathcal{G}$ the T'-typed grammar defined recursively by letting $\sigma \cdot \mathcal{G} = (T', \sigma \cdot G_{in}, \sigma \cdot P)$, with $\sigma \cdot P = \{\sigma \cdot p \mid p \in P\}$ where $\sigma \cdot p : (\sigma \cdot L_T \xleftarrow{l} \sigma \cdot K_T \xrightarrow{r} \rho_{\sigma,T_p} \cdot \mathcal{G}_{T_p})$ for any $p : (L_T \xleftarrow{l} K_T \xrightarrow{r} \mathcal{G}_{T_p})$.*

Note that $\rho_{\sigma,T_p} \cdot \mathcal{G}_{T_p}$ is a T_{σ,T_p}-typed grammar and $\sigma \cdot p$ is a T'-typed production.

To define the behaviour of DynGGs, note that the type graph and the available productions can change over time: as the computation progresses new items and productions can be spawn. Hence, as it is typical of reflexive systems, the actual configuration must comprise data (i.e., the underlying graph), their typing and the control (i.e., available productions). This means that configurations are themselves DynGGs.

In DynGGs, productions are nested inside (the right-hand sides of) other productions, but only top-level productions can be applied, by finding a matching of their left-hand sides into the initial graph. When such a production p is applied, then fresh instances of the productions P_p, nested one level below p, become available at the top-level, and can be unwound themselves in successive steps. Given a DynGG $\mathcal{G} = (T, G_{in}, P)$, a production $p : (L_T \xleftarrow{l} K_T \xrightarrow{r} \mathcal{G}_{T_p}) \in P$ with $\mathcal{G}_{T_p} = (T_p, G_{T_p}, P_p)$, and a matching $m : L_T \to G_{in}$, we proceed as follows (see Figure 4):

- We check that m and $l : K_T \to L_T$ satisfy the gluing conditions.
- We build the pushout complement of $\langle m, l \rangle$, obtaining a T-typed graph D with morphisms $k : K_T \to D$ and $b : D \to G_{in}$.

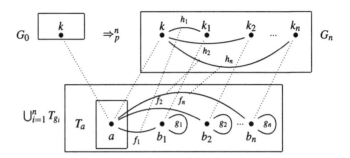

Fig. 5. A derivation where p is applied n times

- We build the standard type graph T_{T,T_p} associated with $\sigma = id_T : T \to T$ and $\iota_p : T \hookrightarrow T_p$. Note that $\iota_{T,T_p} = \rho_{T,T_p} \circ \iota_p$. Fresh items of the underlying graph produced by the application of p must be typed over T_{T,T_p}.
- We build the retyped graphs $D' = \iota_{T,T_p} \cdot D$ and $K' = \iota_{T,T_p} \cdot K_T$ and take the morphism $k' : K' \to D'$ induced by k.
- We build the retyped DynGG $\mathcal{G}'_p = \rho_{T,T_p} \cdot \mathcal{G}_{T_p} = (T_{T,T_p}, G'_p, P')$ with $G'_p = \rho_{T,T_p} \cdot G_{T_p}$ and $P' = \rho_{T,T_p} \cdot P_p$ and take the morphism $r' : K' \to G'_p$ induced by $r : \iota_p \cdot K_T \to G_{T_p}$.
- We take the pushout of k' and r', resulting in a T_{T,T_p}-typed graph H with morphisms $d : D' \to H$ and $h : G'_p \to H$.
- Finally, we build the DynGG $\mathcal{G}' = (T_{T,T_p}, H, P' \cup \iota_{T,T_p} \cdot P)$.

When all the above is applicable, we say that there is a *direct dynamic derivation* α from $\mathcal{G}$ to $\mathcal{G}'$ using p (based on m) and write $\alpha : \mathcal{G} \Rightarrow_p \mathcal{G}'$. A *dynamic derivation* is a sequence of direct dynamic derivations starting from the initial graph.

Example 3.1. Let us consider the T_a-typed grammar $\mathcal{G}_a$ presented in the Introduction (see productions p and q in Figures 1 and 2). The configuration after n applications of p is shown in Figure 5: the type graph has evolved from T_a to $\bigcup_{i=1}^{n} T_{g_i}$ (with $T_a = \bigcap_{i=1}^{n} T_{g_i}$) and there are $n+1$ available productions $p', q_1, ..., q_n$ at the top level that are suitable retyped instances of p and q. However, it is not possible to find a valid matching for any q_i, while there is (always) exactly one valid matching for the application of p.

3.1   About Shallow Graph Grammars

In the case of shallow grammars, the definition of derivation boils down to classic DPO derivation. This can be easily proved by noting that the fresh retyping leads to $T_{T,T_p} = T$ (i.e., the typing is vacuous) and that $P' = \emptyset$ (by definition of shallow grammars).

Proposition 3.1. *DynGGs are a conservative extension of GGs.*

The proof takes any T-typed graph grammar $\mathcal{G} = (T, G_{in}, P)$ and constructs the corresponding T-typed shallow graph grammar $\mathbf{Sh}(\mathcal{G})$. By what said above, it is then immediate to prove that $\delta : G \Rightarrow_p G'$ iff $\delta : \mathbf{Sh}(T, G, P) \Rightarrow_p \mathbf{Sh}(T, G', P)$.

Since the flattening $\mathbf{F}(\mathcal{G}) = (\mathbf{T}(\mathcal{G}), \iota_T \cdot G_{in}, \mathbf{P}(\mathcal{G}))$ of a DynGG $\mathcal{G} = (T, G_{in}, P)$ is also shallow, an obvious question is "how are the behaviours of $\mathbf{F}(\mathcal{G})$ and $\mathcal{G}$ related?"

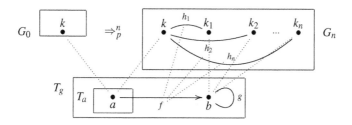

Fig. 6. A flattened derivation where p is applied n times

Proposition 3.2. *Let $\mathcal{G}_0$ be a DynGG, then $\{\delta_i : \mathcal{G}_{i-1} \Rightarrow_{p_i} \mathcal{G}_i\}_{i \in \{1,...,n\}}$ implies $\{\delta'_i : \mathcal{G}'_{i-1} \Rightarrow_{\mathsf{flat}(p'_i)} \mathcal{G}'_i\}_{i \in \{1,...,n\}}$. where $\mathcal{G}'_0 = \mathbf{F}(\mathcal{G}_0)$ and each p_i is instance of $p'_i \in (\mathcal{G})$.*

The proof shows that there is a standard mapping from the dynamically evolving type graph of any $\mathcal{G}$ to the static representative $\mathbf{T}(\mathcal{G})$ and also a mapping from productions dynamically originated by $\mathcal{G}$ to the productions of $\mathbf{F}(\mathcal{G})$. In particular, any two different freshly-generated instances q', q'' of the same production q are mapped to $\mathsf{flat}(q)$. We remind that, contrary to $\mathcal{G}$, all productions in $\mathbf{F}(\mathcal{G})$ are always available and cannot change over time. Thus, any valid match for the dynamic graph remains valid in its flattening (via the retyping) and any direct derivation using the instance p_i of $p'_i \in (\mathcal{G})$ can be simulated using $\mathsf{flat}(p'_i) \in \mathbf{P}(\mathcal{G})$.

The counterexample below shows that $\mathbf{F}(\mathcal{G})$ has possibly more derivations than $\mathcal{G}$.

Example 3.2. Let us take the flattening $\mathbf{F}(\mathcal{G})$ of the T_a-typed DynGG $\mathcal{G}$ in Example 3.1. The configuration after n applications of p is in Figure 6. Note that the type graph remains $T_g = \mathbf{T}(\mathcal{G})$ and that there are only two available productions p, q at any time. Compare the situation with that in Figure 5: in the flattened version it is now possible to apply q to any pair of (distinct) arcs h_i, h_j!

4   Case studies: Dynamic Nets and Join Calculus

Dynamic nets [1,2] are an extension of Petri nets where firings can add fresh places and transitions. In this sense, any DynGG $\mathcal{G} = \langle T, G_{in}, P \rangle$ where $\mathbf{T}(\mathcal{G})$ is a discrete graph can be seen as a dynamic net. Nevertheless, not all dynamic nets can be represented by DynGGs over discrete type graphs because tokens may be coloured with the names of places in the net, and transitions may use such colours to designate places where to spawn new tokens. Since dynamic nets are in one-to-one correspondence with processes of the Join calculus [2], we present the encoding of the latter. Let $\mathcal{N}$ be an infinite set of names ranged by $u, v, x, y, z, \ldots$. The syntax of Join is given by the grammar

$$P ::= 0 \mid x\langle y \rangle \mid \mathbf{def}\, D\, \mathbf{in}\, P \mid P|Q \qquad D ::= J \triangleright P \mid D \wedge D \qquad J ::= x\langle y \rangle \mid J|J$$

The occurrences of x and u in $x\langle u \rangle$ are *free*. Differently, x and y occur bound in $P = \mathbf{def}\, x\langle u \rangle | y\langle v \rangle \triangleright P_1\, \mathbf{in}\, P_2$, while u and v occur bound in $D = x\langle u \rangle | y\langle v \rangle \triangleright P_1$. The sets of free and bound names of P are written respectively $fn(P)$ and $bn(P)$. Moreover, x and y are the defined names of D (written $dn(D)$).

The semantics of the Join calculus relies on the *reflexive chemical abstract machine* model [9]. In this model a solution is roughly a multiset of active definitions $J \triangleright P$ and messages $x\langle u \rangle$ (separated by comma). Moves are distinguished between *structural*, which heat or cool processes, and reductions $\rightarrow$, which are the basic computational steps. The multiset rewriting rules for Join are as follows:

$$\frac{0}{\mathbf{def}\, D\, \mathbf{in}\, P} \quad \frac{P \mid Q \qquad P, Q}{D\sigma_{dn(D)}, P\sigma_{dn(D)}} \quad \frac{D \wedge E \qquad D, E}{(range(\sigma_{dn(D)}) \text{ globally fresh})}$$

$$J \triangleright P, J\sigma \;\rightarrow\; J \triangleright P, P\sigma$$

Structural moves allow for the rearrangement of terms inside a solution. Note that the term denoting a process with local definitions can be represented by two terms (one for the definitions and other for the process) only when the locally defined ports are renamed by fresh names (this rule stands for the dynamic generation of new names). A reduction can take place when the solution contains a rule $J \triangleright P$ and an instance $J\sigma$ of the Join pattern J: when such a match is found, $J\sigma$ is replaced by $P\sigma$. We write $P \mapsto P'$ for $P \quad {}^* Q \rightarrow Q' \quad {}^* P'$.

Join processes as DynGGs. For simplicity we assume definitions not to share names. Any process P is encoded as a DynGG $\mathcal{G}_P = \langle T_P, G_{in}, Q \rangle$. Generally speaking, a channel x will be encoded as node n but the fact that the channel is named x is denoted by an arc $x : n \rightarrow n$. A message $x\langle y \rangle$ is represented with the arc $m : n_1 \rightarrow n_2$, where n_1 corresponds to x and n_2 to y. Any firing rule $J \triangleright P$ will be encoded as a production. More formally, the initial type graph T_P is shown in Figure 7(a), where $fn(P) \cup bn(P) = \{x_1, \ldots, x_n\}$. T_P has a unique node n standing for channels, one arc m for denoting messages, and one arc x_i for any free or bound name of P. We call the *context* of P the T_P-typed graph C_P with one node n_{x_i} and one arc $x_i : n_{x_i} \rightarrow n_{x_i}$ for each $x_i \in fn(P) \cup bn(P)$. Then, the initial graph G_{in} and the set of productions Q are inductively defined as follows:

- $P = 0$. $G_{in} = C_P$ is the empty graph and $Q = \emptyset$.
- $P = x\langle y \rangle$. If $x \neq y$, then $G_{in} = C_P \cup \{m : n_x \rightarrow n_y\}$ is the graph shown in Figure 7(b), with the typing morphism mapping both nodes to n and being the identity on arcs. Otherwise, $G_{in} = C_P \cup \{m : n_x \rightarrow n_x\}$ is as in Figure 7(c). In both cases $Q = \emptyset$.
- $P = \mathbf{def}\, J_1 \triangleright P_1 \wedge \ldots \wedge J_n \triangleright P_n \mathbf{in}\, P'$. Let $\mathcal{G}_{P'} = \langle T_{P'}, G'_{in}, Q' \rangle$ be the encoding of P', then $\mathcal{G}_P = \langle T_P, C_P \cup G'_{in}, Q' \cup \bigcup_{1 \leq n \leq n} \{p_i\} \rangle$, where $T_P \supseteq T_{P'}$ and p_i encodes $J_i \triangleright P_i$. Assuming $J_i = x_1 \langle u_1 \rangle | \ldots | x_k \langle u_k \rangle$, then p_i is shown in Figure 7(d), where $\mathcal{G}'_{P_i}$ is the extension of $\mathcal{G}_{P_i} = \langle T_i, G_{in_i}, Q_i \rangle$ over the type graph $T_i \cup T_P$ and whose initial graph is the union of G_{in_i} with the items preserved by the production. The self-loop arcs naming the nodes u_i are not present in p_i because the identities of formal parameters are not known a priori and they will be provided by valid matchings. Moreover, the left-hand-side and the interface contain a node n_{y_h} and an arc y_h for any free name y_h of P_i not in $\{x_1, \ldots, x_k, u_1, \ldots, u_k\}$. In this way the context of the initial graph of $\mathcal{G}_{P_i}$ is bound to the names of the left-hand-side of the production.
- $P = P_1 | P_2$. Let $\mathcal{G}_{P_1} = \langle T_1, G_{in_1}, Q_1 \rangle$ and $\mathcal{G}_{P_2} = \langle T_2, G_{in_2}, Q_2 \rangle$ be the encoding of P_1 and P_2, then the initial graph is the pushout object of the span $C_P \cup G_{in_1} \hookleftarrow C_P \hookrightarrow C_P \cup G_{in_2}$, and Q is the union of Q_1 and Q_2 (upon production retyping over T_P).

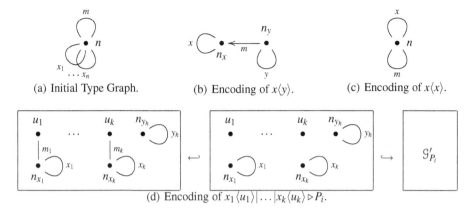

(a) Initial Type Graph.   (b) Encoding of $x\langle y\rangle$.   (c) Encoding of $x\langle x\rangle$.

(d) Encoding of $x_1\langle u_1\rangle|\ldots|x_k\langle u_k\rangle \triangleright P_i$.

Fig. 7. Join processes as Dynamic Graph Grammars

Example 4.1. Let $P = \mathbf{def}\, x\langle u\rangle \triangleright (\mathbf{def}\, y\langle v\rangle \triangleright v\langle y\rangle \,\mathbf{in}\, y\langle u\rangle | x\langle y\rangle)\,\mathbf{in}\, x\langle z\rangle$. The corresponding grammar is $\mathcal{G}_P = \langle T_P, G_{in}, \{p\}\rangle$, where T_P and G_{in} are shown in Figure 8(a). The unique top-level production p is in Figure 8(b) (for space reasons we omit the representation of the typing). The right-hand-side of p is typed over the graph T' that adds two fresh arc types $y : n \to n$ and $u : n \to n$ to T_P. Production p describes the consumption of a message sent to the channel x regardless of the name contained by the message (note that the particular name of the port n_u is not fixed by the production). When fired, p generates two fresh types $y' : n \to n$ and $u' : n \to n$ and modifies the underlying graph by removing m_1 and by adding (*i*) a new node $n_{y'}$, (*ii*) a new arc of the fresh type y', (*iii*) a new arc of the fresh type u' that works as an alias for the actual name of the actual parameter n_u, and (*iv*) two new messages m_2 and m_3. Moreover, p spawns a new production q (Figure 8(c)), which handles the messages sent to the fresh port y'.

The encoding of Join processes establishes a tight correspondence between derivations in the two frameworks. The following results hold up to aliasing of names (i.e., by removing aliasing from grammars).

Proposition 4.1. *For any Join process P we have:*

- *If $P \mapsto P'$ using $J_i \triangleright P_i$, then $\exists Q$ s.t. $\mathcal{G}_P \Rightarrow_{p_i} \mathcal{G}_Q$ and $Q \quad{}^* P'$;*
- *If $\mathcal{G}_P \Rightarrow_{p_i} \mathcal{G}'$, then $\exists P'$ s.t. $P \mapsto P'$ using $J_i \triangleright P_i$ and $\mathcal{G}' = \mathcal{G}_{P'}$.*

5   Encoding Dynamic Graph Grammars as Graph Grammars

In this section we show that DynGGs can be encoded back in GGs. The encoding of a Dynamic Graph Grammar $\mathcal{G}$ relies on the definition of a unique type graph expressive enough for distinguishing all the types generated dynamically by $\mathcal{G}$. As a first step, we show how to describe a chain of types $\mathcal{T}$ ordered by inclusions $T_1 \subset T_2 \subset \ldots \subset T_n$ with a unique type graph $\{[T_n]\}_{\mathcal{T}}$, called the *refined type graph*. Informally, any item (i.e., node or arc) of T_n is mapped to a node in $\{[T_n]\}_{\mathcal{T}}$. Every graph T_i in the chain $\mathcal{T}$ is also

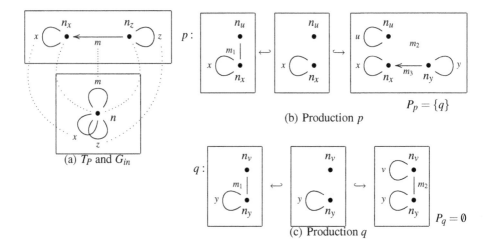

(a) T_p and G_{in} (b) Production p

(c) Production q

Fig. 8. A Join process as a DynGG

represented in $\{[T_n]\}_{\mathcal{T}}$ by a node n_{T_i}. Moreover, any node w corresponding to an item k of T_n has an arc t_w to the node n_{T_i} if T_i is the minimal type in $\mathcal{T}$ that includes k. We call T_i the *type of k* in $\mathcal{T}$. Formally, for any $k \in T_n$, the type of k is $\mathcal{T}(k) = T_i$ if $k \in T_i \setminus T_{i-1}$.

Definition 5.1 (Refined type graph). *Given a type graph T and a chain of types $\mathcal{T} = T_1 \subset \ldots \subset T_n$, with $T_n = T$, the* refined type graph *is $\{[T]\}_{\mathcal{T}} = \langle N_R, E_R, s_R, t_R \rangle$, where:*

- $N_R = N_T \cup E_T \cup \{n_{T_i} | T_i \in \mathcal{T}\}$ *where n_{T_i} are fresh names, i.e., the nodes of $\{[T]\}_{\mathcal{T}}$ are the nodes and arcs of T plus one extra node for any type in $\mathcal{T}$;*
- $E_R = \{e_0, e_1 | e \in E_T\} \cup \{t_w | w \in N_T \cup E_T\} \cup \{s_{i,i+1} | 0 < 1 < n-1\}$ *where all edge names are fresh. Source and target functions are defined s.t. the following holds:*
  - $e_0 : s_T(e) \to e$ *and* $e_1 : e \to t_T(e)$*, i.e., e_0 connects the node $e \in N_R$ to its original source in T while e_1 connects e to its target;*
  - $t_w : w \to n_{\mathcal{T}(w)}$*, i.e., t_w connects w to the node representing its type;*
  - $s_{i,i+1} : n_{T_i} \to n_{T_{i+1}}$ *denotes the inclusion of types $T_i \subset T_{i+1}$.*

Example 5.1. Consider the type graph T_g depicted in the bottom part of the Figure 9(a). The refined type graph for the chain $T_a = \{a\} \subset T_g = T_a \cup \{f, b, g\}$ is shown at the bottom of Figure 9(b). The original arc f (resp. g) of T_g is represented by the homonymous node f (resp. g) and the pair of fresh arcs f_0 and f_1 (resp. g_0 and g_1). The types T_a and T_g are represented by the fresh nodes n_{T_a} and n_{T_g}, while the inclusion relation $T_a \subset T_g$ is denoted by the arc $s_{1,2}$. Finally, for any item w, t_w connects w to its type node.

Definition 5.2 (Refined T-Typed Graph). *Given a T-typed graph $G = \langle |G|, \tau_G \rangle$ and a chain $\mathcal{T} = T_1 \subset \ldots \subset T_n = T$, the $\{[T]\}_{\mathcal{T}}$-typed graph $\{[G]\}_{\mathcal{T}} = \langle |H|, \tau_H \rangle$ is defined as:*

- $N_H = N_G \cup E_G \cup \{n_{T_i} | T_i \in \mathcal{T}\}$*, i.e., N_H has all items of G plus nodes denoting types;*
- $E_H = \{e_0, e_1 | e \in E_G\} \cup \{t_w | w \in N_G \cup E_G\} \cup \{s_{i,i+1} | 0 < 1 < n-1\}$*, where:*

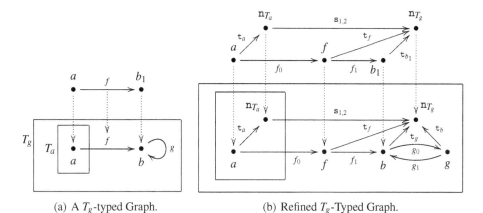

(a) A T_g-typed Graph.   (b) Refined T_g-Typed Graph.

Fig. 9. A refined T-Typed Graph

- $e_0 : s_T(e) \to e$ and $e_1 : e \to t_T(e)$;
- $t_w : w \to n_{\mathcal{J}(\tau_G(w))}$, *i.e.*, t_w *connects w to the node representing its type in* $\mathcal{J}$, *which is obtained by using the typing morphism* $\tau_G(w)$;
- $s_{i,i+1} : n_{T_i} \to n_{T_{i+1}}$, *for the inclusion of types.*
- *The typing morphism* τ_H *is defined as follows*

$$\tau_H(k) = \tau_G(k) \text{ if } k \in G \qquad \tau_H(n_{T_i}) = n_{T_i}$$
$$\tau_H(e_i) = \tau_G(e)_i \qquad \tau_H(t_w) = t_{\tau_G(w)} \qquad \tau_H(s_{i,i+1}) = s_{i,i+1}$$

Example 5.2. Consider the T_g-typed graph G in Figure 9(a). Its refined version for the chain $T_a = \{a\} \subset T_g = T_a \cup \{f, b, g\}$ is shown in Figure 9(b) (we omit the representation of the obvious typing of arcs).

We refer to the nodes n_{T_i} and the arcs t_w and $s_{i,i+1}$ of a refined type graph (resp., a refined T-typed graph) as the *location of the type graph* (resp., *location of the graph*).

Definition 5.3 (Refined T-Typed DynGG). *Let* $\mathcal{G} = (T, G_{in}, P)$ *be a DynGG, and* $\mathcal{J} = T_1 \subset \ldots \subset T_n$, *with* $T_n = T$ *be a chain of types. Then, the refined version of* $\mathcal{G}$ *is defined as* $\mathcal{G}_{\mathcal{J}} = (\{[T]\}_{\mathcal{J}}, \{[G_{in}]\}_{\mathcal{J}}, \{[P]\}_{\mathcal{J}})$, *where* $\{[P]\}_{\mathcal{J}} = \{\{[p]\}_{\mathcal{J}} | p \in P\}$ *is obtained by encoding any production* $p : (L \xleftarrow{l} K \xrightarrow{r} (T', G'_{in}, P'))$ *in P as follows:*

$$\{[p]\}_{\mathcal{J}} : (\{[L]\}_{\mathcal{J}} \xleftarrow{l'} \{[K]\}_{\mathcal{J}} \xrightarrow{r'} \{[(T', G'_{in}, P')]\}_{\mathcal{J} \subset T'})$$

where morphisms l' *and* r' *are the obvious extensions of l and r with the identity over the location of the graph.*

Example 5.3. Consider the production p in Figure 1. Its refined version is in Figure 10. The type graphs are the refined versions of the original type graphs, while the left-hand-side, the interface, and the right-hand-side are the refined version of the original ones. In particular, the left-hand-side is typed over the refined version of T_a, while the right-hand-side grammar is typed over the refined version of T_g. Moreover, the production

$\{[p]\}_{T_a}$:

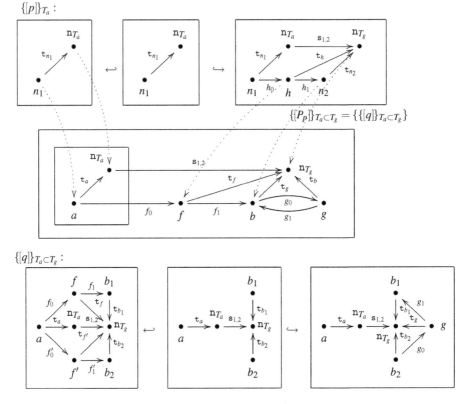

Fig. 10. A refined production p

$\{[q]\}_{T_a \subset T_g}$ created by the reduction corresponds to the refined versions of the original q (for clarity we do not draw the typing morphism, which is the obvious one).

The refined version of a grammar $\mathcal{G}$ recreates the static tree of types $\mathbf{T}(\mathcal{G})$. In fact, any production $p : (L_T \xleftarrow{l'} K_T \xrightarrow{r'} \mathcal{G}_{T'})$ is encoded by considering the path $\mathcal{T}$ of $\mathbf{T}(\mathcal{G})$ starting from the root of $\mathbf{T}(\mathcal{G})$ to T. Moreover, since previous definitions can be straightforwardly extended to consider the whole tree instead of a path, we will use $\{[\_]\}_{\mathcal{T}}$ to denote also the refined versions obtained by considering the tree of types $\mathcal{T}$. Given any tree $\mathcal{T}$ describing type inclusions, the tree $\mathcal{T}' = \mathcal{T}, T \hookrightarrow T'$ stands for the tree $\mathcal{T}$ with the addition of the type T' as a child of T (if T' is already in the tree, then $\mathcal{T}' = \mathcal{T}$).

Remark 5.1. For simplicity, we assume the name of any production to be decorated with the types of its left and right-hand-sides, i.e., $p_{T \hookrightarrow T'} : (L_T \xleftarrow{l} K_T \xrightarrow{r} \mathcal{G}_{T'})$.

The result below shows that a refined grammar behaves like the original one.

Lemma 5.1. *Let* $\mathcal{G}_0 = (T_0, H_0, P_0)$ *and* $\mathcal{G}_n = (T_n, H_n, P_n)$ *be DynGGs, then*

$$\{\delta_i : \mathcal{G}_i \Rightarrow_{p_{i_{T_{p_i} \hookrightarrow T'_{p_i}}}} \mathcal{G}_{i+1}\}_{i \in \{0,...,n-1\}} \quad \textit{iff} \quad \{\delta'_i : \{[\mathcal{G}_i]\}_{\mathcal{T}_i} \Rightarrow_{p_i} \{[\mathcal{G}_{i+1}]\}_{\mathcal{T}_{i+1}}\}_{i \in \{0,...,n-1\}}$$

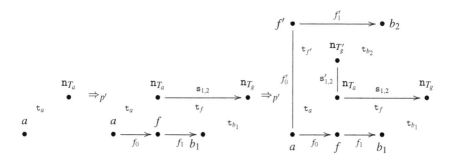

Fig. 11. A flattened, refined derivation

where $\mathcal{I}_0 = T_0$ and $\mathcal{I}_{i+1} = \mathcal{I}_i, T_{p_i} \hookrightarrow T'_{p_i}$ for $1 \le i \le n$.

Proof (Sketch). Consider $\mathcal{G}_i \Rightarrow_{p_{T \hookrightarrow T'}} \mathcal{G}_{i+1}$ and $p_{T \hookrightarrow T'} : (L_T \overset{l}{\leftarrow} K_T \overset{r}{\rightarrow} \mathcal{G}_{T'})$. Then, the derivation $\mathcal{G}_i \Rightarrow_{T \hookrightarrow T'} \mathcal{G}_j$ is analogous to that one in Figure 4. By construction of $\{[\mathcal{G}_i]\}_{\mathcal{I}_i}$, there exists $m : L_T \to G_{in}$ iff there exists $m' : \{[L_T]\}_{\mathcal{I}_i} \to \{[G_{in}]\}_{\mathcal{I}_i}$ in the refined grammar. Since $\{[K_T]\}_{\mathcal{I}_i}$ preserves the "same" elements (up-to suitable encoding) as K_T plus the location of the graph, then D' obtained as the pushout complement of $\langle m', l' \rangle$ coincides with $\{[D]\}_{\mathcal{I}}$. Since $\{[R_T]\}_{\mathcal{I}_i}$ preserves the location of already existing items and generates a new node $n_{T'}$ and a new arc $s : n_T \to n_{T'}$ for typing fresh items, then H' coincides with $\{[H]\}_{\mathcal{I}_i, T \hookrightarrow T'}$. The correspondence among fresh productions is straightforward.

Definition 5.4 (Encoding). *Let* $\mathcal{G} = \langle T, G_{in}, P \rangle$ *be a DynGG. Then, the equivalent graph grammar* $\{[\mathcal{G}]\}$ *is defined as* $\{[\mathcal{G}]\} = \mathbf{F}(\{[\mathcal{G}]\}_T)$.

Example 5.4. Consider the DynGG $\mathcal{G} = \langle T_a, G_{in}, \{p\} \rangle$ with T_a and p as in Figures 1 and 2 and its encoding $\{[\mathcal{G}]\} = \langle T', \{[G_{in}]\}_{T_a}, \{p', q'\} \rangle$, where T' is the type graph in Figure 10, and p' and q' are analogous to $\{[p]\}_{T_a}$ and $\{[q]\}_{T_a \subset T_g}$ in Figure 10. Figure 11 shows a derivation that applies twice the rule p' over the initial graph consisting of a unique node typed a. Although the final graph contains two arrows of type f with same source of type a, there is not a matching for q', since the left-hand-side of q' requires the targets of the two arrows to have the same location. Hence, the encoding does not confuse different instantiations of the same type, as formalised by the following result.

Theorem 5.1 (Correspondence). *Let* $\mathcal{G}_0$ *be a dynamic graph grammar, then*

$$\{\delta_i : \mathcal{G}_{i-1} \Rightarrow_{p_i} \mathcal{G}_i\}_{i \in \{0, \dots, n\}} \quad iff \quad \{\delta'_i : \mathcal{G}'_{i-1} \Rightarrow_{\{[p_i]\}} \mathcal{G}'_i\}_{i \in \{0, \dots, n\}}$$

where $\mathcal{G}'_0 = \{[\mathcal{G}_0]\}$ *and* $\{[p]\}$ *is the encoding of the rule* p *in* $\mathbf{F}(\mathcal{G}_0)$.

Proof. $\Rightarrow$) Follows immediately by Lemma 5.1 and Proposition 3.2. $\Leftarrow$) It remains to prove that the matchings on the encoded version are the same as those of the original one. In fact, the encoding of a rule assures the location of any graphs to identify items with the same type while differentiating items with distinct types.

6   Concluding Remarks

We have proposed the original framework of Dynamic Graph Grammars, as a conservative extension of Graph Grammars that offers a convenient level of abstraction for modeling reflexive systems. Our main result proves that Dynamic Graph Grammars can be simulated by ordinary Graph Grammars, though a non-trivial encoding is necessary.

When compared to the vast literature of theoretical foundations and applications of graph transformation systems, our investigation on reflexive productions is still preliminary under many aspects. A fully extensive development and assessment is therefore a very ambitious programme, along which we foresee several promising directions: (1) to express suitable notion of independent derivations, parallelism, process semantics, unfolding semantics and event structure semantics so to fully develop a true concurrent semantics of DynGGs; (2) to show that concurrency is preserved by our encoding of DynGGs in GGs; (3) to consider other flavours of dynamic productions, like the SPO [12,7]; (4) to exploit the encoding in § 5 to reuse verification tools developed for GGs for systems modeled using DynGGs.

Acknowledgement. The authors want to thank Ivan Lanese for many helpful discussions on the encoding of DynGGs back to GGs.

References

1. A. Asperti and N. Busi. Mobile Petri nets. Technical Report UBLCS 96-10, Computer Science Department, University of Bologna, 1996.
2. M. Buscemi and V. Sassone. High-level Petri nets as type theories in the Join calculus. *Proc. of FoSSaCS'01, Lect. Notes in Comput. Sci.* 2030, pp. 104–120. Springer, 2001.
3. A. Corradini, P. Degano, and U. Montanari. Specifying highly concurrent data structure manipulation. *Proc. of Computing'85: A Broad Perspective of Current Developments.* 1985.
4. A. Corradini, U. Montanari, F. Rossi, H. Ehrig, R. Heckel, and M. Löwe. Algebraic approaches to graph transformation I: Basic concepts and double pushout approach. In [13].
5. A. Corradini, U. Montanari, and F. Rossi. Graph processes. *Fund. Inf.*, 26:241–265, 1996.
6. F. Drewes, H.-J. Kreowski, and A. Habel. Hyperedge replacement graph grammars. In [13].
7. H. Ehrig, R. Heckel, M. Korff, M. Löwe, L. Ribeiro, A. Wagner, and A. Corradini. Algebraic approaches to graph transformation II: Single pushout approach and comparison with double pushout approach. In [13].
8. J. Engelfriet and G. Rozenberg. Node replacement graph grammars. In [13].
9. C. Fournet and G. Gonthier. The reflexive chemical abstract machine and the Join calculus. *Proc. of POPL'96*, pp. 372–385. ACM Press, 1996.
10. I. Lanese. *Synchronization Strategies for Global Computing Models.* PhD thesis, Department of Computer Science, University of Pisa, 2006.
11. D. Hirsch. *Graph Transformation Models for Software Architecture Styles.* PhD thesis, Departamento de Computación, Universidad de Buenos Aires, 2003.
12. M. Löwe. Algebraic approach to single-pushout graph transformation. *Theoret. Comput. Sci.*, 109:181–224, 1993.
13. G. Rozenberg, editor. *Handbook of Graph Grammars and Computing by Graph Transformation. Vol. 1: Foundations.* World Scientific, 1997.

Autonomous Units and Their Semantics — The Sequential Case*

Karsten Hölscher, Hans-Jörg Kreowski, and Sabine Kuske

University of Bremen, Department of Computer Science
P.O. Box 330440, D-28334 Bremen, Germany
{hoelscher, kreo, kuske}@informatik.uni-bremen.de

Abstract. In this paper, we introduce the notion of a community of autonomous units as a rule-based and graph-transformational device to model processes that run interactively but independently of each other in a common environment. The emphasis of the approach is laid on the study of the formal semantics of a community as a whole and of each of its member units separately. We concentrate on the sequential case where only one unit can act at a time and the rule applications of the involved units are interleaved with each other.

1 Introduction

Data processing of today (like communication networks, multiagent systems, swarm intelligence, ubiquitous, wearable and mobile computing) is often distributed and comprises various components that run partially independent of each other, but may access and update the same information structures, communicate with each other and interact in various ways. They may cooperate to reach a common goal or may compete with each other to achieve their individual aims. Typical examples of this kind are logistic processes and systems like transport and production networks where many actors from different companies come together and cooperate to a certain degree. But they are usually still competitors who are not willing to transfer their control to others or to a central entity. On the more technical level, transport networks, for example, comprise many transport vehicles, lots of goods to be shipped, various further components for storing, loading, reloading, etc. It is not meaningful to model such a network as a centralized system with a single control. The same applies to production networks with respect to the involved machines, materials, storage areas, etc.

The main idea of this paper is to provide a formal graph-transformational and rule-based framework for the modeling of such systems composed of a variety of highly self-controlled components that make their decisions on their own depending on the information they get from their environment.

* Research partially supported by the EC Research Training Network SegraVis (Syntactic and Semantic Integration of Visual Modeling Techniques) and the Collaborative Research Centre 637 (Autonomous Cooperating Logistic Processes: A Paradigm Shift and Its Limitations) funded by the German Research Foundation (DFG).

A. Corradini et al. (Eds.): ICGT 2006, LNCS 4178, pp. 245–259, 2006.

The basic notion is that of a community of autonomous units which exist in a common environment (assumed to be a graph). There are initial environments to start computational processes, and there is an overall goal. Each autonomous unit in a community has its own individual goal in addition. To reach its goal, the unit can apply its rules or ask imported units for help. Moreover, each unit has a control mechanism to decide which rule is applied next or which imported unit is used next. This establishes the autonomy of a unit.

In this paper, we concentrate on the sequential semantics of communities of autonomous units. The semantics is given by all sequential processes - finite and infinite - that start in an initial environment, are composed of rule applications of autonomous units and calls of imported units, and follow, in each step, the control of the active unit. From the point of view of a single unit, this means that its own actions (being rule applications or calls of imported units) take place interleaved with other changes of the dynamic environment caused by the coexisting units. Clearly, the sequential semantics is only adequate if one deals with systems in which activities take place one after the other. Examples of this kind are card and board games, sequential algorithms, single-processor systems and such. Moreover, there are many modeling approaches the semantics of which assumes one action at a time. But even sequential systems may consist of self-controling components that decide about their own activities independently of the others like the examples of card and board games with several players show.

Autonomous units generalize our former modeling concept of graph transformation units (see, e.g., [1]). While the latter apply their rules and call imported units without any interference from the outside, an autonomous unit works in a dynamic environment which may change because of the activities of other units in the community. This makes a tremendous difference because the running of the system is no longer controlled by a central entity.

The benefit we expect of using autonomous units is to obtain an easy-to-use and visually well-understandable formal framework with a precise semantics that allows to model systems of interacting components so that on the one hand external control structures are set aside and on the other hand string-based representation is replaced by graph- and rule-based representation that allows to visualize and specify the systems more like they are. Nevertheless, the presented concepts are not restricted to graphs and graph transformation but can be used for any rule-based mechanism where rules modify some kind of configurations (cf. also [2]).

The paper is organized as follows. In Sect. 2 we briefly recall the notion of a graph transformation approach. In Sect. 3 autonomous units are introduced and a sequential semantics for them is given. Section 4 presents communities of autonomous units and how they interact within a common environment. Section 5 compares communities of autonomous units with the original transformation units introduced and studied in [1]. In Sect. 6 we present a case study modeling the players of the board game *Ludo* as autonomous units. The conclusion is given in Sect. 7. For reasons of space limitations proofs are omitted in this paper.

2   Graph Transformation Approaches

Whenever one has to do with dynamic graph-like structures, graph transformation (see also [3]) constitutes an adequate formal specification technique because it supports the visual and rule-based transformation of such structures in an intuitive and direct way. The ingredients of graph transformation are provided by a so-called graph transformation approach. In this section, we recall the notion of a graph transformation approach as introduced in [1] but modified with respect to the class of control conditions.

Two basic components of every graph transformation approach are a class of graphs, and a class of rules that can be applied to these graphs. In many cases, rule application is highly nondeterministic — a property that is not always desirable. Hence, graph transformation approaches can also provide a class of control conditions so that the degree of nondeterminism of rule application can be reduced. Moreover, graph class expressions can be used in order to specify for example sets of initial and terminal graphs of graph transformation processes.

Formally, a graph transformation approach is a system $\mathcal{A} = (\mathcal{G}, \mathcal{R}, \mathcal{X}, \mathcal{C})$ the components of which are defined as follows.

- $\mathcal{G}$ is a class of *graphs*.
- $\mathcal{R}$ is a class of *graph transformation rules* such that every $r \in \mathcal{R}$ specifies a binary relation on graphs $SEM(r) \subseteq \mathcal{G} \times \mathcal{G}$.
- $\mathcal{X}$ is a class of *graph class expressions* such that each $x \in \mathcal{X}$ specifies a set of graphs $SEM(x) \subseteq \mathcal{G}$.
- $\mathcal{C}$ is a class of *control conditions* such that each $c \in \mathcal{C}$ specifies a set of sequences $SEM_{E, Change}(c) \subseteq SEQ(\mathcal{G})$ where $E \colon ID \rightarrow 2^{SEQ(\ )}$, for some set ID of names and $Change \subseteq \mathcal{G} \times \mathcal{G}$.[1] As we will see later the mapping E is meant to associate a semantics to rules and imported autonomous units. The relation $Change$ defines the changes that can occur in the environment of an autonomous unit. Hence, control conditions have a loose semantics which depends on the semantics associated to rules and imported units via the mapping E and on the changes of the environment given by $Change$.

For technical simplicity we assume in the following that ID is an arbitrary but fixed set with $\mathcal{R} \subseteq ID$ and that $\mathcal{A} = (\mathcal{G}, \mathcal{R}, \mathcal{X}, \mathcal{C})$ is an arbitrary but fixed graph transformation approach.

3   Autonomous Units

Autonomous units act within or interact on a common environment which is modeled as a graph. An autonomous unit consists of a set of graph transformation rules, a control condition, and a goal. Moreover, it can import other units to which it may delegate auxiliary tasks. The graph transformation rules contained

---

[1] For a set A 2^A denotes its powerset and $SEQ(A)$ the set of finite and infinite sequences over A.

in an autonomous unit *aut* and the imported units of *aut* specify all transformations the unit *aut* can perform. Such a transformation comprises for example a movement of the autonomous unit within the current environment, the exchange of information with other units via the environment, or local changes of the environment. The control condition regulates the application process. For example, it may require that a sequence of rules be applied as long as possible or infinitely often. In this first approach the goal of a unit is a graph class expression determining how the transformed graphs should look like.

Definition 1 (Autonomous unit). An *autonomous unit* is a system $aut = (g, U, P, c)$ where $g \in \mathcal{X}$ is the *goal*, U is a set of imported autonomous units, $P \subseteq \mathcal{R}$ is a set of graph transformation rules, and $c \in \mathcal{C}$ is a control condition. The components of *aut* are also denoted by g_{aut}, U_{aut}, P_{aut}, and c_{aut}, respectively.

In the following we consider only autonomous units with acyclic import structure. Moreover, for technical simplicity we assume that in addition to the rules all autonomous units are contained in the set *ID*.

An autonomous unit modifies an underlying environment while striving for its goal. Its semantics consists of a set of transformation processes being finite or infinite sequences of environment transformations. An environment transformation comprises the application of a local rule or a transformation process performed by an imported unit or an environment change typically performed by another autonomous unit that is working in the same environment. These environment changes are given as a binary relation of environments. Hence, in this sequential approach a transformation process of an autonomous unit interleaves local rule applications with transformation processes of imported units and environment changes specified by other components. This implies that an environment transformation of an imported unit cannot be interrupted by changes of the importing unit but it can be interleaved with the change relation induced by other components. Hence, every autonomous unit has exactly one thread of control. Autonomous units regulate their transformation processes by choosing in every step only those rules or imported units that are allowed by its control condition.

The definition of the sequential semantics of autonomous units makes use of the sequential composition of sequences. Let $s = (x_0, \ldots, x_n)$ and $s' = (x'_0, x'_1, \ldots)$ be sequences such that s is finite, s' is finite or infinite, and $x'_0 = x_n$. Then the *sequential composition* of s and s' is equal to $s \circ s' = (x_0, \ldots x_n, x'_1, \ldots)$. Moreover, the number of elements of a finite sequence $s = (x_0, \ldots, x_n)$ is equal to $n + 1$ and is denoted by $|s|$. For an infinite sequence s its number of elements is $|s| = \infty$. The first element of a sequence s is denoted by $first(s)$ and its last element by $last(s)$ in the case where s is finite.

Definition 2 (Sequential semantics). Let $aut = (g, U, P, c)$ be an autonomous unit and let $Change \subseteq \mathcal{G} \times \mathcal{G}$. Let $s \in SEQ(\mathcal{G})$. Then $s \in SEM_{Change}(aut)$ if

- there is a sequence $seq = (s_0, s_1, \ldots) \in SEQ(SEQ(\mathcal{G}))$ such that
  - for $0 \leq i < |seq|$ if seq is finite and for all $i \in$ if seq is infinite, s_i is finite and $last(s_i) = first(s_{i+1})$;

- $s = s_0 \circ s_1 \circ \cdots;^2$
- for $i = 0, \ldots, |seq|$ if seq is finite and for all $i \in$ if seq is infinite

$$s_i \in \bigcup_{p \ P} SEM(p) \cup Change \cup \bigcup_{u \ U} SEM_{Change}(u).$$

- $s \in SEM_{E(aut),Change}(c)$ with $E(aut)(id) = SEM(id)$ if $id \in P$, $E(aut)(id) = SEM_{Change}(id)$ if $id \in U$, and $E(aut)(id) = \emptyset$, otherwise.

It is worth noting that the semantics of autonomous units is inductively defined meaning that it covers the case where no unit is imported and in the case where the set of imported units is not empty the semantics of every imported unit is recursively computed.

Every autonomous unit induces a set of atomic (i.e. not interruptible) environment transformations that consist of the semantic relation of all local rules plus the atomic transformations of the imported autonomous units.

Definition 3 (Atomic transformations). The set of *atomic transformations* of an autonomous transformation unit $aut = (g, U, P, c)$ is defined as $AT(aut) = \bigcup_{p \ P} SEM(p) \cup \bigcup_{u \ U} AT(u)$.

As one would expect a transformation process of an autonomous unit consists of a sequence of atomic transformations interleaved with other changes of the environment.

Observation 1. Let $aut = (g, U, P, c)$ be an autonomous unit and let $s = (G_0, G_1, \ldots) \in SEM_{Change}(aut)$ for some relation $Change \subseteq \mathcal{G} \times \mathcal{G}$. Then for $i = 1, \ldots, |s|$ if s is finite and for all $i \in$ $^+$ if s is infinite $(G_{i\ 1}, G_i) \in AT(aut) \cup Change.^3$

4 Communities of Autonomous Units

Autonomous units are meant to work within a community of autonomous units that modify the common environment together. In the sequential case these modifications take place in an interleaving manner. Every community is composed of an overall goal that should be achieved, an environment specification that specifies the set of initial environments the community may start working with, and a set of autonomous units. The overall goal may be closely related to the goals of the autonomous units in the community. Typical examples are the goals admitting only graphs that satisfy the goals of one or all autonomous units in the community.

Definition 4 (Community). A *community* is a triple $COM = (Goal, Init, Aut)$, where $Goal, Init \in \mathcal{X}$ are graph class expressions called the *overall goal* and the *initial environment specification*, respectively, and Aut is a set of autonomous units.

---

^{2} Please note that $s = s_0 \circ s_1 \circ \cdots$ stands for $s = s_0 \circ s_1 \circ \cdots \circ s_{|seq|}$ if seq is finite.
^{3} $^+$ denotes the set of all positive natural numbers.

In a community all units work on the common environment in a self-controlled way by applying their rules or their imported units. The change relation integrated in the semantics of autonomous units makes it possible to define an interleaving semantics of a community in which every autonomous unit may perform its transformation processes. For this purpose it is necessary to know for every autonomous unit the set of atomic transformations of all other units in the community.

Definition 5 (Change relation). Let $COM = (Goal, Init, Aut)$ be a community. Then for each $aut \in Aut$ the *change relation* w.r.t. aut is defined as $Change(aut) = \bigcup_{aut' \ Aut \ aut} AT(aut')$.

Every transformation process of a community must start with a graph specified as an initial environment of the community. Moreover, it must be in the sequential semantics of every autonomous unit participating in the community. A finite transformation process of a community is successful if its last environment satisfies the overall goal. Every infinite transformation process of a community is successful if it meets infinitely many environments that satisfy the overall goal.

Definition 6 (Sequential community semantics)

1. Let $COM = (Goal, Init, Aut)$. Then the *sequential community semantics* of COM consists of all finite or infinite sequences $s = (G_0, G_1, \ldots) \in SEQ(\mathcal{G})$ such that $G_0 \in SEM(Init)$ and $s \in SEM_{Change(aut)}(aut)$ for all $aut \in Aut$.
2. The sequence s is called a *successful transformation process* if s is finite and $G_s \in SEM(Goal)$ or if for all $j \in$ there is a finite sequence $s_j = (G_{j,0}, \ldots, G_{j,n_j})$ with $G_{j,n_j} \in SEM(Goal)$ such that $s = s_0 \circ s_1 \circ \cdots$. The set of all successful transformation processes of COM is denoted by $STP(COM)$.

As the definition of the community semantics shows, there is a strong connection between the semantics of a community $COM = (Goal, Init, Aut)$ and the semantics of an autonomous unit $aut \in Aut$. More precisely, the semantics of COM is a subset of the semantics of aut w.r.t. the change relation $Change(aut)$.

For the community semantics we can also show in a straightforward way that only atomic transformations of the participating units are applied in every transformation process.

Observation 2. Let $COM = (Goal, Init, Aut)$ be a community and let $s = (G_0, G_1, \ldots) \in SEM(COM)$. Then for $i = 1, \ldots, |s|$ if s is finite and for all $i \in {}^+$ if s is infinite $(G_{i\ 1}, G_i) \in \bigcup_{aut\ Aut} AT(aut)$.

5   Comparison with Transformation Units

In this section we compare communities of autonomous units with transformation units that have an acyclic import structure (see e.g. [1]). Autonomous units are up to a certain degree similar to transformation units because both concepts are graph- and rule-based, use control conditions, import other units, and employ graph class expressions. Nevertheless there are some fundamental differences: (1) While an autonomous unit interacts with other autonomous units, an ordinary transformation unit runs its computations without interference of other units except those imported. Hence the semantics of transformation units is not defined with respect to possible environment changes. (2) All transformations are solely controlled by the transformation units whereas an autonomous unit controls its own actions, but not those of the other units in the community. (3) The semantics of a transformation unit and consequently also the semantics of control conditions used in transformation units are binary relations on graphs whereas the semantics of autonomous units consists of finite and infinite transformation processes, i.e. finite and infinite sequences of graphs. Because of these differences communities have the advantage that systems such as logistic processes or games consisting of many automomously and perhaps infinitely long acting components can be modeled in a more realistic way.

   If one considers only the semantic relation induced by all finite transformation sequences of communities and if there exist appropriate control conditions in the underlying graph transformation approach, one can translate transformation units into communities. To this aim we define for every finite sequence $s = (G_0, \ldots, G_n)$ its *induced pair* as $pair(s) = (G_0, G_n)$ and for every set S of sequences its *induced binary relation* as $rel(S) = \{pair(s) \mid s \in S'\}$ where S' is the set of all finite sequences in S. Moreover, for a set $\mathcal{G}$, its identity relation is the set $\Delta\mathcal{G} = \{(G, G) \mid G \in \mathcal{G}\}$. Finally, for a binary relation $R \subseteq \mathcal{G} \times \mathcal{G}$ the set $\overline{R}$ denotes the set of all finite sequences obtained from sequentially composing pairs of $R \cup \Delta\mathcal{G}$, i.e. $\overline{R} = \{(r_0, \ldots, r_n) \mid (r_{i-1}, r_i) \in R \cup \Delta\mathcal{G} \text{ for } i = 1, \ldots, n, n \geq 1\}$.[4]

5.1   Transformation Units

A *transformation unit* is a system $tu = (I, U, P, C, T)$ where $I, T \in \mathcal{X}$, U is a set of imported transformation units, $P \subseteq \mathcal{R}$, and C is a control condition that specifies for every mapping $E \colon ID \to 2^{\overline{}}$ a binary relation on graphs. Please note that analogously to autonomous units, transformation units are also inductively defined i.e. they have an acyclic import structure.[5] The set of *directly and indirectly imported transformation units* of tu is inductively defined by $IMP(tu) = U \cup \bigcup_{u' \in U} IMP(u')$. A pair $(G, G') \in SEM(I) \times SEM(T)$ is in the *interleaving semantics* $SEM(tu)$ of tu if there is a sequence $s = (G_0, \ldots, G_n)$ of graphs such that $G_0 = G$, $G_n = G'$, for $i = 1, \ldots n$ $(G_{i-1}, G_i) \in \bigcup_{p \in P} SEM(p) \cup \bigcup_{u \in U} SEM(u)$, and $(G, G') \in SEM_{E(tu)}(C)$ where $E(tu)(id) = SEM(id)$ if $id \in P \cup U$ and $E(tu)(id) = \emptyset$, otherwise. Analogously to autonomous units the set of *atomic transformations* of tu is defined by $AT(tu) = \bigcup_{p \in P} SEM(p) \cup \bigcup_{u \in U} AT(u)$.

5.2   Translating Transformation Units into Communities

For comparing transformation units with communities we first translate every transformation unit tu into two sets of autonomous units namely $TRANS1(tu)$

---

[4] Obviously, $rel(R^*)$ corresponds to the reflexive and transitive closure of R.
[5] Transformation units with an arbitrary import structure are studied in [4].

and $TRANS2(tu)$. The units in both sets are inductively defined such that the rule set of every autonomous unit in $TRANS1(tu) \cup TRANS2(tu)$ is equal to the rule set of tu, the goal is equal to the terminal graph class expression of tu, and the control condition can be anyone satisfying a certain property that is different for $TRANS1(tu)$ and $TRANS2(tu)$.

Definition 7 (Translations). Let $tu = (I, U, P, C, T)$ be a transformation unit.

1. The *first translation* of tu is the set $TRANS1(tu)$ consisting of all autonomous units $aut(tu) = (T, \{aut(u) \mid u \in U\}, P, c)$ such that for all $s \in AT(tu)$ with $pair(s) \in SEM_{E(tu)}(C)$, $s \in SEM_{E(aut(tu)),\Delta}(c)$, and $aut(u) \in TRANS1(u)$ for each $u \in U$.

2. The *second translation* of tu is the set $TRANS2(tu)$ consisting of all autonomous units $aut(tu) = (T, \{aut(u) \mid u \in U\}, P, c)$ such that

$$rel(SEM_{E(aut(tu)),\Delta}(c)) \subseteq SEM_{E(tu)}(C),$$

and $aut(u) \in TRANS2(u)$ for each $u \in U$.

It can be shown that the first translation preserves the behaviour of the original transformation unit tu (but can do more), and the second only performs such transformations that can also be done by tu, provided that the graph class expressions occurring in tu do not restrict the class $\mathcal{G}$ to some proper subclass. As a consequence we get that every autonomous unit in the intersection of $TRANS1(tu)$ and $TRANS2(tu)$ behaves as tu, if tu satisfies the mentioned property with respect to the graph class expressions.

These facts imply the following observation that relates transformation units with communities. In particular, for every autonomous unit aut in $TRANS1(tu) \cup TRANS2(tu)$ let $COM(aut)$ be the community with aut as its only autonomous unit, the initial graph class expression of tu as the initial environment specification, and the terminal expression of tu as the goal. Then the interleaving semantics of tu is contained in the binary relation induced by the transformation processes of $COM(aut)$ if aut belongs to the first translation of tu. Moreover, the binary relation induced by the successful transformation processes of $COM(aut)$ are contained in the interleaving semantics of tu if aut belongs to the second translation of tu. Consequently, the interleaving semantics of tu is equal to the binary relation induced by the successful transformation processes of COM if aut belongs to both translations. In the last two cases, the graph class expressions of all directly and indirectly imported transformation units of tu must specify the class of all graphs. It is worth noting that this condition concerning the graph class expressions can be dropped by requiring additionally that the control condition of every $aut(tu)$ in $TRANS2(tu)$ admits only sequences from initial into terminal graphs of tu.

Observation 3. Let $tu = (I, U, R, C, T)$ be a transformation unit and let $COM = (T, I, \{aut\})$. Then the following holds.

1. $SEM(tu) \subseteq rel(SEM(COM))$ if $aut \in TRANS1(tu)$.
2. $rel(STP(COM)) \subseteq SEM(tu)$ if $aut \in TRANS2(tu)$ and if $SEM(I_u) = SEM(T_u) = \mathcal{G}$ for all $u \in IMP(tu)$.
3. $SEM(tu) = rel(STP(COM))$ if $aut \in TRANS1(tu) \cap TRANS2(tu)$ and if $SEM(I_u) = SEM(T_u) = \mathcal{G}$ for all $u \in IMP(tu)$.

6   Modeling Ludo Players as Autonomous Units

Board games are a typical example of communities of autonomous units with sequential semantics where the board provides the common environment and the players are the autonomous units. As a concrete example we consider in this section the game *Ludo*.[6]

The graph transformation approach used in this example consists of labeled directed graphs and double-pushout rules (cf. [5]). The control conditions used are regular expressions and priorities. As graph class expressions we use subgraph conditions and the graph class expression specifying the class of all graphs. A *subgraph condition* is a graph G that admits all graphs that have (an isomorphic copy of) G as subgraph. Please note that in order to verify the presented case study we have implemented it based on the AGG system [5].

A possible environment graph of *Ludo* is the initial game situation where four players of different colours have all their tokens at the start place and there is one die showing an arbitrary number between one and six. This graph is depicted in Fig. 1. Every player is drawn as a kind of actor labeled with a colour out of b(lue), y(ellow), r(ed), and g(reen) so that every player has a different colour. Technically, a player is a labeled node. The players are connected via some directed edges indicating the playing direction. The game board consists of a start node and four home nodes for every player and a set of round nodes. The start node of a c-labeled player is depicted as a c-labeled polygon with six corners. The home nodes are drawn as rhombuses. Every c-labeled player has four c-labeled tokens that are all situated at her/his start node at the beginning of a match. The fact that a token of colour c is situated at a node v is visualized with a c-labeled token that is connected to v via an undirected edge. Technically, this can be modeled by means of a c-labeled loop connected to the node v. The directed edges between the nodes of the game board indicate where and in which direction the tokens can move around the game board.

Every round node and every directed edge between round and home nodes are labeled with a set $M \subseteq \{b, y, g, r\}$. The label of every round node contains all colours that can visit this node. Since at the beginning of a game all round nodes are vacant, i.e. they can be visited by all colours, they are all labeled with $\{b, y, g, r\}$. The labels of the edges connecting home and round nodes contain also all colours the tokens of which can move via these edges. For example, only yellow tokens can move to a home node of a yellow player. Moreover, no yellow token may go over the edge labeled with $N\_y = \{b, g, r\}$, because it has to enter

---

[6] There exist several distinct versions of the game *Ludo*. In this paper we consider one of the standard german versions.

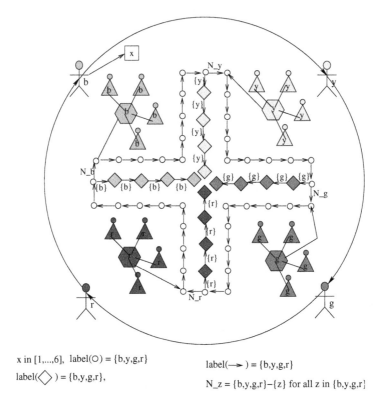

x in [1,...,6], label(○) = {b,y,g,r}
label(◇) = {b,y,g,r},

label(—➤) = {b,y,g,r}
N_z = {b,y,g,r}–{z} for all z in {b,y,g,r}

Fig. 1. An environment of *Ludo*

its home. Please note that the labels of most of the nodes and edges of Fig. 1 are depicted below the graph in order to keep the graph easy to read.

The goal of every player is to have all four tokens at home, one in each home node. To reach a home place, a token must go from the start place over the round fields in the indicated direction. To move a token, a die must be thrown. If a six is thrown the current player must move one of her/his tokens from the start node to the first round node, i.e. to the round node connected to the start node. If there is no token left at the start node, the player can take any other of her/his tokens. A six allows for throwing again. We assume here that the blue player starts to play. This is why the *b*-labeled player is holding the die (represented by the edge from the player to the die). Afterwards it is the turn of the yellow player.

Every player of *Ludo* can be realized as the autonomous unit depicted in Fig. 2. The goal of a player is to have all of her/his tokens at home, one at each home node. The rules and the imported units model all possible actions of a player. A rule is depicted by an arrow pointing from the left-hand side of the rule to the right-hand side. The possible values of the variables occurring in the left- or right-hand side are put under the arrow. If the label of a node or an edge is not significant it is omitted in the rule, i.e. an item without a label can be matched to an item with any label. The rule *go-to-startpoint* of the unit

player(c:[b,g,y,r])

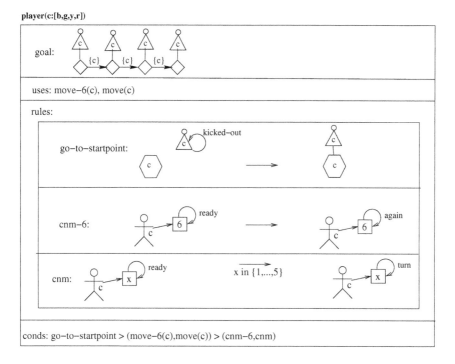

Fig. 2. A ludo player

player moves a token that has been kicked out to its home node. As the control condition prescribes this rule has the highest priority, i.e. it should be always applied if possible. The rule *cnm-6* and the rule set *cnm* are applied if no token can be moved by the player, i.e. they have the lowest priority. The rule *cnm-6* asks the die to throw again and *cnm* asks the die to turn to the next player if a number between one and five was thrown.

Every player imports the two units *move-6* and *move*. The unit *move* is depicted in Fig. 3. It models all moves of a token if no six is thrown. The moves corresponding to a six are contained in the unit *move-6*. For reasons of space limitations it is not depicted. The unit *move* contains four rules. The first, *mf* moves a token from the first round node (the one connected to its start node) x nodes ahead where $x \in \{1, \dots, 5\}$ is the number thrown by the die. This move can only be performed if the target node is not occupied and if there is still a token at the start node. Moreover, the token can only be moved if it is the turn of its player. This is indicated by the arrow pointing from player c to the die. On the left-hand side the die has a *ready*-loop which means that the die has already thrown itself. On the right-hand side the die is asked to turn. The rule *mfko* is similar to *mf*. The difference is that another token is kicked out. The rule *go* moves from a round or a home node to another round or home node. The rule *goko* does the same but it additionally kicks out another token. The rules *go* and *goko* can only be applied if the first two rules are not applicable. This is why the first round move(c:[b,y,g,r])

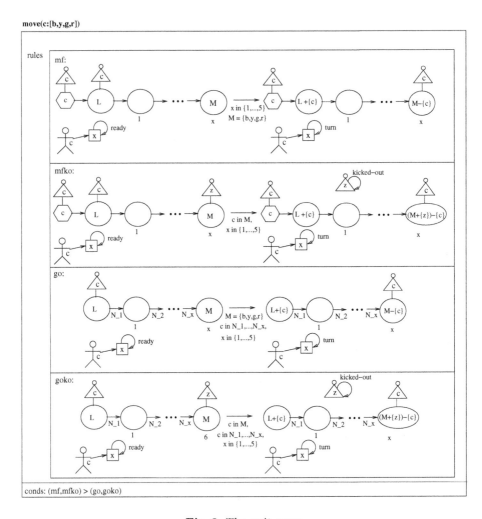

Fig. 3. The unit *move*

node must be left if it is occupied by a token of colour c and if there is still a token at the start node. If this move is not possible any other token can be taken.

Please note that players select their tokens nondeterministically. More sophisticated rules would allow to decide whether it is appropriate to choose a token that can kick out another one, etc. The rules for making such decisions possible are more complicated, because they have to consider a wider context of the environment (e.g. such a rule could check whether the kicking out of another token brings the own token into a "dangerous" position). For reasons of space limitations they are kept simple in this paper.

The last autonomous unit of *Ludo* models the die and is depicted in Fig. 4. The unit *die* has no special goal, i.e. it admits every graph as a goal. The only functionality of *die* is to throw itself and to move to the next player. The first

die

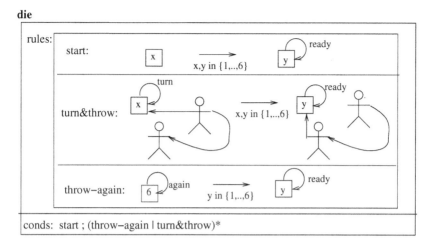

Fig. 4. The unit *die*

rule throws the die. The second rule throws and turns the die in the case where the die gets a corresponding *turn*-message from the player. With the third rule the die throws itself again without moving to the next player. This can be only done if a six was thrown before. The control condition requires that the *start* rule be applied once at first. Afterwards any of the two remaining rules can be applied arbitrarily often.

The game *Ludo* (including the board, up to four players, one die and the game rules) can be modeled as the community *Ludo* the goal of which specifies all graphs in which at least one player unit has reached its goal, the initial environment specification specifies all possible start situations of *Ludo*, and the set autonomous units consists of the unit *die* and one player unit for every colour.

7 Conclusion

In this paper, we have introduced communities of autonomous units as a means for modeling systems in which different components interact in a rule-based, self-controlled, and goal-driven manner within a common environment. We have compared communities of autonomous units with the original transformation units and we have illustrated the notion of communities with a case study modeling the board game *Ludo* in which every player as well as the die can act as an autonomous unit.

The underlying formal framework for communities of autonomous units has been graph transformation which is highly adequate if the common environment can be represented in a natural way as a graph as for example in the case of board games and logistic applications. Nevertheless, it is worth noting that the graphs and the graph transformation rules the units are working with are not further specified in the underlying graph transformation approach so that in general, one can take as formal basis any rule-based mechanism that provides a set of configurations and a set of rules specifying a binary relation on such configurations.

The presented paper has been concentrated on the sequential semantics of communities of autonomous units which seems to be suitable for a series of applications. For applications in which unit actions can happen in parallel or concurrently, the semantics can be defined accordingly as is done in a very first attempt in [6] where one can find also a case study from the area of transport logistics.

There are at least the following interesting points for future work. (1) Communities of autonomous units should be compared with related approaches such as agent systems [7,8], swarm intelligence [9] and distributed graph transformation [10]. (2) Communities of autonomous units should be implemented in order to be able to elaborate and to verify case-studies of realistic size. In order to verify the case study presented in Sect. 6 we have implemented it based on the AGG system. For this pupose the community *Ludo* had to be translated into a single flat graph transformation system because AGG does not support the concept of transformation units or communities. Currently there is being done some work towards the implementation of communities at the University of Bremen which has as one aim to allow to plug in other already existing graph transformation tools (cf. [11]). (3) Up to now, the goal of an autonomous unit is defined as a graph class expression. Since for some applications this may not be sufficient, other adequate classes of goals should be studied. It should also be studied which concrete classes of control conditions are the right ones for autonomous units. In particular, in the case of games different playing strategies should be employed for the different units in a communities. (4) More case studies should be specified. In this context it would be meaningful to elaborate for one application several case studies with a different degree of control distribution. For example in the case of the game *Ludo* one could compare the presented case study with another in which also the tokens are modeled as autonomous units with their own control and rules. (5) Communities of autonomous units that can perform parallel and concurrent transformations should be further investigated taking into account concepts and results from concurrent, parallel and distributed graph transformation (cf.,e.g., [12,10]).

References

1. Kreowski, H.J., Kuske, S.: Graph transformation units with interleaving semantics. Formal Aspects of Computing **11**(6) (1999) 690–723
2. Kreowski, H.J., Kuske, S.: Approach-independent structuring concepts for rule-based systems. In Wirsing, M., Pattison, D., Hennicker, R., eds.: Proc. 16th Int. Workshop on Algebraic Development Techniques (WADT 2002). Volume 2755 of Lecture Notes in Computer Science. (2003) 299–311
3. Rozenberg, G., ed.: Handbook of Graph Grammars and Computing by Graph Transformation, Vol. 1: Foundations. World Scientific, Singapore (1997)
4. Kreowski, H.J., Kuske, S., Schürr, A.: Nested graph transformation units. International Journal on Software Engineering and Knowledge Engineering **7**(4) (1997) 479–502

 5. Ehrig, H., Ehrig, K., Prange, U., Taentzer, G., eds.: Fundamentals of Algebraic Graph Transformation. Springer (2006)
 6. Hölscher, K., Klempien-Hinrichs, R., Knirsch, P., Kreowski, H.J., Kuske, S.: Regelbasierte Modellierung mit autonomen Transformationseinheiten. Technical Report 1, University of Bremen (2006)
 7. Weiss, G., ed.: Multiagent Systems — A Modern Approach to Distributed Artificial Intelligence. The MIT Press (1999)
 8. Wooldridge, M., Jennings, N.R.: Intelligent agents: Theory and practice. The Knowledge Engineering Review **10**(2) (1995)
 9. Kennedy, J., Eberhart, R.C.: Swarm Intelligence. Morgan Kaufmann (2001)
10. Ehrig, H., Kreowski, H.J., Montanari, U., Rozenberg, G., eds.: Handbook of Graph Grammars and Computing by Graph Transformation, Vol. 3: Concurrency, Parallelism, and Distribution. World Scientific, Singapore (1999)
11. Ehrig, H., Engels, G., Kreowski, H.J., Rozenberg, G., eds.: Handbook of Graph Grammars and Computing by Graph Transformation, Vol. 2: Applications, Languages and Tools. World Scientific, Singapore (1999)
12. Heckel, R.: Open Graph Transformation Systems: A New Approach to the Compositional Modelling of Concurrent and Reactive Systems. PhD thesis, Technical University of Berlin (1998)

Termination Analysis of Model Transformations by Petri Nets [*]

Dániel Varró[1], Szilvia Varró–Gyapay[1], Hartmut Ehrig[2],
Ulrike Prange[2], and Gabriele Taentzer[2]

[1] Budapest University of Technology and Economics
Department of Measurement and Information Systems
{varro, gyapay}@mit.bme.hu
[2] Technical University of Berlin
{ehrig, uprange, gabi}@tfs.cs.tu-berlin.de

Abstract. Despite the increasing relevance of model transformation techniques in model-driven software development, research is mainly conducted to the specification and the automation of such transformations. However, since the transformations themselves may also contain conceptual flaws, it is essential to formally analyze them prior to executing them on user models. In the current paper, we focus on a central validation problem of trusted model transformations, namely, *termination* and propose a Petri net based analysis method that provides a sufficient criterion for the termination problem of model transformations captured by graph transformation systems.

Keywords: graph transformation, termination, model transformation, Petri nets.

1 Introduction

Many researchers and practitioners have recently revealed that model driven software development relies not only on the precise definition of modeling languages taken from different domains, but also on the unambiguous specification of transformations between these languages. To provide a standardized support for capturing queries, views and transformations (QVT) between modeling languages defined by their standard MOF metamodels, the Object Management Group (OMG) is soon to issue QVT [17] as a standard. QVT provides a declarative rule-based, model transformation language where control structures are restricted to embedding transformation rules into each other.

Graph transformation (GT) [8,20] has been applied successfully to many model transformation (MT) problems. Many success stories were in the field of model analysis which aim at projecting high-level UML models into mathematical domains by model transformations to carry out formal analysis.

---

[*] This work was partially supported by the Segravis Research Training Network. Dániel Varró was also supported by the János Bolyai Scholarship.

A. Corradini et al. (Eds.): ICGT 2006, LNCS 4178, pp. 260–274, 2006.

As revealed in a recent study [14], graph transformation and QVT-like declarative techniques show a close correspondence. A first precise formulation of this correspondence has been studied in [19]. As a consequence the theoretical background of graph transformation is expected to provide useful results for QVT.

Problem statement. A core problem, which is very vaguely addressed by QVT, is related to the *correctness of model transformations*, namely, to guarantee that certain semantic properties hold for a *trusted model transformations*. For instance, when transforming UML models into mathematical domains, the results of a formal analysis can be invalidated by erroneous model transformations as the systems engineers cannot distinguish whether an error is in the design or in the transformation. In case of QVT, it is possible that the embedded transformation rules interfere with each other and thus they may cause semantic problems, which is not acceptable for trusted model transformations.

Most typical correctness properties of a trusted model transformation are termination, uniqueness (confluence) and behaviour preservation. In [11], we proposed a set of sufficient criteria that guarantees the termination of model transformations specified by so-called layered graph transformation systems (GTS). While this technique was applicable to various practical model transformation problems, further experiments have revealed that these sufficient criteria exclude model transformations where rules are causally dependent on themselves.

Objectives and Approach. In the current paper, we provide a Petri Net based technique for the termination analysis of model transformations specified by GTSs. As termination is undecidable for graph grammars in general [18], we propose a sufficient criterion, which either proves that a GTS is terminating, or it yields a "maybe nonterminating" (do not know) answer.

The essence of our technique is to derive a simple Petri net which simulates the original GTS by abstracting from the structure of instance models (graphs) and only counting the number of elements of a certain type. If we manage to prove by algebraic techniques that the Petri net runs out of tokens in finitely many steps regardless of the initial marking, then we can conclude that the original GTS is terminating due to simulation. In order to handle graph transformation systems with negative application conditions as well, we introduce the notions of forbidden and permission patterns, and overapproximate how different rules influence each other when generating permissions.

As the derived Petri net model is of managable size (comparable to the number of elements in the metamodels), our technique can yield positive results for judging the termination of various model transformation problems captured by graph transformation techniques.

*Structure of the paper.*The rest of the paper is organized as follows: Sec. 2 presents a running example where we specify (with graph transformation rules) a transformation from UML class diagrams to relational databases. Sec. 3 provides an overview on graph transformation systems and Place / Transition (P/T) nets. Sec. 4 proposes a P/T net abstraction of GTS with rules having negative application conditions (NAC). Sec. 5 presents sufficient conditions for termination of

GTSs by solving algebraic inequalities. Sec. 6 discusses related work and finally Sec. 7 presents our conclusions and proposals for future work.

2   Motivating Example: The Object-Relational Mapping

As the motivating example of the current paper, we map simple UML class diagrams into relational database tables by using one of the standard solutions. This transformation problem (with several variations) is frequently used as a model transformation benchmark of high practical relevance [15].

The source and target languages (UML and relational databases, respectively) are captured by their corresponding metamodels in Fig. 1. In Sec. 3.1, metamodels will be represented formally by means of type graphs [9], while instance models will be graphs typed over a type graph.

UML class diagrams in our paper consist of classes arranged into an inheritance hierarchy (by *parent* edges). Classes have attributes (*attrs*), which are typed over classes (*type*). Directed associations are leading from a source (*src*) class to a destination (*dst*) class.

Relational databases consist of tables, which are composed of columns (*tcols*). Each table has a single primary key column (*pkey*). Foreign key (*FKey*) constraints

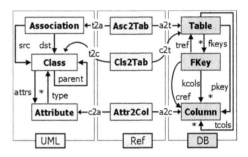

Fig. 1. Metamodels (type graphs): Source, reference, target

can be assigned to tables (*fkeys*). A foreign key refers to certain columns (*cref*) of a table (*tref*), and it is related to the columns *kcols* of (local) referencing table.

These metamodels (adapted from [15]) are extended by a *reference metamodel* to interconnect the elements of the source and the target language. This way it defines the main guidelines of (this variant of) the object-relational mapping itself, which can be summarized as follows:

- Each top-level UML class (i.e. a top-most class in the inheritance tree) is projected into a database table. Two additional columns are derived automatically for each top-level class: one for storing a unique identifier (primary key), and one for storing the type information of instances.
- Each attribute of a UML class will appear as columns in the table related to the top-level ancestor of the class. For the sake of simplicity, the type of an attribute is restricted to user-defined classes. The structural consistency of valid object instances in columns is maintained by foreign key constraints.
- Each UML association is projected into a table with two columns pointing to the tables related to the source and the target classes of the association by foreign key constraints.

3 Introduction to Graph Transformation and Petri Nets

Now we provide a brief overview on the formal background of graph transformation and Petri nets. Only those concepts will be introduced which are essential for presenting our main results in Sec. 4 and 5.

3.1 Typed Graph Transformation

Type and Instance Graphs. The metamodels of different modeling languages are frequently formalized as type graphs and instance models are typed over this type graph. The traditional instance-of relationship between metamodels and models is captured formally by a typing morphism.

A *graph* $G = (N, E, src, trg)$ is a 4-tuple with a set N of nodes, a set E of edges, a source and a target function $src, trg : E \rightarrow N$. A *type graph* TG is an ordinary graph. An *instance graph* G is typed over TG by a typing morphism $type : G \rightarrow TG$. Let $card(G, x)$ denote the *cardinality* (i.e. the number of graph objects) of a type $x \in TG$ in graph G. Formally, $card(G, x) = |\{n \mid n \in N \cup E \wedge type(n) = x\}|$.

For the current paper, we assume that there is a unique edge of a certain type between two nodes, i.e., if $src(e_1) = src(e_2) \wedge trg(e_1) = trg(e_2) \wedge type(e_1) = type(t_2) \Rightarrow e_1 = e_2$, which simplifies the proofs of our theorems.

Graph Transformation. Graph transformation (GT) [8] provides a rule-based manipulation of graph models. A *graph transformation rule* $r = (L \xleftarrow{l} K \xrightarrow{r} R)$ typed over a type graph TG is given by triple where L (left-hand side, LHS), K (context) and R (right-hand side, RHS) graphs are typed over TG and graph morphisms l, r are injective and assumed to be type preserving.

The *negative application conditions* (NACs) of a GT rule are a (potentially empty) set of pairs (N, n) with N being a graph also typed over TG and $n : L \rightarrow N$ being an injective graph morphism. A GT rule with NACs is denoted shortly as $r = (L \xleftarrow{l} K \xrightarrow{r} R, \{L \xrightarrow{n_i} N^i\})$ $(i = 1 \ldots k)$. Moreover, we assume that no rules exist where all L and N are empty.

Application of a Rule. The *application* of a rule to a *host model graph* G alters the model graph by replacing the pattern defined by L with the pattern defined by R. This is performed by (i) *finding an injective matching* $m : L \rightarrow G$ of the L pattern in model graph G; (ii) *checking the negative application conditions* N which prohibit the presence of certain model elements, i.e. for each NAC $n : L \rightarrow N$ of a rule no injective graph morphism $q : N \rightarrow G$ exists with $m = q \circ n$; (iii) *removing* a part of the model graph M that can be mapped to L but not to R yielding an intermediate graph D; (iv) *adding* new elements to the intermediate graph D which exist in R but not in L yielding the derived graph H. A GT step is denoted formally as $G \xRightarrow{r,m} H$, where r and m denote the applied rule and the matching along which the rule was applied, respectively. In the paper, we follow the *Double Pushout Approach* [8].

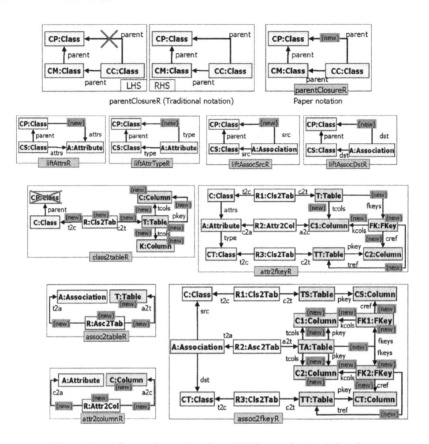

Fig. 2. Model transformation from UML to relational databases

Example 1. A sample graph transformation rule calculating the transitive closure of the parent relation is depicted in the top rule (*parentClosureR*) of Fig. 2. The rule prescribes that if class *CP* is parent of class *CM* (i.e. there is a *parent* edges between them), and *CM* is a parent of class *CC*, but there is no *parent* edge from *CC* to *CP*, then such an edge should be created.

For a more compact presentation of the rules, we abbreviate the *L*, *N* and *R* graphs of a rule into one, and we only mark the (images of) graph elements to be removed (*del*), or created (*new*). We assume that all elements in *R* marked as *new* are implicitly present in the negative application condition *N* as well. In case of rule *class2TableR* we use crossed lines to denote the second negative application condition (that is not part of *R*).

Example 2. The object-relational mapping is captured by the set of graph transformation rules in Fig. 2. The entire transformation starts with a preprocessing phase when the transitive closure of *parent* relations is calculated (*parentClosureR*), and then all attributes and associations are lifted up to the top-level classes in the inheritance (*parent*) hierarchy (rules *liftXYZ*). Then the main model transformation (Fig. 2) proceeds as described in Sec. 2 by transforming classes into tables (*class2tableR*), associations into tables (*assoc2tableR*), attributes into columns (*attr2columnR*), attribute types and destination class of associations into foreign key constraints (*attr2fkeyR* and *assoc2fkeyR*).

A *graph transformation system* $GTS = (R, TG)$ consists of a type graph TG and a finite set R of graph transformation rules typed over TG. A *graph grammar* $GG = (GTS, G_0)$ consists of a graph transformation system $GTS = (R, TG)$ and a so-called *start (model) graph* G_0 typed over TG.

The *state space* $Sem(GG)$ generated by a graph grammar $GG = (GTS, G_0)$ is defined as a graph where nodes are model graphs, and edges are graph transformation steps $G \xRightarrow{r,m} H$ such that the source and target nodes of the edge are graphs G and H, respectively. Starting from G_0 the *state space* (i.e. the reachable model graphs) of the GG is represented taking into account all applicable rules from a given model graph for all possible matchings.

A *graph grammar* $GG = (G_0, GTS)$ is *terminating* if there are no infinite sequences of rule applications starting from G_0. A *graph transformation system* $GTS = (R, TG)$ is called *terminating* if for all G_0, the corresponding graph grammar $GG = (G_0, GTS)$ is terminating.

3.2   Place/Transition Nets

In the current section we give a short introduction into the theory of Place/Transition nets based on [16].

A *Place/Transition net* (or shortly P/T net) is a 4-tuple $PN = (P, T, E, w)$ where P is a set of *places* (represented graphically as circles), T is a set of *transitions* (represented as horizontal bars), $E \subseteq (P \times T) \cup (T \times P)$ is the set of *arcs* (where no arcs connect two places or two transitions), and the *weight function* $w : E \to {}^+$ maps arcs to positive integers.

Places may contain tokens. The distribution of tokens at different places is called a *marking* $M : P \to$, which maps places to non-negative integers. The initial marking is denoted as M_0.

The token distribution can be changed in the net by firing transitions. A transition t is *enabled* (i.e. it may fire), if each of its input places contain at least as many tokens as it is specified by the weight function. The *firing* of an enabled transition t removes a $w(p, t)$ amount of tokens from the input places, and $w(t, p)$ tokens are produced on each output place p. As a result, the marking M changes to M' (denoted as $M \xRightarrow{t} M'$) according to $\forall p \in P : M'(p) = M(p) - w(p, t) + w(t, p)$.

The *incidence matrix* W of a (finite) net describes the net token flow (of the P/T net) when firing a transition. Mathematically, W is a $|P| \times |T|$–dimensional matrix of non-negative integers   such that $w_{ij} = w(t_j, p_i) - w(p_i, t_j)$, where $1 \le i \le |P|, 1 \le j \le |T|$.

After firing a transition t in marking M, the result marking M' can be computed with the incidence matrix: $M' = M + W \cdot \underline{e}_t$, where $\underline{e}_t$ is a $|T|$-dimensional unit vector, where the t-th component is 1 and the others are 0.

A *(transition) firing sequence* $s = \langle t_1, t_2, \dots, \rangle$ is a sequence of transition firings starting from state M_0 such that $M_0 \overset{t_1}{\Longrightarrow} M_1, \overset{t_2}{\Longrightarrow} \dots$, i.e. for all $1 \leq j$ t_j is enabled in M_{j-1} and M_j is yielded by the firing of t_j in M_{j-1}.

The marking of the net after executing the first k steps of the firing sequence s can be calculated by the state equation: $M_k = M_0 + W \cdot \underline{\sigma}$, where $\underline{\sigma}$ is the *transition occurrence vector* or *Parikh–vector* of the trajectory s counting the number of occurrences of individual transitions in the firing sequence.

4   A Petri Net Abstraction of Graph Transformation

4.1   Definition of the Core Abstraction

First we map a graph transformation system *without negative application* conditions into a Petri net (which is called *cardinality (P/T) net* in the sequel) by only keeping track of the number of objects in the instance graph (separately for each node and edge in the type graph) but abstracting from the structure of the instance graph.

Informally speaking, since the LHS of a GT rule requires the presence of nodes and edges of certain types, the derived transition removes tokens from all the places storing the instances of the corresponding types. Furthermore, the RHS of a GT rule guarantees the presence of nodes and edges of certain types, thus the derived transition generates tokens for the places storing the instances of such types. Later we show that this is a proper abstraction, i.e. the derived P/T net simulates the original GTS, i.e. when a GT rule is applicable, the corresponding transition in the P/T net can be fired as well.

This mapping $\mathcal{F}(GTS) = (\mathcal{F}_{TG}, \mathcal{F}_G, \mathcal{F}_R) \to PN$ (where $GTS = (R, TG)$ and $PN = (P, T, E, w)$ with initial marking M_0) is formally defined as follows:

- $\mathcal{F}_{TG} : TG \to P$: *Types into places.* For each node and edge $y \in N_{TG} \cup E_{TG}$ in the type graph TG, a corresponding place $p_y = \mathcal{F}(y)$ is defined in the cardinality P/T net.
- $\mathcal{F}_G : G \to M_0$: *Instances into tokens.* For each node and edge $x \in N_G \cup E_G$ in an instance graph G with type $y = type(x)$, a token is generated in the corresponding marking $M_G = \mathcal{F}(G)$ of the target P/T net. Formally, for all places $p_y = \mathcal{F}(y)$, the marking of the net is defined as $M_G(p_y) = card(G, y)$.
- $\mathcal{F}_R : R \to (T, E, w)$: *Rules into transitions.* For each rule r in the graph transformation system GTS, a transition $t_r = \mathcal{F}(r)$ is generated in the cardinality P/T net such that
  - *Left-hand side:* If there is a graph object x in L with $y = type(x)$, then an incoming arc (p_y, t_r) is generated in the P/T net where $p_y = \mathcal{F}(y)$ and the weight of the arc $w(p_y, t_r)$ is equal to the number of graph objects in L of the same type y. Formally, if $\forall x, y : x \in L \wedge y = type(x) \wedge \mathcal{F}(y) = p_y \Rightarrow (p_y, t_r) \in E \wedge w(p_y, t_r) = card(L, y)$.
  - *Right-hand side:* If there is a graph object x in R with $y = type(x)$, then an outgoing arc (t_r, p) is generated in the P/T net where $p_y = \mathcal{F}(y)$ and the weight of the arc $w(t_r, p_y)$ is equal to the number of graph objects in R of the same type y. Formally, if $\forall x, y : x \in R \wedge y = type(x) \wedge \mathcal{F}(y) = p_y \Rightarrow (t_r, p_y) \in E \wedge w(t_r, p_y) = card(R, y)$.

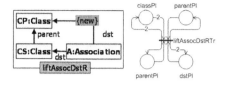

Fig. 3. Transition corresponding to rule liftAssocDstR

In Fig. 3 rule *liftAssocDstR* of the example in Fig. 2 is shown on the left together with the corresponding transition *liftAssocDstR* (on the right) of the P/T net abstraction of the example. Note that indices of $\mathcal{F}()$ will be omitted for simplicity.

As the GT rule *liftAssocDstR* contains two *Class* nodes, one *Association* node, one *parent* edge and one *dst* edge, the corresponding transition is enabled if the corresponding type places (with identical labels) contain at least 2, 1, 1, and 1 tokens, respectively. Since the application of the rule preserves all items and creates one *dst* edge, the firing of transition *liftAssocDstR* puts 2, 1, 1, and 2 tokens to these places, respectively.

Note, however, that the transition of Fig. 3 is always enabled and thus, it would directly cause non-termination. Therefore, we now extend our abstraction technique to handle graph transformation rules with negative application conditions as well, which are frequently used in model transformation problems.

4.2 Extensions for Negative Conditions

Permission Places. In order to cope with NACs, the P/T net is extended with so-called *permission places* to restrict the firing of a transition. We add one permission place for each NAC in the GTS, and the idea of a permission place is to count how many times the GT rule can be applied to the current instance graph (such that the corresponding NAC does not violate these matchings).

- *Start graph.* The initial marking of permission places shall enable the firing of a transition as many times as the corresponding GT rule is applicable to the start graph by giving a permission token.
- *Removing permissions.* If a new matching of some NAC N_i of a GT rule r is generated or an existing matching of the LHS of the same rule r is destroyed by the application of some GT rule r' then one or more tokens should be removed from the permission place corresponding to N_r^i.
- *Creating permissions.* If an existing matching of the NAC of a GT rule r is destroyed or a new matching of the LHS of the same rule r is generated by the application of some GT rule r' then one or more tokens should be generated to the permission place corresponding to N_r^i.

Unfortunately, the exact number of tokens created for or removed from a permission place depends on the actual graph structure. Therefore, we cannot derive a constant weight *a priori* for the corresponding arcs in the P/T net; instead we write $w(G)$ on such arcs to denote that the weight of the arc is dependent on graph G. However, we know that such an arc weight $w(G)$ is finite, i.e. we can only generate and remove a finite number of new permissions for any permission place.

Overapproximation for Permissions. Therefore, we need to define an overapproximation of the potential number of rule applications, which still simulates the GTS, yet it is precise enough to detect termination for a certain class of model transformation problems.

- In our proposal, we only remove one token from a permission place when it is absolutely guaranteed (by analyzing the original GT rule) that a *permission should be destroyed* each time the rule is applied. In case of GT rules with NAC, such a situation is when a GT rule cannot be applied on the same matching twice due to a NAC.
- In case of *generating a permission*, we should consider all possible values for the arc weight $w_i(G)$, thus we create a new variable c_i which runs over positive integers.

Permission and Forbidden Patterns. An initial idea for granting permissions is to consider the causalities of GT rules, i.e. when a rule generates a new matching for another rule, a new permission is generated as well. However, this solution is unable to handle cases when GT rules are generating a bounded number of new matchings for themselves (i.e., when a rule is causally dependent on itself).

For instance, each application of rule *liftAssocDstR* (in Fig. 3) generates a new *dst*, thus a new matching for itself, which seems to be a direct cause for non-termination. On the other hand, if the meaning of a permission is related to the number of *Class-Association* pairs not connected by a *dst* edge, we notice that this number is strictly decreasing, thus no new permission is granted by GT rule *liftAssocDstR* for itself. This insight is captured formally by *forbidden and permission patterns*.

Definition 1 (Forbidden and permission pattern). *Let $GTS = (R, TG)$ be a graph transformation system. A forbidden pattern fp_r^i is defined for each NAC N_r^i of rule r as the smallest subgraph of N_r^i that contains $N_r^i \setminus L_r$ (also called as the context of $n^i : L_r \to N_r^i$).*

The permission pattern pp_r^i (of the same NAC N_r^i) is defined as smallest subgraph of fp_r^i that contains $N_r^i \setminus L_r$ (also called as the boundary of $n_r^i : L_r \to N_r^i$), which is defined formally as $fp_r^i \setminus (N_r^i \setminus L_r)$.

Informally, the permission pattern can be interpreted as an LHS pattern having a NAC with the forbidden pattern. The exact number of permissions for a rule is calculated as the number of matchings of the permission pattern having the forbidden pattern as a NAC.

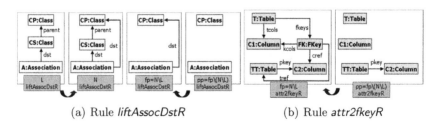

(a) Rule *liftAssocDstR*   (b) Rule *attr2fkeyR*

Fig. 4. Forbidden and permission patterns

Example 3. The concepts of forbidden and permission patterns are demonstrated in Fig. 4(a). The forbidden pattern (*FP*) of rule *liftAssocDstR* contains a *dst* edge leading from Association *A* to Class *CP*. Here $N \setminus L$ contains the single *dst* edge while the two nodes are added to guarantee that the forbidden pattern forms a graph. In order to obtain the permission pattern (*PP*), we simply remove this *dst* edge from the forbidden pattern.

Definition of cardinality P/T with permission places. The *cardinality P/T net with permission places* of GTS is a $PN = (P, T, E, w)$ derived by the mapping $\mathcal{F}_{pp}(GTS)$ by extending $\mathcal{F}(GTS)$ in the following way:

- *Variables as weight functions.* We extend the weight function of a P/T net to $w : E \to \mathbb{N}^+ \cup V$ where V is a set of variables ranging over $\mathbb{N}^+$.
- *NACs into permission places.* For each NAC N^i of a rule r a corresponding permission place $p_{r_{N^i}} = \mathcal{F}_{pp}(r_{N^i})$ is defined in the cardinality net.
- *Matchings of permission patterns into tokens (initial marking).* For each NAC N^i of a rule r as many tokens are generated in the corresponding permission place as the number of injective matchings m of permission pattern pp_r^i in the instance graph G which satisfies the derived NAC $pp_r^i \to fp_r^i$, (i.e., there is no injective matching of the forbidden pattern fp_r^i to G along m).
- *NACs into pre arcs.* For each rule r with NACs $N^1, \ldots, N^k$, if there is an injective morphism $k_i : N^i \to R$ compatible with r for some NAC N^i (informally, everything included in the NAC N^i exists or it is created by the RHS), an incoming arc $(p_{r_{N^i}}, t_r)$ is generated in the P/T net with weight 1.

- *Rule actions into post arcs.* For each pair of rules $r = (L_r \xleftarrow{l} K_r \xrightarrow{r} R_r)$ with NACs $N^1, \ldots, N^k$ and $r' = (L_r' \xleftarrow{l} K_r' \xrightarrow{r} R_r')$, an outgoing arc $(t_{r'}, p_{r_{N^i}})$ ($i : 1 \le i \le k$) is generated in the P/T net (i.e. from the transition of rule r' to the permission place of r_{N^i}) with a *variable* arc weight $v_{r', r_{N^i}}$ if
  1. at least one graph object o is deleted by r' (from the forbidden pattern fp_r^i of r) such that there exists a graph object $o' \in N^i \setminus L_r$, and $type(o) = type(o')$ or
  2. at least one graph object o is created by r' such that there exists a graph object $o' \in pp_r^i$, and $type(o) = type(o')$.

Informally, instead of regarding the causality between two rules based upon the RHS of rule r' and the LHS of r, we define causality between the effects of a rule r' and the permission pattern of r.

Furthermore, in order to overapproximate the graph dependent arc weights $w(G)$, we introduce variables as weights for such arcs. As a consequence, for each step of the P/T net, we can substitute the variables with proper values to simulate the original GTS in a step-wise way. In order to prove termination later in Sec. 5, we will show that any substitution of these variables fulfill certain algebraic properties.

The *incidence matrix* of the P/T net abstraction of GTS with NACs is denoted as $W(\underline{v})$, which notation emphasizes that W contains variables at locations where new permissions are generated for a rule.

	UML								Ref								
	Assoc	Class	Attr	src	dst	attrs	type	par	t2a	t2c	c2a	A2T	C2T	A2C	a2t	c2t	a2c
parentClosureR	0	0	0	0	0	0	0	1	0	0	0	0	0	0	0	0	0
liftAttrsR	0	0	0	0	0	1	0	0	0	0	0	0	0	0	0	0	0
liftAttrTypeR	0	0	0	0	0	0	1	0	0	0	0	0	0	0	0	0	0
liftAssocSrcR	0	0	0	1	0	0	0	0	0	0	0	0	0	0	0	0	0
lisftAssocDstR	0	0	0	0	1	0	0	0	0	0	0	0	0	0	0	0	0
class2tableR	0	0	0	0	0	0	0	0	0	1	0	0	1	0	0	1	0
assoc2tableR	0	0	0	0	0	0	0	0	1	0	0	1	0	0	1	0	0
attr2columnR	0	0	0	0	0	0	0	0	0	0	1	0	0	1	0	0	1
attr2fkeyR	0	0	0	0	0	0	0	0	0	0	0	0	0	0	0	0	0
assoc2fkeyR	0	0	0	0	0	0	0	0	0	0	0	0	0	0	0	0	0

	DB									permission places (for each NAC)										
	Tab	Key	Col	tref	fkeys	pkey	tcols	cref	kcols	parCl	lAttr	lAttrT	lAssSrc	lAssDst	cl2ta1	cl2ta2	as2tab	attr2c	at2fk	as2fk
parentClosureR	0	0	0	0	0	0	0	0	0	-1	0	0	0	0	0	0	0	0	0	0
liftAttrsR	0	0	0	0	0	0	0	0	0	0	-1	0	0	0	0	0	0	0	0	0
liftAttrTypeR	0	0	0	0	0	0	0	0	0	0	0	-1	0	0	0	0	0	0	0	0
liftAssocSrcR	0	0	0	0	0	0	0	0	0	0	0	0	-1	0	0	0	0	0	0	0
lisftAssocDstR	0	0	0	0	0	0	0	0	0	0	0	0	0	-1	0	0	0	0	0	0
class2tableR	1	0	2	0	0	1	2	0	0	0	0	0	0	0	0	-1	0	0	v_1	v_2
assoc2tableR	1	0	0	0	0	0	0	0	0	0	0	0	0	0	0	0	-1	0	v_3	v_4
attr2columnR	0	0	1	0	0	0	0	0	0	0	0	0	0	0	0	0	0	-1	v_5	v_6
attr2fkeyR	0	1	0	1	1	0	1	1	1	0	0	0	0	0	0	0	0	0	-1	0
assoc2fkeyR	0	2	2	1	2	2	2	2	2	0	0	0	0	0	0	0	0	0	0	-1

Fig. 5. Incidence matrix of the P/T net abstraction

The incidence matrix of the example GTS is given in Fig. 5. The places (columns) refer to the type places corresponding to the type graph of Fig. 1, while transitions (rows) refer to corresponding rules of Fig. 2. The right-most columns of the matrix denote permission places. Note that the incidence matrix is independent of the initial marking of the cardinality P/T net, thus our termination result is valid for any initial marking.

It is worth pointing out that the proposed abstraction highly relies on the fact that a RHS contains at least one of its NACs. Note that this is typical for model transformation problems where NACs are frequently used to prevent the application of a rule multiple times on the same matching.

Example 4 (Cardinality P/T net with permission patterns). Rule *parentClosureR* generates new *parent* edges, which are required for the matching of rule *liftAssocDstR* (see Fig. 2), thus the two rules are causally dependent. However, no new permissions are generated for the latter, since rule *parentClosureR* should remove a *dst* edge (see the forbidden pattern) or create new *Class* or *Association* nodes (see the permission pattern) for a new permission to be generated (see the permission and forbidden patterns in Fig. 4(a)).

On the other hand, rule *class2tableR* generates new permissions for rule *attr2fkeyR*, since the tables created by the former are present in the permission pattern of the latter (which consists of tables *T*, *TT* and columns *C1* and *C2*, see Fig. 4(b)). Consequently, a variable v_1 is used as the weight of the corresponding arc leading from the transition of *class2table* to the permission place of *attr2fkeyR*.

5 Termination Analysis of Graph Transformation

Now we propose a termination analysis for GTS using a generalization of non-repetitiveness results from P/T nets [16].

A P/T net is *partially repetitive* if there exists a marking M_0 and a firing sequence s from M_0 such that some transition occurs infinitely many times in s. Furthermore, a main result from P/T net theory states that a P/T net with the incidence matrix W is partially repetitive if and only if there exists a Parikh–vector $\underline{\sigma} \geq \underline{0}, \underline{\sigma} \neq \underline{0}$ such that $W^T \cdot \underline{\sigma} \geq \underline{0}$. As a consequence, if a P/T net is not partially repetitive (i.e., no Parikh–vector $\underline{\sigma} \geq \underline{0}, \underline{\sigma} \neq \underline{0}$ exists that satisfies $W^T \cdot \underline{\sigma} \geq \underline{0}$), then only finite firing sequences exist from any intial marking M_0, which proves termination.

Our generalization lies in the fact we do not require the existence of the incidence matrix W. Instead we state that if sequences of state vectors fulfill the condition that at least one component of the state vector is decreasing (wrt. each previous state vector in the sequence) in each step it guarantees that the $\underline{0}$ state is reached in finite steps. Our reason for this generalization is that W may contain variables at permission places.

Theorem 1. *If for all infinite sequences* $\{M_i\} = M_0, M_1, \ldots$ *of n-dimensional (state) vectors of nonnegative integer values with* $M_j - M_{j\ 1} < \infty$ *for all j*

$$(1) \quad \forall i, \forall j : j > i, M_i \not\equiv \underline{0} \Rightarrow \exists k : M_j[k] - M_i[k] < 0, \text{ and}$$

$$(2) \quad \forall i, \forall j : j > i, M_i \equiv \underline{0} \Rightarrow M_j \equiv \underline{0}$$

then $M \equiv \underline{0}$ in finitely many steps, i.e. $\exists s : M_s \equiv \underline{0}$ (where $M_j[k]$ denote component k in vector M_j).

Then, we claim that mapping $\mathcal{F}()$ is a proper abstraction in the sense that the derived P/T net *without permission places* simulates the original GTS. In other terms, whenever a rewriting step is executed in the GTS on an instance graph, then the corresponding transition can always be fired in the corresponding marking in the P/T net, furthermore, the result marking is an abstraction of the result graph.

Theorem 2 (Cardinality P/T net simulates GTS). *Let $GTS = (R, TG)$ be a graph transformation system and $PN = (P, T, E, w)$ be a cardinality P/T net derived by the mapping $\mathcal{F}(GTS)$. Furthermore, let G, H be instance graphs typed over TG. Then PN simulates GTS, formally*

$$\forall G, H, r, o : (G \overset{r,o}{\Longrightarrow} H) \Rightarrow (M_G \overset{t_r}{\Longrightarrow} M_H),$$

where $\mathcal{F}(G) = M_G$, $\mathcal{F}(H) = M_H$, and $\mathcal{F}(r) = t_r$.

Finally, as a termination "oracle", we solve quadratic inequalites based on the incidence matrix of the P/T net with variables as defined in Sec. 4.1-4.2. If there are no solutions for the inequality for any evaluation of variables in the incidence matrix, we state that the original GTS is terminating.

Theorem 3 (Termination). *Let $W(\underline{v})$ be the incidence matrix of a cardinality P/T net $PN = \mathcal{F}_{pp}(GTS)$ derived as the abstraction of a GTS.*
If $\exists \underline{\sigma} \exists \underline{v}\ \ W(\underline{v}) \cdot \underline{\sigma} \geq \underline{0}$ has no solutions with $\underline{v} \geq \underline{1}, \underline{\sigma} \geq \underline{0}, \underline{\sigma} \neq \underline{0}$ (thus $\forall \underline{\sigma} \forall \underline{v}\ \ \exists k : (W(\underline{v}) \cdot \underline{\sigma})[k] < 0$), then GTS is terminating.

In order to show that the quadratic inequality $W(\underline{v}) \cdot \underline{\sigma} \geq \underline{0}$ has no solutions for proving the termination of GTSs with negative application conditions, we used a symbolic optimization toolkit (GAMS [12]) which supports mixed integer nonlinear programming.

6   Related Work

Relation of Graph Transformation and Petri Nets. The main idea of this paper
is to analyze graph transformation systems via Petri nets. In fact, there is a long
tradition concerning the relationship of both areas. The basic observation is that
a P/T net is essentially a rewriting system on multisets, which allows to encode
the firing of P/T nets as a direct graph transformation in the Double Pushout approach using discrete graphs and empty interfaces for the productions only (see
[7]). Taking into account general graphs and nonempty interfaces graph transformation systems are closer to some generalizations of Petri nets, like contextual
nets. This relationship has been used in [2] to model concurrent computations of
graph grammars.

Vice versa the existence of powerful analysis techniques for P/T nets motivates
to simulate graph transformation by P/T nets [3], which allows to conclude correctness properties of graph grammars from properties of corresponding P/T nets.
The main novelty of this paper wrt. [3] (and subsequent papers of the authors) is
that (i) we take into account also negative application conditions of graph transformations and (ii) the size of the derived P/T is dependent on the type graph and
not to the instance graph. The price we have to pay for a more efficient termination
analysis is that our P/T net can be too abstract to verify all the safety properties
investigated in [3].

Termination of Graph Transformation Systems. Termination of graph transformation systems is undecidable in general [18], but several approaches have been
considered to restrict a graph transformation system such that termination can
be shown. The classical approach of proving termination is to construct a mono-
tone function that measures graph properties, and to show that the value of such a
function decreases with every rule application. Concrete criteria such as the number of nodes and edges of certain types have been considered by Aßman in [1].
However, he sticks to these concrete criteria, while Bottoni et.al. [5] developed a
general approach to termination based on measurement functions.

With respect to termination for graph transformation systems, the current work
generalizes and formalizes the work begun at [13]. This, in fact, is an extension of
the layering conditions for deleting grammars proposed in [6], which were used
for parsing. A main advantage of our approach with respect to the termination
requirements of this parsing algorithm is that we do not require to partition the
rules (and the alphabet) into layers.

As pointed out already in the introduction, we have presented termination criteria for graph transformation systems in [11], which allow to prove termination
of several practical relevant model transformations. However these criteria are not
applicable to model transformations where rules are causally dependent on themselves (e.g. transitive closure) like our motivating example. Since each layer of [11]
can be treated separately by our current techniques, furthermore, the termination
criteria proposed in [11] imposes a special structure on the derived incidence matrix of the P/T net, it is possible to show that our termination analysis technique
based on P/T nets subsumes our former results in [11].

7   Conclusion

In this paper, we have presented a termination analysis technique for model transformations expressed as graph transformation systems using an abstraction into Petri nets. This way, the termination problem of (a special class of) graph transformation systems can be proved by its Petri net abstraction using algebraic techniques. Since the termination of graph transformation systems is undecidable in general, our approach yields a sufficient criterion: either it proves that a GTS is terminating, or gives a "do not know" answer.

We believe that our results can also be useful for proving the termination of QVT-based model transformations, which also uses a very limited set of control structure. For instance, triple graph grammars (TGG) [21] provide a declarative means to specify model transformations, and show a strong conceptual correspondence with bidirectional QVT mappings. Moreover, a pair of traditional (operational) graph transformations can be easily derived for each TGG rule, and then our termination criteria become directly applicable.

Although not mentioned explicitly, the termination criteria presented can also be used for graph transformation with node type inheritance, since a flattening to graph transformation without inheritance is available in [4]. Thus, the termination analysis can always be done and need not be translated back.

Acknowledgments. The authors are grateful to András Pataricza for initial ideas, and Paolo Baldan for fruitful discussions. Valuable comments of anonymous reviewers were also highly helpful.

References

1. Aßmann, U. 2000. *Graph Rewrite Systems for Program Optimization.* ACM TOPLAS, vol. 22(4), pp. 583–637, ACM Press, New York.
2. Bardohl, P., 2000. *Modelling Concurrent Computations: From Contextual Petri Nets to Graph Grammars.* PhD thesis, University of Pisa.
3. Baldan, P., Corradini, A., and König, B. 2001. *A Static Analysis Technique for Graph Transformation Systems* In *Proc. CONCUR 2001*, LNCS 2154, pp. 381–395. Springer.
4. Bardohl, R., Ehrig, H., de Lara J., and Taentzer, G. 2004. *Integrating Meta Modelling with Graph Transformation for Efficient Visual Language Definition and Model Manipulation.* In *Proc. FASE'04*, LNCS 2984, pp. 214–228. Springer.
5. Bottoni, P., Koch, M., Parisi-Presicce, F., Taentzer, G. 2004. *Termination of High-Level Replacement Units with Application to Model Transformation.* In proceedings of VLFM'04, ENTCS.
6. Bottoni, P., Taentzer, G., Schürr, A. 2000. *Efficient Parsing of Visual Languages based on Critical Pair Analysis and Contextual Layered Graph Transformation.* In Proc. Visual Languages 2000 IEEE Computer Society. pp.: 59-60.
7. Corradini, A. 1996 *Concurrent Graph and Term Graph Rewriting.* In Proc. CONCUR'96, LNCS 1119, pp. 438–464, Springer.
8. Corradini, A., Montanari, U., Rossi, F., Ehrig, H., Heckel, R., and Löwe, M. *In [20]*, chap. Algebraic Approaches to Graph Transformation — Part I: Basic Concepts and Double Pushout Approach, pp. 163–245. World Scientific, 1997.

9. Corradini, A., Montanari U., Rossi, F. *Graph Processes*. Fundamenta Informaticae 26(3/4):241-265.
10. Ehrig, H., and Ehrig, K., and Prange, U., and Taentzer, G. Fundamentals of Algebraic Graph Transformation. *Monographs in Theoretical Computer Science. An EATCS Series*, Springer-Verlag New York, Inc., 2006.
11. Ehrig, H., Ehrig, K., de Lara, J., Taentzer G., Varró, D., Varró-Gyapay, Sz. *Termination Criteria for Model Transformation*. In FASE 2005: Internation Conference on Fundamental Approaches to Software Engineering (Edinburgh, UK), LNCS 3442, pp. 49-63, Springer, 2005.
12. GAMS: General Algebraic Modeling System. http://www.gams.com.
13. de Lara, J., Taentzer, G. 2004. *Automated Model Transformation and its Validation with AToM³ and AGG*. In DIAGRAMS'2004 (Cambridge, UK). Lecture Notes in Artificial Intelligence 2980, pp.: 182–198. Springer.
14. Küster, J., Sendall, S., Wahler, M.. *Comparing two model transformation approaches*. In OCL and Model Driven Engineering 2004.
15. Model Transformations in Practice (Satellite Workshop of MODELS 2006) http://sosym.dcs.kcl.ac.uk/events/mtip.
16. T. Murata. Petri nets: Properties, analysis and applications. In *Proc. IEEE*, vol. 77, pp. 541–580. 1989.
17. Object Management Group. *QVT: Request for Proposal for Queries, Views and Transformations*. http://www.omg.org.
18. Plump, D. 1998. *Termination of Graph Rewriting is Undecidable*. Fundamenta Informaticae 33(2):201-209.
19. Rensink, A., and Nederpel R. 2006. Graph transformation semantics for a QVT language. In *Proc. Fifth Intern. Workshop on Graph Transformation and Visual Modelling Techniques (GT-VMT 2006)*, ENTCS, pp. 45–56. Elsevier. In Press.
20. Rozenberg, G. (ed) 1997. *Handbook of Graph Grammars and Computing by Graph Transformation*. World Scientific. Volume 1.
21. Schürr, A. Specification of Graph Translators with Triple Graph Grammars. In *Proc. WG94: Int. Workshop on Graph-Theoretic Concepts in Computer Science*, vol. 903 of *LNCS*, pp. 151–163. Springer-Verlag, 1994.

Non-functional Analysis of Distributed Systems in Unreliable Environments Using Stochastic Object Based Graph Grammars

Odorico Machado Mendizabal[1] and Fernando Luis Dotti[2]

[1] Faculdade de Ciências, Universidade de Lisboa
Bloco C6, Campo Grande 1749-016, Lisboa - Portugal
omendizabal@lasige.di.fc.ul.pt
[2] Faculdade de Informática, Pontifícia Universidade Católica do Rio Grande do Sul
Avenida Ipiranga, 90619-900, Porto Alegre - Brazil
fldotti@inf.pucrs.br

Abstract. In unreliable environments, *e.g.* wireless networks, often there are messages lost, connection and process crashes, among other undesirable fault occurrences. Mechanisms to enhance the dependability of these systems can be employed, but with a performance cost. Analytical approaches are useful to predict performance and dependability values, guiding the system developer to adjust bounds for specific requirements in complex systems. In this paper we use non-functional analysis of Stochastic Object-Based Graph Grammars (SOBGG) models considering classical fault behaviors in distributed systems, allowing the developer to predict performance and dependability values for high performance and resilient systems. The specific contributions of this paper are: (i) revisit the notion of fault representation to allow non-functional analysis, more specifically, steady-state analysis; (ii) discuss the specification of rates associated to SOBGG rules, describing an adequate approach to distributed systems; (iii) show the suitability of the proposed techniques through their application to a case study.

Keywords: Object-based graph grammars, distributed systems, fault-tolerance, non-functional analysis, dependability.

1 Introduction

In unreliable environments, *e.g.* wireless networks, often there are messages lost, connection and processes crashes, network partitions, among other undesirable fault occurrences. Such deficiencies obligate applications running in these environments to use some type of support, like atomic broadcast, fault detection, reconfiguration, recovery actions, among others.

These mechanisms can lead to some resilience level, but with a performance cost. Analytical approaches are useful to predict the cost in terms of performance as well as the dependability levels achieved with the use of such mechanisms, guiding the system developer to construct high performance or resilient systems.

A. Corradini et al. (Eds.): ICGT 2006, LNCS 4178, pp. 275–290, 2006.

In [5] OBGG (Object Based Graph Grammars) were introduced as a specification language suited for asynchronous distributed systems communicating via message passing. Validation of functional aspects of the OBGG models has been accomplished via simulation and model checking [4,7,18]. In [4,7] the functional analysis of OGBB models in the presence of classical fault models in distributed systems (e.g. crash, omission) was discussed. Nevertheless, performability and dependability analysis were not addressed in those works. In [15] we introduced the notion of non-functional analysis of OBGG models, proposing SOBGG (Stochastic Object-Based Graph Grammars).

In this paper we would like to show the applicability and relevance of joining these results through the use of non-functional analysis of SOBGG models considering classical fault behaviors in distributed systems, allowing the model developer to predict performance and dependability values for high performance and resilient systems. As a case study we start from the *Token Ring* model presented in [15] and extend it to consider also fault behaviors. The performance parameters of the model are based on experimental analysis by [9,19].

The specific contributions of this paper are: (i) revisit the notion of fault representation to allow non-functional analysis, more specifically, steady-state analysis; (ii) discuss the specification of rates associated to SOBGG rules, describing an adequate approach to distributed systems; (iii) show the suitability of the proposed techniques through their application to a case study.

This work is organized as follows. In Section 2 the SOBGG formalism is described together with the Token Ring case study which is used in the other sections of the paper. Section 3 and Section 4 briefly present the SAN (Stochastic Automata Network) formalism and the translation from SOBGG to SAN. The fault specification approach is discussed in Section 5 and in Section 6 the Token Ring model is analysed. Final remarks are in Section 7.

2   Stochastic Object-Based Graph Grammars

In [12] a first step towards the stochastic analysis of graph transformation systems is given. In that contribution, the authors associate occurrence rates (corresponding to exponential distributions of probabilities) to rules. With this, the transition system obtained from the graph grammar gives raise to a Continuous Time Markov Chain that can be analysed with existing tools.

In [15] Stochastic OBGG (SOBGG) were proposed. SOBGG is a stochastic extension to OBGG, where rates, likewise to [12], are associated to rules, allowing one to derive probabilities associated to the reachable states of the Graph Grammar. OBGG is a restricted form of graph grammar and therefore the results of [12] apply to OBGG as well. However, due to the state-space explosion problem, in [15] Markov Chains are avoided and an equivalent method with better scalability is preferred. Stochastic Automata Networks (SAN) [16] - see Section 3 - is a Markov Chain equivalent formalism having as advantage its modularity in terms of representation and the compact mathematical solution, allowing the analysis of models with larger state space, if compared to Markov Chains [10]. Once a

system is represented in SAN, it is possible derive the probabilities associated to
the states using the PEPS tool (Performance Evaluation of Parallel Systems)[17].
In [15] a translation from SOBGG to SAN is proposed - see Section 4.

SOBGG, as OBGG, is a restricted form of Graph Grammars [8] that offers
objects which communicate through asynchronous message passing as main abstractions. The specifications are done in an object-based style that is quite
familiar to most of the users, and therefore easy to construct, understand and
use. An SOBGG system is modular, once is composed of independent entities
(objects).

In SOBGG, each object may be composed by the vertices and edges shown
in Figure 1(a). The vertices represent classes and ADTs (Abstract Data Types),
whereas messages and attributes of classes are modeled as hyperedges (edges with
one destination and many source vertices). We defined a distinguished graphical
representation for these graphs to increase the readability of the specifications.
This representation is shown in Figure 1(b). Elements of ADTs are allowed as
attributes of classes and/or parameters of messages. Note that the graph in
Figure 1 defines only a scheme of the kinds of vertices and edges that may occur
in a specification, and does not oblige objects or messages to have attributes.
For example, this graph specifies that, if a class has attributes, they must be
either of type ADT or of type Class.

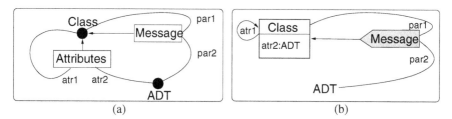

Fig. 1. (a) Object-Based Graph Scheme (b) Graphical Representation of Object-Based
Graphs

A rule will express the reaction of an object to the receipt of a message. A
rule of an object-based graph grammar consists of:

- *a left-hand side L*: describes the items that must be present in the current
  state to enable the application of the rule. The restrictions imposed to left-
  hand sides of rules are:
    - There must be exactly one message vertex, called *trigger message* (this
      is the message treated by this rule).
    - Only attributes of the object that is the target of the trigger message
      may appear.
  These restrictions represent characteristics of the Object-based style, namely
  reaction to messages and encapsulation (only attributes of the instance which
  receives the message can be used in the rule application).
- *a right-hand side R*: describes the items that will be present after the application of the rule. It consists of:

- Objects: all objects and attributes present in the left-hand side of the rule. Although OBGG does not restrict object creation, the analysis method presented in this paper does not support dynamic object creation. The values of attributes may change, but attributes cannot be deleted;
- Messages created by the rule application.

- *a condition*: that must be satisfied for the rule to be applied. This condition is an equation over the attributes of left- and right-hand sides.
- *an occurrence rate:* each rule has an occurrence rate associated. The inverse of the occurrence rate is the mean value of the exponential distribution function that regulates the sojourn time in the state before the rule application. For instance, let g be a state where rule r may be applied and r has rate $t = 5$. Then, the system will stay, in average, $1/5$ time units in state g. However, observing a specific application of r, according to the exponential distribution the system could stay any time in g.

Formally, we use typed attributed hypergraphs and a rule is a (partial) graph homomorphism with application conditions. A SOBGG model is composed of: *(i) a Type Graph*: a graph containing information about all attributes of all classes involved in this system and messages sent/received by each kind of object; *(ii) a set of Rules*: these rules specify how the objects behave when receiving messages. For the same kind of message, there may be many rules specifying the intended behavior. The behavior of an object when receiving a message is not specified as a series of steps that shall be executed, but rather as an atomic change of the values of the object attributes together with the creation of new messages to other (or the same) objects; *(iii)an Initial Graph*: this graph specifies the initial values of attributes of the objects, as well as messages that must be sent to these objects when they are created. The messages in this graph can be seen as triggers of the execution of the object.

The behavior of an OBGG is given by the state transition system generated by applying rules of the grammar starting in the initial state. The computations of a SOBGG are the same as the underlying OBGG (grammar without the rates associated). Note that since the occurrence rate of a rule is given by an exponential distribution, we may have any positive value of delay for the rule application (including ∞). Therefore, the computations of the SOBGG do not exclude any computation of the underlying OBGG.

Provided that the behavior of a SOBGG is given by a state transition system which is finite and irreducible, i.e. from any state of the system it is possible to reach any other one [20], a SOBGG as discussed above defines a discrete state continuous time stochastic process, and has the *memoryless property*. This property is assured by the use of exponential distributions associated to the transitions, meaning that the time to the transition to the next state depends only on the current state of the system and not on the previous ones. With this it is possible to associate a probability to each state, which is the probability of the system being in that state in a steady state situation.

2.1   SOBGG Token Ring Model

The token ring protocol is used to control the access of various stations to the shared transmission medium in a ring topology network [21]. According to the protocol, a special bit pattern, called token, is transmitted from station to station in only one direction. When a station wants to send some content through the network, it awaits for the token, holding it, and sends the message on the ring. The message circulates the ring and the destination station may copy its contents. When the message completes the cycle, it is received by the originating station. The originating station then removes the it from the ring and sends the token to the next station, which then may act as already described. Having only one token, only one station may be transmitting in a given time.

Figure 2(a) is a Type Graph and defines the type *Node*. Instances of *Node* have one boolean attribute called *sent* and may receive two kinds of messages: *Msg* meaning a frame of data and *Token* meaning the token. The *link* to the next *Node* is given by the object reference *next* [1].

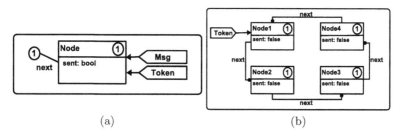

(a)   (b)

Fig. 2. (a) Type Graph for *Node* and (b) Initial Graph for the Token Ring model

The rules that define the behavior of this model are presented in Figure 3. If a *Node* receives the *token* it may send a *Msg* (rule *Send*) or pass the *Token* (rule *Token Pass*). If the *Node* decides to send a *Msg*, the attribute *sent* is assigned to *true*. When a *Msg* is received by a *Node* and it is the originating *Node* (if its attribute *sent* is *true*) then rule *Complete* is applied, removing *Msg* from the ring and generating the *Token* to the (next) *Node*. If the receiving *Node* is not the originating one (its attribute *sent* is *false*) then rule *Transmit* is applied and *Msg* is passed to the next *Node*.

The Initial Graph is shown in Figure 2(b), defining the various instances, attributes and messages of the start situation. A ring with four nodes is defined, called *Node1*, *Node2*, *Node3* and *Node4*. The attribute *next* of each instance refers to the next *Node*. All *sent* attributes are initially *false* and only one *Node1* has the *token*.

---

[1] Graphical notation: in Figure 2(a) rectangles are vertices and the numbers inside circles are used to indicate the type of each vertex in Figures 2(b) and 3. The items within a vertex are the vertex attributes. Messages that appear in Figures 2(a) and 3 are hyperarcs.

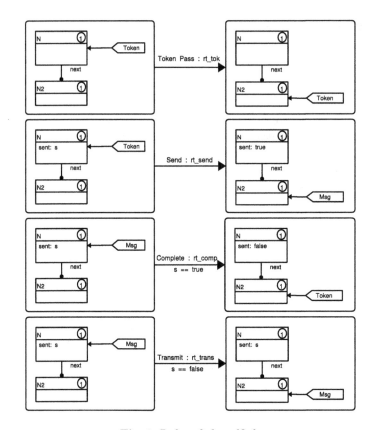

Fig. 3. Rules of class *Node*

3   Stochastic Automata Networks

In the Stochastic Automata Network (SAN) formalism, a system is modeled by interacting subsystems which, in turn, are represented by automata that may behave independently or may have dependencies. According to [1], SAN has exactly the same application scope as Markov Chains, with the advantage that models are constructed componentwise [16].

An automaton is composed by states and transitions labeled with event names. A SAN model is composed by various automata. These automata may evolve independently with local events (that may affect only the local state of the automata participating in this event), whereas synchronizing events are used to model joint evolution of two or more automata. With the association of distribution probabilities to the events, the labeled transition system generated by a SAN gives raise to a Markov Chain and it is possible to calculate the steady state probability of each state of a SAN. More concretely, to each event there is an occurrence rate associated. The inverse of the occurrence rate is the mean value of the exponential distribution function that regulates the time interval between two occurrences of the event.

A SAN defines the set of events that are used to synchronize the different automata during the execution. The state changes of SANs are possible when all different automata that may be engaged in some event are in some state in which a transition labeled with this event is possible. Note that since there may be different transitions labeled with the same event, there may be different reachable state starting with the same state and executing the same event.

In this paper we do not present more details about SAN, but the reader may refer to [1,16] for more information about SAN. In [15] a formal presentation of the subset of SAN used in [15] and in this paper is provided. As an example, in Section 4.1 a SAN model representing the translation of the SOBGG Token Ring model from Section 2.1 is presented.

4   Translation from SOBGG into SAN

To associate probabilities to the states of the behavior semantics of a SOBGG we have to solve the respective stochastic model. To do this, we translate SOBGG to SAN and solve the resulting SAN model using the PEPS tool[17]. The translation is briefly discussed here, but a more complete presentation is done in [15].

According to our translation approach, each object in a SOBGG initial graph originates a SAN. The composition of the various objects is possible by the composition of different SANs, resulting again in a SAN. For each object:

- each attribute of the internal state is represented by a separate automaton in SAN;
- for each input message type for that object:
  - for each possible configuration of concrete parameters of the message: an automaton is generated, representing how many messages of that type, with that configuration of parameters, is stored in the state of the system;
- for each rule:
  - for each possible occurrence of the rule: a SAN event having the same rate as the originating SOBGG rule is generated, as well as a series of transitions associated to that event.
    These transitions represent all the possible state changes in a rule application and they are synchronized by the same event. Thus the state changes are made in an atomic way, maintaining the SOBGG semantic. Due to the encapsulation of object-based systems, these transitions affect only: the automata representing the internal states of the object; the automaton representing the input message consumed by the rule; and the automata representing input messages of objects that receive messages due to this rule application.

From the initial graph the initial state of the SAN is derived. More concretely, for the PEPS tool the initial graph is translated into a partial reachability function that specifies the states of all automata of the SAN according to the values of the attributes, messages and parameters in the initial graph.

4.1   SAN Token Ring Model

This section presents the SAN model obtained from the Token Ring SOBGG model in Section 2.1 through the translation.

Figure 4 depicts partially the SAN model. Note that *sent_Node1* and *next_Node1* automata represent the *sent* and *next* attributes for the object *Node*1 of the Token Ring SOBGG model (the initial state of each automata is the gray circle). Since the topology is static, the automaton *next_Node1* is actually not needed. To represent the possibility of referring to different nodes during system functioning, such an automaton would be needed. The names of events used to label transitions are composed of the name of the rule applied, a list of attribute names and the respective value needed to build the match (in case of attributes that are object references), and the object that receives the message. In this figure, only the automata corresponding to *Node1* are presented. The complete model is composed by automata for all *Node* instances in the initial graph.

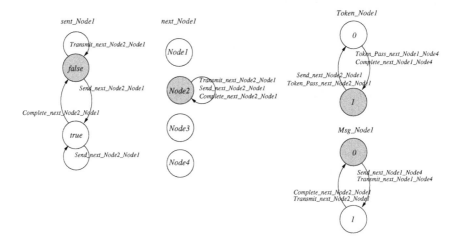

Fig. 4. Token Ring model translated

Token_Node1 and *Msg_Node1* are message automata for object *Node*1, corresponding to messages *Token* and *Msg*, respectively. For example, when the rule *Transmit* is applied by *Node4*, a message *Msg* to *Node1* is generated. This transformation is given by event *Transmit_next_Node1_Node4*. When the rule *Transmit* is applied by *Node1*, one message *Msg* to *Node1* is consumed. This transformation is given by event *Transmit_next_Node2_Node1*.

To complete the SAN model, we translate the initial graph depicted in Figure 2(b) into the following partial reachability function in PEPS:

```
partial reachability =
(st sent_Node1==false) && (st Token_Node1==1) && (st Msg_Node1==0) &&
(st sent_Node2==false) && (st Token_Node2==0) && (st Msg_Node2==0) &&
(st sent_Node3==false) && (st Token_Node3==0) && (st Msg_Node3==0) &&
(st sent_Node4==false) && (st Token_Node4==0) && (st Msg_Node4==0);
```

The partial reachability function in PEPS can be used to define one or more reachable states. Here we use it to define the initial state of the SAN.

A token ring network with four nodes was modeled, having a static topology, *i.e.* a node does not change its neighbors. Due to the static topology of the example, the automata *next_Node1* to *next_Node4*, representing the *next* attribute of the instances are not necessary[2]. Therefore, each node was modeled with three automata, one for the attribute *sent* and two for the possible input messages. There are two states in each automaton. This results in 12 automata and a product state space of 4096 states. However, considering the initial state as described, only 20 states are reachable.

5   Modeling Fault Behavior

In [4,6,7] the specification and analysis, through simulation and model checking, of OBGG systems in the presence of classical faults were discussed. The types of faults assumed are benign, such as crash and omission. The representation of fault behavior is suggested by [3], that states that a system may change its state based on two event classes: normal system operation and fault occurrences. Based on this observation a fault can be modelled as an undesired (but possible) state transition of a system [11], *i.e.* a fault is just another kind of (programmable) behavior. [11] further provides a straightforward transformation of a model of the system into another model, representing the system in the presence of a selected type of fault.

An interesting observation is that due to the declarative and reactive characteristics of OBGG, the same ideas are applicable to OBGG. A fault behavior F is thus represented by the transformation of an OBGG model M into an OBGG model M_F and this transformation can be automated. The same results are valid for SOBGG.

In the **crash fault behavior** a process fails by halting and processes that maintain communication with the halted process are not warned about the fault. This section describes how to transform a SOBGG model M into a model M_F that incorporates the behavior of a crash fault. In order to represent the behavior of a given object in the presence of a crash fault, a transformation procedure is applied:

- a boolean attribute, called *down* is inserted in the *type graph* of the respective class. Depending on the value of this attribute, the object may exhibit the fault behavior (*down* is *true*) or not (*down* is *false*).
- new rules are added to activate the fault behavior and to cease it (changing the values of the attributes (*down*). To illustrate, the rules *Crash* and *Uncrash* for the *Node* class are presented in Figure 5(a).
- for each original rule in M, add in the rule ¬*down* as condition;
- for each original rule in M, generate a new rule in M_F. This rule will have the same name as the original, but preceded by F_-, have as condition *down*, and will consume the input message but without any effect, i.e. without

---

[2] This and other optimizations on the translation approach are found in [15].

changing attributes or sending messages. The desired (faulty) effect is that the input message is simply lost. See for instance Figure 6 where rule *F_Send* is generated from *Send*.

- one or more rules specifying the recovery actions are described in reaction to message *Recovery*. Note that since the recovery depends on the application, these rules have to be defined by the system designer.

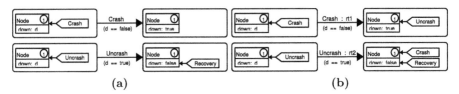

(a)   (b)

Fig. 5. Rules *Crash* and *Uncrash* used to activate/deactivate the crash fault behavior: (a) for qualitative analysis; (b) for quantitative analysis

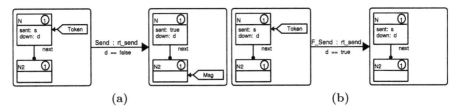

(a)   (b)

Fig. 6. Rule *Send*: (a) correct behavior; (b) fault behavior

In this work we apply a steady state analysis over the model. Thus, we can analyse situations where fault occurrence and recovery take place cyclically. Although this is a restriction, the quantification of fault and recovery actions is typically given in terms of MTBF (Mean Time Between Failures) and MTTR (Mean Time To Repair), which can be well captured in our model. Thus we change the rules *Crash* and *Uncrash* in order to always generate one message *Uncrash* when the rule *Crash* is applied and generate one message *Crash* when the rule *Uncrash* is applied (Figure 5(b)). The rates to rules *Crash* and *Uncrash* can be assigned to represent MTBF and MTTR, respectively. The resulting SOBGG with crash representation has 11 rules.

The rules reacting to *Recovery* describe the actions of the faulty node when it is correct again. A possible recovery procedure for a node in a Token Ring is to await for a period, which is enough for the message with maximum length to circulate the Ring, and, in the meanwhile if a message or token is received, simply follow the protocol or, if no packet is received, generate a new token in the Ring.

From the SOBGG with crash representation we apply the translation into SAN. Figure 7 depicts the automata representing *down*, *Crash* and *Uncrash*, together with the events generated by the rules specifying the fault behavior (beginning with *F_*). Note that the attributes are not updated by these rules, only the original rules (correct behavior) can update attributes. However, the messages are consumed by both types of rules (see rules *Send_Node2_Node1* and *F_Send_Node2_Node1*). In addition, *Crash* and *Uncrash* messages are alternated, according to the rules in Figure 5(b).

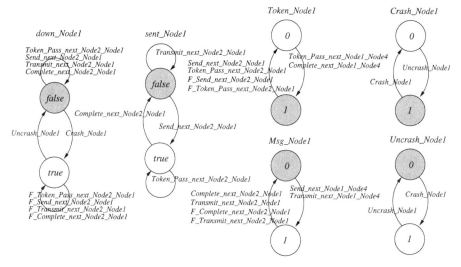

Fig. 7. Translated SAN model for crash fault behavior

This approach to fault specification increases the state space of the model due to new messages (*Crash*, *Uncrash* and *Recovery*) and attributes (*down*). Each of these automata has two states, thus the new product state space, to represent one faulty node in the ring, $prod(M_F) = 2 \times 2 \times 2 \times 2 \times prod(M)$. As stated in Section 2.1, $prod(M) = 4096$, so $prod(M_F) = 65536$ states. Considering the same initial state as in section 2.1, the reachable state space is 16384 states.

6   Model Analysis

This section presents results obtained from the steady state analysis of the Token Ring translated model. We discuss the representation of rates for the rules and the quantitative analysis of the fault-free and faulty models, considering crash.

6.1   Representation of Rates

For each rule r_i of the model, an occurrence rate rt_i is associated. The inverse of the occurrence rate is the mean value of the exponential distribution function that describes the sojourn time in the state before the rule application. In other words, $t_i = 1/rt_i$ is the mean value, in time units, that the system stays in the state before the rule application.

We consider the influence of three different time measures in the model, based on [19,22], to calculate t_i: time to send a message t_{send}, which is the time spent on local processing in the node until the message is considered to be posted in the network; network contention t_{net}, which is the time spent by the message in the path to the destination; and time to receive a message $t_{receive}$, which is the time spent since the message arrived at the destination node in the communication devices until it is processed by the application.

The left-hand side of a rule r_i describes the delivery of the activation message of this rule. The delivery time consists of $t_{net} + t_{receive_i}$. The right-hand side specifies attribute changes and messages sent. If one or more messages are created, it is necessary to include t_{send_i} in t_i in order to represent the time to send the set of messages sent by r_i. When rule r_i does not send any message we consider $t_{send_i} = 0$ and t_i will represent only the time for delivery of the activation message of r_i. In this way, the time t_i will be represented as follows:

$$t_i = t_{receive_i} + t_{net} + t_{send_i}$$

With the analysis so far, each rule r_i, having a t_i calculated as above, would have a rate $rt_i = 1/t_i$. This would model the sojourn time t_i in the state before the rule application. However, when various rules can be applied to a same state, it is interesting to be able to represent different probabilities to the choice of the different rules. This allows to study different conditions (e.g. workload) of the system. For each set of n rules with the same activation message and which can be enabled by the same states, we assign probabilities $p_1, \ldots, p_n$, where $p_1 + \ldots + p_n = 1$, representing the probability of choice for each rule. The identification of these sets of rules is left to the designer since it is dependent on the application.

Consider that in a given state, rules r_i with i from 1 to n, can be applied with respective probabilities p_i and sojourn times t_i associated. Then, to associate rates rt_i to the rules in order to represent this probability of choice, we have to:
- calculate the mean sojourn time mt considering p_i and t_i as follows:
  $mt = \sum_{i=1}^{n} t_i \times p_i$
- calculate the rate to rule r_i to represent the probability p_i as follows:
  $rt_i = p_i/mt$

For example, in the Token Ring, one can observe that both rule *TokenPass* and rule *Send* can be applied to the same states (see Figure 3) while the other rules are applied in exclusive situation. Thus, for this case study we define probabilities associated to *TokenPass* and *Send*. Suppose that the $t_{tokenPass} = 570\mu s$ and $t_{send} = 630\mu s$. Imagine that we may choose *tokenPass* and *Send* in various configurations of probabilities in order to model various scenarios: 9:1; 7:3, 1:1, 3:7, and 1:9. Thus, the configurations of rates for rules *tokenPass* and *send* would be as in Table 1.

Table 1. Configuration of rates according to probabilities

Properties	Scenarios (proportion token:message)				
	9:1 (tok:msg)	7:3 (tok:msg)	1:1 (tok:msg)	3:7 (tok:msg)	1:9 (tok:msg)
rate tokenPass	0.9/576	0.7/588	0.5/600	0.3/612	0.1/624
rate send	0.1/576	0.3/588	0,5/600	0.7/612	0.9/624

6.2   Evaluation of Results

In this section we analyse the Token Ring model presented in Section 2.1 based on the rates used in [19]. In that work, simulation and measurement of a real network was performed in order to analyse the performance of failure detection mechanisms. Since we are considering an analogous environment in this paper we adopt the measures presented in [19].

We consider that the time spent to receive a token is $230\mu s$, to receive a message is $260\mu s$, to send a token is $240\mu s$ and to send a message is $300\mu s$. Further, we consider a network latency of $100\mu s$ to deliver any type of message. In Section 6.1 we discussed how to define the rates associate to rules. With the communication parameters described above we achieve the following time values: $t\_tok = 570\mu s$ or $t\_tok = 230\mu s + 100\mu s + 240\mu s = 570\mu s$, $t\_send = 630\mu s$, $t\_comp = 600\mu s$ and $t\_trans = 660\mu s$.

In addition, we model the probability of sending a message or a token (rules *Token Pass* and *Send*). We solve the model assuming various configurations of probabilities: the same probability, *i.e.* 1:1 (token:msg); representing higher probabilities of sending data instead of passing the token with 3:7 (token:msg) and 1:9 (token:msg); and representing situations of lower workloads with 7:3 (token:msg) and 9:1 (token:msg), increasing the network idleness.

Evaluating the Fault-Free Environment. In the fault-free environment there are no crash occurrences. For this environment we measured the probability of the network being busy or idle. The properties are described informally and as SAN integration functions, as follows:

- **busy network:** the probability of having a data message in the ring, (*i.e.* the probability of any automata *Msg_Node1* to *Msg_Node4* being in state *1*)
    busy = (nb [Msg_Node1 .. Msg_Node4] 1) > 0;
- **idle network:** the probability of having a token message in the ring
    idle = (nb [Token_Node1 .. Token_Node4] 1) > 0;

The operation *(nb [list of automatonName] stateName)* returns a vector of probabilities mapping from 0 to the number of listed automata to a probability of having that number of automata in the state mentioned. Thus, the function *busy* determines the probability of more than 0 automata, out of *Msg_Node1* to *Msg_Node4*, being in the state 1. Table 2 presents the results achieved.

Table 2. Quantitative analysis of the Token Ring model for a fault-free environment

Properties	Scenarios (proportion token:message)				
	9:1 (tok:msg)	7:3 (tok:msg)	1:1 (tok:msg)	3:7 (tok:msg)	1:9 (tok:msg)
busy	30.93 %	55.82 %	68.25 %	74.69 %	78.81 %
idle	69.06 %	43.17 %	31.74 %	25.31 %	21.18 %

Note that when messages and tokens are sent proportionaly (1:1), the probability of the network to be busy is higher. This occurs because the time to deliver messages is higher than the time to deliver tokens (since messages circulate the whole ring) so most of the time *Msg* is being transmitted on the network.

Evaluating Faulty Environment. In this section we solve the Token Ring model augmented with fault behavior according to changes described in Section 5. To solve models considering fault behavior, we define values for MTBF and MTTR, representing how often a node will crash and the time to recover a crashed node. The values of MTBF and MTTR can be configured to represent and analyse different and limit situations. To exemplify their use here we consider MTBF of $1s$ and MTTR of $100\mu s$ (see Table 3).

Table 3. Quantitative analysis of the Token Ring model for a faulty environment (MTBF $= 1s$ and MTTR $= 100\mu s$)

Properties	Scenarios (proportion token:message)				
	9:1 (tok:msg)	7:3 (tok:msg)	1:1 (tok:msg)	3:7 (tok:msg)	1:9 (tok:msg)
busy	28.90 %	52.14 %	62.40 %	68.16 %	71.72 %
idle	63.15 %	39.70 %	29.30 %	23.45 %	19.81 %
unavailable	7.95 %	8.16 %	8.30 %	8.39 %	8.47 %
node crashed	9.10 %	9.10 %	9.10 %	9.10 %	9.10 %

Moreover, we can perform other analyses since we have more elements. For instance, we may compute the probability of having a Node crashed (st down_Node1 == true) [3] and, more interestingly, we can observe that there is a non zero probability of having the network unavailable, i.e. with no token and no message in the network. This is described as: $unavailable = 1 - busy - idle$.

Note that the probability of a node being crashed and the network being unavailable is not the same. This occurs since the communication between non-crashed nodes remains even some node has failed. The network is unavailable only when a message is sent to a crashed node and, consequently it is lost. The network unavailability is dependent on MTBF and MTTR.

7   Final Remarks

In this paper we briefly presented Stochastic Object-Based Graph Grammars (SOBGG), an extension of Object-Based Graph Grammars (OBGG) to represent stochastic systems. We also briefly discussed the translation from SOBGG to Stochastic Automata Network (SAN), allowing one to perform steady state analysis on the model, associating probabilities to the reachable states [15].

Various stochastic formalisms could have been chosen for the translation, like Stochastic Process Algebras (SPA) [13] and Generalized Stochastic Petri Nets (GSPN) [14]. Considering the data driven characteristic of Graph Grammars, it is more natural to adopt a target formalism with explicit state representation.

---

[3] The function *st aut == x* returns the probability of an automaton *aut* being in state *x*.

We consider that this eases both the translation as well as the analysis of the translated model. Thus SPA were not preferred. As far as GSPN are concerned, even GSPN representations tend to be more compact than SAN due to the use of tokens [1,2], we consider that the resulting GSPNs to represent SOBGG models would have a high number of transitions involving many places, hindering legibility. Nevertheless, the use of GSPN should be considered in future steps mainly due to tool support. In the case studies carried out so far, we could notice that SOBGG models, when translated to SAN, tend to generate a large product state space but a reduced reachable state space. The PEPS tool, in the current version, first calculates the product state space and then solves the system, assigning probabilities to the reachable states. Therefore, using PEPS, our models are restricted in the product state space. Existing GSPN tools would not impose such a restriction. A new version of the PEPS tool is being developed whereby the the reachable state space is calculated directly. With the ideas presented in [15] and here, this enhancement should allow the stochastic analysis of SOBGG models with considerable size.

A particular interesting result achieved for OBGG, which is highly desired to have also in SOBGG is the possibility of representing classical fault models for distributed systems [4,6,7]. With this it is possible to evaluate important aspects such as availability and performance impact of fault-tolerance mechanisms. Therefore, in this work we revisited the previous ideas on representing faults to be used with SOBGG. More concretely, we discussed the crash fault model and its representation in SOBGG. Moreover, we discussed the assignment of rates to SOBGG rules in order to represent meaningful performance aspects of distributed systems using message passing. These ideas were applied to a case study and the numerical results achieved are coherent with the model analysed.

As could be noticed, in order to analyse the SOBGG model the user has to know the generated SAN model to extract results. One future work should be to allow the analysis of the model based on the SOBGG abstractions and not on the generated SAN, which should be hidden from the user.

References

1. L. Brenner, P. Fernandes, and A. Sales. The need for and the advantages of generalized tensor algebra for kronecker structured representations. *International Journal of Simulation: Systems, Science & Technology*, 6(3-4):52 – 60, 2005.
2. Ming-Ying Chung, Gianfranco Ciardo, Susanna Donatelli, N. He, Brigitte Plateau, William J. Stewart, E. Sulaiman, and J. Yu. A comparison of structural formalisms for modeling large markov models. In *IPDPS*. IEEE Computer Society, 2004.
3. F. Cristian. A rigorous approach to fault-tolerant programming. *IEEE Trans. on Soft. Eng.*, 11(1):23–31, 1985.
4. F. L. Dotti, O. M. Mendizabal, and O. M. Santos. Verifying fault-tolerant distributed systems using object-based graph grammars. In *Lecture Notes in Computer Science*, volume 3747, pages 80 – 100. Springer-Verlag GmbH, October 2005.
5. F. L. Dotti and L. Ribeiro. Specification of mobile code systems using graph grammars. In *4th International Conference on Formal Methods for Open Object-Based Distributed Systems*, volume 177, pages 45–63. IFIP Conference Proceedings, Kluwer Academic Publishers, 2000.

6. F. L. Dotti, L. Ribeiro, and O. M. Santos. Specification and analysis of fault behaviours using graph grammars. In *2nd International Workshop on Applications of Graph Transformations with Industrial Relevance*, volume 3062 of *Lecture Notes in Computer Science*, pages 120–133, USA, 2003. Springer-Verlag.

7. F. L. Dotti, O. M. Santos, and E. T. Rödel. On use of formal specification to anayse fault behaviors of distributed systems. In *1st Latin-American Symposium on Dependable Computing*, volume 2847, pages 341–360. LNCS, 2003.

8. H. Ehrig. Introduction to the algebraic theory of graph grammars. In *1st International Workshop on Graph Grammar and Their Application to Computer Science and Biology*, volume 73, pages 1–69. Lecture Notes in Computer Science, 1979.

9. L. Falai and A. Bondavalli. Experimental evaluation of the qos of failure detectors on wide area network. In *DSN '05: Proceedings of the 2005 International Conference on Dependable Systems and Networks (DSN'05)*, pages 624–633, Washington, DC, USA, 2005. IEEE Computer Society.

10. P. Fernandes, B. Plateau, and W. J. Stewart. Numerical evaluation of stochastic automata networks. In *Proceedings of the Third International Workshop on Modeling, Analysis, and Simulation of Computer and Telecommunication Systems*, pages 179–183, 1995.

11. F. C. Gärtner. Fundamentals of fault-tolerant distributed computing in asynchronous environments. *ACM Computing Surveys*, 31(1):1–26, 1999.

12. R. Heckel, G. Lajios, and S. Menge. Stochastic graph transformation systems. In *Proc. 2nd Intl. Conference on Graph Transformations (ICGT 2004)*, volume 3256 of *Lecture Notes in Computer Science*, pages 210–225. Springer, 2004.

13. Hermanns, Herzog, and Katoen. Process algebra for performance evaluation. *TCS: Theoretical Computer Science*, 274, 2002.

14. M. Ajmone Marsan, G. Balbo, and G. Conte et al. *Modelling with Generalized Stochastic Petri Nets*. Wiley series in parallel computing. Wiley, New York, 1995.

15. O. M. Mendizabal, F. L. Dotti, and L. Ribeiro. Stochastic Object-Based Graph Grammars. In *Proceedings of the Brazilian Symposium on Formal Methods (SBMF 2005)*, pages 128–143, 2005. http://bibliotecadigital.sbc.org.br/?module=Public &action=PublicationObject& subject=147&publicationobjectid=7.

16. B. Plateau. On the stochastic structure of parallelism and synchronization models for distributed algorithms. In *SIGMETRICS*, pages 147–154, 1985.

17. B. Plateau and K. Atif. Peps: a package for solving complex Markov models of parallel systems. In *Proceedings of the 4th International Conference on Modelling Techniques and Tools for Computer Performance Evaluation*, 1988.

18. O. M Santos, F. L. Dotti, and L. Ribeiro. Verifying object-based graph grammars. *Eletronic Notes in Theoretical Computer Science*, 109:125–136, 2004.

19. N. Sergent, X. Défago, and A. Schiper. Impact of a failure detection mechanism on the performance of consensus. In *Proc. 8th IEEE Pacific Rim Symp. on Dependable Computing (PRDC'01)*, Seoul, Korea, December 2001.

20. W.J. Stewart. *Introduction to the numerical solution of Markov chains*. Princeton University Press, 1995.

21. A. S. Tanenbaum. *Computer Networks*. Prentice-Hall, Englewood Cliffs, New Jersey, third edition, 1996.

22. P. Urbán, X. Défago, and A. Schiper. Contention-aware metrics for distributed algorithms: Comparison of atomic broadcast algorithms. In *Proc. 9th IEEE Int'l Conf. on Computer Communications and Networks (IC3N)*, pages 582–589, 2000.

Temporal Graph Queries to Support Software Evolution⋆

Tobias Rötschke and Andy Schürr

Real-Time Systems Lab
Darmstadt University of Technology
{rotschke, schuerr}@es.tu-darmstadt.de

Abstract. Graph transformation techniques have already been used successfully by several research groups to support re-engineering of large legacy systems. Where others often aim at transforming the system to improve it, we advocate an evolutionary approach that embeds transformations within the ordinary development process and provides tool support to monitor the ongoing progress regularly. In this paper, we discuss how temporal graph queries based on Fujaba story diagrams can provide a natural means to express trend-oriented metrics and consistency rules that we identified in our industrial case studies. To this end, we discuss a first-order logic rather than operational interpretation of a graph queries and show how well-known temporal logic operators can be added to express rules over consecutive states of the same instance graph.

1 Introduction

After successfully developing and enhancing a software intensive product for many years, developers often realize that it becomes harder and harder to add new features to the system. Subsequent reverse engineering usually shows that the software architecture needs some major restructuring of the existing code. In some cases, these restructurings can be quickly performed by automated re-engineering tools. In other cases, there are technical obstacles or rigid development processes prohibiting the use of transformation tools. Manual restructuring of the complete system however can delay the development of new features for several weeks or months, which is not acceptable from an economic point of view.

Under these circumstances, less invasive tool support can provide means to achieve long-term goals such as restructuring without sacrificing business-critical short term goals. The idea is to improve the software architecture in small steps as part of the ordinary development process and monitor the progress towards long-term goals with specially designed analysis tools. In our approach, such tools can be generated from domain-specific meta models describing architectural concepts, design tool data and source code structure as well as appropriate graph transformations and queries defining metrics and consistency rules.

⋆ Work supported in part by the European Community's Human Potential Programme under contract HPRN-CT-2002-00275, SegraVis.

A. Corradini et al. (Eds.): ICGT 2006, LNCS 4178, pp. 291–305, 2006.

Naturally, the initial situation is usually pretty bad in the sense, that it is quite different from the desired situation. When the existing problems cannot be solved at once or their dynamics are not yet fully understood, a first goal might be that things don't become worse than they already are. In later stages of the restructuring process, improvements should take place at a certain pace to ensure that the long-term goals are finally met. Hence, our analysis tools focus on trends in metric values and rule violations rather than absolute numbers.

Consequently, we came to the conclusion, that it would be beneficial to be able to specify rules with respect to analysis data of the same system sampled at different points of time. One could think of the following kind of temporal rules:

- A file with less than 1000 lines of code (LOC) *now* should have less than 1000 LOC *henceforth*.
- If a file includes a header file, but its block does not see an adequate exporting interface *now*, the block has *never* seen such an interface.
- *After* the next milestone (01-05-07), the number of invalid includes must be reduced by 100 every week.
- Each block must provide at least one interface *before* the next release.

During our research, we found out that temporal logic provides a means to express a number of these rules we had in mind. However, we had to investigate, how temporal logic and graph transformation fit together. As we point out in Section 5, some work had already be done to bring graph transformation and time together, but did not fully serve our purposes. The main contribution of this paper is the definition of a temporal graph query language derived from story diagrams used by Fujaba [1]. To define the semantics, we construct a pair grammar [2] mapping temporal graph queries on equivalent temporal first order logic formulae.

The rest of the paper is structured as follows: In Section 2, we introduce an example that is referred to throughout the paper. In Section 3, we introduce a subset of Fujaba story diagrams referred to as *graph queries*. Next, in Section 4, we show how well-established time operators from temporal logic can be added to graph queries so that they can be interpreted by temporal first order logic. Section 5 discusses related work with particular emphasis on graph transformation-based re-engineering and graph transformation considering time. We summarize our results in Section 6, and provide some insight into ideas to continue this work.

2   Running Example – Motivated by Case Study

Throughout the paper, we use a simple example motivated by a more complex industrial case study described in [3]. For this example, we assume that we start with an existing system programmed in C. The project started quite small more than a decade ago, with only a couple of developers. Back then, nobody anticipated the growth of the system and hence no-one cared about "architecture". In typical C-style, the system was divided into several libraries with globally

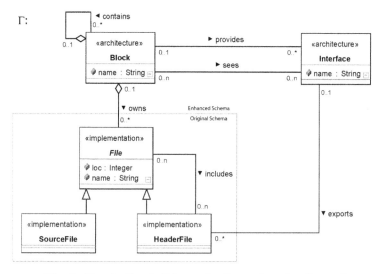

Fig. 1. Simple Graph Schema for Re-engineering Example

available header files. Over the years, the system has become bigger and now consists of several million lines of code, which are maintained and improved by more than 200 developers.

Due to the complexity of the dependencies, it has become more and more difficult to modify the existing code without breaking something. As a solution, the software architects defined, in their terms, architecture concepts – actually a proprietary module interconnection language – consisting of a nested building blocks and associated export interfaces. Fig. 1 presents a graph schema Γ^1 for these concepts and shows the dependencies between new *architecture* and already existing *implementation* concepts.

Given the right parsers, which are out of scope of this paper, it is possible to periodically create an instance graph of this schema representing the architecture and the related implementation at the time of analysis. Existing Graph queries (and transformations) can be defined to check (and transform) each single instance with respect to certain consistency rules. As one goal was to avoid *new* violations in the architecture, we found that we would like to have rules that enabled us to reason about certain objects in *different* instance graphs. The contribution of this paper is to propose the syntax and semantics of a temporal language extention for graph queries to make this possible. The exact implementation of these queries is out of the scope of this paper, but some basic ideas are sketched in Section 4.3.

The case study discussed in [3] involved several metrics and 23 consistency rules which used to be checked every night. The system consisted of 500 building blocks, 8000 files and 40000 include relationships. Most of the analysis tool was hand-coded and implemented with the specific system in mind. The code

---

[1] A formal definition for graph schemas can be found in [4].

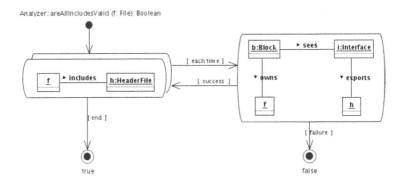

Fig. 2. Consistency Rule for All Included Header Files

for checking consistency rules however, was already generated from Fujaba 3 specifications, with a strong emphasis on changes in metrics and consistency violations, rather than trying to interpret absolute figures.

Ever since, our research effort has been put into generalizing the approach to make it easily available to other development projects as well. Recently started case studies indicate that our approach will be useful in other industrial application areas as well. This research is performed as part of the MOFLON project [5], which is a MOF 2.0-based meta modeling framework on top of the Fujaba Toolsuite [1]. Hence, our suggestions for temporal graph queries are based on the Story Driven Modeling (SDM) language, which is the graph transformation language featured by Fujaba.

3   First Order Logic Interpretation of Graph Queries

In the following, we describe the systax and semantics of a subset of SDM referred to as "graph queries". As we restrict ourselves to queries, the semantics can be descibed by well-known first order logic (FOL) formulae. First we discuss an example query that serves as consistency rule and discuss its logic interpretation in Section 3.1. Then we approach the issue more systematically in Section 3.2 and show by construction, how any FOL formula can be translated into a normalized graph query. We omit formulae for non-normalized graph queries, as they would be too complex considering the size restrictions of a paper. However, we discuss in Section 3.3, under which conditions more user-friendly story diagrams can be translated into FOL formula.

3.1   Example Interpretation

Fig. 2 shows a Fujabagraph query representing a consistency rule to check whether all[2] outgoing *include* relationships between a given file f and header

---

[2] Indicated by the doube-boxed "forall" activity.

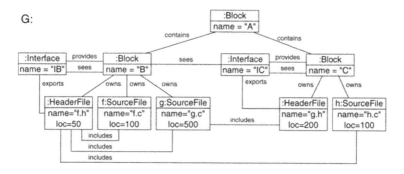

Fig. 3. Instance Graph for Schema in Fig. 1

files h are correct with respect to the architecture due to an adequate *sees* relationship between the related block b and the interface i.

$$areAllIncludesValid(f) \Leftrightarrow \forall h($$
$$(type(h, HeaderFile) \land \neg(f \equiv h) \land rel(f, includes, h))$$
$$\rightarrow \exists b \exists i (type(b, Block) \land type(i, Interface)$$
$$\land \neg(b \equiv f) \land \neg(b \equiv h) \land \neg(b \equiv i) \land \neg(f \equiv h) \land \neg(f \equiv i) \land \neg(h \equiv i)$$
$$\land rel(b, owns, f) \land rel(b, sees, i) \land rel(i, exports, h))). \tag{1}$$

Equation 1 is a FOL interpretation of this query, where type, rel and eval are predicate symbols and attr is a function symbol with the following interpretation:

- *type(x, T)*: object x is an instance of type T.
- *rel (x,E,y)*: between objects x and y exists a link of association E.
- *eval (x, a, Ω, e)*: the value of attribute a of object x compares to expression e by operator $\Omega \in \{=, \neq, <, \leq, \geq, >\}$.
- $x \equiv y$: x is the same object as y.

Let G_Γ be an instance graph consistent with the schema Γ. A valuation for a given G is a function $\mathcal{A}_G : \bar{A} \longrightarrow \{0, 1\}$, where $\bar{A}$ is the set of FOL formulae. As a shorthand, we write $G \models F$ for $\mathcal{A}_G(F) = 1$. Fig. 3 represents an instance graph G consistent with the schema Γ introduced in Fig. 1. As one can easily reproduce, the following holds: $G \models areAllIncludesValid(f)$, $G \models areAllIncludesValid(g)$, and $G \not\models areAllIncludesValid(h)$.

3.2 From Formulae to Graph Queries

As we have seen, it is possible to interpret the graph query in Fig. 2 as a FOL formula, but is there an equivalent graph query for any given FOL formula? Indeed, we were able to construct a graph query language comprising a useful subset of Story Diagrams defined in [1], which essentially are UML activity diagrams containing visually notated transformation rules of the instance graph. To this end, we defined the pair grammar shown in Fig. 4 and 5.

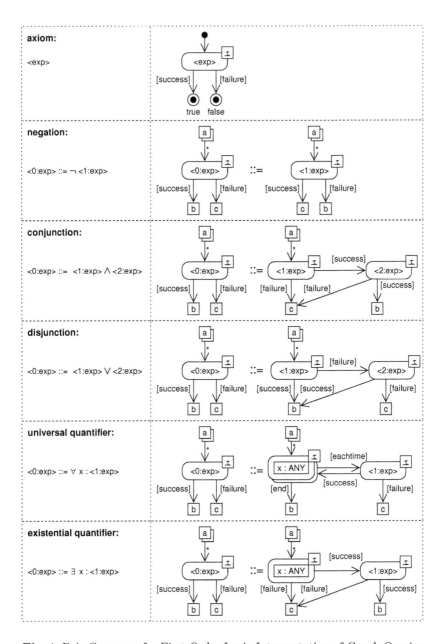

Fig. 4. Pair Grammar for First Order Logic Interpretation of Graph Queries

Pair grammars have been invented by Pratt for the precise specification of text-to-graph and graph-to-text translations [2]. For this purpose, they combine a string and a graph grammar which derives corresponding string and graph

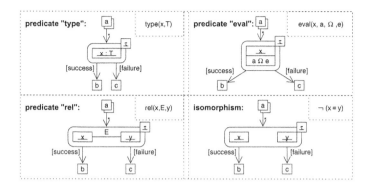

Fig. 5. Pair Grammar Right Hand Sides of Atomic Rules

sentences simultaneously. Note, that we define the pair grammar based on the concrete rather than the abstract syntax of SDM which would be correct, but less readable.

Each row in Fig. 4 contains one rule that is divided into two sub rules that are applied simultaneously. The left sub rule describes the construction of the FOL formula, while the right sub rule describes the construction of the related graph query. Non-terminals are denoted by angle brackets (e.g. <exp>). Corresponding non-terminals in both sub rules have the same name.

In the right sub rule, non-terminals are embedded in rounded boxes representing *activities*. We always consider the edge context of the activity extended by the rule and embed activities on the rule's right hand side in the context of the left hand side. Every activity may have any number of incoming transitions with arbitrary labels. The incoming edge context is marked by a box named *a*. Apart from *forall*-activities, each activity has exactly one outgoing *success* and *failure* transition. The success edge context is named *b* and the failure edge context *c*. As every word in this grammar is derived from the first rule, the resulting story diagram has one start activity, exactly one stop activity marked with *true*, and exactly one stop activity marked with *false*.

Fig 4 contains definitions of formulae containing the classical FOL operators negation, conjunction, disjunction, universal quantifier and existential quantifier. Further operators like implication and equivalence can be derived according to the usual substitutions. The *forall* activity is used to reflect the universal quantifier and must have exactly one outgoing *eachtime* and *end* transition instead of *success* and *failure*.

Fig 5 defines the atomic propositions that might occur in a graph query, namely node matching, edge matching, attribute evaluation and isomorphism. Rectangular boxed inside activies represent nodes of the instance graph. Note that this figure only contains the right hand side of each sub rule with the textual part in the upper right corner. The missing left-hand sides are the same as in Fig. 4 as e.g. for the rule "negation".

To save some space, the rules are already prepared for the temporal extensions proposed in Section 4. Therefore, each activity carries a temporal expression τ. In the case of static graph queries, i.e. without temporal operators, the temporal expression can be ignored.

Using this pair grammar, we can parse any FOL formula[3] using the context-free string grammar rule. We also can compute the equivalent depiction of an SDM graph using its corresponding graph grammar rules.

3.3   From Graph Queries to Formulae

As we have argued in Section 3.2, we can construct a graph query for every FOL formula. Obviously, we can do the reverse, and construct a FOL formula for any graph query that can be constructed by the grammar. However, Story Diagrams usually contain a relative small number of Story Patterns consisting of multiple objects, links and attribute-value pairs. We consider these story diagrams as shorthands for those that can be created by the pair grammar.

Because of size restrictions, we can not provide a tranformation system that is able to translate each "nice" story diagram into a normalized graph query to be able to find the FOL formula. This system would also translate additional features such as negative nodes, negative edges and optional nodes[4].

Though we do not provide the complete transformation, we at least want to provide an example how the graph query in Fig. 2 looks like in the normalized form. Fig. 6 represents this equivalent graph query. To keep the diagram layout clear, we merged some failure transitions using gray circles which are not actually part of the concrete syntax of our graph queries. The normalized form is obviously much more complex than the original form. On the other hand, it is easy to verify that the normalized form corresponds to formula 1.

Note, that the first order logic interpretation of graph queries does only work, if the operational semantics of story diagrams is modified with respect to the original definition in [1]: Atomic propositions contained in the same activity are connected by conjunction. Transitions between activities however effectively describe a *cut* as for instance in Prolog. Thus, if the selection of possible matches is non-deterministic, a story pattern might have different valuations even on the same instance graph, depending on the actual selection order.

To solve this problem, the operational semantics of the graph queries proposed in this paper includes backtracking as in PROGRES [4]. Only then all possible matches are tried if necessary and the graph queries can be fully described by FOL formulae. Minor changes to the operational semantics are necessary to allow objects with unspecified type *ANY* and to permit nesting of forall-activities with outgoing success and failure transitions from enclosed activities. Syntactically, our graph queries are distinguished by the attached temporal expressions.

---

[3] Rules for parentheses and precedence have been omitted due to lack of space.

[4] Set nodes may not be used in these graph queries, as second order logic is required to describe their semantics, which would complicate things unnecessarily and are of little use in graph queries: If a set node can be matched, so can be a single node from the set and the result of the formula is the same.

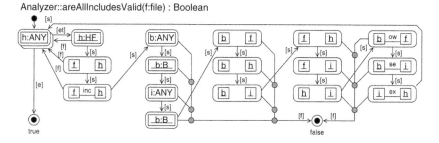

Fig. 6. Equivalent Normalized Graph Query for Fig. 2

4 Adding Time to Graph Queries

Having defined how FOL formulae can be expressed as graph queries, we now extend the definition to include temporal operators. In Section 4.1, we elaborate on the concept of time used in our approach. Next, we enhance the pair grammar from Section 3 with unary temporal operators.

4.1 The Notion of Time

In this paper we have to deal with different concepts of time. As we make use of temporal logic we need to treat time as an infinitive sequence of states, i. e. instance graphs in our case. Naturally, these states coincide with measurements of the system under analysis. However, queries might be expressed in terms of "real" time, and the specificator might not be aware of the corresponding states. In the Unix-world, time is expressed as an 64-bit natural number $t \in \quad = \quad \cap [0; 2^{64} - 1]$, being the number of seconds that have passed since January 1st, 1970. This concept of time would serve our purpose well.

Fig. 7 provides an overview, how the different concepts of time correspond to each other. In this example, we assume that the initial measurement takes place on Monday, March 29th of any given year after January 1st, 1970. Consecutive measurements are scheduled once for each working day. Thus, the gaps between April 2nd and April 5th as well as April 9th and April 12th indicate weekends. On April 8th, there is an exceptional gap, e.g. because of a power failure. On April 5th, there is an additional measurement, e.g. because some changes had been applied to the analysis scripts this day.

Let $\mathcal{V}$ be a set of atomic propositions as defined in Fig. 5. Each measurement i taken at a certain point of time $t_i \in \quad$ corresponds to a state s_i, which is a mapping $s_i : \mathcal{V} \longrightarrow \{0, 1\}$. The *Temporal Structure* for $\mathcal{V}$ is an arbitrary long, but finite sequence $T = (s_0, s_1, \dots, s_\omega)$.

As indicated by the brackets on the time axis, a state s_i is valid from the moment of the measurement to the second before the next measurement at t_{i+1}. The initial state s_0 is valid by definition between January 1st, 1970 and $t_1 - 1$. As queries may refer to any point of time $t \in \quad$, but the corresponding state might be unknown a priori, we define a mapping function $\mu : \quad \longrightarrow T$, where

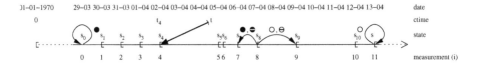

Fig. 7. Time line for Temporal Graph Queries

$$\mu(t) = \begin{cases} s_0 & : \quad t < t_0 \\ s_\omega & : \quad t \geq t_\omega \\ s_i & : \quad t_i \leq t < t_{i+1} \end{cases}$$

As we are mainly interested in changes of instance graphs over time, pretending that the unknown situation before the first measurement was always the same as at t_0 is more useful than assuming that an empty instance graph existed before that time. With the same argument, we project the last known state s_ω for eternity into the future. As a consequence, queries referring to future states might produce different results depending on the current s_ω at query time.

4.2   A Pair Grammar for Temporal Graph Queries

As shown in Fig. 8, we extend the pair grammar with additional rules to cover unary temporal logic operators. As in Fig. 5, we only provide the right hand sides of both sub rules. The left column defines future time operators, while the right column defines past time operators. Binary operators like "until", "unless", "atnext", "before" are not yet supported but could be added easily. An nice summary of first order temporal logic and related operators can be found in [6].

When designing the extensions, we tried to keep as close to ordinary story diagrams as possible. The basic idea is to provide each activity with a time object, i.e. a state $\tau \in T$, where T is the temporal structure defined in Section 4.1. The time object is treated in a way similar to regular objects.

The *henceforth* and *hitherto* operators can be seen as universal quantifiers over unbound time variables and hence are visualized by double boxes around the time expression. Therefore, we use the same pattern with *eachtime* and *end* transitions as for universal quantifiers. The quantified time variable σ is successively bound to τ and, depending on the operator, to every state defined in T before or after τ.

The *eventually* and *once* operators are consequently treated as existential quantifier over unbound time variables. The time variable σ is successively bound to every state in T. If a consecutive story pattern fails, backtracking is used to assign another state. If there is no state so that consecutive story patterns match, the rule finally fails.

The *next* and *previous* operators also define a new time variable σ, but immediately assign a value based on the enclosing variable τ. As shown in Fig. 7, the *next* and *previous* operators refer to states rather than points in time. Note that our temporal structure is not infinite as usual, because there is always a last

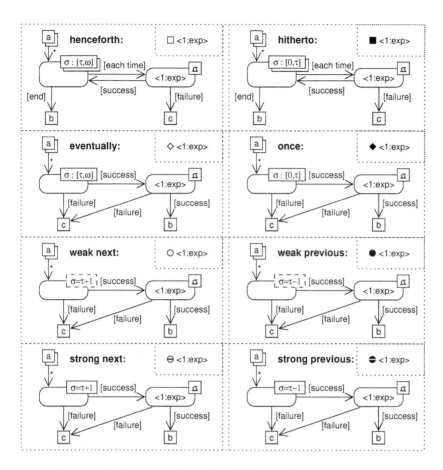

Fig. 8. Pair Grammar Right Hand Sides for Temporal Operators

state σ_ω. The original definition of the *weak previous* operator, i.e. that $\bullet A = 1$ at the inital state s_0 for any given formula A, is also problematic: The value "1" means that the related story pattern succeeds, but the new time variable σ cannot be bound although it is used in consecuting story patterns.

We solve this problem by the following deviations from ordinary temporal logic: We define *strong* and *weak previous* and *next* operators with symmetric semantics. To distinguish between *weak* and *strong* operators, we adopted the notation for *optional* and *obligate* nodes of Fujaba story diagrams. For instance, *strong next* can only be true, if there really is a next state, i.e. $s_\tau \neq s_\omega$. *Weak next* in contrast, is also defined for $s_\tau = s_\omega$. In this case, σ is bound to the last known state ω. Weak operators are visualized by dashed boxes.

For convenience, real time expressions like "$\tau = $ '01-05-07'" or "$\sigma = \tau + 20$ days" may be used. They are implicitly transformed into states using the

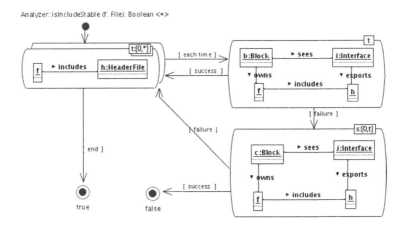

Fig. 9. Example for Temporal Graph Query

mapping function μ defined in Section 4.1. Besides, normalized temporal graph queries as defined by our pair grammar can be condensed by defining activities containing more than one atomic proposition. Again, we have to skip the exact transformation rules.

4.3   Example Revisited

Fig. 9 represents a non-normalized temporal graph query based on the static query in Fig. 2. Rather than checking whether there is a valid *sees* relationship for every *include* relationship, the rule ensures that valid include relationships never become invalid. As a result, only changes after the initial measurement are taken into account and *deterioration* of the system is detected as soon as it occurs so that it can be quickly countered.

The temporal graph query contains the equivalent of the future time operator *henceforth* and the past time operator *once*. The free time variable t is sequentially bound to all states in T. For each t, the remaining part of the first story pattern is evaluated. For every match, the second story pattern is evaluated for the current state t. If the second story pattern matches for the given t, the next iteration begins with the following state in T. If the second story pattern fails, the third story pattern is evaluated. It has the same body as the second, but represents a *once* operator. If s can be bound to any state up to the current t so that the third pattern succeeds, the whole query fails. If there is no possible match for any s, the outer iteration continues.

Note that this procedure does not necessarily describe the actual implementation. Our idea is to equip every graph element with a *created* and *deleted* timestamp so that we are able to avoid iterations in many cases. However, a detailed discussion of the implementation is out of the scope of this paper.

5   Related Work

The idea of applying graph transformation in the field of software re-engineering has already been proposed by others. There have been several approaches to re-engineer industrial software systems based on PROGRES [7]. Another example is the Varlet project [8] that deals with re-engineering of information systems. To our knowledge, all graph transformation-based approaches deal with snapshots of the analyzed systems and hence do not require language features that allow to consider time.

There are some approaches that add time aspects to graph transformations. In [9], a notion of time is added to graph transformations which is adapted from time environment-relationship nets. The approach introduces logical clocks and allows the user to specify the age of nodes in the instance graph and durations for transformations so that the age of involved nodes can be updated. Time can be used to determine the order of productions in transformation sequences so that the age of related nodes does never increase in the course of its transformations. However, this approach allows to reason about *durations* rather than to reason about multiple instance graphs at different times as we do in this paper. The same holds for real-time automatons as described for instance in [10].

The contribution related closest to this paper is described in [11,12]. These documents introduce graphically denoted temporal invariants to graph transformations. In this approach, states are the result of runs, i.e. sequences of atomic graph transformations. Among other things, the temporal invariants allow to specify the correct order of transformations if the choice of transformations is non-deterministic. The main differences are, that we completely embed temporal operators into graph queries while they are expressed as constraints in the context of concrete matches in [11]. Before the extension, the former already require first order logic to describe. Hence we need to support temporal first order logic, while propositional logic suffices for the latter. Besides we provide a more explicit and complete definition of the syntax of temporal graph queries.

There are other approaches that allow to visually define constraints. For instance, [13] provides a visual representation of arbitrary OCL expressions. The operators "not", "implies", "or", etc. are represented as nested sub diagrams with the corresponding keywords as attachments, i.e. the structure of OCL expression (first order logic expression) is preserved, in contrast to our approach that avoids nesting of boxes in favour of the usage of "flat" control flow diagrams. There is no support for temporal logic expressions. An alternative for graphical FOL constraints are spider diagrams [14] that provide no support for temporal operators either.

In [15,16], temporal extensions for OCL are described. The former work allows to specify the behavior of business software components using linear time logic in OCL syntax including operators such as initially, until, etc. It is a straightforward extension of OCL with just the temporal operators added. The latter provides a computational tree logic extension of OCL used for the definition of temporal invariants of state charts to model the behavior of real-time systems. Both approaches rely on a textual definition of temporal logic formulas and could not be integrated with story diagrams without destroying the essential property of story diagrams of being a visual notation.

While many papers deal with textual notations for propositional temporal logic, there are some that also discuss first order temporal logic. The main problem is that first order logic might lead to an infinite number of models for the formulae that have to be considered. In our case, the number of models is always finite, as our temporal structure is defined by an arbitrary large though finite sequence of measurements. We would like to mention the work described in [6], where a normal form for first order temporal logic is suggested. We are currently investigating, if this normal form provides a better means to reason about the differences in expressive power of temporal graph queries and first order temporal logic.

6   Conclusions and Future Work

In this paper, we have motivated why temporal graph queries can be beneficial in the area of evolutionary software restructuring. As main contributions of this paper, we provide a temporal first order logic interpretation of a well-defined subset of temporal graph queries by means of a pair grammar. We further define a mapping function to translate real time expressions into states of the underlying temporal structure. Equipped with these new concepts, we are able to specify visual consistency rules over a temporal sequence of instance graph, i.e. models of the analyzed system.

There are several possible continuations of the work described in this paper: First, we need to broaden the set of graph queries, that we are able to construct FOL formulae for, ideally so that we can support at least all features that are already available in story diagrams. Next, we have to integrate the temporal graph queries into our meta modeling framework, MOFLON, and therefore enhance our editor and code generator. While doing so, we will have to turn special attention on the compact representation of graph instances, probably based on versioned graphs, as we have to deal efficiently with large amounts of data. Finally, we intend to applying our approach to new industrial case studies and learn how we can benefit from temporal graph queries in practice.

References

1. Zündorf, A.: Rigorous Object Oriented Software Development. University of Paderborn (2001) Habilitation Thesis.
2. Pratt, T.W.: Pair Grammars, Graph Languages and String-to-Graph Translations. Journal of Computer and System Sciences **5** (1971) 560–595
3. Rötschke, T.: Re-engineering a Medical Imaging System Using Graph Transformations. In: Applications of Graph Transformations with Industrial Relevance. Volume 3062 of LNCS., Springer (2003) 185–201
4. Schürr, A., Winter, A.J., Zündorf, A.: The PROGRES Approach: Language and Environment. In Ehrig, H., Engels, G., Kreowski, H.J., Rozenberg, G., eds.: Handbook of Graph Grammars and Computing by Graph Transformation. Volume 2. World Scientific Publishing (1999) 487–550

5. Amelunxen, C., Königs, A., Rötschke, T., Schürr, A.: MOFLON: A Standard-Compliant Metamodeling Framework with Graph Transformations. In: Proc. European Conference on Model-Driven Architecture. (2006) Accepted for publication.
6. Dixon, C., Fischer, M., Barringer, H.: A Graph-Based Approach to Resolution in Temporal Logic. In Gabbay, D.M., Ohlbach, H.J., eds.: Temporal Logic, ICTL '94. Number 827 in LNAI, Springer (1994) 415–429
7. Cremer, K., Marburger, A., Westfechtel, B.: Graph-Based Tools for Re-Engineering. Journal of Software Maintenance **14**(4) (2002) 257–292
8. Jahnke, J., Wadsack, J.: Integration of Analysis and Redesign Activities in Information System Reengineering. In: Proc. of the 3rd European Conference on Software Maintenance and Reengineering (CSMR'99), IEEE Press (1999) 160–168
9. Gyapay, S., Heckel, R., Varró, D.: Graph Transformation with Time: Causality and Logical Clocks. Fundamenta Informaticae **58**(1) (2003) 1–22
10. Burmester, S., Giese, H., Hirsch, M., Schilling, D., Tichy, M.: The FUJABA Real-Time Tool Suite: Model-Driven Development of Safety-Critical Real-Time Systems. In: Proc. 27th ICSE, ACM Press (2005) 670–671
11. Koch, M.: Integration of Graph Transformation and Temporal Logic for the Specification of Distributed Systems. PhD thesis, TU Berlin (1999)
12. Gadducci, F., Heckel, R., Koch, M.: A Fully Abstract Model for Graph-Interpreted Temporal Logic. In Ehrig, H., Engels, G., Kreowski, H.J., Rozenberg, G., eds.: Proc. 6th International Workshop on Theory and Application of Graph Transformation (TAGT'98). Number 1764 in LNCS, Springer (2000) 310–322
13. Bottoni, P., Koch, M., Parisi-Presicce, F., Taentzer, G.: A Visualization of OCL Using Collaborations. LNCS **2185** (2001) 257–271
14. Howse, J., Molina, F., Taylor, J.: On the Completeness and Expressiveness of Spider Diagram Systems. In: [17]. (2000) 26–41
15. Conrad, S., Turowski, K.: Specification of Business Components Using Temporal OCL. In Favre, L., ed.: UML and the Unified Process, IRM Press/IDEA Group Publishing (2003) 48–65
16. Flake, S., Müller, W.: An OCL Extension for Real-Time Constraints. In Clark, T., Warmer, J., eds.: Object Modeling with the OCL, The Rationale behind the Object Constraint Language. Volume 2263 of LNCS., Springer (2002) 150–171
17. Anderson, M., Cheng, P., Haarslev, V., eds.: Theory and Application of Diagrams, First International Conference. In Anderson, M., Cheng, P., Haarslev, V., eds.: Diagrams 2000. Volume 1889 of LNCS., Springer (2000)

On the Use of Alloy to Analyze Graph Transformation Systems

Luciano Baresi and Paola Spoletini

Politecnico di Milano
Dipartimento di Elettronica e Informazione
piazza Leonardo da Vinci 32, 20133 Milano, Italy
{baresi, spoleti}@elet.polimi.it

Abstract. This paper proposes a methodology to analyze graph transformation systems by means of Alloy and its supporting tools. Alloy is
a simple structural modeling language, based on first-order logic, that
allows the user to produce models of software systems by abstracting
their key characteristics. The tools can generate instances of invariants,
and check properties of models, on user-constrained representations of
the world under analysis. The paper describes how to render a graph
transformation system —specified using AGG— as an Alloy model and
how to exploit its tools to prove significant properties of the system.
Specifically, it allows the user to decide whether a given configuration
(graph) can be obtained through a finite and bounded sequence of steps
(invocation of rules), whether a given sequence of rules can be applied
on an initial graph, and, given an initial graph and an integer n, which
are the configurations that can be obtained by applying a sequence of n
(particular) rules.

1   Introduction

Graphs provide the underlying structure for many artifacts produced during the
development of software systems. No matter of the actual process we follow,
we always end up with diagrams and models that can easily be conceived as
suitably annotated graphs. For example, graphs provide a sufficiently general
infrastructure to model the topology of object-oriented and component-based
systems, as well as the architecture of distributed applications. This means that
graph transformations are often needed —either explicitly or implicitly— to
specify how these models are built and interpreted, and how they can evolve
over time.

Graph transformation [5] originated in reaction to shortcomings in the expressiveness of classical approaches to rewriting, like Chomsky grammars and
term rewriting, to deal with non-linear structures. The theory behind it has
been evolving for some thirty years, but only recently there have been attempts
to analyze modeled transformation systems. After many different modeling notations and theoretical approaches for specifying graph transformation systems,
part of the research community is addressing the problem of analyzing such sets
of rules. Besides approaches, like *critical pair analysis* [1], that provide a means

A. Corradini et al. (Eds.): ICGT 2006, LNCS 4178, pp. 306–320, 2006.
c Springer-Verlag Berlin Heidelberg 2006

to discover conflicts among rules, tools like *CheckVML* [20] and *GROOVE* [11] start from conventional analysis techniques (model checking, in these cases) and exploit them to prove reachability properties on the modeled sets of rules. Typically, these proposals allow the designer to understand if a given target graph is reachable from an initial graph, through a finite number of transformation rules.

The importance of these approaches is twofold. Besides providing interesting insights from a purely theoretical point of view, they are also valuable for analyzing and discovering properties in the host domain. For example, if we used a graph transformation system to model the dynamic behavior of a software architectural style [6], or we used it to model the operational semantics of a visual notation [17], the capability of analyzing the graph transformation system allows us to discover properties on designed software architectures or on the evolution of produced models.

When the analysis approach exploits model checking, it must face and cope with the typical problems of this analysis technique: state explosion and thus limited capability of rendering the peculiarities of the modeled domain. In contrast, Alloy [16] provides a viable compromise between the richness of models and their analyzability. Alloy is a simple structural modeling language, based on first-order logic, that allows the user to produce models of software systems by abstracting their key characteristics. The SAT-based analyzer can generate instances of invariants, and check properties of models, on user-constrained representations of the world under analysis. Alloy does not address infinite words and it copes with state explosion by asking the user to specify the maximum cardinality of the worlds under analysis. The interesting features of Alloy led us to investigate the possibility of encoding a graph transformation system and use its tools to analyze reachability properties and to study the applicability of sequences of transformation rules.

This paper presents the first results of this encoding. It highlights the translation process and exemplifies it on a simple case study, taken from [1]. The paper also describes the properties on the graph transformation system that can be checked with Alloy.

The rest of the paper is organized as follows. Section 2 briefly introduces Alloy. Section 3 explains the translation of graph transformation systems into Alloy and Section 4 discusses the properties we can check on these models and how we can verify them. Section 5 surveys similar proposals and Section 6 discusses the positive and negative aspects of the approach and concludes the paper.

2   Alloy

Alloy is a formal notation based on relational logic, that is, a logic with clear semantics based on relations. In this section, we only introduce the key characteristics of the notation through an example; interested readers can refer to [15] for an in-depth presentation. The example models finite state automata and specifies some basic properties.

```
module Automa
```

is the module declaration.

```
sig Event{}
sig State{}
```

are empty signatures used to introduce the concepts of Event and State, respectively.

```
sig Transition{
startingState: State,
arrivalState: State,
trigger: Event
}
```

defines a Transition as three relations: startingState and arrivalState identify the source and target States, while trigger defines the Event that triggers the transition. Relations are similar to the fields of an object in the classical object-oriented paradigm.

```
sig FiniteStateAutomaton{
states: set State,
transitions: set Transition,
initialState: states,
finalStates: some states,
dangerousStates: set(states-initialState-finalStates)
}
```

states and transitions are the sets of States and Transitions that belong to the automaton. initialState is one of the states, while finalStates are a non-empty subset (some) of the states. The dangerousStates are those states that are neither initial nor final.

After the signatures, we have the facts that constrain the instantiation of the signatures previously defined and of their relations. A fact is an explicit constraint on the model. It is possible to express constraints on the relations of a signature directly after the signature body without using the keyword fact.

```
fact Determinism{
all a:FiniteStateAutomaton| no disj t1, t2: a.transitions{
t1.startingState=t2.startingState
t1.trigger=t2.trigger}}
```

imposes that, for any FiniteStateAutomaton a, there does not exist a pair of disjoint transitions in a such that they have the same startingState and the same trigger.

```
fact CorrectTransition{
all a:FiniteStateAutomaton |
all t: a.transitions |
t.startingState in a.states && t.arrivalState in a.states}
```

imposes that for any `FiniteStateAutomaton` a, the `startingState` and `arrivalState` of its `transitions` must be contained in its `states`.

Alloy comes with dedicated analysis tools: a consistency checker and a counterexample extraction tool. Both the analyses are fully automated and based on SAT solvers[1]. The Alloy model is translated into a boolean formula, which is then passed to the SAT solver that tries to find an assignment for all the variables in the formula to satisfy it. If the assignment exists, it is translated back into Alloy.

The consistency checker evaluates **pred**icates on defined models. Predicates are like facts, but they do not affect the structure of the world under analysis; they impose "temporary" constraints whose validity is limited to the predicate they belong to.

```
pred example(){}
```

If the **predicate** is empty, as the above **example**, the Alloy analyzer checks the consistency of the model itself (with no further constraints).

A predicate is verified through a **run**, which tries to find an assignment that satisfies the model, along with the constraints of the predicate under analysis.

```
run example for 1 FiniteStateAutomaton, 2 Transition, 3 State, 2 Event
```

When we **run** a model, we must specify the maximum cardinality of the sets of the world under analysis. We can supply a unique cardinality for all the sets, but we can also associate special values with particular sets. In this case, **example** considers exactly 1 `FiniteStateAutomaton`, 2 `Transitions`, 3 `States`, and 2 `Events`.

The identification of counterexamples is performed through **assert**s, which claim that something must be true due to the behavior of the model.

```
assert isolation{
all a:FiniteStateAutomaton| #a.states>1 =>
   all s:a.states| some t:Transition|
   s = t.startingState || s =  t.arrivalState}
```

```
check isolation for 5
```

states that there is no state in any automaton with at least two states that is not the initial or the final state of a transition of the automaton. Assertions are checked by searching for counterexamples, using **check** commands and again we need to set the upper-bound for the cardinalities of the sets that define the world under analysis. In this case, we use a single value, and we only consider sets of five elements. The assertion generates a counterexample since there is no constraint on the automaton that limits the number of isolated states.

Alloy is targeted to describing and analyzing structural properties of systems, but Jackson et al. [16] introduce also a mechanism to represent traces. A more

---

[1] Alloy works with different SAT solvers with different characteristics.

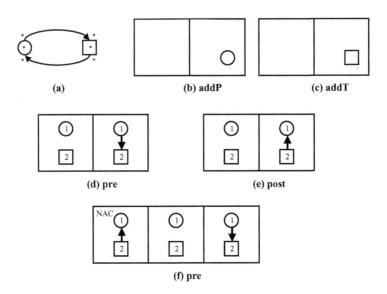

Fig. 1. Graph transformation system to create place-transition nets

general approach to deal with execution traces is presented in [10]: DynAlloy is a dynamic version of Alloy for modeling the execution of operations and reasoning about execution traces. Alloy specifications can also be verified by using theorem proving techniques: Arkoudas et al. [2] present Prioni, a tool that uses the semi-automatic theorem prover Athena to prove properties regarding Alloy specifications.

3   Encoding

This paper uses the notation proposed by AGG [7] to model graph transformation systems. For example, Fig. 1 shows the simple rules needed to build correct place-transition nets, a particular instance of finite state automata. More precisely a place-transition net is a tuple (P, T, π, τ), where P is a set of places, T is a set of transitions, $\pi : P \to T$ and $\tau : T \to P$ are functions that associate a transition to a place and a place to a transition, respectively.

Fig. 1(a) shows the type graph, i.e., the meta-model of the system, and identifies two components: places, the circle, and transitions, the square; places can be connected to a finite number of transitions and vice versa. All the other parts of the figure represent the rules: Fig. 1(b) represents the rule to add a place to the net, Fig. 1(c) the rule to add a transition, and Fig. 1(d) and Fig. 1(e) the rules to connect a place to a transition and a transition to a place, respectively. Since the presented example does not contain negative application conditions, Fig. 1(f) shows the modification of the rule of Fig. 1(d) to add the constraint that to add a relation from a place to a transition, we require that no inverse relations exist.

Our use of Alloy aims at representing the evolution of a system, described as a graph, through the application of the rules that compose the transformation system. To this end, we introduce the following signature —borrowed from [16]— to represent graph paths:

```
sig Path{
    elem: set Graph,
    first, last: elem,
    next:(elem-last) one -> one (elem-first)
} {
    first!=last
}
```

The path itself represents the evolution of the system. A path is a set of graphs that goes from an initial graph, `first`, to the graph, denoted as `last`, reached by applying $|elem| - 1$ transformation rules.

The relation `next` assigns exactly one element to each graph different from the final graph . The `fact`[2] (`first!=last`) ensures that the path is composed of at least a transformation, imposing that the initial and final graphs must be different. This restriction guarantees that all the elements in the path are connected; in fact if initial and final graphs were the same, the path would be disconnected.

This signature is independent of the context and it is added to the Alloy model for any graph transformation system. The composition of the path and the rules used to build the system, instead, are defined by means of a fixed translation process that depends on the particular graph transformation system.

Specifically, we introduce a signature `Graph` along with a signature for all the elements that compose the meta-model. The newly introduced `Graph` contains a relation to each of these signatures. The cardinality of this relation depends on the cardinalities that the elements have in the *type graph*. In our example of Fig. 1(a), the meta-model contains two types of elements, nodes and transitions, both with multiple cardinality. Hence, the `Graph` is defined in Alloy as follows:

```
sig Graph{
    places: set Place,
    transitions: set Transition
}
```

The signatures representing the elements are composed of two parts: attributes and connections. The first contains a unary relation for each attribute of the element. Appropriate signatures are introduced to render the types ascribed to the different elements. Particular attention must be paid to integers and booleans: the former are a *built-in* concept in Alloy, while the latter can be represented using a `lone` (zero or one) relation to a mirror signature with the empty relation that corresponds to *false*. The connection part contains a set of

---

[2] Notice that in this case, we use the compact form to represent facts within the signatures they belong to.

relations, one for each connection in the *type graph*. The relations are added only to the source elements of the associations in the meta-model. The cardinality of these relations is given based on the cardinality of the target association end. The cardinality of the source association end is used to constrain the system. For instance, if the cardinality of the target association end between elements e_1 and e_2 is 1, this means that we cannot have two instances of e_1 related to the same instance of e_2. These constraints are added as facts.

In our example, the elements have no attributes, but they have outgoing arcs, and thus each of them has a relation. The arcs (associations) are marked with an $*$ on both sides, which become sets in the Alloy model, with no constraints added to the signatures:

```
sig Place{
    enable: set Transition
}
sig Transition{
    fireTo: set Place
}
```

The rules of the graph translation system, which represent the rules to build the edges in signature Path, are modeled in Alloy as predicates. For each graph transformation rule, we introduce a predicate with the following parameters: a) Two graphs, g_1 and g_2, represents the LHS and the RHS of the production, b) two parameters for each element that is modified correspond to the old and new values, c) one parameter for each element that is deleted from the LHS or added to RHS. The RHS graph g_2 is obtained as follows:

$$g_2.r_i = g_1.r_i - \Sigma_j eLHS_j^{mod} - \Sigma_k eLHS_k^{canc} + \Sigma_j eRHS_j^{mod} + \Sigma_h eRHS_h^{add}$$

where $eLHS_j^{mod}$, $eLHS_k^{canc}$, $eRHS_j^{mod}$, and $eRHS_h^{add}$ are all the elements of the predicates of same type of the right-hand side of relation r_i. $eLHS_j^{mod}$ represents the element that has to be modified as it appears in the LHS, $eRHS_j^{mod}$ the element modified as it appears in the RHS, $eLHS_k^{canc}$ the element in the LHS that is cancelled in the RHS, and $eRHS_h^{add}$ the element added in the RHS.

We also add a constraint to ensure that all the elements that appear in the RHS and not in LHS, are not part of the graph that represents the left hand side. The predicates also contain the characteristics that the elements have in terms of attributes and relations, and the connections between $eLHS_j^{mod}$ and $eRHS_j^{mod}$.

The representation in Alloy of the AGG rules of Fig. 1 is the following.

```
pred addP(g1,g2:Graph, p:Place){
    g2.transitions=g1.transitions &&
    g2.places=g1.places+p &&
    p not in g1.places &&
    #p.enable=0
}
```

Rule `AddP` adds a place to the RHS, hence it is translated into a predicate that has the initial and the final graphs and the added place as parameters. Since this place has no relations with any transition, it is imposed that the number of elements (#) in relation `enable` for p is zero. Then, since the only other parameter, besides the graphs, is place p, the transitions of the two graphs are the same, while the places of the final graph are the union of those of the initial graph and p, which must not be present in `g1`. Rule `addT` is the same as `addP`, but it adds a transition.

```
pred addT(g1,g2:Graph, t:Transition){
    g2.transitions=g1.transitions+t &&
    g2.places=g1.places &&
    t not in g1.transitions &&
    #t.fireTo=0
}
```

Rules `pre` and `post` are more complex.

```
pred pre(g1,g2:Graph, p1,p2:Place,t:Transition){
    g2.transitions=g1.transitions &&
    g2.places=g1.places-p1+p2 &&
    p2 not in g1.places &&
    p2.enable=p1.enable+t
}
```

```
pred post(g1,g2:Graph, p:Place, t1,t2:Transition){
    g2.transitions=g1.transitions-t1+t2 &&
    g2.places=g1.places &&
    t2 not in g1.transitions &&
    t2.fireTo=t1.fireTo+p
}
```

Since the two rules are symmetric, we only consider the first one. This rule involves a place and a transition, the place is modified, while the transition is not. Hence, the parameters are the two graphs, two places, and a transition. No particular constraints are imposed on t, while the set of places in the final graph must differ from the one in the initial graph for a place. In fact, place p1 is replaced by p2, which is related to all the transitions in relation `p1.enable`, with the addition of t. Moreover p2 must not appear in the original graph.

When a graph transformation rule has a negative application condition, this constraint must be represented in the Alloy predicate. The NAC is a precondition on the elements involved in the transformation. Generally, these elements are already parameters of the predicate; if not, they are added. Then, for each element in the negative application condition, we add a formula to constrain involved elements not to assume the forbidden configuration. As example, we can consider rule `pre` of Fig. 1(f), where the NAC imposes that we cannot add a relation from a place to a transition if the opposite relation already exists. Notice that this does not mean that place-transition nets are acyclic or they have not cycle of length 2 (place – trantition – place), but that, if such a cycle exists, it was built by creating the association between a place and a transition before its opposite. The Alloy predicate is modified as follows:

```
pred pre(g1,g2:Graph, p1,p2:Place, t:Transition){
    g2.transitions=g1.transitions &&
    g2.places=g1.places-p1+p2 &&
    p2 not in g1.places &&
    p2.enable=p1.enable+t &&
    p1 not in t.fireTo
}
```

where the formula `p1 not in t.fireTo` is the representation of the NAC.

After defining the predicates for all the rules of the graph transformation system, we want to impose that these rules are the only way to move from a configuration of the graph to another, that is, they are the only way to relate two elements in the `Path` with respect to relation `next`. To model this constraint in Alloy, we add an additional constraint to the facts regarding signature `Path` to impose that for relation `next`, one of the predicates holds, that is, the edge is obtained by applying one of the rules of the system. In our example the constraint is:

```
all e,e':elem| (e in e'.next =>
    ((one p:e.places| addP(e',e,p))||
     (one t:e.transitions| addT(e',e,t))||
     (one p1:e'.places| one p2:e.places|
            one t:e.transitions| pre(e',e,p1,p2,t))||
     (one t1:e'.transitions| one t2:e.transitions|
                one p:e.places| post(e',e,p,t1,t2))))
```

and states that for each pair of adjacent graph configurations, there exists exactly one place that is added to the first configuration to obtain the second through production `addP`, or exactly a transition is added to the first configuration to create the second through `addT`, or the second configuration is obtained by adding a relation from a place to a transition, or from a transition to a place, to the first configuration.

Even if presented through an example, the encoding is general and algorithmic, and thus can be applied on other and more complex transformation systems.

4   Verification

The tools allow us to check the reachability of given configurations of the host graph through a finite sequence of steps (invocations of rules), to verify whether given sequences of rules can be applied on an initial graph, and to show all the configurations that can be obtained by applying a sequence of n (particular) rules on a given initial graph. Moreover, particular systems might motivate us to verify the validity of particular user-defined properties (constraints) on valid configurations.

Alloy allows us to try whether a property can be satisfied or to show that it does not hold. The property is translated into a predicate, in the former case, or its negation is translated into an assertion, in the latter case.

The reachability of configurations allows us to check whether a given configuration can be reached through a finite set of transformation rules. This property can be applied to any user-defined configuration; the default one is the empty graph.

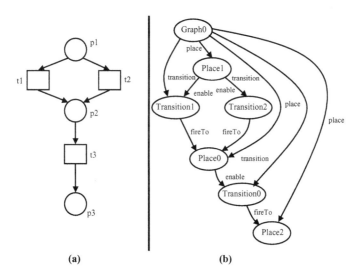

(a)   (b)

Fig. 2. Example initial graph configuration and a solution produced by Alloy

If we consider the graph of Fig. 2(a) and we want to verify that it is reachable from the the empty graph, we can build the following predicate:

```
pred reachConfig(){
    some p:Path| initialGraph(p.first) && DefConfig(p.last)
}
```

where `initialGraph(p.first)` and `defConfig(p.last)` are two predicates representing the initial and desired final configurations, respectively. In our example, they are defined as follows (but they might be much more complex):

```
pred initialGraph(g:Graph){
    #g.places=0 && #g.transitions=0
}
```

```
pred defConfig(g:Graph){
    #g.places=3 && #g.transitions=3 &&
    one p1,p2,p3:g.places| some disj t1,t2,t3:g.transitions|
    p1.enable=t1+t2 && t1.fireTo=p2 && t2.fireTo=p2 &&
    p2.enable=t3 && t3.fireTo=p3 && #p3.enable=0
}
```

Fig. 2(b) shows an instance of the model that satisfies `pred defConfig()`. The model is a set of nodes that represent the different instances of the signature; edges represent the relations among them. For example, `Graph0` embeds place `Place1`, which is related, through `enable`, to `Transition1` and `Transition2`.

If we want to check that a given configuration is not reachable, we must check the following assertion:

```
assert unreachConfig(){
    no p:Path| initialGraph(p.first) && defConfig(p.last)
}
```

The validation of a sequence of rules allows us to verify that, starting from an initial graph, a given sequence of rules is applicable on it. To verify the property, we have to introduce a predicate with the following structure:

```
pred validRules(){
  some p:Path| some disj t1,...,tk:Type1| ... | some disj f1,...,fh:Typej|
  givenConfig(p.first) &&
  R1(p.first, p.next[p.first], parR1) &&
  ...
  Rn(p.next[p.next...[p.first]],p.next[p.next[p.next...[p.first]]],parRn)
}
```

where `givenConfig()` is a predicate that represents the starting graph, `some disj t1,...,tk:Type1| ... | some disj f1,...,fh:Typej` identifies the variables needed, besides the graph, in the predicates that represent the rules to be applied and `R1, ..., Rn` are the rules of the sequence we want to verify.

As an example, the verification of the sequence `addT`, `addP`, `pre` from the initial graph is performed through the following predicate:

```
pred validRules(){
  some p:Path| some disj t1,t2:Transition| some disj p1,p2,p3:Place
  givenConfig(p.first) &&
  addT(p.first, p.next[p.first], t1) &&
  addP(p.next[p.first], p.next[p.next[p.first]], p1) &&
  pre(p.next[p.nex[p.first]],p.next[p.next[p.next[p.first]]], p2, p3, t2)
}
```

Finally, to analyze reached configurations, that is, to show all the possible configurations obtained by means of a generic allowed path of a specified length from an initial state, we need to use one of the SAT solvers embedded in Alloy that allows us to find multiple solutions (e.g., MCHAFF). Moreover, since this SAT solver shows all the possible configurations that satisfy the property, it is necessary to constrain the assertion we use to obtain "only" the paths we want. The general structure of this property is the following:

```
pred configLength_n(){
    some p:Path| p.elem=n+1 && givenConfig(p.first) &&
    one Path && no(A1-p.elem.a1) && ... &&
    no(Am-p.elem.am)
}
```

where n is the desired length for the path, a1, ..., am represent the relations in **Graph** that are used for the path, and A1, ...", Am are the corresponding signatures.

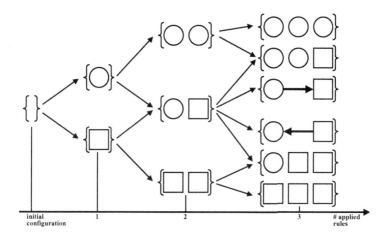

Fig. 3. All possible paths

Hence, if we wanted to extract all the possible paths of length 3 from the initial configuration in our example, the predicate is the following:

```
pred ConfigLength_3(){
    some p:Path| #p.elem=4 &&
    one Path &&  initialGraph(p.first) &&
    no(Transition-p.elem.transitions) &&
    no(Place-p.elem.places)
}
```

The **run** of this property allows us to find all the possible paths of length 3 in the given graph transformation system starting from the empty configuration. The result is presented in Fig. 3.

5   Related Work

The analysis of graph transformation systems can be performed in different ways. Heckel in [13] gives the theoretical foundations for the verification of graph transformation systems through model checking: graphs are interpreted as states and rule applications as transitions. This idea is exploited both by GROOVE [18] and by checkVML [20].

GROOVE [18] is based on the idea of using the core concepts of graphs and graph transformations all the way through during model checking. States are represented as graphs, while transitions are represented as applications of graph transformation rules. This way, properties are specified in a graph-based logic to apply graph-specific model checking algorithms. Hence, in this approach, only some ideas of traditional model checkers can be applied immediately, since the most basic concept, namely the underlying model, has been extended drastically.

On the other side, CheckVML [20] exploits off-the-shelf model checker tools, like SPIN [14], to verify graph transformation systems. More thoroughly, Check-VML takes as input a graph transformation system parameterized with a type graph and an initial graph (represented by means of an abstract transition system) and gives an equivalent model in Promela, the input language of SPIN. Property graphs are also translated into their temporal logic equivalents.

In [19], Rensink et al. propose a comparison between these two tools, and conclude that CheckVML always performs better when dynamic allocation and symmetries are limited, while for dynamic problems GROOVE is preferable.

VIATRA [8] is another graph transformation tool with interesting capabilities for controlling the transformation and composition of complex transformations. The graph transformations are driven by abstract state machines, as specification formalism; extended hierarchical automata represent the model. In the end, VIATRA checks statecharts with SPIN.

Baldan and König [4] describe a different theoretical framework. It aims at analyzing a special class of hypergraph rewriting systems by a static analysis technique based on approximative foldings and unfoldings of a special class of Petri nets. Baldan et al. [3] extend this work by providing a precise (McMillan-style) unfolding strategy. Dotti et al. [9] use object-based graph grammars for modeling object-oriented systems and define a translation into the input language of SPIN to use model checking. The authors allow a restricted structure for graph transformation rules tailored to the message call mechanism of object-oriented systems. Even if the chosen representation in SPIN only supports a restricted problem, the structure of the generated code, in general, results in better run-time performance.

6   Conclusions and Future Work

The paper presents a proposal for exploiting the formal language Alloy to analyze graph transformation systems. We present the first ideas behind the encoding process and we demonstrate them on a simple case study. Besides the example in the paper, we applied the proposed methodology on the Concurrent Append example presented in [19], and on the Shopping example shown in [12], and we obtained encouraging results.

The main drawback of this approach, with respect to GROOVE and Check-VML, is the need for tailoring the search space. In fact, these methods utilize model checking techniques, and the whole space of the modeled system is searched automatically —if the model checker does not go out of memory. In our case, with Alloy, we only analyze the system for a finite scope, whose size is user-defined. Even if this feature may look like a limitation, it is true that this way users have much more freedom to tailor the details embedded in the Alloy models they want to analyze with respect to the size of the spaces they want to deal with. It is also true that the SAT-based analysis techniques produce interesting results faster and by using less memory.

We do not want to say that the proposed approach is better than those that employ model checking; we only want to let the reader know of other options for analyzing designed graph transformation systems, fostering the capability of specifying reach models, which are analyzable because of the finite size of the search space, and the SAT-based techniques.

We think that our method is useful to find instances and counterexamples for models and graph transformation systems. In general, we are not interested in infinite paths, but it is interesting to analyze that our properties are verified by applying a certain number of transformation rules and with a "big enough" model. Heuristics and practical cases suggest us that it is always possible to identify a scope that is big enough to let us explore the interesting parts of systems.

In the future, we plan to improve the approach and implement a component to automatically translate a graph transformation system described using AGG into an Alloy model. This would be the natural conclusion of the first phase of the research, which is investigating the best encoding heuristics. It is the only way to conduct more significant experiments and refine the properties we are interested in. We also have plans to integrate our approach with Prioni, and thus exploit the analysis capabilities of a theorem prover.

References

1. AGG. http://tfs.cs.tu-berlin.de/agg/.
2. K. Arkoudas, S. Khurshid, D. Marinov, and M. Rinard. Integrating model checking and theorem proving for relational reasoning. In *Proceedings of the Seventh International Seminar on Relational Methods in Computer Science (RelMiCS 2003)*, volume 3015 of *Lecture Notes in Computer Science (LNCS)*, pages 21–33, 2003.
3. P. Baldan, A. Corradini, and B. König. Verifying finite-state graph grammars: An unfolding-based approach. In *Proceedings of CONCUR 2004*, pages 83–98, 2004.
4. P. Baldan and B. König. Approximating the behaviour of graph transformation systems. In *Proceedings of ICGT*, pages 14–29, 2002.
5. L. Baresi and R. Heckel. Tutorial Introduction to Graph Transformation: A Software Engineering Perspective. In *Proceedings of the First International Conference on Graph Transformation (ICGT 2002)*, volume 2505 of *Lecture Notes in Computer Science*, pages 402–429. Springer-Verlag, 2002.
6. L. Baresi, R. Heckel, S. Thöne, and D. Varró. Modeling and validation of service-oriented architectures: Application vs. style. In *Proceedings of European Software Engineering Conference and ACM SIGSOFT Symposium on the Foundations of Software Engineering*, pages 68–77. ACM Press, 2003.
7. M. Beyer. *AGG1.0 - Tutorial.* Technical University of Berlin, Department of Computer Science, 1992.
8. G. Csertán, G. Huszerl, I. Majzik, Z. Pap, A. Pataricza, and D. Varró. Viatra - visual automated transformations for formal verification and validation of uml models. In *Proceedings of ASE*, pages 267–270, 2002.
9. F.L. Dotti, L. Foss, L. Ribeiro, and O. Marchi dos Santos. Verification of distributed object-based systems. In *Proceedings of FMOODS*, pages 261–275, 2003.

10. M. Frias, C. Lopez Pombo, G. Baum, N. Aguirre, and T. Maibaum. Reasoning about static and dynamic properties in alloy: A purely relational approach. *ACM Trans. Softw. Eng. Methodol.*, 14(4):478–526, 2005.
11. GROOVE. `http://groove.sourceforge.net/groove-index.html`.
12. J.H. Hausmann, R. Heckel, and G. Taentzer. Detection of conflicting functional requirements in a use case-driven approach: a static analysis technique based on graph transformation. In *Proceedings of ICSE*, pages 105–115, 2002.
13. R. Heckel. Compositional verification of reactive systems specified by graph transformation. In *Proceedings of FASE*, pages 138–153, 1998.
14. G.J. Holzmann. The model checker spin. *IEEE Trans. Software Eng.*, 23(5):279–295, 1997.
15. D. Jackson. *Software Abstractions : Logic, Language, and Analysis*. The MIT Press, 2006.
16. D. Jackson, I. Shlyakhter, and M. Sridharan. A micromodularity mechanism. In *Proceedings of the 8th European software engineering conference held jointly with 9th ACM SIGSOFT international symposium on Foundations of software engineering*, pages 62–73. ACM Press, 2001.
17. S. Kuske. A formal semantics of UML state machines based on structured graph transformation. In M. Gogolla and C. Kobryn, editors, *Proceedings of UML 2001*, volume 2185 of *Lecture Notes in Computer Science*. Springer-Verlag, 2001.
18. A. Rensink. The GROOVE simulator: A tool for state space generation. In *Applications of Graph Transformations with Industrial Relevance (AGTIVE)*, volume 3062 of *Lecture Notes in Computer Science*, pages 479–485. Springer-Verlag, 2004.
19. A. Rensink, Á. Schmidt, and D. Varró. Model checking graph transformations: A comparison of two approaches. In *Proc. ICGT 2004: Second International Conference on Graph Transformation*, volume 3256 of *LNCS*, pages 226–241. Springer, 2004.
20. Á. Schmidt and D. Varró. CheckVML: A tool for model checking visual modeling languages. In *Proc. UML 2003: 6th International Conference on the Unified Modeling Language*, volume 2863 of *LNCS*, pages 92–95, San Francisco, CA, USA, October 20-24 2003. Springer.

Non-materialized Model View Specification with Triple Graph Grammars

Johannes Jakob, Alexander Königs, and Andy Schürr

Real-time Systems Lab
Darmstadt University of Technology
D-64283 Darmstadt, Germany
{jakob, koenigs, schuerr}@es.tu-darmstadt.de
www.es.tu-darmstadt.de

Abstract. Model-based tool data transformation and integration are crucial tasks in software and system development relying on model-driven development (MDD). Since the tool-specific meta models of the involved system development tools are often too generic and lack the desired level of abstraction, it is inappropriate to specify model transformation and integration rules on top of them. Domain-specific views on tool-specific meta models are needed which provide meaningful information on a higher level of abstraction. Current approaches usually consider a view as a separate model which has to be kept consistent with the tool's model and, thus, duplicate the data. In this paper we discuss different implementations of our declarative view specification approach called VTGG that are based on modified triple graph grammars. As a result we come up with an implementation with non-materialized views that avoids the duplication of data.

1 Introduction

Typically, software development and system engineering projects involve lots of tools each specialized in a number of development phases as requirements elicitation, system modeling, coding, testing, etc. Manually keeping the data stored in these tools consistent is time-consuming and error-prone. Thus, there is a need for automatic data transformation and integration support. Model transformations are a powerful way to realize the needed support. Current approaches including OMG's Query/View/Transformation (QVT) [20] standard and our own triple graph grammar approach [17] allow for the specification and application of model transformation or model integration rules based on the metamodels of the considered tools. Most tools are designed for multiple types of projects. Thus, their APIs and, therefore, their metamodels are rather generic on the one hand. On the other hand these metamodels reflect tool-specific and technical details which usually results in quite complex APIs. This makes it difficult and intricate to specify model transformation or model integration rules based on tool metamodels. Therefore, it is desirable to define views on tool metamodels in order to add project-specific information on the one hand and to suppress technical details on the other hand. In this paper we focus on the latter issue.

A. Corradini et al. (Eds.): ICGT 2006, LNCS 4178, pp. 321–335, 2006.

Current model transformation and integration approaches either disregard views at all or realize them as separate models that are kept consistent with the viewed model by applying common model transformation techniques. This results in an awkward duplication of data. In contrast, we want to realize logical views as functional interface adapter layers that do not replicate any data at all.

To summarize, we propose the following list of requirements that, in our opinion, should be supported by a view specification approach:

- Abstraction of tool-specific data, like mapping of several objects or links to one object or link (includes the mapping of links to objects).
- Domain-specific modeling; for instance, to define project class specific constraints on and adaption for tool metamodels.
- Views on top of views to support different abstraction levels.
- Multiple views for one base model, for example, to realize viewpoint approaches.
- Metamodel based scheme definition of view and tool data structures.
- Avoiding the duplication of data; view creation should not result in coexisting view and tool data representations.
- Updateable views. Updates should be incrementally propagated in both directions, between a view and its base model.

In this paper we introduce a new view specification approach for MDD that fulfills the listed requirements. The paper is based on our initial ideas outlined in [14]. The main enhancements to the latter are the usage of parameters for attribute value assignments and the definition of an implementation metamodel on which the implementation of the operational rules are based.

In Section 2 we discuss related view specification approaches. Thereafter, in Section 3 we present a running example that we use for illustration purposes in the following. In Section 4 we introduce our declarative view specification approach called VTGGs. The implementation of view specifications by applying the class adapter pattern of the Gang of Four [8] is described in Section 5. Finally, Section 6 concludes this paper, discusses open issues, and future work.

2   Related Work

Within the context of *view specification*, we have to distinguish different kinds of *views*. In the following we distinguish between three main categories.

The first category describes approaches that can be classified as *visual representations* of models. Meta-case tools like Pounamu [25] or MetaEdit [18] use "model view" as a short-hand for visualization of a model. In the same way the MViews/JViews [9,10] approach for the construction of design environments supports a transformation approach that generates visual representations. AToM^3 is another approach that is based on metamodeling and a special form of triple graph grammars to generate environments for visual languages (VLs) supporting multiple-views [12,13,4]. It mainly supports propagation of updates from a base model to its materialized views and relies on a single metamodel for all views.

The animated simulation of a model's behavior with GenGED [6,7] proposes the use of animation views instead of simulating the model behavior. Thus, GenGED defines a view as an incomplete specification of a system like a VL model that is part of a larger VL model. The relation between the view and the larger system is captured by graph transformations in the form of materialized views. For our approach, a visual representation is out of scope. We aim at logical model views that are again "models" and not just visualizations.

The second category comprises approaches dealing with *logical views*. Naturally, in respect of having a self-contained view model, this part includes "regular" model transformation approaches like AGG, VIATRA2, VMTS that are compared in [5]. Of course, these approaches have other intentions, but they can also be utilized for view specification purposes with the weakness of generating materialized views. The OMG's QVT standard [20] explicitly excludes in its current version the part of view specification. However, as far as we know they plan a definition of view specification that is also based on model transformation. The viewpoint-oriented approach of [24] also provides an own model transformation language called ArchiMate with an underlying repository in which the models and their views are made persistent. A fully materialized view is one of the main disadvantages of all these related approaches, which results in an intricate duplication of data together with the inherent view update problem. Another drawback of the approaches above is the disability of mapping view associations onto base model objects (or vice versa).

Views in the world of databases either belong to the second category of logical views or consitute a third category. Usually, database views are defined as query results for relational databases. Incremental propagation of updates is often not supported and views for object-oriented data models is usually out-of-scope. For a more detailed discussion of research related to database views the reader is referred to [14].

As far as we can estimate, there are no tools or approaches that fulfill our requirements of view specification.

3   Running Example

In this section we present a running example that we use throughout this paper. On the one hand we aim at integrating the data of a model-based software development project stored in quite a number of different tools. On the other hand we aim at manipulating the data of a single tool by applying model transformation rules. Since the tools' metamodels are too generic and, thus, offer too little project-specific abstraction on the one hand and reflect unnecessary tool-specific details on the other hand it is desirable to specify the model integration and transformation rules on a higher abstraction level. This abstraction can be realized by creating views on the tools' data as shown in Figure 1a. The depicted tool adapter provides a standard compliant interface (e.g. Java Metadata Interface (JMI) [15]) on the tool's data by adapting its proprietary tool interfaces (e.g. APIs). Thereby, the adapter reflects the internal data structure of the tool, which correspond to the tool-specific metamodel (TMM). On

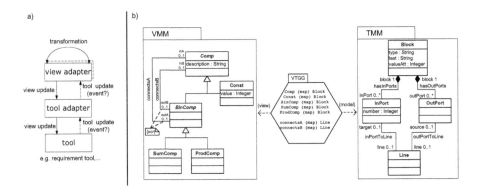

Fig. 1. a. Adapter layers, **b.** Metamodels of view, correspondence graph, and tool

top of these tool-specific interfaces we specify view adapters which provide read and write access to the tool's data through view-specific interfaces reflecting the view-specific metamodels (VMMs)[1]. Please note that we will introduce later on an implementation of this 3-layer architecture that avoids duplication of data by translating all read/write operations on one layer into read/write operations of the next lower layer.

In the running example we want to perform a model transformfation on a Matlab/Simulink[2] model (cf. Figure 2a). Two `Constant` blocks carrying the values 1 and 2 respectively are used as input for a `Sum` block which calculates the sum of its inputs. The result as well as a third `Constant` block carrying the value 2 are used as input for a `Product` block which returns the product of its inputs. We aim at specifying a model transformation rule that substitutes a `Sum` block that has two `Constant` blocks as input by a single `Constant block` whose value is set to the sum of the values of the existing `Constant` blocks.

Figure 2b shows the abstract representation of the Matlab/Simulink model as an object graph (a *model* according to OMG's metamodeling approach). On the one hand the object graph contains undesirable tool-specific details as the `Line`, `InPort`, or `OutPort` objects for instance. On the other hand the object graph lacks preferable abstract information as dedicated `Constant`, `Sum`, and `Product` objects instead of general `Block` objects carrying type attributes. The reason for these deficiencies can be found in the simplified tool-specific graph schema (*metamodel*; cf. Figure 1b on the right-hand side) to which the object graph complies. The graph schema states that a Matlab/Simulink model basically consists of `Blocks` that carry a `type` and a `text` attribute. Furthermore, each `Block` is provided with an arbitrary number of `InPorts` and `OutPorts`. For identification purposes each `InPort` carries a `number` attribute. In- and `OutPorts` can be connected by `Lines`.

---

[1] In principle we can specify further view adapters on top of existing view adapters in order to realize different levels of abstraction.

[2] http://www.mathworks.com

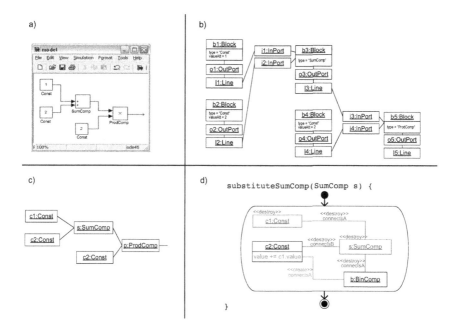

Fig. 2. a. Matlab/Simulink model, **b.** Tool-specific object graph, **c.** View-specific object graph, **d.** Model transformation rule

We aim at an object graph without these deficiencies as depicted in Figure 2c. This graph consists of dedicated `Constant`, `Sum`, and `Product` objects and omits the low level `Line`, `InPort`, and `OutPort` objects. Therefore, it provides the desired level of abstraction. The graph complies to the graph schema shown in Figure 1b on the left-hand side. According to this schema an object graph consists of components `Comp`. Components either are binary components `BinComps` or constants `Const`. Each `BinComp` can be connected with up to two `Comps` through its input links `inA` and `inB`. Correspondingly, a `Comp` can be connected with up to one `BinComp` through either output port `outA` or `outB`. This is expressed by the attached {`XOR`} constraint. Finally, a `BinComp` either can be a sum component `SumComp` or a product component `ProdComp`.

Based on the view metamodel, we can easily specify the model transformation rule as depicted in Figure 2d[3]. We denote the model transformation rule as a graph rewriting rule according to the Story Driven Modeling (SDM) language introduced in [26]. The method `substituteSumComp` is provided with an input parameter `s` of type `SumComp`. The graph transformation rule checks whether `s` is connected to two `Const` blocks `c1` and `c2` as input and an `BinComp` block `b` as output. If this pattern is matched on the regarded model the rule destroys `c1` and `s` as well as their links of type `connectsA` and `connectsB` respectively to connected objects. Furthermore, the rule changes the value of the attribute

---

[3] Due to the consistency of view and tool data, dangling edges are not possible.

value of c2 to the sum of the value of c1 and c2. Finally, it replaces the link of type connectsA (or connectsB) between s and b by a link of type connectsA (or connectsB) between c2 and b. The same graph transformation rule specified based on Simulink's tool-specific metamodel is considerably more complex, and, therefore, less understandable. Due to lack of space we have to omit this rule.

In order to realize the object graph from Figure 2b as a view on the graph from Figure 2b we have to map them to each other. We declare the needed mapping dependencies in the graph schema (cf. Figure 1b in the middle - metamodel of correspondence graph). For clarity reasons we introduce these dependencies only textually. In fact these dependencies are declared visually as well. Besides the mapping dependencies we need rules that describe which objects of the one graph are actually mapped to which objects of the other graph. To this end we rely on the triple graph grammar approach as pointed out in the following section.

4   VTGGs

In this section we describe the basics of our model view specification approach called VTGGs. VTGGs are a special kind of triple graph grammars (TGGs). TGGs have been introduced in [22] about ten years ago. Hitherto, they only have been implemented in prototypical ways [2,3,11,16]. Currently, we are working on an implementation [17] that adopts recent OMG standards as the Meta Object Facility (MOF) 2.0 [19] and Query/View/Transformation (QVT). Thereby, MOF 2.0 plays the role of a graph schema definition language, whereas QVT acts as a model integration specification language. Basically, a TGG is a regular graph grammar consisting of a set of graph rewriting rules taking the empty graph as the axiom. Each graph that has been derived by applying triple graph grammar rules can be divided into three related subgraphs. Two subgraphs represent a pair of corresponding graphs, whereas the third keeps track of correspondences by means of traceability relationships. In our context one graph represents the tool-specific object structure. The second graph represents the corresponding view-specific object structure. Finally, the third graph keeps track of the mapping dependencies between the first and the second graph. It is one of the main advantages and, thus, a reason for using TGGs, that a set of regular graph rewriting rules implementing model integration tasks as *forward transformation, backward transformation, traceability link creation, change propagation,* and so on, can be automatically derived from each triple graph grammar rule. For more details the reader is referred to [17].

Regular TGGs are used for model integration and transformation purposes. Thereby, changes on one model are propagated to the other model by applying forward and backward model transformation rules. In the context of view creation we want to materialize neither the view nor the correspondence graph. They should only exist virtually. In the following we present how we modify TGGs to this end. Thereby, we refer to the running example we introduced in Section 3. Particularly, we come up with a new rule derivation strategy that derives a set of view implementing graph rewriting rules from a declarative view specification. According to regular TGGs, the creation of new objects or links are denotated with the {new} tag. The mapping relationships from virtually existing VMM objects (left-hand side) to really existing TMM objects (right-hand side) are modeled as links between objects combined with the tag {new}[4]. Furthermore, we modify TGGs as follows. First of all, we allow that the creation of a link in one graph simultaneously creates a subgraph in the other graph. Additionally, a VTGG rule may only create one new object or link on the view side. Otherwise it would be very difficult to unambiguously propagate changes on the view to the tool model, i.e. to translate declarative VMM rules into operational TMM rules automatically.

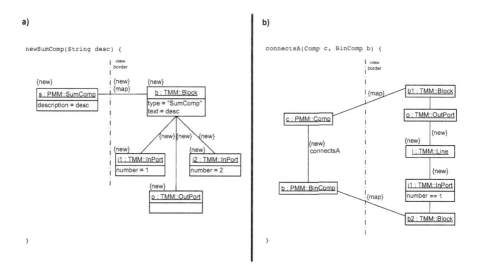

Fig. 3. Examples of declarative VTGG rules

Figure 3 depicts examples of declarative VTGG rules. Both rules describe the simultaneous evolution of the view and the tool object graph. The rule from Figure 3a simultaneously adds a new SumComp object s to the view and a new Block object b to the tool graph. Moreover, this rule adds secondary objects InPort i1, InPort i2, and OutPort o to the tool graph (to match the tool specific structure) and connects them with b. The rule is provided with an input parameter desc of type String which is used to set the attributes description of s and text of b[5]. The value of the type attribute of b is set to "SumComp". The values of the number attributes of i1 and i2 are set to 1 and 2,

---

[4] Normally, the mapping dependencies are represented as dedicated objects in declarative rules. We omit these objects for the sake of clarity.

[5] In former TGG papers OCL constraints have been used for attribute assignments. It's part of ongoing research that this is superseded by input parameters.

respectively. Finally, s and b are linked by a new mapping dependency to reflect their correspondence.

Rule 3b depicts a rule that is an enhancement of hitherto existing TGG rules. It is provided with two input parameters Comp c and BinComp b and connects both objects on the view side by a new connectsA link. Simultaneously, the rule connects the OutPort o of the Block b1 that corresponds to c with the InPort i1 of the Block b2 that corresponds to b with a new Line l object on the tool side. Thereby, the rule matches if the number attribute of InPort i1 has been set to 1. In the same way, the rule for adding a new link of association connectsB matches if the number attribute of an InPort instance has been set to 2.

Rules that create new Const and ProdComp blocks on the view side look similar to rule 3a. The entire set of rules enables us to simultaneously create the view and tool object graphs from Figure 2 as well as their mapping dependencies. Actually, the simultaneous evolution of both object graphs is not intended. Instead, we want to manipulate a virtually existing view on the tool graph. Therefore, we derive regular graph rewriting rules based on Fujaba SDM diagrams [23,26]. These rules implement basic operations (e.g. creation and deletion of objects and links, navigating on links, manipulating attribute values) on the view by translating them into corresponding operations on the underlying TMM.

5   Implementation

Basically, we have three alternatives how to realize views based on the declarative view specification we introduced in the preceding section. First of all, we can use regular model integration/graph transformation approaches. That means that we realize the view as a fully materialized model (graph) by applying a model-to-model integration between tool and view model. This approach suffers from the fact that we duplicate the number of graph objects as well as the whole data (e.g. attribute values) stored in the tool. Furthermore, changes on the tool model must be propagated to the materialized view. For this purpose a regarded tool's API must offer event notification mechanisms that are not (yet) supported by a majority of system engineering tools. A better solution is the view implementation based on the object adapter pattern [8]. This means that we still duplicate the number of graph objects but do not replicate the tool's data anymore. Rather, queries on attributes on the view level are delegated to the tool level. The most elegant solution is the application of the class adapter pattern [8]. This results in having only one single graph, the resulting view adapter, whose objects implement the view interfaces as well as the tool interfaces[6]. Thereby, we even avoid the duplication of graph objects and do not have to propagate tool model modifications to a materialized view.

Since we want to adopt the latest OMG standards such as MOF 2.0 and plan to implement our approach as part of the MOFLON [21] project which is written in Java, it is appropriate to rely on Sun's JMI specification for defining the view and tool Java interfaces. According to JMI each class in a MOF-compliant

---

[6] Calls to the view interface are internally translated into calls to the tool interface.

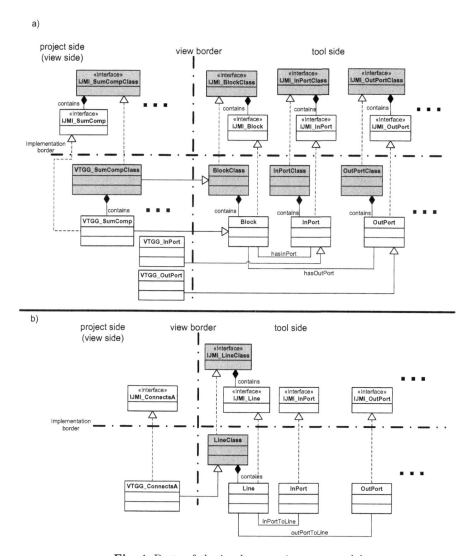

Fig. 4. Parts of the implementation metamodel

class diagram is translated into two Java interfaces. The first interface represents instances of the regarded class and provides among others methods for querying and manipulating their attribute values. The second interface is called the *proxy* which allows for the creation of new instances and keeps track of the set of already existing instances. Furthermore, each association in a MOF-compliant class diagram is translated into one Java interface. This interface represents a singleton[7] which keeps track of linked objects and provides query and manipulation methods on the internally maintained link set.

---

[7] There exists at most one instance at runtime.

5.1   Implementation Metamodel

In order to realize our view specification approach we have to translate the view and the tool metamodels into JMI-compliant Java interfaces and come up with an implementation of them. We assume that the tool's metamodel has already been translated into JMI interfaces and implemented as an adapter to the tool's API beforehand [1].

Using our MOFLON framework we can automatically generate the needed JMI interfaces for the view metamodel. As pointed out above we want to implement the resulting interfaces as class adapters [8]. Basically, a class adapter maps a desired interface (e.g. view interface) to an existing interface (e.g. tool interface). To this end the class adapter inherits from an implementation of the existing tool JMI interface. It implements the desired interface by internally translating calls to the desired interface into calls to the inherited implementation. Figure 4 depicts parts of the implementation metamodel that results from the JMI interface generation[8]. Figure 4a shows the part of the metamodel which deals with the declaration of the VTGG_SumComp class adapter that represents the mapping of a SumComp object on the view side to a Block object on the tool side. The top of the figure depicts the JMI interfaces, the bottom shows their corresponding implementations. The interfaces as well as the implementation of JMI proxy classes are gray. The left-hand side of the figure shows the interfaces and implementations of the view side, the right-hand side represents the tool side. On the tool side there are interfaces and implementations for the tool-specific classes Block, InPort, and OutPort as well as their corresponding proxies. According to the class adapter pattern VTGG_SumComp implements the interface of IJMI_SumComp and inherits from the Block implementation. Therefore, a single instance of VTGG_SumComp simultaneously represents an instance of IJMI_SumComp as well as of Block without replicating the data. VTGG_SumComp realizes the IJMI_SumComp interfaces by delegating method calls to the inherited Block implementation. For delegation purposes and to support a complete tool adapter, secondary objects on tool side are inherited by utility classes on view side (e.g. VTGG_InPort).

Correspondingly, Figure 4b introduces the VTGG_ConnectsA class adapter. This class adapter implements the JMI interface of the connectsA association on the view side and inherits from the proxy of Line since connectsA and LineClass represent singletons that internally keep track of links and Line instances, respectively.

5.2   Derived Graph Rewriting Rules

We automatically generate implementations of the JMI interfaces by deriving regular graph rewriting rules from our declarative triple graph grammar rules from Section 4, which in turn are automatically translated into executable Java

---

[8] It's just an elaborated scheme that describes the implementation of operational rules. So, it is not a substitute for the VTGG schema (cf. Figure 1b).

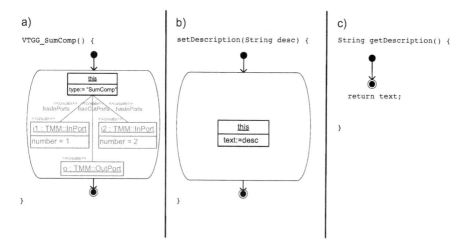

Fig. 5. a. VTGG_SumComp constructor, **b.** Write attribute operation, **c.** Read attribute operation

code by our MOFLON framework. Thus, the derived graph rewriting rules represent the methods that are declared by the JMI interfaces. Figure 5 gives examples of such derived regular graph rewriting rules for VTGG_SumComp that correspond to the declarative rule from Figure 3a. Figure 5a depicts the constructor of VTGG_SumComp which is invoked by the corresponding **create** method of the proxy VTGG_SumCompClass. According to the declarative rule from Figure 3a the constructor additionally creates secondary objects i1, i2 of type InPort, and o of type OutPort and sets their attribute values. Secondary objects are objects that are only visible on the tool side and have no representation on the view side. Rule 5b illustrates the implementation of the method setDescription from the

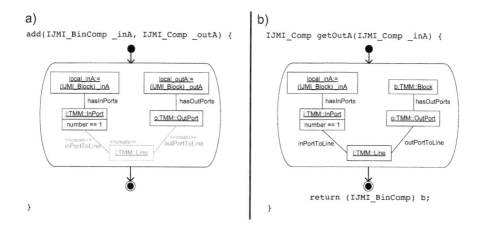

Fig. 6. a. Add new VTGG_ConnectsA link operation, **b.** Navigate link operation

IJMI_SumComp interface. The rule sets the value of the text attribute which has been inherited from the Block implementation to the value of the input parameter desc. The implementation of the method getDescription from the IJMI_SumComp interface is shown in Figure 5c. This rule just returns the value of the Block's attribute text. In general, an attribute reading operation requires navigation on secondary tool objects resulting in more complex rules.

Figure 6 illustrates some of the methods of VTGG_ConnectsA that are derived from the declarative rule in Figure 3b. The rule from Figure 6a represents the add method of the IJMI_ConnectsA interface. The rule is provided with two input parameters _inA of type IJMI_BinComp and _outA of type IJMI_Comp which are supposed to be linked to each other. Internally, the rule casts the input parameters to IJMI_Block. This may be done since the view JMI interfaces are implemented by our VTGG class adapters which implement the tool JMI interfaces as well. This has to be done in order to access the corresponding

a)

```
public class VTGG_SumComp extends Block implements IJMI_SumComp {
  public VTGG_SumComp(RefObjectm etaObject, RefPackage ImmediatePackage,
      RefPackage outermostPackage) {
    super(metaObject, immediatePackage, outermostPackage);
    this.setType("SumComp");
    this.setDesc("");

    IJMI_TMMPackage tmmPackage = (IJMI_TMMPackage) super.refImmediatePackage();
    IJMI_HasInPortsh asInPortsAssoc = tmmPackage.getHasInPorts();
    IJMI_HasOutPorts hasOutPortsAssoc= tmmPackage.getHasOutPorts();

    IJMI_InPorti 1 = tmmPackage.getInPort().createInPort();
    IJMI_InPorti 2 = tmmPackage.getInPort().createInPort();
    IJMI_OutPort o = tmmPackage.getOutPort().createOutPort();

    i1.setNumber(1);
    i2.setNumber(2);

    hasInPortsAssoc.add(i1, this);
    hasInPortsAssoc.add(i2, this);
    hasOutPortsAssoc.add(o, this);
}
/* Additional methodso mitted duet o lack of space. */
```

b)

```
public class VTGG_connectsA extends LineClass implements IJMI_ConnectsA {
  public boolean add(IJMI_BinComponent_ inA, IJMI_Component _outA) throws JmiException {
    IJMI_TMMPackage tmmPackage = (IJMI_TMMPackage) super.refImmediatePackage();
    IJMI_HasOutPorts hasOutPortsAssoc= tmmPackage.getHasOutPorts();
    IJMI_HasInPortsh asInPortsAssoc = tmmPackage.getHasInPorts();

    Collection outports = hasOutPortsAssoc.getOutPort((IJMI_Block) _outA);
    IJMI_OutPort outport= (IJMI_OutPort) outports.iterator().next();

    Collection inports= hasInPortsAssoc.getInPort((IJMI_Block) _inA);
    IJMI_InPorti nport= null;

    for(Objectt mpInport : inports) {
      if(((IJMI_InPort) tmpInport).getNumber() == 1) {
        inport = (IJMI_InPort) tmpInport;
        break;
      }
    }
    if(inport! = null && outport! = null) {
      this.createLine(outport, inport);
      return true;
    }
    return false;
}
/* Additional methodso mitted duet o lack of space. */
```

Fig. 7. a. Executable Java code that corresponds to Figure 5a, **b.** Executable Java code that corresponds to Figure 6a

In- and OutPort objects on the tool side. Both Port objects are then linked to a new Line object on the tool side that represents the desired association link on the view side. Similarly, the rule 6b returns the IJMI_Comp object that is connected to the given IJMI_BinComp _inA object by a connectsA link. Again the rule internally casts the input parameter for the same reasons mentioned above. Thereafter, the rule identifies the InPort object i of _inA. Navigating via the Line object that is attached to i the rule determines the connected OutPort object o. Finally, the rule returns the casted Block object b that owns o. A rule that deletes an existing connection between a given IJMI_BinComp and a IJMI_Comp object looks similar to the one depicted in Figure 6a. The <<create>> tags are just replaced by <<delete>> tags.

5.3   Executable Code

From the regular graph rewriting rules we now generate executable Java code using our MOFLON framework. Figure 7a illustrates the Java code that results from the regular graph rewriting rule depicted in Figure 5a. First of all, the constructor calls the constructor of its superclass Block. Afterwards, the values of the attributes Type and Desc are initialized. Using the corresponding proxies the constructor creates two IJMI_InPort objects and an IJMI_OutPort object. Finally, the ports are connected to the regarded VTGG_SumComp using the corresponding associations.

The Java code corresponding to the rule presented in Figure 5a is shown in Figure 7b. Initially, the add method casts the provided input parameter _outA of type IJMI_Comp to IJMI_Block as demanded by the rule. After this the method determines the attached OutPort object. Correspondingly, the method determines the InPort object which number attribute has the value 1 and is connected to _inA that also has been cast to IJMI_Block beforehand. Finally, add links both ports by a new Line object.

6   Conclusion

In this paper we have outlined a more detailed definition of our view specification approach realized by a modified version of triple graph grammars called VTGG. Based on the initial ideas presented in [14], the paper introduces three possible solutions on how to implement the automatically derived operational rules (e.g. for instantiating, read/write access, deleting, ...). The mainly presented realization based on class adapters avoids the creation of two coexisting object graphs. Furthermore, it reuses MOFLON's TGG specification and translation framework as well as its JMI compliant Java code generator backend.

By using our VTGG approach for view specification purposes we are able to support tool integration, model transformation as well as checking tailored design rules at various levels of abstraction. VTGGs fulfill all requirements listed in section 1. Thus, our approach allows for the specification of multiple views such as complementary or overlapping views to one underlying tool model by weaving together the metamodels using the class adapter pattern multiple times.

Furthermore, we are able to create views on top of views while having again just one object graph . Since our approach is realized as a single object graph, the resulting adapter, we are not facing the view update problem.

As a drawback while realizing the class adapter layer, we have to know and access the tool-specific adapter classes. Moreover, applying quite a number of views on top of views, the resulting class hierarchy becomes quite complex. Due to the resulting view adapters, an intricate control mechanism is necessary if multiple complementary views have to exist at the same time.

An implementation of the VTGG approach as well as our regular TGGs are currently under development. The result will be a plug-in of the MOFLON meta modeling environment. It is subject of ongoing research activities to adapt the ideas presented here to the syntax of OMG's model transformation language standard QVT. In this context, we will extend our VTGG approach with the definition of abstract mapping rules. Moreover, we want to make a proposal for a uniform treatment of model, view, and transformation specifications in QVT.

References

1. F. Altheide et al. An Architecture for a Sustainable Tool Integration. In Dörr and Schürr, editors, *TIS'03: Workshop on Tool Integration in System Development*, pages 29–32, 2003.
2. S. Becker, T. Haase, and B. Westfechtel. Model-Based A-Posteriori Integration of Engineering Tools for Incremental Development Processes. *SoSym*, 4(2):123–140, 2005.
3. S. Burmester, H. Giese, J. Niere, M. Tichy, J. P. Wadsack, R. Wagner, L. Wendehals, and A. Zündorf. Tool Integration at the Meta-Model Level within the FUJABA Tool Suite. *STTT*, 6(3):203–218, August 2004.
4. J. de Lara, E. Guerra, and H. Vangheluwe. A Multi-View Component Modelling Language for Systems Design: Checking Consistency and Timing Constraints. In *VMSIS'05*, 2005.
5. K. Ehrig, E. Guerra, J. de Lara, L. Lengyel, T. Levendovszky, U. Prange, G. Taentzer, D. Varró, and S. Varró-Gyapay. Model Transformation by Graph Transformation: A Comparative Study. In *MTiP'05*, 2005.
6. C. Ermel and K. Ehrig. View Transformation in Visual Environments applied to Algebraic High-Level Nets. In *PNGT'04*, volume 127(2), pages 61–86, 2005.
7. C. Ermel, K. Hölscher, S. Kuske, and P. Ziemann. Animated Simulation of Integrated UML Behavioral Models based on Graph Transformation. In *VL/HCC'05*, 2005.
8. E. Gamma, R. Helm, R. Johnson, and J. Vlissides. *Design Patterns*. Addison-Wesley Professional Coumputing Series. Addison-Wesley Publishing Company Inc., 1995.
9. J. Grundy and J. Hosking. Constructing Integrated Software Development Environments with MViews. In *IJAST Vol.2*, 1996.
10. J. Grundy, R. Mugridge, J. Hosking, and M.k Apperley. Tool Integration, Collaboration and User Interaction Issues in Component-based Software Architectures. In *TOOLS'98*, 1998.
11. L. Grunske, L. Geiger, A. Zündorf, N. VanEetvelde, P. VanGorp, and D. Varro. Using graph transformation for practical model driven software engineering. In *Research and Practice in Software Engineering*, volume 2, pages 91 – 119, 2005.

12. E. Guerra and J. de Lara. Event-Driven Grammars: Towards the Integration of Meta-modelling and Graph Transformation. In *ICGT'04*, pages 54–69, 2004.
13. E. Guerra, P. Diaz, and J. de Lara. A Formal Approach to the Generation of Visual Language Environments Supporting Multiple Views. In *VL/HCC'05*, 2005.
14. J. Jakob and A. Schürr. View Creation of Meta Models by Using Modified Triple Graph Grammars. In *GT-VMT'06*, ENTCS, pages 175–185, 2006.
15. SUN JCP: Java Metadata Interface(JMI) Specification, 2002.
16. E. Kindler, V. Rubin, and R. Wagner. An Adaptable TGG Interpreter for In-Memory Model Transformation. In A. Schürr and A. Zündorf, editors, *Fujaba Days 2004*, volume tr-ri-04-253 of *Technical Report*, pages 35–38. University of Paderborn, September 2004.
17. A. Königs and A. Schürr. Tool Integration with Triple Graph Grammars - A Survey. In *SegraVis*, ENTCS, London, 2005. Academic Press.
18. MetaCase. *MetaEdit+*, 2005. http://www.metacase.com.
19. OMG: Meta Object Facility (MOF) 2.0 Core Specification, 2004.
20. OMG: Meta Object Facility (MOF) 2.0 Query/View/Transformation Specification, 2005.
21. Real-Time Systems Lab, Darmstadt University of Technology. *MOFLON*, 2006. http://www.moflon.org.
22. A. Schürr. Specification of graph translators with triple graph grammars. In Mayr and Schmidt, editors, *WG'94: Workshop on Graph-Theoretic Concepts in Computer Science*, pages 151–163, 1994.
23. Software Engineering Group, University of Paderborn. *FUJABA*, 2005. http://www.fujaba.de.
24. M.W.A. Steen, D.H. Akehurst, H.W.L. ter Doest, and M.M. Lankhorst. Supporting Viewpoint-Oriented Enterprise Architecture. In *EDOC'04*, 2004.
25. N. Zhu, J. C. Grundy, and J. G. Hosking. Pounamu: A Meta-Tool for Multi-View Visual Language Environment Construction. In *VL/HCC'04*, pages 254–256, 2004.
26. A. Zündorf. *Rigorous Object Oriented Software Development*. University of Paderborn, 2001. Habilitation Thesis.

Model-Driven Monitoring: An Application of Graph Transformation for Design by Contract

Gregor Engels[1,2,3], Marc Lohmann[1], Stefan Sauer[1,3], and Reiko Heckel[4]

[1] University of Paderborn, Department of Computer Science
Warburger Str. 100, 33098 Paderborn, Germany
{engels, mlohmann}@uni-paderborn.de
[2] sd&m AG, software design & management
Am Schimmersfeld 7a, 40880 Ratingen, Germany
[3] Software Quality Lab, University of Paderborn,
Warburger Str. 100, 33098 Paderborn, Germany
sauer@s-lab.upb.de
[4] University of Leicester, Department of Computer Science
University Road, Leicester, United Kingdom
reiko@mcs.le.ac.uk

Abstract. The model-driven development (MDD) approach for constructing software systems advocates a stepwise refinement and transformation process starting from high-level models to concrete program code. In contrast to numerous research efforts that try to generate executable function code from models, we propose a novel approach termed *model-driven monitoring*. Here, models are used to specify minimal requirements and are transformed into assertions on the code level for monitoring hand-coded programs during execution.

We show how well-understood results from the graph transformation community can be deployed to support this model-driven monitoring approach. In particular, models in the form of visual contracts are defined by graph transitions with loose semantics, while the automatic transformation from models to JML assertions on the code level is defined by strict graph transformation rules. Both aspects are supported and realized by a dedicated Eclipse plug-in.

1 Introduction

Object-oriented technology provided us with a better handle on complexity than previous technologies. Nevertheless, the growing size of applications and the demand for shorter time-to-market entail that many issues remain. In recent years, the paradigm of a model-driven development (MDD) approach has been introduced and discussed heavily. In particular, the Object Management Group (OMG) favored a model-driven approach to software development and pushed its Model-Driven Architecture (MDA) [1] initiative as well as standards such as the Unified Modeling Language (UML) that provides the foundation for MDA.

However, model-driven development is still in its infancy compared to its ambitious goals of having a (semi-)automatic, tool-supported stepwise refinement

A. Corradini et al. (Eds.): ICGT 2006, LNCS 4178, pp. 336–350, 2006.

process from vague requirements specifications to a fully-fledged running program. A lot of unresolved questions exist for horizontal *modeling* tasks as well as for vertical *model transformation* tasks.

In principle, models provide an abstraction from the detailed problems of implementation technologies. They allow software designers to focus on the conceptual task of modeling static as well as behavioral aspects of the envisaged software system. Unfortunately, abstraction naturally conflicts with the desired automatic code generation from models. To enable the latter, fairly complete and low-level models are needed. Today, a complete understanding of the appropriate level of detail and abstraction of models is still missing.

Horizontal modeling levels are interrelated by vertical model transformations. Here, too, a complete understanding is missing how such a transformation might be specified and implemented. A number of model transformation approaches have been proposed and discussed, in particular, as answer to the Query-View-Transformation (QVT) RFP of the OMG [2].

The graph transformation community has been investigating and discussing since years graph-based approaches for specifying structure and behavior of software components as well as for specifying transformations.Thus, graph transformation provides well-defined and well-investigated candidate solutions for the mentioned open issues in the MDD realm.

In our work, we employ results from research on graph transformation to offer solutions for horizontal modeling as well as vertical model transformation problems. In particular, we introduce a novel modeling approach. We do not follow the usual approach that models should operate as source for an automatic code generation step that produces the executable function code of the program. Rather, we restrict the modeling task to providing structure information and minimal requirements towards behavior for the subsequent implementation. We expect that only structural parts of an implementation are automatically generated, while the behavior is manually added by a programmer.

As a consequence it can not be guaranteed that the hand-coded implementation is correct with respect to the modeled requirements. Yet, we will show how models can be used to generate assertions which monitor the execution of the hand-coded implementation. Herewith, violations of the modeled requirements will be detected at runtime and reported to the environment. We call this novel approach *model-driven monitoring*.

Model-driven monitoring (MDM) is based on the idea of Design by Contract (DbC) [3], where so-called contracts are used to specify the desired behavior of an operation. Contracts consist of pre- and post-conditions. Before an operation is executed, the pre-condition must hold, and in return, after the execution of an operation, it has to be guaranteed that the post-condition is satisfied.

The DbC approach has been introduced for textual programming languages and is supported by appropriate tools, e.g. for the Eiffel language [4]. Recently, the same approach has been put into effect for the Java programming language. For instance, the Java Modeling Language (JML) extends Java with Design by Contract concepts [5]. JML assertions are based on Java expressions and are annotated to the source code. During the execution of such an annotated Java program, any violation of an assertion is monitored. An exception is raised as soon as a violation is detected.

We lift this idea of contract specifications to the level of visual models and reuse the concept of graph transformation to specify pre- as well as post-conditions of an operation in a graphical, UML-like way. As those *visual contracts* define minimal requirements towards an operation, the semantic concept of loose graph transitions, formalized by the double-pullback (DPB) approach [6], is deployed to provide the semantics of the contract-based approach.

Besides this novel modeling approach, we deploy graph transformation results for defining the automatic transformation step from visual contract specifications to textual JML assertions. In contrast to the modeling of minimal requirements illustrated above, we provide a complete specification of this transformation step here. Thus, the semantic concept of strict graph transformations is deployed which is formalized by the double-pushout (DPO) approach [7].

The complete approach of contract-based modeling of a software system is supported by a tool chain that we implemented as Eclipse plug-in. The presented method for specifying software components by contracts has been studied in an industrial setting [8,9].

We give an overview of the model-driven monitoring approach in the following section. Section 3 explains our method of modeling with visual contracts based on the concepts of graph transitions. The translation from visual contracts to JML assertions is described in Sect. 4. There we use graph transformation rules for specifying the translation. The tools that we provide to support our method are introduced in Sect. 5. Finally, we summarize the achievements and sketch future perspectives.

2   Towards Model-Driven Monitoring

Model-driven monitoring (MDM) constitutes a novel strategy for model-driven software development beyond the classical idea of model-driven development (MDD) centered upon the automatic generation of function code and model-driven testing (MDT) focussing on automatically deriving test cases from models. We enable model-driven monitoring by embedding visual contracts in a model-driven software development process according to Fig. 1. Visual contracts are interpreted as models of behavior from which code for testing and runtime assertion checking can be generated. The visual contracts also specify the behavior which is then manually implemented by programmers.

On the design level, a software designer has to specify a model of the system under development. This model consists of class diagrams and visual contracts. The class diagrams describe the static aspects of the system. Each visual contract specifies the behavior of an operation. The behavior of the operation is given in terms of data state changes by pre- and post-conditions, which are modeled by a pair of UML composite structure diagrams as explained in Sect. 3. Both the pre- and post-condition of a visual contract are typed over the class diagram.

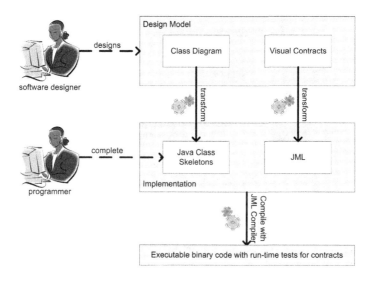

Fig. 1. Towards Model-Driven Monitoring

In the next step, we generate Java code from the design model. This generation process consists of two parts. First, we generate Java class skeletons from the design class diagrams. Second, we generate JML assertions from every visual contract and annotate each of the corresponding operations with the generated JML contract. The JML assertions allow us to check the consistency of models with manually derived code at runtime. The execution of such checks has to be transparent in that, unless an assertion is violated, the behavior of the original program remains unchanged. Thus, our transformation rules (see Sect. 4) for generating JML assertions from the UML design model only generate assertions that behave accordingly.

Then, a programmer uses the generated Java fragments to fill in the missing behavioral code in order to build a complete and functional application. Her programming task will emanate from the design model of the system. Particularly, she will use the visual contracts as reference for implementing the behavior of operations. She has to code the method bodies, and may add new operations to existing classes or even completely new classes, but she is not allowed to change the JML contracts. The latter guarantees that the JML contracts remain consistent with the visual contracts. Integrity of visual contracts can be technically assisted by separating Java class skeletons and JML assertions into two different files and prohibiting access to the JML assertions file. Programmers do not need to see the JML annotations; rather they should use the more intuitive visual contracts as the starting point for their programming.

When a programmer has implemented the behavioral code, she uses the JML compiler to build executable binary code. This binary code consists of the programmer's behavioral code and additional executable runtime checks which are generated by the JML compiler from the JML assertions. This leads to a runtime

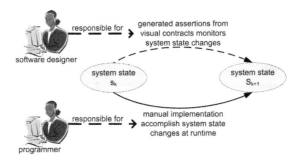

Fig. 2. Behavior at runtime

behavior as shown in Fig. 2. The manual implementation of a programmer leads to a system state change. The generated runtime checks monitor the pre- and post-conditions during the execution of the system. They monitor whether the manually coded behavior of an operation fulfills its JML specification. Thus, we indirectly monitor whether the system state change performed by the manual implementation complies with the visual contract specification of the design model since the JML annotations are purely generated from the visual contracts. Thus, we support model-driven monitoring of implementations by transforming our visual contracts into contracts in JML.

Our visual contracts are given in a UML-like notation of graph transformation rules. However, the classical interpretation of graph transformation rules based on the double-pushout approach (DPO) [7] is not adequate for the representation of a contract. In this approach it is assumed that during the execution of an operation nothing is changed beyond the specification in the rule. This would mean that we have to describe the behavior of an operation completely on the model level, which would lead to the drawbacks mentioned in the introduction. Rather, our method builds upon the loose semantic interpretation of visual contracts. They are interpreted as a minimal description of the data state transformation which has to be implemented by the programmer. Thus, a visual contract specifies only what at least has to happen on a system state, but it allows the programmer to implement additional effects. This loose interpretation is necessary both to give the programmer the opportunity for optimizing her code, e.g. by adding new classes or methods, and to generate assertions from partial, incomplete models. Therefore, we have to interpret our visual contracts as graph transitions. In the double-pullback (DPB) [6] approach graph transitions allow additional changes that are not encoded in the transformation rules.

3   Modeling with Visual Contracts

We show how to specify a system with visual contracts by the example of an online shop. We distinguish between a static and a functional view.

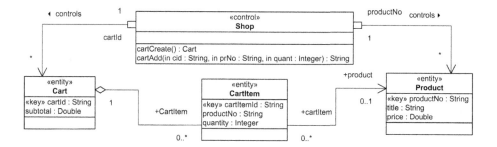

Fig. 3. Class Diagramm specifying static structure of online shop

UML class diagrams are used to represent the *static view* of a system specification. Figure 3 shows the class diagram of the sample online shop. We use the stereotypes `control` and `entity` as introduced in the Unified Process [10]. The stereotype `key` indicates a unique identifier for each of a set of objects of the same type. Qualifiers on associations (e.g. `productNo` on the association `controls` between `Shop` and `Product`) designate an attribute of the referenced class and provide direct access to a specific object.

The *functional view* of the system is described by visual contracts (i.e., graph transformations) for a selected set of operations. It integrates static and dynamic aspects to describe the effect of an operation on the data state of the system. Therefore, visual contracts take an operation-wise view on the internal behavior.

On the functional level a designer has different degrees of freedom to decide how detailed a model is. At first a designer can decide which of the operations to specify by visual contracts. If an operation is not detailed by a visual contract then the only consequence is that the operation is not monitored at runtime.

Further, if a designer describes an operation by a visual contract, she has the freedom to decide how detailed the specification shall be. The less detailed an operation is specified by a contract, the more freedom has a developer in implementing an operation. This is possible due to the assumption that the contracts are an incomplete description of the system state changes by an operation. A contract only specifies what at least has to happen, but it allows a developer to implement additional effects. For example, the implementation of the visual contract of Fig. 4 can additionally calculate the total costs of a cart and assign this value to the attribute `subtotal` of `Cart`. This interpretation is supported by the loose semantics of open graph transformation systems [6].

Structurally, a visual contract consists of two graphs, representing the precondition and the post-condition, respectively, like the left- and a right-hand side of a graph transformation rule (compare Fig. 4). The graphs are visualized by UML composite structure diagrams. Each of the diagrams is typed over the design class diagram.

Additionally, we may extend the pre- or post-condition of a visual contract by negative pre-conditions (i.e., negative application conditions [11]) or respectively

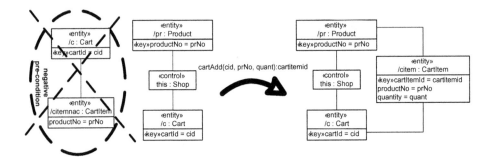

Fig. 4. Visual contract for operation `cartAdd`

by negative post-conditions. A slashed ellipse marks them. The negative pre-condition specifies object structures that are not allowed to be present before the operation is executed. The negative post-condition identifies object structures that are not allowed to be present after the execution of the operation.

Beside the different graphs, a visual contract contains the operation name, a parameter-list and a return-result. The variables of the parameter-list and the return-result are used in the visual contracts to further qualify the objects.

The visual contract in Fig. 4 specifies the operation `cartAdd`. This operation adds a new `CartItem`, which references an existing `Product`, to an existing `Cart`. The variables of the parameter-list and the return-value are used to determine values of attributes of different objects. For a successful execution of the operation, the object `this` must know two objects: an object of type `Cart` that has an attribute `cartId` with the value `cid`, and an object of type `Product` that has an attribute `productNo` with the value `prNo`. The actual argument values are bound when the client calls the operation. The `Cart` object is reused in the negative pre-condition (compare object identifiers). The negative pre-condition extends the pre-condition by the requirement that the `Cart` object is not linked to any object of type `CartItem` that has an attribute `productNo` with the value `prNo`. This means, it is not permitted that the product is already contained in the cart. As a result, the operation creates a new object of type `CartItem` with additional links to previously identified objects. The return value of the operation is the content of the attribute `cartItemId` of the newly created object.

4   Transformation of the Design Model to Java Code with JML Assertions

After describing the modeling of a software system with visual contracts, we now present how the model-driven software development process continues from the design model. A transformation of visual contracts to JML constructs provides for model-driven monitoring of the contracts. The contracts can be automatically evaluated for a given state of a system, where the state is given by object configurations. The generation process as well as the kind of code that is generated from a class diagram and the structure of a JML assertion that is generated from a visual contract are described in detail in [12, 13]. Here we describe the transformation more generally and from a methodical perspective and explain the formalization by graph transformation rules which underlies the transformation.

4.1 Transformation of Class Diagrams to Java

Each UML class is translated to a corresponding Java class. Attributes and associations are complemented by the corresponding access methods (e.g., get, set). For multi-valued associations we use classes that implement the Java interface Set. Qualified associations are provided by classes that implement the Java interface Map. We add methods like getProduct(int productNo) that use the attributes of the qualified associations as input parameters. Operation signatures that are specified in the class diagram are translated to method declarations in the corresponding Java class.

4.2 Transformation of Visual Contracts to JML

For operations that are specified by a visual contract, the transformation of the contract to JML yields a Java method declaration that is annotated with a JML assertion. The pre- and post-conditions of the generated JML assertions are interpretations of the graphical pre- and post-conditions of the visual contract. When any of the JML pre- and post-conditions is evaluated, an optimized breadth-first search (compare [14]) is applied to find an occurrence of the pattern that is specified by the pre- or post-condition in the current system data state. The search starts from the object this which is executing the specified behavior. If the JML pre-condition or post-condition finds a correct pattern, it returns true, otherwise it returns false.

4.3 Specifying the Contract Transformation

After demonstrating the transformation in principle, we explain in the following how we have defined a precise specification of the transformation from visual contracts to JML. The declarative specification in [12] abstracts from representation details of the visual contracts and leaves out different details of the mapping between visual contracts and JML. In contrast, we present an operational specification of the transformation from visual contracts to JML here. The provision of the operational model transformation is the prerequisite for an automated translation of visual contracts to JML as implemented in our development tools.

The operational specification is the second application of graph transformation concepts in our method towards model-driven monitoring. Other than the first application for specifying visual contracts that state minimal requirements on a single horizontal modeling level, we need a complete specification of the transformation behavior to support the automation of the model transformation in the vertical direction. The operational specification is based upon an extension of the UML 2 metamodel for visual contracts. The metamodel represents the source language of the model transformation and provides the type graph on which the graph transformation rules operate, i.e., the graph transformation rules are specified on the metamodel level, and the concrete models are viewed as metamodel instances when they are transformed.

4.4   Extended UML 2 Metamodel

Our visual contracts integrate with the UML 2 metamodel. Mainly we use elements from the UML 2 metamodel packages *InternalStructures* and *Collaborations*. The *InternalStructure* subpackage provides mechanisms for specifying structures of interconnected elements, representing runtime instances, which collaborate over communication links to achieve some common objectives. A *collaboration* represents how elements of the model cooperate to perform some essential behavior. Among others, the participating elements may include classes and objects, associations and links as well as attributes and operations. Collaborations allow us to describe only the relevant aspects of the cooperation of a set of instances by identifying the specific roles that the instances will play.

Figure 5 provides a view on the metamodel for our visual contracts. *Visual-Contract* specializes *Collaboration*. A collaboration defines a set of cooperating entities to be played by instances (its roles) as well as a set of connectors that define communication paths between the participating instances. The roles are represented by *ConnectableElements*, which are referenced by a *Collaboration*. *ConnectableElement* is a *TypedElement*, which references a *Type*. *Class* is a *Classifier*, which is a *Type*. Consequently, the *ConnectableElement* can define a role that classes have to play in order to accomplish the behavior of a collaboration (visual contract, respectively). *ConnectableElements* are linked by a *Connector* with *ConnectorEnds*. A *Connector* specifies a link that enables communication between two or more instances. This link may be an instance of an association. In contrast to associations, which specify links between any instance of the associated classifiers, connectors specify links between instances playing the roles of the connected parts only. Additionally, the UML 2 metamodel offers specializations of *ConnectableElement* for representing parameters and variables.

We also define attribute values that an instance must provide in order to play one of the defined roles. According to the UML metamodel, you cannot specify the content of the features (properties) of a role in more detail. Therefore, we have introduced a specialization of a *ConnectableElement* named *VCElement* and a class *Constraint* to restrict possible attribute values. The class *Constraint* groups a feature (which represents an attribute of a class) and a permitted value. The permitted value of a feature can be a simple value (*ValueSpecification*) or another *VCElement*. Since the value of a feature can change from the pre- to the post-condition, we distinguish in the meta-model by association whether the reference value belongs to the pre- or post-condition.

We have to define whether a *VCElement* is part of the pre- or post-condition. To specify the absence of certain structures, both pre- and post-conditions may contain negative conditions. Therefore, we have added three metamodel classes to the UML metamodel: *Precondition*, *Postcondition*, and *NegativeCondition*.

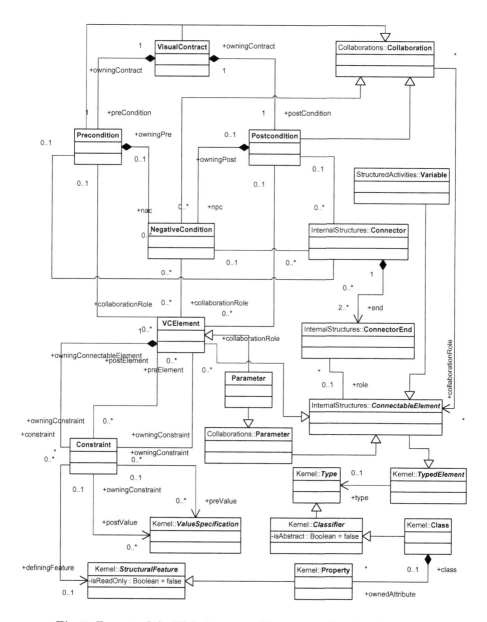

Fig. 5. Extract of the UML 2 metamodel extension for visual contracts

4.5   Operational Transformation with Compound Rules

For the operational specification of our transformation from visual contracts to JML, we assume that the source model is syntactically correct according to our metamodel. We define the transformation by a set of compound rules as introduced in [15].

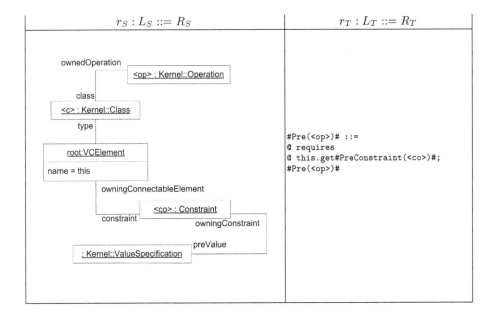

Fig. 6. Compound rule that starts the generation of JML assertions for checking attribute values of object *this*

The basic idea of compound rules is that a model transformation from a source language to a target language can be defined by a synchronized model transformation on the source and the target language. Such a synchronized model transformation can be specified by a set of model transformation rules, consisting of two parts for transforming both the source and target model.

Figure 6 depicts a sample compound rule that starts the generation of the JML assertions for checking the attribute values of the object this in the precondition. A compound rule $r : (r_s, r_t)$ consists of two parts, a UML part and a JML part. Both r_s and r_t can be viewed as graph transformation rules. In general, the source transformation rule $r_s : L_S ::= R_S$ describes the transformation of the source model, the target transformation rule $r_t : L_T ::= R_T$ specifies the transformation of the target model. Note that in Fig. 6 r_s is an identical transformation with $L_S = R_S$, which is visualized by the left-hand side only. Source and target rules are coupled by the ability of using shared variables. Such variables are denoted by $\#variable\#$.

When applying a compound rule for the transformation of a source to a target model, at first an occurrence of the left-hand side L_s of the source transformation rule is searched within the source model (source match). In Fig. 6 the left-hand side of the source rule matches, if a VCElement (part of a visual contract added to an operation) this has a constraint with a value specification. If a source match is found, the variables are instantiated. This means, that a value is assigned to each variable according to the source match. Then, an occurrence of the left-hand side of the target transformation rule L_T (using the instantiated variables—in our example there is only one variable op) is searched within the target model (target match). Then the target match is replaced by the right-hand side R_T of the target transformation rule. In our example, the target transformation rule prepares the code for testing the content of an attribute of the object this.

In order to specify model transformations with control, the approach in [15] provides support for assembling compound rules into transformation units. Such units consist of a set of compound rules with control. Each compound rule is contained in a rule set. The rule sets are then organized in a sequence of rule sets where each rule set can be considered as a layer. Within a rule set, rules may be applied non-deterministically. A transformation unit consists of a set of compound rules together with a control expression specifying the organization of rules into rule sets, layers and determining whether a rule should be applied once or as long as possible.

For defining the transformation of our models consisting of class diagrams and visual contracts to Java classes and JML we need round about 95 compound rules. These compound rules have to be organized in approximately 25 transformation units.

5   Tool Support

In the previous sections, we have shown how to use visual contracts in models of software systems for specifying operations and how to translate visual contracts to JML. This enables model-driven monitoring. We can monitor the correctness of a manual implementation with respect to its specification.

Existing CASE tools or graph transformation tools do not support the use of visual contracts for specifying software systems as described in our approach. As a proof of concept and for showing the practical feasibilty of our approach, we have developed an integrated development environment for using visual contracts in a software development process. This development environment allows software developers to model class diagrams and specify the behavior of operations by visual contracts. It further supports automatic code generation as described in Sect. 4, the manual implementation to get a complete application, and the compilation of the generated assertions by a JML compiler.

Figure 7 shows the user interface of our tool for modeling visual contracts. The central workspace of the visual contract editor is divided into four sectors. A software designer can specify the pre- and post-condition in the bottom-left sector (labeled with LHS) and the bottom-right sector (labeled with RHS), respectively. An object this (the active object executing the operation) is added automatically in both sectors when a new visual contract for an operation is created. Every object added to the pre- or post-condition must be within reach of the object this by links. Additionally, the top sector allows for specifying negative conditions. The top-left sector (labeled with NAC for negative application condition) is for specifying object structures that are not allowed to be present

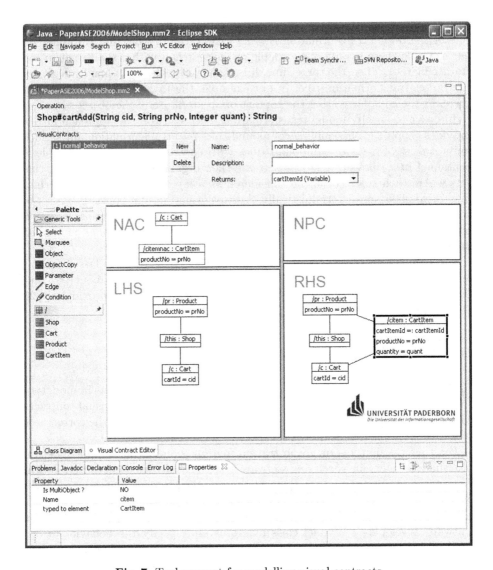

Fig. 7. Tool support for modelling visual contracts

before the operation is executed. The top-right sector (labeled with *NPC* for negative post-condition) is for specifying object structures that are not allowed to be present after the execution of the operation.

The development environment is implemented as an Eclipse plug-in. We mainly used the Graphical Editor Framework (GEF) [16] and the Eclipse Modeling Framework (EMF) [17] for the implementation of the plug-in. The code generation was implemented using Eclipse JET [18], which is a part of the EMF.

6   Conclusion

We have shown in this paper, how we have been employing results from research on graph transformation in model-driven software development processes. Addressing horizontal modeling issues, we have lifted the Design by Contract idea to the visual model level. Visual contracts use graph transformation concepts for the specification of pre- and post-conditions of operations. Since they only define minimal requirements towards the implementation of an operation, we use the loose semantics of graph transitions of the double-pullback approach.

For the vertical direction of model transformations, we use compound (graph transformation) rules to define a transformation of our visual contracts to the Java Modeling Language JML, a Design by Contract extension for Java. To automate this model transformation, we need the strict semantic interpretation of graph transformation rules as formalized by the double-pushout approach.

Altogether, we have introduced model-driven monitoring as a new and practically useful amalgamation of graph transformation and Design by Contract concepts. In contrast to the automatic generation of function code, we generate assertions from contracts that are monitored and automatically checked while the actual and manually implemented function code is executed.

To support our model-driven monitoring method, we provide an editor that allows developers to coherently model class diagrams and visual contracts. The editor is complemented by code generation facilities for Java classes with JML assertions for their operations.

In an industrial case study [9,8], we have successfully applied visual contracts for specifying the interfaces of Web services. Our method and tools are currently considered by an industrial partner software company of the Software Quality Lab (s-lab) for deployment in their software development projects.

References

1. Meservy, T., Fenstermacher, K.D.: Transforming software development: An MDA road map. Computer **38**(9) (2005) 52–58
2. OMG (Object Management Group): Request for proposal: Mof 2.0 query / views / transformations rfp (2002)
3. Meyer, B.: Applying "Design by Contract". IEEE Computer **25**(10) (1992) 40–51
4. Meyer, B.: Eiffel: The Language. second printing edn. Prentice-Hall (1992)
5. Leavens, G.T., Baker, A.L., Ruby, C.: Preliminary design of JML: A behavioral interface specification language for Java. Technical Report 98-06-rev27, Department of Computer Science, Iowa State University (2005)
6. Heckel, R., Ehrig, H., Wolter, U., Corradini, A.: Double-pullback transitions and coalgebraic loose semantics for graph transformation systems. APCS (Applied Categorical Structures) **9**(1) (2001) 83–110
7. Ehrig, H., Pfender, M., Schneider, H.: Graph grammars: an algebraic approach. In: 14th Annual IEEE Symposium on Switching and Automata Theory, IEEE (1973) 167–180

8. Engels, G., Güldali, B., Juwig, O., Lohmann, M., Richter, J.P.: Industrielle Fallstudie: Einsatz visueller Kontrakte in serviceorientierten Architekturen. In Biel, B., Book, M., Gruhn, V., eds.: Software Enginneering 2006, Fachtagung des GI Fachbereichs Softwaretechnik. Volume 79 of Lecture Notes in Informatics., Köllen Druck+Verlag GmbH (2006) 111–122
9. Lohmann, M., Richter, J.P., Engels, G., Güldali, B., Juwig, O., Sauer, S.: Abschlussbericht: Semantische Beschreibung von Enterprise Services - Eine industrielle Fallstudie. Technical Report 1, Software Quality Lab , Unversity of Paderborn (2006)
10. Jacobson, I., Booch, G., Rumbaugh, J.: The Unified Software Development Process. Addison-Wesley Professional (1999)
11. Habel, A., Heckel, R., Taentzer, G.: Graph grammars with negative application conditions. Fundamenta Informaticae **26**(3,4) (1996) 287–313
12. Lohmann, M., Sauer, S., Engels, G.: Executable visual contracts. In Erwig, M., Schürr, A., eds.: 2005 IEEE Symposium on Visual Languages and Human-Centric Computing (VL/HCC'05). (2005) 63–70
13. Heckel, R., Lohmann, M.: Model-driven development of reactive informations systems: From graph transformation rules to JML contracts. International Journal on Software Tools for Technology Transfer (STTT) (2006) accepted for publication.
14. Zündorf, A.: Graph pattern matching in progres. In Cuny, J., Ehrig, H., Engels, G., Rozenberg, G., eds.: 5th. Int. Workshop on Graph-Grammars and their Application to Computer Science. LNCS 1073 (1996)
15. Heckel, R., Küster, J.M., Taentzer, G.: Towards automatic translation of UML models into semantic domains. In Kreowski, H.J., Knirsch, P., eds.: Proceedings of the Appligraph Workshop on Applied Graph Transformation. (2002)
16. Eclipse Consortium: Eclipse graphical editing framework (GEF) - version 3.1.1. http://www.eclipse.org/gef/ (2006)
17. Eclipse Consortium: Eclipse modeling framework (EMF) - version 2.1.2. http://www.eclipse.org/emf/ (2006)
18. Eclipse Consortium: Java emitter templates (JET). Eclipse Modeling Framework (EMF) - Version 2.1.1, http://www.eclipse.org/emf/ (2006)

Model View Management with Triple Graph Transformation Systems

Esther Guerra[1] and Juan de Lara[2]

[1] Computer Science Department,
Universidad Carlos III de Madrid (Spain)
eguerra@inf.uc3m.es
[2] Polytechnic School,
Universidad Autónoma de Madrid (Spain)
jdelara@uam.es

Abstract. In this paper, we present our approach for model view management in the context of Multi-View Visual Languages (MVVLs). These are made of a number of diagram types (or viewpoints) that can be used for the specification of the different aspects of a system. Therefore, the user can build different *system views* conform to the viewpoints, which are merged in a repository in order to perform consistency checking. In addition, the user can define *derived views* by means of *graph query patterns* in order to extract information from a base model (a system view or the repository). We have provided automatic mechanisms to keep synchronized the base model and the derived view when the former changes. Predefined queries by the MVVL designer result in so-called *audience-oriented views*. Finally, *semantic views* are used for analysing the system by its translation into a semantic domain.

Our approach is based on meta-modelling to describe the syntax of the MVVL and each viewpoint, and on triple graph transformation systems to synchronize and maintain correspondences between the system views and the repository, as well as between the derived, audience-oriented and semantic views and the base models. We illustrate these concepts by means of an example in the domain of security for web systems.

1 Introduction

Visual Languages (VLs) are extensively used in many engineering activities for the specification and design of systems. As these become more complex, there is a need to split specifications into smaller, more comprehensible parts that use the most appropriate notation. Thus, there are VLs (such as UML) made of a family of diagram types, which can be used for the description of the different aspects of a system. We call such VLs Multi-View Visual Languages (MVVLs) [5]. In these languages, the user has a need of building models (using the different diagram types) and check their consistency; of querying models to obtain partial models; and of transforming models into other formalisms. All these artefacts can be considered different kinds of views of the system. This necessity has been recently recognised by the OMG by defining a standard language for expressing Queries, Views and Transformations (QVT, see [11]).

A. Corradini et al. (Eds.): ICGT 2006, LNCS 4178, pp. 351–366, 2006.

Our approach for model view management is based on meta-modelling and graph transformation. With meta-modelling, we define the syntax of the complete MVVL. Each diagram type (or viewpoint) has its own meta-model too, which is a part of the complete MVVL meta-model. At the model level, the user builds different *system views* conform to a viewpoint meta-model. System views are merged together in a unique model called *repository*. Triple Graph Transformation Systems (TGTSs) automatically derived from the meta-models perform the merging, allow incremental updates and relate the system views and the repository. They also provide syntactic consistency and change propagation from one view to the others (i.e. they are bidirectional). In addition, it is possible to generate TGTSs modelling different behaviours for view management (e.g. cascading deletion vs. conservative deletion of elements).

We also present *graph query patterns* as a declarative visual query language to obtain *derived views* (in the sense of QVT [11]) from a base model. Starting from the patterns, a TGTS is automatically generated to build the derived view and maintain it consistent with respect to changes in the base model (i.e. derived views are incremental). If the query is predefined by the MVVL designer and later used by a specific kind of user, we call it *audience-oriented view*.

Finally, the system views (or the repository) can be translated into another formalism for dynamic semantics checking, analysis and simulation. We call the target model *semantic view*. The MVVL designer defines the translation by means of a TGTS that establishes correspondences between the elements in the source model and its semantic view. Thus, the results of the analysis can be back annotated and shown in the base model, likely more intuitive for the user.

The main contribution of this paper is the use of a uniform specification of the different kinds of views by means of meta-modelling and TGTSs. Moreover, we propose graph query patterns to specify derived views, together with mechanisms to automatically obtain TGTSs that build the view. In [5] we presented the first steps towards the definition of MVVLs, where we only considered system views. We have extended previous work, as we now consider other kinds of views, and improve system views by allowing configurable behavioural patterns. This work is founded in an extension of the classical notion of Triple Graph Grammars (TGGs) by Schürr [13]. In its original sense, a TGG is a grammar that generates a language of triple graphs, from which triple rules implementing forward or backward translations are derived. In our case, we generate the TGTSs that implement translations and propagation of updates from meta-models or queries. In addition, our TGTSs are formally defined in the double pushout approach (DPO), and extend triple graphs with inheritance and more flexible morphisms in the correspondence graph (see [6] for details).

The paper is organized as follows. Section 2 presents an overview of our formalization of TGTSs. Section 3 shows our approach for defining the syntax of MVVLs and handling system views. In subsection 3.1, we describe several ways of configuring the behaviour of a modelling environment to manage the system views. Section 4 describes graph query patterns and how TGTSs to build the derived view are obtained from them. Section 5 shows how to define a semantic view. Section 6 compares with related research. Finally, section 7 ends with the conclusions and further research. In all sections, we illustrate the concepts with an example of MVVL in the domain of security for web systems.

2   Triple Graph Transformation Systems

TGTSs model transformations of triple graphs, which are made of three separate graphs: source, target and correspondence. As originally defined [13], nodes in the correspondence graph had morphisms to nodes in the source and target graphs. We have extended the notion of triple graph by allowing attributes on nodes and edges. Moreover, the relation between the source and target graphs is more flexible, as we allow morphisms from nodes in the correspondence graph to nodes and edges in the other two graphs, as well as being undefined. Finally, we also provide triple graphs with a typing by a triple type graph (similar to a triple meta-model) which may contain inheritance relations between nodes or edges. We follow the DPO approach [3] for the formalization of triple graph rules (see [6] for details). Here, for space limitations, we only present a brief summary.

Our triple graphs are based on the notion of $E - graph$ [3], which extends regular graphs with node and edge attribution. Attribute values are indeed data nodes, while attributes are edges connecting graph nodes and edges with data nodes. We define a $TriE - graph$ as three E-graphs (source, correspondence and target) and two correspondence functions c_1 and c_2. The correspondence functions are defined from the nodes in the correspondence graph to nodes and edges in the other two graphs. In addition, the functions can be undefined. This is modelled with a special element in the codomain (named "·"). Visually, this is denoted as a correspondence graph node from which some of the correspondence functions are missing (see for example rules in Fig. 7). TriE-graph objects and morphisms form category **TriEGraph**. In order to structure the data values in sorts and make operations available, TriE-graphs are provided with an algebra over a suitable signature, resulting in category **TriAGraph**. Finally, we provide TriAGraphs with a typing by defining a triple type graph (similar to a meta-model triple). This is an attributed triple graph where the algebra is final. The typing is a TriAGraph-morphism from the graph to the type graph. Indeed, attributed typed triple graphs (short ATT-graphs) can be modelled as objects in the slice category **TriAGraph/TriATG**, which we write as **TriAGraph$_{\text{TriATG}}$**. In [6] we extend type graphs and triple graph rules with inheritance for nodes and edges.

Fig. 1 shows an ATT-graph that relates a source Role Based Access Control model [4] (up) and a target Coloured Petri Net [9] (down). Its meta-model triple is not shown in the paper for space constraints, although the part that corresponds to the source type graph is shown on the upper part of Fig. 4.

We manipulate ATT-graphs by means of DPO triple rules. In the DPO approach, rules are modelled using three components: L, K and R. The L component (or left hand side, LHS) contains the elements to be found in the structure (a graph, a Petri net, etc.) where the rule is applied. K (the kernel) contains

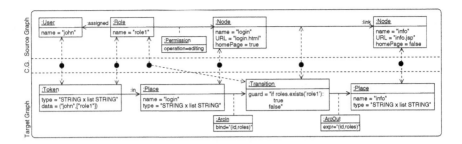

Fig. 1. An Attributed Typed Triple Graph

the elements preserved by the rule application. Finally, R (the right hand side, RHS) contains the elements that should replace the part identified by L in the structure. Therefore, $L - K$ are the elements that should be deleted by the rule application, while $R - K$ are the elements that should be added. Note that DPO transformation has been lifted to work not only in the **Graph** category, but also with any (weak) adhesive HLR category [3]. In [6] we show that category **TriAGraph$_{\text{TriATG}}$** is an adhesive HLR category. Therefore, in our case, L, K and R are ATT-graphs.

In addition, we provide triple rules with a set of application conditions that restrict their applicability. An application condition $c = (X, X \xrightarrow{y_i} Y_i)$ has a premise ATT-graph X, a set of consequent ATT-graphs Y_i, and morphisms y_i from X to each Y_i. In order to apply a rule, if a match of X is found in the host ATT-graph, then a match of some Y_i has also to be found. If the application condition does not have any consequent ATT-graph, finding a match of the premise forbids the rule application. This is a special case of condition called *negative application condition* (NAC). On the contrary, if the premise is isomorphic to the LHS, then this is a *positive application condition* (PAC). We use a shortcut notation for application conditions: the subgraph of L (resp. X) that does not have an image in X (resp. Y_i) is isomorphically copied into X (resp. Y_i) and appropriately linked with their elements.

Fig. 2 shows a triple rule. It creates a place in the target graph (lower part) and relates it with an existing node in the source graph (upper part). The NAC forbids the rule application if the node is related to a place. The K component is omitted for clarity. It contains the common elements of LHS and RHS (i.e. node labelled "1"). We will use such notation throughout the paper.

3   Multi-View Visual Languages: System Views

MVVLs are made of a set of diagram types, each one of them defined by its own meta-model and dealing with a different viewpoint of the system [5]. However, all these separate definitions are based on a unique meta-model that relates their abstract syntax concepts. This is for example the approach of UML2.0. The different viewpoints may overlap in this unique meta-model. It is key to

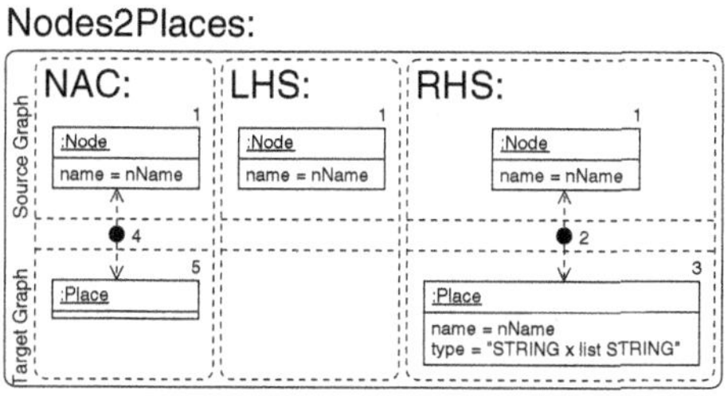

Nodes2Places:

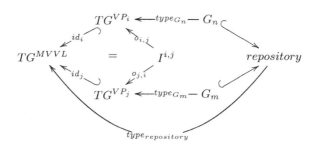

Fig. 2. An Attributed Typed Triple Graph Rule

identify the overlapping parts for each pair of viewpoints in order to be able to maintain them coherent through their common elements. Fig. 3 expresses this situation in categorical terms. Thus, a MVVL is defined by means of an attributed type graph TG^{MVVL} (i.e. its meta-model). The different viewpoints TG^{VP} are inclusions of it, although in a more general approach they can be any function. For each two viewpoints TG^{VP_i} and TG^{VP_j}, the overlapping part $I^{i,j}$ is calculated as the pullback of its respective type graphs and TG^{MVVL}. Thus, $id_i \circ o_{i,j} = id_j \circ o_{j,i}$. At the model level, the user builds *system views* conform to some viewpoint (i.e. a typing morphism exists from each system view to a viewpoint). Note how there might be more than one system view for the same viewpoint. In order to guarantee syntactic consistency, a *repository model* is built by merging all the system views. The repository is the colimit of the views, and there is a typing morphism from it to the MVVL type graph.

$$TG^{MVVL} = \quad TG^{VP_i} \xleftarrow{type_{G_n}} G_n \quad repository$$

Fig. 3. Multi-View Visual Language Definition in Categorical Terms

In our approach, the merging operation is performed by TGTSs automatically derived from the meta-model information. Note how updating the repository (because of a system view modification) may leave other views in an inconsistent state. At this point, other automatically generated TGTSs update the rest of the views. The identification of the overlapping of each two viewpoints helps to minimize the rules that must be tried in this change propagation. In this way, we have TGTSs that propagate changes in the two directions in consecutive steps: first from the views to the repository, and then the other way round (if necessary). This is similar to the Model-View-Controller pattern. For static semantics consistency the MVVL designer may provide additional triple rules. In this way, both syntax and static semantics can be checked in a uniform way.

In order to illustrate these concepts, as well as the ones presented in the following sections, we introduce a case study for modelling a Role Based Access Control (RBAC) [4] for web systems[1]. Its meta-model is shown in the upper part of Fig. 4. Briefly, a web application is made of nodes with a name and a URL, where one of the nodes is the home page. Navigation between nodes is modelled by means of relation "link". In addition, roles can be defined with a set of permissions for accessing nodes. Roles are nested in hierarchies through relation "contains", and inherit all permissions of reachable roles through such relation. Finally, users can be assigned a set of roles, from which they obtain the permissions for interacting with the system.

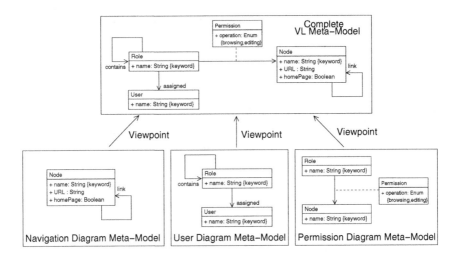

Fig. 4. Definition of a Role Based Access Control for Web-Based Systems

This meta-model comprises the structure of the web system as well as its security policy. However, it is more suitable to handle each aspect in separate diagrams. Thus, we specify three viewpoints on the MVVL meta-model. The first one (*Navigation Diagram*) is used for specifying the web structure, and it only contains nodes and links. The second one (*User Diagram*) is used for specifying role hierarchies and their assignment to users. It contains roles, users, and relations "contains" and "assigned". The third viewpoint (*Permission Diagram*) allows assigning permissions to roles. It contains roles, nodes and permissions. Note how in this view we are only interested in identifying the node, therefore only its attribute "name" is relevant.

---

[1] We do not consider the internal structure of web pages for simplicity.

From this definition, a TGTS is automatically generated between each viewpoint and the repository. Their purpose is building the repository model from the system views. The transformation systems contain the rules shown in Fig. 5 for each concrete class and association[2] in the viewpoint. We only show the rules for class "Role". The first creation rule adds a role to the repository if, given a new role in a view, it does not exist in the repository yet (NAC1). NAC2 is needed to prevent creating a new role in the repository when changing the name of an existing role in the view (i.e. with an associated role in the repository). The second creation rule relates a new role in a view with the same role in the repository. This means that the role was previously created and added to the repository from another view. Attribute "refcount" counts how many times a role appears in different views. When a role is added to the repository, the counter is set to 1; each time the same role is added to any view, the counter is incremented. The first deletion rule detects when a role is removed from a view (i.e. the correspondence function to the view is undefined for a role in the repository), and decrements the "refcount" attribute. When the counter reaches zero, this means that the role does not appear in any view, so it is removed from the repository by the second deletion rule. Finally, the editing rule propagates changes in the attributes from the roles in the views to the corresponding roles in the repository.

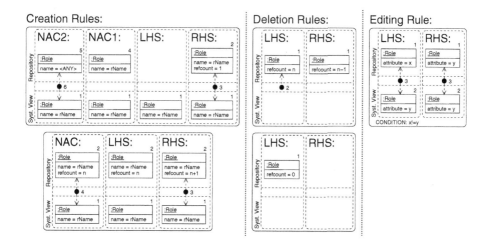

Fig. 5. Triple Graph Rules for Building the Repository

Other TGTSs (one for each viewpoint) are automatically generated which propagate changes in attribute values from the repository to the other system views. These are made of just one kind of rule, like the one shown in Fig. 6.

---

2 An association is not unambiguously identified by its source and target. Thus, we relate associations in views with the corresponding ones in the repository for change propagation. This is possible as we allow correspondence functions to edges [6].

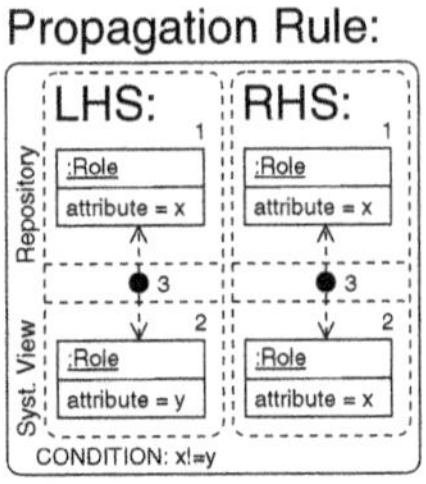

Propagation Rule:

Fig. 6. Triple Graph Rules for Change Propagation

These transformation systems are applied only when the repository has changed due to the execution of one of the former TGTSs.

3.1   Configurable Behavioural Patterns

The presented TGTSs determine the behaviour of the MVVL modelling tool. However, a different behaviour can be more appropriate for a given MVVL. For this reason, we provide a catalogue of different behavioural patterns to configure the behaviour of the modelling tool. Different sets of automatically generated rules are added to the consistency TGTSs depending on the desired behaviour.

For example, the deletion rules in Fig. 5 perform a conservative deletion (i.e. a role is deleted from the repository only when the user has deleted all its instances from the views). On the contrary, cascading deletion implies that when an element is removed from a view, it is also removed from any other view and the repository. Such behaviour can be provided by replacing the previous deletion rules by the ones shown in Fig. 7. The first two rules belong to the TGTS for building the repository. The first one detects when a role has been removed from a view (i.e. the correspondence function to the view for the role in the repository is undefined), and then removes it from the repository. The dangling condition forbids the rule application when the role has incoming or outgoing relations. The second rule handles the deletion of one of such relations. Similar rules are generated for each possible incoming and outgoing relation to a role. Note how the rule is applied to the triple graph relating a particular view and the repository; therefore, correspondence relations from the same node to elements in other views are not part of the graph.

Other rules propagate the deletion of elements from the repository to the other views. Some of them are shown to the right of Fig. 7. The first rule deletes a role from a view if it is not in the repository. The dangling condition forbids the rule application if the role has some incoming or outgoing relation. The following rule (and other similar ones for each possible type of relation) handles this. It can be noted how this behavioural pattern does not need attribute "refcount".

The presented TGTSs are asynchronous. They are executed when the user finishes editing a system view and validates the changes. In addition, there are also synchronous behavioural patterns that execute a TGTS in response to a user

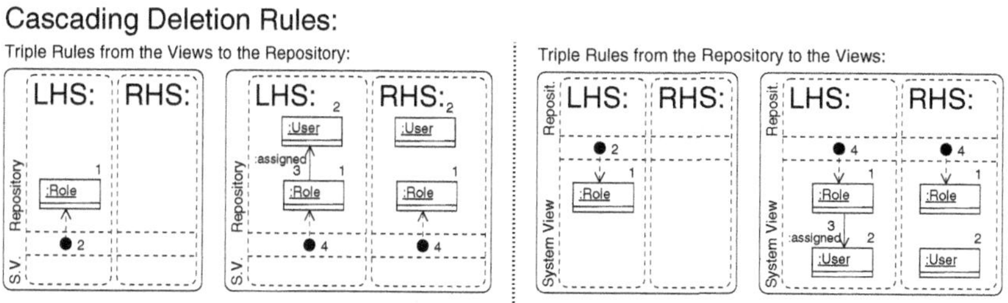

Cascading Deletion Rules:

Fig. 7. Cascading Deletion by means of Triple Graph Rules

action (e.g. a creation). One example is the behavioural pattern for intelligent creation. This creates one rule for each type in the complete MVVL meta-model that is immediately executed after creating a new element of such type in a view and specifying its keyword. The rule copies the value of the attributes from the same element in the repository (if exists) to the element in the view.

4   Derived and Audience-Oriented Views

In addition to adding information to the system, there is a necessity of extracting information from it. In this way, a *derived view* is defined as a sub-model that contains part of another model, called *base model*. The users of a modelling tool can define derived views. However, there is also the possibility for the MVVL designer to predefine derived views oriented to certain type of final user. We call them *audience-oriented views*. For both kinds of views, we propose the use of a kind of declarative, visual queries called *graph query patterns*. These are evaluated on a base model G (a system view or the repository) to obtain the derived view V^Q. A query pattern $Q = (TG^Q, \{(P_i^Q, P_i)\}_{i \in I}, \{(N_j^Q, N_j)\}_{j \in J})$ is made of:

- a meta-model TG^Q. It is a restriction of the base model's meta-model TG.
- a set of positive restrictions P_i. They are patterns that have to be present in the base model for an element to be included in the derived view. P_i^Q contains the element of P_i to which the restriction is applied. Whereas P_i is typed over TG, P_i^Q is typed over TG^Q. Several positive restrictions applied on the same type have a meaning of "or"; if the restrictions apply on different types, they have a meaning of "and".
- a set of negative restrictions N_j. They are patterns that must not be fulfilled by the elements included in the derived view. As before, a subgraph N_j^Q contains the element where the restriction is applied. Several negative restrictions applied on the same or different types have a meaning of "and" (all have to be fulfilled).

Fig. 8 shows a diagram with a query pattern evaluated on a base model G (the typing of the restriction graphs has been omitted for clarity). In a first step, the pullback object V^G of $id_{TG^Q}: TG^Q \rightarrow TG$ and $type_G: G \rightarrow TG$ is calculated (square (1) in the figure). V^G is the restriction of G by meta-model TG^Q. In a second step, graph V^G is further restricted to take into account the restrictions of the query pattern. Thus, for each match p_{il} of a positive restriction P_i, the element identified by P_i^Q has to be included in V^Q, such that square (2) commutes[3]. Moreover, if a matching n_{jk}^Q from a negative restriction N_j^Q is found on V^Q, then no matching n_{jk} must be found from N_j to G such that square $id_{V^G} \circ id_{V^Q} \circ n_{jk}^Q = n_{jk} \circ id_{N_j^Q}$ commutes. Thus, for an element x with type $type_{V^Q}(x)$ to be included in V^Q, it has to fulfill some of the positive restrictions and all the negative ones defined in the query pattern for such type[4].

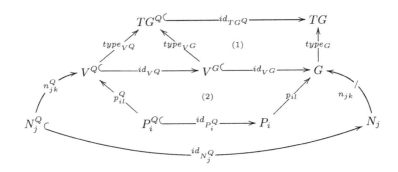

Fig. 8. Query Pattern Evaluated on graph G to obtain the Derived View V^Q

As an example, Fig. 9(a) shows a graph query pattern to be executed on the repository model. It defines a derived view, which should contain nodes and links (modelled by graph TG^Q in the query pattern). The derived view should include nodes for which no role is allowed to access (negative restriction N_1), and which are source or target of a link (positive restrictions P_1 and P_2). That is, the derived view contains those nodes that are not isolated in the navigation design but for which nobody has been granted access permission. On the other hand, Fig. 9(b) contains a graph query pattern defining an audience-oriented view with the role hierarchy extracted from the repository. In this case, it is enough to express the TG^Q component, since no additional restriction is imposed.

In order to evaluate a graph query pattern on a base model, a TGTS is automatically generated from the pattern. This TGTS creates the derived view, as well as correspondences between its elements and the base model elements. Afterwards, if the base model changes, the TGTS also propagates the changes to the derived view, taking into account the negative and positive restrictions.

---

[3] In general, P_i^Q is not the pullback object of TG^Q, P_i and TG.

[4] By now, we only allow subgraphs P_i^Q and N_j^Q to contain either a node for specifying restrictions on the node, or an arrow between two nodes for specifying restrictions on the arrow. Restrictions that apply to more complex graphs are up to future work.

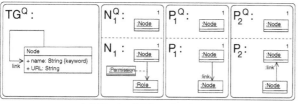

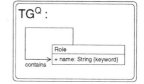

(a) Derived View Definition.   (b) Audience-Oriented
                                      View Definition.

Fig. 9. Graph Query Patterns for the Example

Formally, given a query pattern Q to be applied on a base model typed over TG, the triples rules generated for it can be expressed as $TGTS(Q, TG) = \{C^N, C^E, E^N, E^E, D^N, D^E\}$. Set C^N (resp. C^E) contains the rules that copy the nodes (resp. edges) of the types in TG^Q from the base model to the derived view. Sets E^N and E^E contain the rules that copy the attributes from nodes and edges in the base model to the derived view. Finally, D^N and D^E contain rules that delete nodes and edges from the derived view when they are removed from the base model, or when they (or their context) change and are not consistent with the query pattern restrictions. Positive restrictions in the query pattern are transformed into PACs of the rules in C^i, and into NACs of additional deletion rules in D^i. More precisely, an extra deletion rule is added to D^i for each set of positive restrictions applied on the same type. On the contrary, negative restrictions are transformed into NACs of the rules in C^i, and into PACs of additional deletion rules in D^i. This time one deletion rule is created for each negative restriction, independently of the type where the restriction is applied. Due to the simplicity of our rules, we can easily translate the graph constraints into application conditions [7].

Fig. 10 shows the derived rules for class "Node" from the query pattern in Fig. 9(a). The lower part of the rules corresponds to the base model, and the upper part to the derived view to be created. Creation rules, as the one in the figure, always contain the LHS, RHS and $NAC1$ by default. In addition, for this example, the positive restrictions P_1 and P_2 are transformed into the application condition $(X3, X3 \rightarrow Y3_1, X3 \rightarrow Y3_2)$. Thus, a node is added to the derived view only if it is source $(Y3_1)$ or target $(Y3_2)$ of a link. Besides, the negative restriction N_1 is transformed into the application condition $NAC2$. Thus, a node is not added to the view if some role can access it. The editing rule simply copies the value of the attributes from the nodes in the repository to the nodes in the derived view.

Creation and editing rules are enough for building the derived view. However, a subsequent change in the base model may produce the addition of an element in the view if it fulfills the query pattern; or its deletion if it does not fulfill the query pattern anymore. The former is taken into account by the creation and editing rules. The latter is provided by the deletion rules. The first deletion rule in Fig. 10 removes a node from the view if it does not appear in the base

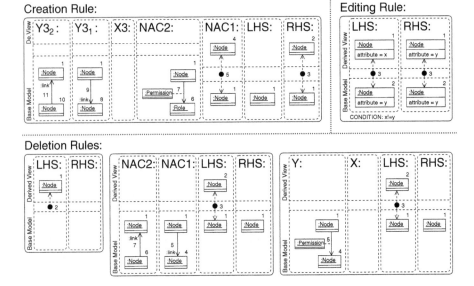

Fig. 10. Triple Graph Transformation System Derived from a Graph Query Pattern

model (i.e. the correspondence function to the base model is undefined). The second rule is derived from the positive restrictions P_1 and P_2. In this way, if a node is neither source nor target of a link, but it appears in the view, it has to be removed from it. The third rule is derived from the negative restriction N_1. Note how a change in any system view is propagated by the consistency TGTSs (Fig. 5) to the repository, and from there, to the derived view by this TGTS.

Fig. 11 shows the derived view that results from the application of the query pattern in Figure 9(a) to a repository model. It contains all the repository nodes except "login" (since it does not satisfy the negative restriction, that is, a role has an editing permission on it) and "admin" (since it does not satisfy the positive restrictions, that is, it is neither source nor target of a link). The links between the nodes are also copied to the view, since they were specified in the TG^Q component of the query pattern. No other types are copied.

5   Semantic Views

Semantic views are parts of the system expressed in other formalism for dynamic semantics checking, analysis and simulation. With this purpose, the MVVL designer can define a TGTS to generate the target model (or semantic view) from a source model (usually the repository, but also audience-oriented and system views). This allows keeping correspondences between the elements of both models, in such a way that the results of analysing the semantic model could be back annotated to the source model.

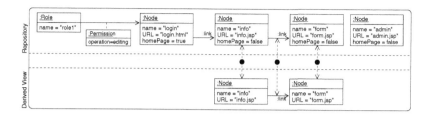

Fig. 11. Derived View from a Repository Model

For example, rules in Figs. 2 and 12 form a TGTS that generates a semantic view of the repository by translating it into a Coloured Petri Net (CPN) [9]. Thus, source graphs in the rules (up) have to conform to the complete MVVL meta-model shown in Fig. 4. Target graphs (down) have to conform to a CPN meta-model, not shown in the paper for space constraints. This contains places with a type, transitions with a guard, tokens with a type and data value, and arcs from places to transitions and the other way round (the former with a binding of the token data values to variables, and the latter with an expression that is evaluated on such data values to change its value). There is also a meta-model for the correspondence graph, which is not shown.

Rule *Users2Tokens* creates a token for each user. The token has type $STRING$ $\times list\ STRING$ and stores the user name and a list with its assigned roles (initially empty). The NAC forbids multiple applications of the rule for the same user. Next, rule *Assignments2DataValues* builds such list of assigned roles. Thus, each role assigned to a user is added to the second component of the data value of the corresponding token. Rule *Nodes2Places* in Fig. 2 creates a place for each node in the source model. Its type is $STRING \times list\ STRING$, since it will contain tokens of such type. Rule *UsersAtHome* puts tokens into the place that corresponds to the homepage. Rule *Links2Transitions* translates each link between two nodes into a transition between the places that correspond to the nodes. The incoming arc to the transition binds the type of tokens to the variables *id* and *roles*. Their value does not change when the transition is fired. However, the transition has a guard that forbids tokens without certain permissions to pass through. Note how, in this rule we associate nodes of type *Transition* with edges of type *link*. Rule *Permissions2Guards* builds the guard expression. Thus, each permission given to a role to access a node is transformed into a condition that is evaluated to true in the guard. If a token has one of the roles allowed by the transition, it can pass through. Finally, rule *InheritedPermissions2Guards* allows inheritance of permissions. Thus, if a role contains another role that can access certain node, then the former can access it as well. The rule modifies the guard of the transition in order to allow the container access the place related to the node. Note how, this TGTS has the potential to be incremental by adding rules for creation, deletion and edition (as previously done for system views) for each element in the source model meta-model.

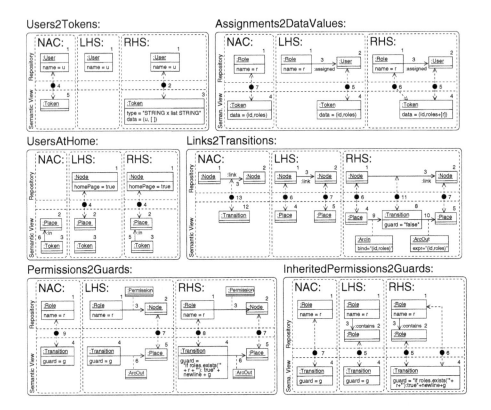

Fig. 12. Semantic View Definition by means of a Triple Graph Transformation System

Fig. 1 shows the ATT-graph resulting from the application of this TGTS to the source model in the upper part. Thus, the lower model is its CPN semantic view. It is possible to analyse the CPN, for example, to check the navigation paths for a certain user or to find unreachable nodes for any user.

6   Related Work

TGTSs are a natural approach to handle views and model transformations. For example, [8] proposes TGGs for view creation. They work with views that can be considered model transformations, similar to our semantic views. They do not consider neither system nor derived views. Moreover, our formalization of triple rules [6] is more expressive as we allow attributes on nodes and edges, more flexible correspondence functions, application conditions and inheritance.

This work is also related to views in database systems [14]. These are virtual relations over a set of real data that are made visible to the user with the purpose of hiding information, or presenting such information in a more adequate way. Such views are defined as the result of a query (in a similar way as our derived views), and present problems for updating the real data as a result of a change in the view [10]. Our work is also related to the problem of consistency of replicated and fragmented data in distributed databases [14]. However, our concept of repository (i.e. we have a centralized control) and the model view controller approach we follow permits an easy solution to the consistency problem.

There are other approaches in the literature to express queries with graphs. For example, in VIATRA [1], queries on graphs can be expressed by generalized (recursive) graph patterns, which may contain a nesting of positive and negative patterns of arbitrary depth. Rensink [12] showed that this possibility is indeed equivalent in expressive power to first order logic. However, incremental transformations are not supported in VIATRA. Other possibility for queries is to use a textual notation, such as OCL [15]. We believe using a graphical approach makes the expression of complex structural patterns easier (in OCL it has to be coded by navigating between the relations), and may be more appropriate for non-computer scientists (patterns use the graphical notation of the given VL). On the other hand, OCL is much more expressive than our patterns (with only one level of nesting), and allows for example to express the absence of cycles. Graph patterns have been discussed extensively in the literature [3], especially its connection with application conditions for rules [7].

The QVT [11] specification includes a facility for queries (in addition to OCL), the *helper*, which allows combining blocks of expressions in a procedural way. Besides, it is also possible to define transformation rules to extract a derived view from a base model in the way we have presented here. Nonetheless, our approach is higher-level and declarative: starting from visual patterns, the rules that perform the transformation are automatically generated. By using QVT, the transformation to extract the view has to be coded by hand. As in our approach, QVT provides a mechanism (similar to the correspondence graph in triple graphs) for leaving traces (mappings) between the source and the target model, which allows a bidirectional update.

7   Conclusions and Future Work

In this paper we have proposed an approach for the uniform specification and handling of the different kind of views in MVVLs. The approach is based on meta-modelling for describing the syntax of the MVVL and its different diagram types. From the meta-models, triple rules are derived in order to build a unique model from the different system views that the user inputs, and to keep them consistent. Several alternative rules allow configuring the behaviour of the MVVL modelling environment. Derived and audience-oriented views are specified through graph query patterns. From these, a TGTS is generated that builds the derived view and keeps it consistent with the base model. Semantic views result from the transformation of a base model into another formalism. Altogether, our approach makes emphasis on using visual, declarative techniques (meta-models, patterns), from which TGTSs are derived. However, for the case of semantic views, the TGTS has to be specified by the MVVL designer.

There is an ongoing implementation in the meta-modelling tool AToM^3 [2] [5]. Up to now, it is possible to define MVVLs and automatically generate the consistency triple rules for several behavioural patterns. Besides, it is also possible for the MVVL designer to define extra static semantics consistency rules as well as semantic views. It is up to future work to implement the graph query patterns. In addition to this, we are studying ways of improving the expressivity of the graph query patterns, and of DPO rules for model transformation.

Acknowledgements. This work has been sponsored by the Spanish Ministry of Science and Education, projects TSI2005-08225-C07-06 and TSI2004-03394. The authors would like to thank the referees for their useful comments.

References

1. Balogh, A., Varró, D. 2006. *Advanced Model Transformation Language Constructs in the VIATRA2 Framework.* To appear in ACM SAC'06.
2. de Lara, J., Vangheluwe, H. 2002. *AToM^3: A Tool for Multi-Formalism Modelling and Meta-Modelling.* FASE'02, LNCS 2306, pp.: 174-188. Springer.
3. Ehrig, H., Ehrig, K., Prange, U., Taentzer, G. 2006. *Fundamentals of Algebraic Graph Transformation.* Monographs in Theoretical Computer Science. Springer.
4. Ferraiolo, D., Cugini, J., Kuhn, D. R. 1995. *Role-Based Access Control (RBAC): Features and Motivations.* Computer Security Applications, pp.: 241-248. Springer.
5. Guerra, E., Díaz, P., de Lara, J. 2005. *A Formal Approach to the Generation of Visual Language Environments Supporting Multiple Views.* Proc. IEEE VL/HCC. pp.: 284-286.
6. Guerra, E., de Lara, J. 2006. *Attributed Typed Triple Graph Transformation with Inheritance in the Double Pushout Approach.* Tech. Rep. Universidad Carlos III. Available at http://www.ii.uam.es/~jlara/investigacion/techRep_UC3M.pdf
7. Heckel, R., Wagner, A. 1995. *Ensuring consistency of conditional graph rewriting - a constructive approach.* Proc. SEGRAGRA'95, ENTCS Vol 2.
8. Jakob, J., Schürr, A. 2006. *View Creation of Meta Models by Using Modified Triple Graph Grammars.* Proc. GT-VMT'06, to appear in ENTCS (Elsevier).
9. Jensen, K. 1992. *Coloured Petri Nets. Basic Concepts, analysis methods and practical use (vol. 1).* EATCS Monogr. on Theoretical Computer Science. Springer-Verlag.
10. Kozankiewicz, H., Subieta, K. 2004. *SBQL Views - Prototype of Updateable Views.* Proc. Advances in Databases and Information Systems (ADBIS).
11. QVT specification by OMG: http://www.omg.org/docs/ptc/05-11-01.pdf.
12. Rensink, A. 2004. *Representing First-Order Logic Using Graphs.* Proc. ICGT'04, LNCS 3256, pp.: 319-335. Springer.
13. Schürr, A. 1994. *Specification of Graph Translators with Triple Graph Grammars.* In LNCS 903, pp.: 151-163. Springer.
14. Silberschatz, A., Korth, H., Sudarshan, S. 2005. *Database System Concepts, 5th Edition.* McGraw Hill.
15. Warmer, J., Kleppe, A. 2003. *The Object Constraint Language: Getting Your Models Ready for MDA, 2nd Edition.* Pearson Education. Boston, MA.

Graph Transformation in Constant Time

Mike Dodds and Detlef Plump

Department of Computer Science
The University of York

Abstract. We present conditions under which graph transformation rules can be applied in time independent of the size of the input graph: graphs must contain a unique *root label*, nodes in the left-hand sides of rules must be reachable from the root, and nodes must have a bounded outdegree. We establish a constant upper bound for the time needed to construct all graphs resulting from an application of a fixed rule to an input graph. We also give an improved upper bound under the stronger condition that all edges outgoing from a node must have distinct labels. Then this result is applied to identify a class of graph reduction systems that define graph languages with a linear membership test. In a case study we prove that the (non-context-free) language of balanced binary trees with backpointers belongs to this class.

1 Introduction

A major obstacle to using graph transformation as a practical computation mechanism is its complexity. Finding a match for a rule r in a graph G requires time $O(\text{size}(G)^{\text{size}(L)})$, where L is the left-hand graph of r. This is too expensive for many applications, even if r is fixed (meaning that $\text{size}(L)$ is a constant). For example, Fradet and Le Metayer [8] and later Dodds and Plump [4] have proposed to extend the C programming language with graph transformation rules to allow the safe manipulation of pointers. To make such a language acceptable for programmers, rules must be applicable in constant time.

In [4], constant-time rule application is achieved by using a form of *rooted* graph transformation which is characterized by the presence of unique root nodes in rules and host graphs. These roots serve as entry points for the matching algorithm and ensure, under further assumptions on left-hand sides and host graphs, that all matches of a rule can be found in time independent of the size of the host graph. The purpose of this paper is twofold: to develop a general approach to rooted graph transformation in the setting of the double-pushout approach, and to demonstrate the expressive power of rooted graph transformation in a case study on graph recognition.

Our contributions are as follows. In Section 3, we present two axiomatic conditions each of which guarantees that rules can be applied in time independent of the size of host graphs. The first condition requires that graphs have a unique root, nodes in left-hand sides of rules are reachable from the root, and nodes in host graphs have a bounded outdegree. Under this condition, we establish a constant upper bound for the time needed to construct all graphs resulting from

A. Corradini et al. (Eds.): ICGT 2006, LNCS 4178, pp. 367–382, 2006.
c Springer-Verlag Berlin Heidelberg 2006

an application of a fixed rule to a host graph. The second condition requires, in addition, that all edges outgoing from a node have distinct labels. We prove that this leads to a greatly reduced upper bound. Then, in Section 4, we introduce rooted *graph reduction specifications* for defining graph languages. We identify a class of graph reduction specifications that come with a linear membership test. In a case study we prove that the non-context-free language of balanced binary trees with backpointers belongs to this class. This is remarkable as there exist context-free graph languages (definable by both edge replacement grammars and node replacement grammars) whose membership problem is NP-complete.

Our approach to rooted graph transformation is similar to Dörr's approach [5] in that he also requires unique root nodes to ensure constant-time application of rules. Instead of limiting outdegree, he aims at avoiding so-called strong V-structures in host graphs. This makes the approaches incomparable in terms of the strength of their assumptions. (We mention a few separating properties in Section 5.) Another major difference is that in [5], all rules are assumed to belong to a graph grammar which produces all host graphs (which allows to analyse the grammar for the impossibility of generating V-structures). We don't require any generation mechanism for host graphs. A final difference is that [5] is based on the algorithmic approach to graph transformation while we work in the setting of the double-pushout approach.

2   Graphs, Rules and Derivations

We review basic notions of the double-pushout approach to graph transformation, using a version that allows unlabelled nodes [12]. Rules with unlabelled nodes allow to relabel nodes and, in addition, represent sets of totally labelled rules because unlabelled nodes in the left-hand side act as placeholders for arbitrarily labelled nodes.

A *label alphabet* is a pair $\mathcal{C} = \langle \mathcal{C}_V, \mathcal{C}_E \rangle$ of finite sets $\mathcal{C}_V$ and $\mathcal{C}_E$. The elements of $\mathcal{C}_V$ and $\mathcal{C}_E$ serve as node labels and edge labels, respectively. For this section and the next, we assume a fixed alphabet $\mathcal{C}$.

A *graph* $G = \langle V_G, E_G, s_G, t_G, l_G, m_G \rangle$ consists of a finite set V_G of *nodes* (or *vertices*), a finite set E_G of *edges*, source and target functions $s_G, t_G \colon E_G \to V_G$, a partial node labelling function $l_G \colon V_G \to \mathcal{C}_V{}^1$ and an edge labelling function $m_G \colon E_G \to \mathcal{C}_E$. The *size* of G, denoted by $|G|$, is the number of its nodes and edges. The *degree* of a node v, denoted by $\deg_G(v)$, is the number of edges incident with v. The *outdegree* of a node v, denoted by $\text{outdeg}_G(v)$, is the number of edges with source v. We write $\text{outlab}_G(v)$ for $m_G(s_G{}^{-1}(v))$, the set of labels of all edges outgoing from v. A node v' is *reachable* from a node v if $v = v'$ or if there are edges $e_1, \ldots, e_n$ such that $s_G(e_1) = v$, $t_G(e_n) = v'$ and for $i = 1, \ldots, n-1$, $t_G(e_i) = s_G(e_{i+1})$.

A *graph morphism* $g \colon G \to H$ between two graphs G and H consists of two functions $g_V \colon V_G \to V_H$ and $g_E \colon E_G \to E_H$ that preserve sources, targets and

---

[1] The domain of l_G is denoted by $\text{Dom}(l_G)$. We write $l_G(v) = \bot$ to express that node v is in $V_G - \text{Dom}(l_G)$.

labels: $s_H \circ g_E = g_V \circ s_G$, $t_H \circ g_E = g_V \circ t_G$, and $l_H(g_V(v)) = l_G(v)$ for all v in $\mathrm{Dom}(l_G)$. A morphism g is *injective* (*surjective*) if g_V and g_E are injective (surjective); it *preserves undefinedness* if $l_H(g(v)) = \perp$ for all v in $V_G - \mathrm{Dom}(l_G)$. Morphism g is an *isomorphism* if it is injective, surjective and preserves undefinedness. In this case G and H are *isomorphic*, which is denoted by $G \cong H$. Furthermore, g is an *inclusion* if $g(x) = x$ for all nodes and edges x in G. (Note that inclusions need not preserve undefinedness.)

A *rule* $r = \langle L \leftarrow K \rightarrow R \rangle$ consists of two inclusions $K \rightarrow L$ and $K \rightarrow R$ such that (1) for all $v \in V_L$, $l_L(v) = \perp$ implies $v \in V_K$ and $l_R(v) = \perp$, and (2) for all $v \in V_R$, $l_R(v) = \perp$ implies $v \in V_K$ and $l_L(v) = \perp$. We call L the *left-hand side*, R the *right-hand side* and K the *interface* of r.

A *direct derivation* from a graph G to a graph H via a rule $r = \langle L \leftarrow K \rightarrow R \rangle$, denoted by $G \Rightarrow_{r,g} H$ or just $G \Rightarrow_r H$, consists of two natural pushouts[2] as in Figure 1, where $g \colon L \rightarrow G$ is injective.

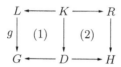

Fig. 1. A direct derivation

In [12] it is shown that for rule r and injective morphism g given, there exists such a direct derivation if and only if g satisfies the *dangling condition*: no node in $g(L) - g(K)$ must be incident to an edge in $G - g(L)$. If this condition is satisfied, then r and g determine D and H uniquely up to isomorphim and H can be constructed (up to isomorphism) from G as follows: (1) Remove all nodes and edges in $g(L) - g(K)$, obtaining a subgraph D'.[3] (2) Add disjointly to D' all nodes and edges from $R - K$, keeping their labels. For $e \in E_R - E_K$, $s_H(e)$ is $s_R(e)$ if $s_R(e) \in V_R - V_K$, otherwise $g_V(s_R(e))$. Targets are defined analogously. (3) For each node $g_V(v)$ in $g(K)$ with $l_L(v) \neq l_R(v)$, $l_H(g_V(v))$ becomes $l_R(v)$.

To keep the complexity considerations below independent of any type system imposed on graphs, we introduce abstract graph classes and rules preserving such classes. A *graph class* is a set $\quad$ of graphs over $\mathcal{C}$. A rule r is $\quad$-*preserving* if for every direct derivation $G \Rightarrow H$, $G \in \quad$ implies $H \in \quad$. We can now state the basic problem we are interested in.

Graph Transformation Problem (GTP)
Given: A graph class $\quad$ and a $\quad$-preserving rule $r = \langle L \leftarrow K \rightarrow R \rangle$.
Input: A graph G in $\quad$.
Output: The set $\{H \mid G \Rightarrow_r H\}$.

---

[2] A pushout is *natural* if it is a pullback, too [12].
[3] D differs from D' in that nodes are unlabelled if they are the images of unlabelled nodes in K that are labelled in L. We do not need D to transform G into H though.

We consider the graphs in $\{H \mid G \Rightarrow_r H\}$ up to isomorphism, which makes this set finite (as there are only finitely many morphisms $L \rightarrow G$). The time complexity of GTP is dominated by the cost of finding the injective graph morphisms $L \rightarrow G$. This is because for each of these morphisms, checking the dangling condition and transforming G into H can be done in time independent of the size of G (assuming a suitable data structure for graphs). This leads us to the core problem to solve.

Graph Matching Problem (GMP)
Given: A graph class $\mathcal{C}$ and a $\mathcal{C}$-preserving rule $r = \langle L \leftarrow K \rightarrow R \rangle$.
Input: A graph G in $\mathcal{C}$.
Output: The set $\{g \colon L \rightarrow G \mid g$ is injective$\}$.

To solve the GMP in general requires time $|G|^{L}$ – better algorithms are not known. If we consider L as part of the input rather than as given, the GMP essentially becomes the *subgraph isomorphism problem* which is NP-complete [11]. (This is the problem to decide whether there exists an injection from L to G. In the worst case, if there is none, this is as expensive as finding all injections.)

3   Rooted Graph Transformation

We present two conditions each of which ensures that the problems GTP and GMP can be solved in time independent of the size of the input graph G. Both conditions put restrictions on the rule r and the graph class $\mathcal{C}$.

Condition I
There are $\varrho \in \mathcal{C}_V$ and an integer $b \geq 0$ such that

(1) L contains a unique ϱ-labelled node from which each node is reachable, and
(2) for every graph G in $\mathcal{C}$,
     (i) there is a unique ϱ-labelled node, and
     (ii) the outdegree of each node is bounded by b.

We call the distinguished node labelled with ϱ the *root*. The next condition differs from Condition I in clause (2)(ii).

Condition II
There exists $\varrho \in \mathcal{C}_V$ such that

(1) L contains a unique ϱ-labelled node from which each node is reachable, and
(2) for every graph G in $\mathcal{C}$,
     (i) there is a unique ϱ-labelled node, and
     (ii) distinct edges outgoing from the same node have distinct labels.

Remark 1 Condition II implies Condition I, which can be seen by choosing bound b as the size of $\mathcal{C}_E$. The converse does not hold in general.

Remark 2 The conditions do not guarantee that r preserves the constraints on   .
To preserve property (2) of Condition I, it suffices that the right-hand side R of r
contains a unique ϱ-labelled node, and that for each node v in K, $\text{outdeg}_R(v) \leq$
$\text{outdeg}_L(v)$. The preservation of property (2) of Condition II is discussed in
Section 4.

Remark 3 Rather than having just one root, one could allow a fixed number
$k \geq 1$ of roots. But this approach can be simulated as follows: one adds to every
graph in   a new root with k outgoing edges that point to the old roots, and
every left-hand side of a rule gets a new root with pointers to the old root(s).

Algorithm 1 below is similar to the matching algorithm described in [5]: it solves
the Graph Matching Problem by constructing the set A of all injections between
the left-hand side of a rule and a host graph. The algorithm starts with the set
A_0 consisting of the partial injections that are defined for the root only. (Under
Condition I, there exists exactly one such morphism). Each iteration of the loop
extends the injections in the previous working set with a single edge and its
target node, or a single edge, until the injections in the set are total. When an
iteration adds some node or edge to the domain of an injection, we speak of the
node or edge being *matched*. A pre-processed enumeration of the edges of the
left-hand side ensures that when an edge is matched, its source must have been
matched in some previous iteration.

In defining the algorithm, we use some extra notions. Given partial functions
$f, f' \colon S \to T$, we write $f \text{ ext } f'$ by Z if $\text{Dom}(f) \supseteq \text{Dom}(f')$, $\text{Dom}(f) - \text{Dom}(f') =$
Z and for each $x \in \text{Dom}(f')$, $f(x) = f'(x)$. A *partial graph morphism* $f \colon G \xrightarrow{par}$
H is a graph morphism from some subgraph of G to H. Given a graph L that
contains a unique ϱ-labelled node from which each node is reachable, an *edge
enumeration* for L is a list of edges $e_1, \ldots, e_n$ such that $E_L = \{e_1, \ldots, e_n\}$ and
for $i = 1, \ldots, n$, $l_G(s_G(e_i)) = \varrho$ or there is some $1 \leq j < i$ with $s_G(e_i) = t_G(e_j)$.

Proposition 1 (Correctness of Algorithm 1). *Algorithm 1 solves the Graph
Matching Problem.*

Proof. We have to show that the returned set A is the set of all total injective
graph morphisms from L to G.

Soundness. We first show that A contains only total injections $L \to G$. By
construction of the sets A_i, it is clear that these consist of partial injections
from L to G, so the same holds for A. It therefore suffices to show that all
morphisms in A are total. Since $e_1, \ldots, e_n$ is an edge enumeration for L, each
edge e_i in L will be considered in the i-th iteration of the loop. At this point
there are three cases: (1) e_i and $t_L(e_i)$ are added to the domain of one or more
morphisms h in A_{i-1}, where h is already defined for $s_L(e_i)$, and A_i becomes
the set of all extended morphisms. (2) e_i is added to the domain of one or more
morphisms h in A_{i-1}, where h is already defined for both $s_L(e_i)$ and $t_L(e_i)$, and
A_i becomes the set of all extended morphisms (3) Matching of e_i fails because
either A_{i-1} is empty or there is no counterpart to e_i in G. Then A_i is empty

Algorithm 1 (Graph Matching Algorithm) The algorithm works for a fixed rule $\langle L \leftarrow K \rightarrow R \rangle$ and an input graph $G \in$, as stated in the Graph Matching Problem, and it assumes an edge enumeration $e_1, \ldots, e_n$ for L.

attach tag to the unique ϱ-labelled node in L
$A_0 \Leftarrow \{h \colon L \xrightarrow{par} G \mid \mathrm{Dom}(h_V) = l_L^{-1}(\varrho) \wedge \mathrm{Dom}(h_E) = \emptyset\}$
for $i = 1$ to n **do**
   if $t_L(e_i)$ is not tagged **then**
      attach tag to $t_L(e_i)$
      $A_i \Leftarrow \{h \colon L \xrightarrow{par} G \mid h \text{ is injective} \wedge \exists h' \in A_{i\ 1} \colon$
           $h_E \text{ ext } h'_E \text{ by} \{e_i\} \wedge h_V \text{ ext } h'_V \text{ by} \{t_L(e_i)\}\}$
   else
      $A_i \Leftarrow \{h \colon L \xrightarrow{par} G \mid h \text{ is injective} \wedge \exists h' \in A_{i\ 1} \colon$
           $h_E \text{ ext } h'_E \text{ by} \{e_i\} \wedge h_V = h'_V\}$
   end if
end for
return $A = A_n$

and hence A will be empty, too. As a consequence, if A is not empty upon termination of the loop, all its morphisms must be defined for all edges in L and their incident nodes. Hence, by the structure of L, they are total morphisms.

Completeness. For $i = 1, \ldots, n$, let L_i be the subgraph of L consisting of the edges $e_1, \ldots, e_i$ and their incident nodes. Also, let L_0 be the subgraph consisting of L's root only. A straightforward induction on i shows that for $i = 0, \ldots, n$,

$$\{h \colon L \xrightarrow{par} G \mid h \text{ is injective and } \mathrm{Dom}(h) = L_i\} \subseteq A_i.$$

Since $L_n = L$ by the structure of L, it follows that A contains all total injections from L to G. $\qquad\square$

Notational convention. In the rest of this section, n always refers to the number of edges in the left-hand side L of the given rule r.

Theorem 1 (Complexity of Algorithm 1). *Under Condition I, Algorithm 1 requires time $\Sigma_{i=0}^{n} b^i$ at most. The maximal size for the resulting set A is b^n.*

Proof. A run of the algorithm involves n iterations of the loop. In each iteration, one of the cases (1) to (3) of the proof of Proposition 1 applies. In finding the maximal running time, case (3) can be ignored as A_i is just set to $\emptyset$.

   Case (1): Both e_i and $t_L(e_i)$ are added to the domain of one or more morphisms h in $A_{i\ 1}$, where h is already defined for $s_L(e)$, and A_i becomes the set of all extended morphisms. By Condition I, there are at most b edges outgoing from $h_V(s_L(e_i))$. It follows $|A_i| \leq b|A_{i\ 1}|$. Hence the maximal time needed to update $A_{i\ 1}$ to A_i is $b|A_{i\ 1}|$.

   Case (2): Only e_i is added to the domain of one or more morphisms h in $A_{i\ 1}$, where h is already defined for both $s_L(e_i)$ and $t_L(e_i)$, and A_i becomes the set of all extended morphisms. For the same reason as in Case (1), the time needed to update A_{i-1} to A_i is $b|A_{i-1}|$.

The initialisation of the algorithm constructs the set A_0 which, by Condition I, contains exactly one morphism and therefore can be constructed in one unit of time (assuming a suitable data structure for graphs in). By the above case analysis, executing the body of the loop takes time $b|A_{i-1}|$ at most. Thus we obtain te following bound for the overall running time:

$$1 + b|A_0| + b|A_1| + \cdots + b|A_{n-1}|.$$

By recursively expanding each term $|A_i|$ to its maximal size, we arrive at the expression

$$1 + b + b^2 + \cdots + b^n = \sum_{i=0}^{n} b^i.$$

This expansion also shows that maximal size of A is b^n.   □

For the rule $r = \langle L \leftarrow K \rightarrow R \rangle$ of the GMP and the GTP, we define size by $|r| = \max(|L|, |R|)$.

Corollary 1 (GTP under Condition I). *Under Condition I, the Graph Transformation Problem can be solved in time $\Sigma_{i=0}^{n} b^i + 4|r|b^n$.*

Proof. Recall from Section 2 that constructing a derivation $G \Rightarrow_r H$ consists of four stages: (1) Finding an injective morphism $L \rightarrow G$ that satisfies the dangling condition. (2) Removing nodes and edges. (3) Inserting nodes and edges. (4) Relabelling nodes. To adapt this to the GTP problem, we extend stage (1) to: Finding the set of all injections $L \rightarrow G$ that satisfy the dangling condition. We then perform stages $(2) - (4)$ for all members of this set.

The dangling condition can be decided for an injection $h \colon L \rightarrow G$ in time $|V_L|$. This is because the condition holds if and only if for all nodes v in $V_L - V_K$, $\deg_L(v) = \deg_G(h(v))$. We assume a graph representation such that the degree of any node can be retrieved in one unit of time, so we can compare $\deg_L(v)$ with $\deg_G(h(v))$ for all nodes v in L in time $|V_L|$.

We construct the set of all injections $L \rightarrow G$ using Algorithm 1 and then complete stage (1) by filtering this set for those morphisms satisfying the dangling condition.

Given a morphism h, it is obvious that stage (2) can be executed in time $|L| - |K|$, and stage (3) can be done in time $|R| - |K|$. Stage (4) requires time $|V_K|$ at most. In the worst case, stages (2) to (4) must be completed for every element in the set A of all injections from L to G. Thus, using Theorem 1, we obtain the time bound

$$\sum_{i=0}^{n} b^i + b^n(|V_L| + |L| - |K| + |R| - |K| + |V_K|).$$

As $|V_L|$, $|L|$, $|R|$, $|K|$ and $|V_K|$ are all bounded by $|r|$, we can estimate the expression from above by $\Sigma_{i=0}^{n} b^i + 4|r|b^n$.   □

Note that according to the GTP and Condition I, n, $|r|$, and b are constants and hence the above time bound is a constant—albeit a possibly large one. The next theorem and the subsequent corollary show that under Condition II, the constant time bounds for the GMP and the GTP decrease from being exponential in n down to being linear in n.

Theorem 2 (Complexity of Algorithm 1 under Condition II). *Under Condition II, Algorithm 1 requires time $n|\mathcal{C}_E| + 1$ at most. The resulting set A contains at most one injection.*

Proof. Under Condition II, outgoing edges from a node must be distinctly labelled. Hence a partial morphism in A_i can be extended by an edge outgoing from a matched node in at most one way. Since $|A_0| = 1$, it follows that $|A_i| \leq 1$ after the i-th iteration of the loop. So in particular $A \leq 1$.

In the time bound $1 + b|A_0| + \cdots + b|A_{n-1}|$ of the proof of Theorem 1, we can therefore replace each term $|A_i|$ with 1. We also replace b with $|\mathcal{C}_E|$, the number of edge labels in the given alphabet. This yields the bound $n|\mathcal{C}_E| + 1$.   □

Corollary 2 (GTP under Condition II). *Under Condition II, the Graph Transformation Problem can be solved in time $n|\mathcal{C}_E| + 4|r| + 1$.*

Proof. By Theorem 2, Algorithm 1 will need time at most $n|\mathcal{C}_E| + 1$ under Condition II. The proof of Corollary 1 shows that applying r for a found morphism can be done in time $|V_L| + |L| - |K| + |R| - |K| + |V_K|$. Similar to the proof of Corollary 1, this results in the overall bound $n|\mathcal{C}_E| + 4|r| + 1$.   □

4   Efficient Recognition of Graph Languages

In this section we apply the results of the previous section to show that graph languages specified by rooted graph reduction systems of a certain form come with an efficient membership test. This is in sharp contrast to the situation for graph grammars where even context-free languages can be NP-complete.

We define graph reduction languages by adapting the approach of [2] to the setting of rooted graph transformation.

Definition 1 (Signature and Σ-graph). A *signature* $\Sigma = \langle \mathcal{C}, \varrho, \text{type} \rangle$ consists of a label alphabet $\mathcal{C} = \langle \mathcal{C}_V, \mathcal{C}_E \rangle$, a *root label* $\varrho \in \mathcal{C}_V$, and a mapping type: $\mathcal{C}_V \rightarrow \mathcal{C}_E$ that assigns to each node label a set of edge labels. A graph G over $\mathcal{C}$ is a *Σ-graph* if it contains a unique ϱ-labelled node, the *root* of G, and if for each node v, (1) $l_V(v) \neq \bot$ implies $\text{outlab}_G(v) \subseteq \text{type}(l_G(v))$ and (2) distinct edges outgoing from v have distinct labels. The set of all Σ-graphs is denoted by $\mathcal{G}(\Sigma)$.

Next we define a class of rules that preserve Σ-graphs.

Definition 2 (Σ-rule). A rule $r = \langle L \leftarrow K \rightarrow R \rangle$ is a *Σ-rule* if L, K and R are Σ-graphs and for each node v in K,

(1) $l_L(v) = \bot = l_R(v)$ implies $\text{outlab}_L(v) = \text{outlab}_R(v)$ and
(2) $l_R(v) \neq \bot$ implies $(\text{outlab}_R(v) \cap \text{type}(l_L(v))) \cup (\text{type}(l_L(v)) - \text{type}(l_R(v))) \subseteq \text{outlab}_L(v)$.

Conditions (1) and (2) ensure that r can add outgoing edges to a node only if the node is relabelled and the edge labels do not belong to the node's old type. Also, outgoing edges of a relabelled node are deleted if their labels do not belong to the node's new type.

Proposition 2 (Σ-rules preserve Σ-graphs). *Let $G \Rightarrow_r H$ such that G is a Σ-graph and r a Σ-rule. Then H is a Σ-graph.*

We now define a "rooted" version of the graph reduction specifications of [2].

Definition 3 (Graph reduction specification). A *graph reduction specification* $S = \langle \Sigma, \mathcal{C}_N, \mathcal{R}, Acc \rangle$ consists of a signature $\Sigma = \langle \mathcal{C}, \varrho, \text{type} \rangle$, a set $\mathcal{C}_N \subseteq \mathcal{C}_V$ of *nonterminal* labels, a finite set $\mathcal{R}$ of Σ-rules and an $\mathcal{R}$-irreducible[4] Σ-graph *Acc*, the *accepting graph*, such that in *Acc* and in all left-hand sides of rules in $\mathcal{R}$, each node is reachable from the root. The graph language specified by S is $L(S) = \{G \in \mathcal{G}(\Sigma) \mid G \Rightarrow_{\mathcal{R}} Acc \text{ and } l_G(V_G) \cap \mathcal{C}_N = \emptyset\}$.

We often abbreviate 'graph reduction specification' by GRS. The following simple example of a GRS specifies cyclic lists as used in pointer data structures.

Example 1 (Cyclic lists). The GRS CL $= \langle \Sigma_{\text{CL}}, \emptyset, \mathcal{R}_{\text{CL}}, Acc_{\text{CL}} \rangle$ has the signature $\Sigma_{\text{CL}} = \langle \{\varrho, E\}, \{p, n\}, \varrho, \{\varrho \mapsto \{p\}, E \mapsto \{n\}\} \rangle$. The accepting graph Acc_{CL} and the rules $\mathcal{R}_{\text{CL}}$ are shown in Figure 2, where the unique ϱ-labelled node is drawn as a small grey node and the label p of its outgoing edge is omitted. Rules are represented by their left- and right-hand sides, the interface consists of the numbered nodes and the root node.

The language $L(\text{CL})$ consists of cyclic lists built up from E-labelled nodes and n-labelled edges, and a distinguished root pointing to any node in the list. For a proof, we have to show soundness (every graph in $L(\text{CL})$ is a cyclic list) and completeness (every cyclic list is in $L(\text{CL})$). Soundness follows from the fact that for every inverse[5] r^{-1} of a rule r in $\mathcal{R}_{\text{CL}}$, and every cyclic list G, $G \Rightarrow_{r^{-1}} H$ implies that H is a cyclic list. For, every reduction $G \Rightarrow Acc$ via $\mathcal{R}_{\text{CL}}$ gives rise to a derivation $Acc \Rightarrow G$ via $\mathcal{R}_{\text{CL}}^{-1}$ and hence G is a cyclic list. Completeness is shown by induction on the number of E-labelled nodes in cyclic lists. The cyclic list with one E-labelled node is Acc_{CL}, which belongs to $L(\text{CL})$. If G is a cyclic list with at least two E-labelled nodes, then there is a unique injective morphism from the left-hand side of either Reduce (if G has more than two E-labelled nodes) or Finish (if G has exactly two E-labelled nodes) to G. Hence there is a step $G \Rightarrow_{\mathcal{R}_{\text{CL}}} H$, and it is easily seen that H is a cyclic list that is smaller than G. Hence, by induction, there is a derivation $H \Rightarrow_{\mathcal{R}_{\text{CL}}} Acc$ and thus $G \in L(\text{CL})$.

---

[4] A graph G is $\mathcal{R}$-*irreducible* if there is no step $G \Rightarrow_{\mathcal{R}} H$.
[5] The *inverse* of a rule is obtained by swapping left- and right hand sides together with the inclusion morphisms.

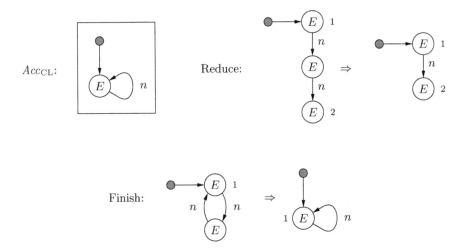

Fig. 2. GRS CL for recognising cyclic lists

Two properties of CL allow to test graphs in $\mathcal{G}(\Sigma_{\mathrm{CL}})$ efficiently for membership in L(CL). Firstly, reduction sequences terminate after a linear number of steps because both rules reduce the size of a graph. Secondly, reduction is deterministic as Σ_{CL}-graphs contain a unique root and the left-hand sides of the two rules do not overlap. Σ_{CL}-graphs can therefore be tested for membership in L(CL) by a straightforward reduction algorithm: apply the rules of CL as long as possible and check if the resulting graph is isomorphic to Acc_{CL}.   □

The properties of CL allowing efficient membership checking can be generalized to obtain a class of GRSs whose languages can be recognised in linear time.

Definition 4 (Linear GRS). A GRS $\langle \Sigma, \mathcal{C}_N, \mathcal{R}, Acc \rangle$ is *linearly terminating* if there is a natural number c such that for every derivation $G \Rightarrow_{\mathcal{R}} G_1 \Rightarrow_{\mathcal{R}} \dots \Rightarrow_{\mathcal{R}} G_n$ on Σ-graphs, $n \leq c|G|$. It is *closed* if for every step $G \Rightarrow_{\mathcal{R}} H$ on Σ-graphs, $G \Rightarrow_{\mathcal{R}} Acc$ implies $H \Rightarrow_{\mathcal{R}} Acc$. A linearly terminating and closed GRS is a *linear* GRS.

The *recognition problem* (or *membership problem*) for GRS languages is defined as follows:
Given: A GRS $S = \langle \Sigma, \mathcal{C}_N, \mathcal{R}, Acc \rangle$.
Instance: A Σ-graph G.
Question: Does G belong to $L(S)$?

Theorem 3 (Linear recognition). *For linear GRSs, the recognition problem is decidable in linear time.*

Proof. Consider a GRS $S = \langle \Sigma, \mathcal{C}_N, \mathcal{R}, Acc \rangle$. Membership of a Σ-graph G in $L(S)$ is tested as follows: (1) Check that G contains no node labels from $\mathcal{C}_N$. (2)

Apply the rules of $\mathcal{R}$ (nondeterministically) as long as possible. (3) Check that the resulting graph is isomorphic to Acc.

Phase (2) of this procedure terminates in a linear number of reduction steps as S is linearly terminating. By Corollary 2, each step can be performed in constant time. So the time needed for phases (1) and (2) is linear. Phase (3) amounts to checking (i) if there is an injective morphism $Acc \to H$, where H is the graph resulting from the reduction of G, and (ii) if $|Acc| = |H|$. Part (i) requires the same time as the Graph Matching Problem under Condition II (note that Acc is a fixed Σ-graph) and hence, by Theorem 2, can be done in constant time. It follows that phase (3) requires only constant time.

The procedure is correct by the fact that S is closed: if $G \Rightarrow_{\mathcal{R}} H$ such that H is $\mathcal{R}$-irreducible and $H \not\cong Acc$, then there is no derivation $G \Rightarrow_{\mathcal{R}} Acc$. This is shown by a simple induction on the length of $G \Rightarrow_{\mathcal{R}} H$.   □

To demonstrate the expressive power of linear GRSs, we show that the non-context-free graph language of *balanced binary trees with back-pointers*, BBTBs for short, can be specified by a linear GRS.[6] A BBTB consists of a binary tree built up from nodes labelled B, U and L such that all paths from the tree root to leaves have the same length. Each node has a back-pointer to its parent node, the back-pointer of the tree root is a loop. Nodes labelled with B have two children to which edges labelled l and r are pointing; nodes labelled with U have one child to which a c-labelled edge points; nodes labelled with L have no children. In addition to this tree, a BBTB has a unique root node whose outgoing edge points to any node in the tree.

The GRS BB $= \langle \Sigma_{\mathrm{BB}}, \{B', U'\}, \mathcal{R}_{\mathrm{BB}}, Acc_{\mathrm{BB}} \rangle$ is shown in Figure 3, where $\mathrm{type}(B) = \mathrm{type}(B') = \{l, r, b\}$, $\mathrm{type}(U) = \mathrm{type}(U') = \{c, b\}$ and $\mathrm{type}(L) = \{b\}$. Note that B' and U' are nonterminal labels. As in Example 1, we draw the root of a BBTB (not to be confused with the tree root) as a small grey node and omit the label of its unique outgoing edge. We also omit the label b of back-pointers and draw them as dashed edges.

Proposition 3 (Correctness). L(BB) *is the set of all balanced binary trees with back-pointers.*

The proof of Proposition 3 is given in the Appendix.

Proposition 4 (Linearity). *GRS* BB *is linearly terminating: the length of any derivation* $G \Rightarrow_{\mathcal{R}_{\mathrm{BB}}} H$ *on* Σ_{BB}-*graphs is at most* $|G| + |V_G|$. BB *is also closed and hence is a linear GRS.*

Proof. For every Σ_{BB}-graph G, define $T(G) = |G| + |l_G^{-1}(\mathcal{C}_V - \mathcal{C}_N)|$ where $|l_G^{-1}(\mathcal{C}_V - \mathcal{C}_N)|$ is the number of nodes not labelled with B' or U'. We show that for every step $G \Rightarrow_{\mathcal{R}_{\mathrm{BB}}} H$ on Σ_{BB}-graphs, $T(G) > T(H)$. This implies the bound in the proposition since $|l_G^{-1}(\mathcal{C}_V - \mathcal{C}_N)| \leq |V_G|$.

---

[6] The language of BBTBs is not context-free in the sense of either hyperedge replacement grammars [6] or node replacement grammars [7].

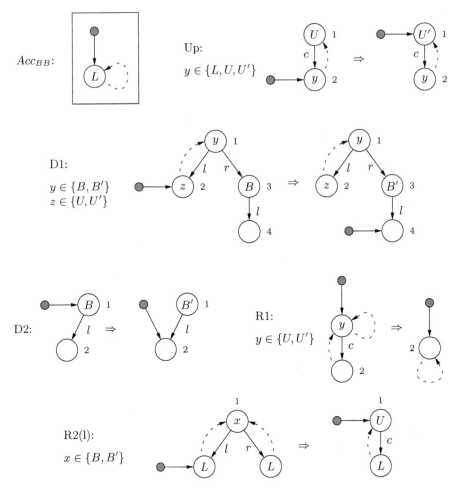

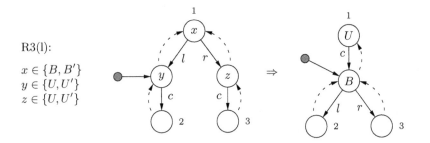

R3(r): as R3(l), but with the left-hand root pointing to the z-node

Fig. 3. GRS BB for recognising balanced binary trees with back-pointers

Rules Up, D1 and D2 preserve the size of graphs but decrease the number of terminally labelled nodes, hence they decrease T's value. Rules R1, R2(l) and R2(r) decrease size without increasing the number of terminal node labels, so they decrease T's value too. Rules R3(l) and R3(r) decrease size by three and increase the number of terminal node labels by at most two, thus they also decrease T's value.

BB is closed because for every Σ_{BB}-graph G, the set

$$\{g\colon L \to G \mid L \text{ is a left-hand side in } \mathcal{R}_{BB} \text{ and } g \text{ is injective}\}$$

contains at most one morphism. This can easily be checked by inspecting the rules of $\mathcal{R}_{BB}$, keeping in mind that distinct outedges of a node always have distinct labels. □

That the non-context-free language of BBTBs is definable by a linear GRS is remarkable — as there exist context-free graph languages with an NP-complete membership problem, and these can even have a bounded node degree [13].

5   Related Work

In addition to the remarks made in the Introduction on the relation of our approach to Dörr's work [5], we mention a few separating properties resulting from the different assumptions. While our approach is restricted to graphs of bounded outdegree, [5] allows an unbounded outdegree as long as outgoing edges form permitted V-structures. On the other hand, we allow parallel edges with the same label which are forbidden in Dörr's approach. Moreover, under Condition I, our only constraint on labels is that there is a uniquely labelled root, while all other items may have the same label. This is not possible with the V-structure approach which needs more structure in the labelling of graphs. Another difference is that rule applications in [5] are always deterministic while rules conforming to our Condition I may be nondeterministic.

Some authors have considered rooted graph matching under severe structural restrictions on host graphs, usually resulting from particular application areas. For example, [9,10] consider graphs representing infinite trees by 'unrolling' from some root. These graphs permit a linear matching algorithm.

There is also work on recognising graph languages which relates to our graph reduction specifications. For example, in [1] it is shown that a graph language expressible in monadic second-order logic (MSOL) can be defined by a reduction system with a linear membership test if the language has bounded treewidth. But our example of balanced binary trees with back-pointers is not expressible in MSOL and hence outside the scope of this result. A linear-time algorithm for recognising graph languages of bounded-treewidth is given in [3]. This algorithm works for so-called special reduction systems, it records which potential application areas for rules in a host graph have already been searched. It is a topic for future work to investigate the relationship between these special reduction systems and our linear GRSs.

References

1. S. Arnborg, B. Courcelle, A. Proskurowski, and D. Seese. An algebraic theory of graph reduction. *Journal of the ACM*, 40(5):1134–1164, 1993.
2. A. Bakewell, D. Plump, and C. Runciman. Specifying pointer structures by graph reduction. *Mathematical Structures in Computer Science*. To appear. Preliminary version available as Technical Report YCS-2003-367, University of York, 2003.
3. H. L. Bodlaender and B. van Antwerpen-de Fluiter. Reduction algorithms for graphs of small treewidth. *Inf. Comput.*, 167(2):86–119, 2001.
4. M. Dodds and D. Plump. Extending C for checking shape safety. In *Proceedings Graph Transformation for Verification and Concurrency*, Electronic Notes in Theoretical Computer Science. Elsevier, 2005. To appear.
5. H. Dörr. *Efficient Graph Rewriting and its Implementation*, volume 922 of *Lecture Notes in Computer Science*. Springer-Verlag, 1995.
6. F. Drewes, A. Habel, and H.-J. Kreowski. Hyperedge replacement graph grammars. In G. Rozenberg, editor, *Handbook of Graph Grammars and Computing by Graph Transformation. Volume I: Foundations*, chapter 2, pages 95–162. World Scientific, 1997.
7. J. Engelfriet and G. Rozenberg. Node replacement graph grammars. In G. Rozenberg, editor, *Handbook of Graph Grammars and Computing by Graph Transformation. Volume I: Foundations*, chapter 1, pages 1–94. World Scientific, 1997.
8. P. Fradet and D. L. Métayer. Shape types. In *Proceedings of the 1997 ACM Symposium on Principles of Programming Languages*, pages 27–39. ACM Press, 1997.
9. J. J. Fu. Linear matching-time algorithm for the directed graph isomorphism problem. In *Proceedings of the 6th International Symposium on Algorithms*, volume 1004 of *Lecture Notes in Computer Science*, pages 409–417. Springer-Verlag, 1995.
10. J. J. Fu. Pattern matching in directed graphs. In *Proc. Combinatorial Pattern Matching*, volume 937 of *Lecture Notes in Computer Science*, pages 64–77. Springer-Verlag, 1995.
11. M. R. Garey and D. S. Johnson. *Computers and Intractability*. W.H. Freeman and Company, 1979.
12. A. Habel and D. Plump. Relabelling in graph transformation. In *Proc. International Conference on Graph Transformation (ICGT 2002)*, volume 2505 of *Lecture Notes in Computer Science*, pages 135–147. Springer-Verlag, 2002.
13. K.-J. Lange and E. Welzl. String grammars with disconnecting or a basic root of the difficulty in graph grammar parsing. *Discrete Applied Mathematics*, 16:17–30, 1987.

Appendix

Proof of Proposition 3. Call a graph an NBBTB if it can be obtained from a BBTB by relabelling any number of B-nodes into B'-nodes, and any number of U-nodes into U'-nodes. Call the single edge with a ϱ-labelled node as its source the *root pointer*.

Soundness. We will show that every Σ_{BB}-graph reducible to Acc_{BB} is an NBBTB, implying that every graph in L(BB) is a BBTB. It suffices to show that the inverses of the rules in $\mathcal{R}_{BB}$ preserve NBBTBs, as then a simple induction on the length of reductions to Acc gives the desired result.

The inverses of the rules Up, D1 and D2 are clearly NBBTB-preserving as they only relabel nonterminal into terminal nodes and redirect the root pointer to some other node in the tree. The inverse of rule R1 can only be applied at the tree root because this is the only node with a loop attached to it. Hence the rule adds a new tree root, which preserves balance. The inverses of rules R2(l) and R2(r) are also NBBTB-preserving: replacing a U-node pointing to a leaf with a B-node pointing to two leaves preserves balance. Similarly, the inverses of R3(l) and R3(r) preserve balance and the other NBBTB properties.

Completeness. Given an NBBTB in which the root pointer does not point to the tree root, call a node a *root-pointer-predecessor* if it is on the unique path of non-back-pointer edges from the tree root to the parent of the root-pointer target. Call a graph an EBBTB if it is an NBBTB satisfying the following conditions: (1) root-pointer-predecessors are not labelled U' and (2) all nodes labelled B' are root-pointer-predecessors. Note that every BBTB it is also an EBBTB. We show that every EBBTB is reducible to Acc_{BB}, implying that every BBTB is in L(BB). In Proposition 4 it is shown that every $\mathcal{R}_{BB}$-derivation sequence terminates, so it suffices to show that (1) Acc_{BB} is the only $\mathcal{R}_{BB}$-irreducible EBBTB and (2) applying any rule in $\mathcal{R}_{BB}$ to an EBBTB results in an EBBTB.

We first show that every non-Acc_{BB} EBBTB is reducible by $\mathcal{R}_{BB}$, by enumerating all possible cases. If the root pointer points to a U or U'-node, we know from Σ_{BB} that it must have an outgoing edge labelled c. If it has no incoming edges, it is the treeroot and we can reduce it using rule R1. If this node has an incoming edge, it must be labelled c, l or r. If c, from the signature and the definition of an EBBTB we know that the source must be a U node, and so rule Up applies. If the incoming edge is labelled l or r, its source must be a B or B' node and it must have a sibling r or l edge. From the graph balance property, the other edge must point to another node with an outgoing edge – either a U, U', B, or B'. B' is excluded by the definition of an EBBTB. If U or U' rule R3(l) or R3(r) applies. If a B, by Σ_{BB} it must have outgoing l and r edges, and so rule D1 applies.

If the root pointer points to a B-node, by Σ_{BB} it must have an outgoing edge labelled l, so the D2 rule applies. By the definition of an EBBTB, the root pointer cannot point to an B'-node.

If the root pointer points to an L-node, either there are no incoming edges (aside from a backpointer loop) so the graph is Acc_{BB}, or it must have a single incoming edge labelled l, r, or c. If c, by Σ_{BB} and definition of EBBTB the source must be a U and the graph can be reduced by the Up rule. If the incoming edge is labelled l or r, its source must be labelled B and it must have a sibling edge labelled r or l. We know by the balance property that the target of this edge must be another L-node, so either R2(l), or R2(r) applies. This completes the proof that every EBBTB apart from Acc_{BB} can be reduced.

We now show that all rules preserve EBBTBs, once again by case enumeration. Rule Up moves the root pointer from its current position to the node's parent and relabels this parent from U to U'. As the rest of the graph is preserved, this preserves EBBTB condition (1). Rules D1 and D2 move the root pointer and relabel a single B-node to B'. In both cases the B'-node is added to the root-pointer-predecessors, so EBBTB condition (2) is satisfied. No B' nodes are added, so EBBTB condition (1) is satisfied. Rules Up, D1 and D2 only relabel nodes and move the root pointer, so balance is preserved.

Rule R1 deletes a U or U'-node and moves the root pointer to the child of the current position. We know from the EBBTB conditions that this child cannot be labelled B', and so the EBBTB conditions are satisfied. Rule R1 can only delete the treeroot, as this is the only non-L-node that can be safely deleted, and so it preserves balance. Rules R2(l) and R2(r) replace a B or B'-node with a U-node and move the root pointer. Replacing a B-node and two leaves with a U-node and one leaf preserves balance. The EBBTB conditions are satisfied, as the new target of the root pointer is a U-node and the root-pointer-predecessors are otherwise preserved. Rules R3(l) and R3(r) replace two U or U'-nodes with a B node. The rule preserves balance because the distance from the 'top' of the rule to the 'bottom' is preserved. These rules replace a single B or B'-node on the path to the treeroot with a U-node, and the root-pointer-predecessors are otherwise unaltered. No B'-nodes are added, so both EBBTB conditions are satisfied. This completes the proof that rules in $\mathcal{R}_{BB}$ are EBBTB-preserving.   □

GrGen: A Fast SPO-Based Graph Rewriting Tool

Rubino Geiß, Gernot Veit Batz, Daniel Grund,
Sebastian Hack, and Adam Szalkowski

Universität Karlsruhe (TH), 76131 Karlsruhe, Germany
rubino@ipd.info.uni-karlsruhe.de
http://www.info.uni-karlsruhe.de/software/grgen/

Abstract. Graph rewriting is a powerful technique that requires graph pattern matching, which is an NP-complete problem. We present GR-GEN, a generative programming system for graph rewriting, which applies heuristic optimizations. According to Varró's benchmark it is at least one order of magnitude faster than any other tool known to us.

Our graph rewriting tool implements the well-founded single-pushout approach. We define the notion of search plans to represent different matching strategies and equip these search plans with a cost model, taking the present host graph into account. The task of selecting a good search plan is then viewed as an optimization problem.

For the ease of use, GRGEN features an expressive specification language and generates program code with a convenient interface.

1 Introduction

Over the last 30 years graph rewriting theory has become mature. The constant rise of applications requires tools that are all *theoretically sound, fast* and *easy to use.* Currently available tools meet these requirements only partially, with varying emphases. Our tool GRGEN, which is presented in this paper, fulfills these requirements [1].

1.1 Graph Rewriting

The concept of *graph rewriting*, as implemented by GRGEN, follows the *single-pushout* (SPO) approach which is a form of rule based graph transformation (see section 2.4). At a basic level a rewrite rule consists of a *pattern graph*, a *replacement graph* and an instruction on what to delete, preserve or insert during rewriting. In order to *apply* a graph rewrite rule to a *host graph* we have to find an instance of the pattern graph in the host graph. Finding such a *match* is called *subgraph matching.*

1.2 Our Contributions

For pattern graphs of potentially unbounded size subgraph matching is an NP-complete problem (see Garey and Johnson, problem GT48 [2]). Hence, the question of performance is essential for the practical relevance of graph rewriting.

A. Corradini et al. (Eds.): ICGT 2006, LNCS 4178, pp. 383–397, 2006.

The multi-purpose graph rewrite generator GrGen allows high-speed graph rewriting. The main features and concepts of GrGen are:

1. *An expressive graph concept.*
   GrGen uses an extension of labeled directed multigraphs, namely *attributed typed directed multigraphs*. The type system features multiple inheritance on node and edge types (see section 2.1).
2. *Separation of meta model and rewrite rules*
   A *meta model* defines the allowed node and edge types as well as the attributes associated with each type. To restrict the set of *well-formed* graphs, the user can give so called *connection assertions*. Meta model and rewrite rules can be specified separately. This enables the developer to utilize different rule sets together with the same meta model description (see section 3.1).
3. *A notion of rewriting close to theory.*
   GrGen implements an extension of the SPO approach to graph rewriting. The differences consist in the use of the extended graph concept, some restrictions regarding the allowed matches and the ability of graph rewrite rules to request the re-labeling (i.e. retyping) of nodes (see section 2.4).
4. *Additional matching conditions and attribute computations.*
   The set of valid matches can be restricted beyond graph patterns by the assignment of *attribute conditions*, *type constraints* and *negative application conditions* (NACs) to every rule. Additionally, *attribute computations* can be associated with each rule (see section 3).
5. *Optimization of the matching process.*
   Subgraph matching is an NP-complete problem. To deal with this challenge in practice, the system is able to optimize the matching process at run time using knowledge about the current host graph (see section 2.2 and 2.3).
6. *Convenient user interface.*
   GrGen features an expressive and concise specification for meta models, rewrite rules, and rule application strategies (see section 3). The generated code can be invoked through an interface, which is easy to use.

We compare GrGen with the most prominent tools, namely PROGRES [3], AGG [4], Fujaba [5], and an approach presented by Varró [6]. Regarding a benchmark also introduced by Varró [7], our graph rewrite engine outperforms all of these tools by at least one order of magnitude (see section 5). While being the fastest graph rewriting system we know, we will show that GrGen is still one of the most expressive ones (see section 4).

2   Fundamental Problems and Their Solutions

Thinking of graph rewriting raises three major questions:

1. What is a graph?
2. How is an occurrence of a pattern graph found?
3. What does rewriting mean in detail?

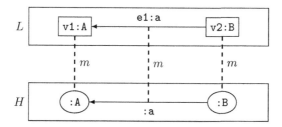

Fig. 1. Named pattern graph L and host graph H together with a match m

2.1 Graphs

The first question concerns two main aspects: which graph concept to use (discussed in this section) and which abilities to specify a meta model we give to the user (see section 3).

GRGEN features *attributed typed directed multigraphs*. These are directed graphs with typed nodes and edges, where between two nodes more than one edge of the same type and direction is permitted. According to its type, each node or edge has a defined set of attributes associated with it. Moreover, the type system features multiple inheritance on node and edge types. A meta model defines the allowed node and edge types as well as the attributes associated with each type. Furthermore it allows to restrict the set of well-formed graphs by so called *connection assertions*. For an example specification see section 3.1.

Throughout this paper graphs are depicted as follows: Nodes are either displayed by rectangles or ellipses. Rectangles are used in pattern graphs, ellipses are used in host graphs. The directed edges are displayed by arrows. Figure 1 shows a pattern graph L and a host graph H. The types of the nodes and edges are represented by node and edge labels with a preceding colon. In case a node or edge is given a name, it is written before the colon.

2.2 Finding a Match

We define a match as a *graph homomorphism* between the pattern graph L and the host graph H. A graph homomorphism is a pair of maps $m = (m_V, m_E)$, where m_V assigns the nodes of L to nodes of H and m_E the edges of L to edges of H. In figure 1 the nodes and edges mapped to each other are connected by dashed lines.

Subgraph matching is known to be NP-complete [2]. So, we propose a heuristically optimizing approach to subgraph matching. Moreover, the optimization is done dynamically at runtime depending on the present host graph (see also section 2.3). The tightest upper bound for the runtime of subgraph matching known to us is $O(|L||H|^L)$, where $|\cdot|$ denotes the sum of the numbers of nodes and edges of a graph. If we consider only fixed size patterns, subgraph matching can be regarded to as polynomial (possibly with a high polynomial degree). This seems to be good news, because we do not have to deal with an exponential runtime. But a runtime of, e.g. $O(|H|^{10})$, is still not feasible even for small constant factors, especially for host graphs H with hundreds or thousands of nodes and edges. Assuming that many application domains provide sparse graphs and a rich type system, we expect that our optimizing approach to subgraph matching leads to acceptable runtimes.

To enable the optimization of the matching process, we perform the subgraph matching according to a so called *search plan*. A search plan is a sequence of *primitive search operations*. Each such operation represents the matching of a single node or edge of the pattern graph to an appropriate node or edge of the host graph. The whole search plan describes the stepwise construction of all (or one) possible matches between L and H. We call a partly constructed match a *candidate*. The runtimes caused by different search plans depend on the present host graph and can vary significantly. Therefore the key idea for finding a match fast is to create a preferably good search plan taking the structure of the present host graph into account. The necessary information is taken from an analysis of the host graph performed at runtime. GRGEN also provides default search plans. They are statically created according to optional user hints.

Consider a search plan $P = \langle s_0, \ldots, s_q \rangle$, i.e., a sequence of primitive search operations s_i. We allow two kinds of search operations: At first there are *lookup* operations. They are denoted by $s_i = \mathsf{lkp}(x_i)$, where x_i is a node or edge of the pattern graph. At second there are *extension* operations $s_i = \mathsf{ext}(v_i, e_i)$, where v_i is a pattern node and e_i is a pattern edge. A lookup operation $\mathsf{lkp}(x_i)$ represents the expansion of a candidate by any node or edge of the host graph, which is suitable for the given x_i. If x_i is a pattern node, an appropriate host graph node must have the same type as x_i or a subtype thereof (we call this an *admissible type*). If x_i is a pattern edge, the incident nodes must also have admissible types (note that GRGEN supports no lookup operations for edges, yet). An extension operation $\mathsf{ext}(v_i, e_i)$ represents the expansion of a candidate by an edge e_i coming from an already matched node v_i. Of course an appropriate host graph edge and the node at its other end must also have admissible types.

The matching of a node can happen explicitly by the execution of a node lookup $\mathsf{lkp}(v)$ or implicitly by the matching of an edge incident to that node. An edge e can also be matched in two different ways (both explicitly): by an edge lookup $\mathsf{lkp}(e)$ or by an extension $\mathsf{ext}(v, e)$. E.g. consider two possible search plans for the pattern graph L shown in figure 2.

$$P_0 = \langle \mathsf{lkp}(\mathsf{v1}), \mathsf{ext}(\mathsf{v1}, \mathsf{e1}), \mathsf{lkp}(\mathsf{v3}), \mathsf{ext}(\mathsf{v2}, \mathsf{e2}) \rangle$$
$$P_1 = \langle \mathsf{lkp}(\mathsf{e1}), \mathsf{lkp}(\mathsf{v3}), \mathsf{ext}(\mathsf{v3}, \mathsf{e2}) \rangle$$

On the execution of a primitive search operation more than one appropriate node or edge may be found. In this case a candidate is replaced by several new candidates, one for every possible node or edge. However, it is not necessary to materialize all candidates at the same time. If a candidate can be expanded by more than one host graph element, we process only one of these. The other alternatives are treated by backtracking.

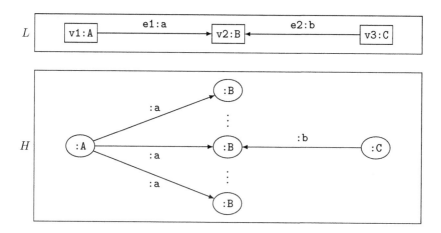

Fig. 2. Operation ext(v1, e1) causes an intense needless splitting of candidates

2.3   Generating Good Search Plans

The runtime of different search plans can vary significantly for a given host graph. For the generation of preferably good search plans, we use an approach originally presented by Batz [8]. It extends a technique invented by Dörr [9] with a cost model that directs a heuristic optimization.

The execution of an operation s_i can cause the splitting of a candidate into several new candidates. If this is the case for a significant ratio of the operations of a search plan, this leads to an exponential growth of the set of candidates. So, if splitting operations could be avoided by a search plan, less runtime would be needed. If the execution of a search plan causes no splitting at all, linear runtime for *sparse* host graphs H is achieved, that is $O(|L|)$.

Consider e.g. the pattern graph L and the host graph H shown in figure 2. In H a single node of type A is connected to a number of nodes of type B (let's say 20), each by an edge of type a. Now let us assume that the search plan

$$P_2 = \langle \mathsf{lkp(v1)}, \mathsf{ext(v1, e1)}, \mathsf{ext(v2, e2)} \rangle$$

is executed. The first operation lkp(v1) leads to the creation of one new candidate. Now the node of type A is incident to 20 outgoing edges of type a, each leading to a node of type B, so in the worst case the candidate splits into 20 new ones. In contrast the execution of the search plan

$$P_3 = \langle \mathsf{lkp(v3)}, \mathsf{ext(v3, e2)}, \mathsf{ext(v2, e1)} \rangle$$

requires no splitting at all. In case of the extension operation for edge e1, the crucial point is that P_3 follows e1 in the opposite direction as P_2 does. That is where Dörr's approach applies to: The direction an edge is followed can determine whether a candidate splits or not. In contrast to extension operations, for lookup operations the splitting depends on the number of present elements having an admissible type.

However, for extension operations, splitting cannot always be avoided. In the following we refer to equally typed edges of equal direction which connect equally typed nodes as *isomorphic*. If there are isomorphic edges present on *both* nodes incident to an edge, splitting occurs inevitably. In such a situation it only remains to choose the direction with *less* splitting. Moreover, we are looking for search plans which cause a low *overall* amount of splitting. Therefore, we extend Dörr's technique by a cost model to direct the optimization of search plans.

For this purpose we assign a cost to every operation which might possibly occur in a search plan: An operation $\mathsf{ext}(v, e)$ gets assigned the average number of splittings for a candidate. A $\mathsf{lkp}(x)$ gets assigned the number of present elements of admissible type. Having done this, we compute the costs of a possible search plan $P = \langle s_0, \ldots, s_q \rangle$ by the formula

$$C_P := c_0 + c_0 c_1 + c_0 c_1 c_2 + \cdots + c_0 c_1 c_2 \cdots c_q$$

where c_i is the cost of the operation s_i.

Essentially the formula estimates the number of host graph elements matched while executing P. If operation s_0 is executed, up to c_0 host graph elements will be matched. This also means that up to c_0 new candidates will be created. If operation s_1 is performed, for all these candidates on average c_1 further elements will be matched. Overall this results in an average amount of up to $c_0 c_1$ matched elements and newly created candidates. Continuing this, one gets the above formula. However, if a candidate fails to complete, no further candidates will be created from it. So, except for constant factors, the above formula yields an *overestimation* of the average number of elements processed while executing P.[1]

We do not know an efficient algorithm yielding a search plan P with minimal costs C_P. So, we use the following heuristic method: In the first step, we minimize the most significant term occurring in the above formula, namely $c_0 c_1 c_2 \cdots c_q$. This is done by choosing a possibly cheap selection from the set of all possible search operations for L. In the second step, we compute an order for the selected operations, such that the cheap operations appear preferably early and the expensive operations as late as possible. This exploits the fact, that a splitting has more impact on C_P, the earlier the according operation occurs in P. The costs of the possible operations are derived from an analysis of H, which can be performed in time $O(|H|)$. A detailed description of this heuristics is given in a technical report [10].

2.4   Meaning of Rewriting

In the literature the meaning of rewriting is treated thoroughly [11,12,3]. Despite this fact it is not a computationally complex problem at all. The approaches differ substantial in understandability and readability of specifications as well as their expressiveness. Also, their degree of theoretical foundation is quite different. We have chosen the well-known SPO approach.

---

[1]   This is due to the assumption, that every node of H has $O(1)$ edges.

Fig. 3. The principle setting of SPO-based graph rewriting

A SPO rewrite rule $p : L \xrightarrow{r} R$ consists of a *pattern graph* L, a *replacement graph* R and a partial graph homomorphism r between L and R. An *application* of a rule p to a host graph H is called a *direct derivation* (see figure 3). It requires a partial graph homomorphism m from L to H called a *match* (GRGEN demands *total* matches). The direct derivation leads to a *result graph* H', see figure 3. For each node or edge x in L there exists a corresponding node or edge in H, namely $m(x)$. Note that m does not need to be injective. The partial *preservation morphism* r determines what happens to $m(x)$: It maps all items from L to R, which are to remain in H during the application of the rule. The images under m of all items in L which have no image under r are to be deleted. The others are retained. Items in R which have no pre-image under r are added to H'. Note that in general ρ is neither surjective nor total. It is partial, because nodes from H may be deleted to get H'. The homomorphism ρ can be non-surjective, because new nodes may be introduced in H'—these nodes are not in the image of ρ but in the image of μ.

The SPO approach is not constructive in a way that it directly gives an algorithm (as sketched above) for obtaining the result graph H'. It rather *characterizes* H' in the set of all graphs using a pushout construction in the category of graphs and partial graph homomorphisms. For conciseness we omit the category theoretical foundations of the SPO approach (for an introduction see [11]). Except for partial matching GRGEN implements the SPO approach to the full extent, but provides additional features not covered by SPO (see also section 3.2). These are: Attribute conditions, type constraints, NACs, node type changes and attribute recalculation. Attribute conditions, type constraints and NACs restrict the set of admissible matches. Retyping and attribute evaluations are performed after the SPO rewrite is done. A formalization of such extensions based on category theory for the DPO approach is presented by Ehrig et al. [13].

3   The Tool

In this section, we present the most important features of GRGEN along with its input language which enables the user to define a meta model for graphs, a set of graph rewrite rules as well as a sequence of rule applications.

The structure of the generated graph rewriters (we call them *graph engines*) yielded by GRGEN arises from the separation of four concerns: defining the type

Listing 1.1. A meta model

```
1   a1 : int;
2   a2 : int;
3 }
4 node class NodeTypeB extends NodeTypeA {
5   a3 :int;
6 }
7 node class NodeTypeC extends NodeTypeA, NodeTypeB;
8
9 edge class EdgeTypeA
10      connect NodeTypeA  [0:1] -> NodeTypeA  [0:1],
11              NodeTypeA  [*]   -> NodeTypeB  [1:5];
12
13 edge class EdgeTypeB extends EdgeTypeA
14      connect NodeTypeB  [4:*] -> NodeTypeA  [1] {
15   a1 : string;
16 }
```

of graph elements, storing the graph data, finding the match, and performing the rewrite. This gives us the freedom to easily change certain aspects of the implementation. The GRGEN(SP) graph engine uses our search plan approach to subgraph matching sketched in section 2.2 and 2.3 (for a technical description see Batz and Szalkowski [8,10,14]). GRGEN(PSQL) is a graph engine variant that uses a Postgres database for storing and matching graphs [15,16].

3.1   Meta Model

The key features of GRGEN's meta model are exemplarily shown in listing 1.1.

Types. Nodes and edges can have types (classes). The syntax is similar to common programming languages (keywords **node class** and **edge class**).

Attributes. Nodes and edges can possess attributes. The set of attributes assigned to a node or edge is determined by its type. The attributes itself are typed, too.

Inheritance. Types (classes) can be composed by multiple inheritance. This eases the way of specifying patterns and improves the expressiveness of graphs. **Node** and **Edge** are the built-in root types of node and edge types, respectively. Moreover, inheritance eases the specification of attributes, because subtypes inherit the attributes of their super types.

Connection Assertions. To specify that certain edge types should only connect specific nodes, we included connection assertions (keyword **connect**). Using these, the system is optionally able to check whether a host graph is well-formed or not. For example, line 12 of listing 1.1 specifies, that nodes of type **NodeTypeA** can have arbitrary outgoing edges of type **EdgeTypeA**. Furthermore these edges must connect to a node of type **NodeTypeB**, whereas one to five such edges may be incoming at a single **NodeTypeB** node.

Listing 1.2. A rewrite rule specification

```
1  rule SomeRule {
2    pattern {
3      node (n1 ~ n2) : NodeTypeA;
4      n1 --> n2;
5      n3 : NodeTypeB;
6      negative {
7        n3 -e1:EdgeTypeA-> n1;
8        if { n3.a1 == 42 * n2.a1; }
9      }
10     negative {
11       node n4 : Node \ NodeTypeB;
12       n3 -e1:EdgeTypeB-> n4;
13     }
14   }
15   replace {
16     n5 : NodeTypeC<n1>;
17     n3 -e1:EdgeTypeB-> n5;
18   }
19   eval {
20     n5.a3 = n3.a1 * n1.a2;
21   }
22 }
```

3.2   Graph Rewrite Rules

For example, consider the graph rewrite rule SomeRule (see listing 1.2). The keyword **pattern** marks the beginning of the pattern graph consisting of a node named n3 of type NodeTypeB as well as two nodes named n1 and n2 of type NodeTypeA. We denote the preservation morphism r implicitly by using named nodes and edges: Identical names in pattern and replacement graph (keyword replace) indicate that this nodes or edges are mapped to each other by r. Anonymous edges are denoted by an arrow (-->). Additionally, we can specify named edges of certain types by annotating the arrows (-EdgeName:EdgeType->). The semantics of the example rule is sketched in the following.

Isomorphic/Homomorphic Matching. The tilde operator (~) between the nodes n1 and n2 specifies that these nodes may be matched homomorphically. In contrast to the default isomorphic matching of morphism m the nodes n1 and n2 *may* be mapped to the same node in the host graph.

Negative Application Conditions (NACs). With negative application conditions (keyword **negative**) we can specify graph patterns which forbid the application of a rule if any of them is present in the host graph (cf. [4]).

Attribute Conditions. The Java-like attribute conditions (keyword **if**) in the pattern part allows for further restriction of the applicability of a rule.

Type Constraints. In general set theoretical operations on types are allowed. By writing n4 : Node \ NodeTypeB we declare a node that is a subtype of Node but not of NodeTypeB.

Replace Part. Because node instances **n1** and **n3** (declared in the pattern part) are used in the replace part (denoting the replacement graph), these nodes are kept. The anonymous edge instance between **n1** and **n2** only occurs in the pattern and therefore gets deleted. The edge **e1** is only declared in the replace part, thus it has to be created. Note that edge **e1** from the replace part and the negative parts are all different, because of their scopes.

Retyping. Node **n5** is a retyped node stemming from node **n1**. This enables us to keep all edges and all attributes stemming from common super types of a node while changing its type.

Eval Part. If a rule is applied, then the attributes of matched and inserted nodes and edges may be recalculated.

3.3   Rule Application

To control the application of rules, we define the set $\mathcal{R}$ of *regular graph rewrite sequences (RGS)*, where $\mathcal{P}$ is a set of rewrite rules:

$$p \in \mathcal{P} \Rightarrow p \in \mathcal{R} \qquad\qquad p \in \mathcal{P} \Rightarrow [p] \in \mathcal{R}$$
$$R_1, R_2 \in \mathcal{R} \Rightarrow R_1 R_2 \in \mathcal{R} \qquad\qquad R \in \mathcal{R} \Rightarrow (R) \in \mathcal{R}$$
$$R \in \mathcal{R} \Rightarrow R* \in \mathcal{R} \qquad\qquad R \in \mathcal{R}, n \in\ \ \Rightarrow R\{n\} \in \mathcal{R}$$

The syntax of RGSs is largely borrowed from regular expressions, but its semantics are only related. The main difference is: Determined and undetermined iteration expressions $R\{n\}$ and $R*$ cause an execution of R until no rule contained in R can be applied (or the iteration count exceeds n, respectively).[2] A subsequence R_2 of a sequence $R_1 R_2$ is executed even if R_1 is not applicable. A single rule application can fail or succeed. In the case of failure nothing happens, except that we carry on with the next step. Please observe that the execution of an RGS does not involve backtracking in any kind. $[p]$ denotes the simultaneous application of all matches of rule p. For the $[\cdot]$ operator GRGEN (or a user supplied application) can sort out overlapping matches or rewrites to maintain desired semantic properties.

E.g. we can express Varró's STS mutex benchmark of size 1000 by the following RGS:

```
newRule{998} mountRule requestRule{1000}
(takeRule releaseRule giveRule){1000}
```

4   Related Work

Over three decades, graph rewrite theory has evolved well. Amongst others, there are two major schools: Firstly, the algebraic rewriting school, which considers graphs as algebraic objects and defines rewriting via mappings. Algebraic rewriting itself has a rich variety of approaches: There is the single-pushout approach (SPO, see section 1.1 and 2.4), the double-pushout approach (DPO) and the

---

[2] The semantics of the RGS is declared operational, starting at the innermost nesting: The execution of $(R*)*$ is always well-defined, but maybe non-terminating.

Table 1. Features of graph rewriting tools

Tool	Semantics	Storage	Matching	Mode	Language
PROGRES	programmed	GRAS	planned LS	int.&comp.	C/C
AGG	SPO&NAC	memory	CSP	interpreted	Java
FUJABA	programmed	memory	LS	compiled	Java/Java
VarróDB	SPO&NAC	RDBMS	SQL	interpreted	Java
GRGEN(PSQL)	SPO&NAC	RDBMS	SQL	compiled	Java/C
GRGEN(SP)	SPO&NAC	memory	planned LS	compiled	Java/C

pullback approach. These approaches are all based on category theory and differ mostly in the fashion of defining the rewrite rules and the behavior when deleting nodes. Regarding the latter, SPO is more powerful then DPO.[3] Secondly, there is the *programmed* approach. It defines rules and rewrites in a more operational style. Its semantics is more complex and hard to define, which on the other hand eases the integration of special application driven needs to the tool. For example, consider the formal definition of a part of PROGRES [17].

In table 1 the most prominent graph rewriting tools are compared. For this purpose we consider five key properties, which give a coarse-grained insight in the theory and implementation of each tool.

Semantics. How is the rewriting described theoretically and how powerful is a single rewriting step? SPO refers to single-pushout approach (see section 1.1 and 2.4). If the tool uses negative application conditions to enhance its expressiveness then we write NAC. By programmed we mean that semantics is rather defined through an operational sequence than a theory.

Storage. The storage property describes how the graph is stored and whether it is persistent: In-memory storage is not necessary persistent. RDBMS and GRAS are both database backed graph storages where the first stands for of the shelf relational database system, the latter is a special graph database implementation.

Matching. The tools vary significantly in the handling of the matching problem. Some transform the matching problem into another well understood and tool supported domain, like constraint satisfaction (CSP) or relational algebra (SQL). Others perform a local search (LS) on the graph structure to find the matchings. This search process can be driven by chance or be planned ahead.

Mode. Does the tool generate code in a conventional programming language, which has to be compiled to perform the matching? Or are the graph rewrite rules just interpreted by the tool, hence no code is generated.

Language. This refers to the implementation languages. For example, GRGEN is implemented in Java and generates matchers implemented in C. For tools with interpreted matching there is only one entry.

---

[3] SPO can delete nodes without specifying its whole context whereas DPO cannot. Moreover, SPO in conjunction with NACs can simulate the dangling edge conditions of DPO.

Table 2. Runtime for several of the Varró benchmarks (in milliseconds)

Benchmark → Tool ↓	STS			ALAP			ALAP simult.			LTS
	10	100	1000	10	100	1000	10	100	1000	1000,1
PROGRES	12	946	459,000	21	1,267	610,600	8	471	2,361	942,100
AGG	330	8,300	6,881,000	270	8,027	13,654,000	–	–	–	$> 10^7$
FUJABA	40	305	4,927	32	203	2,821	20	69	344	3,875
VarróDB	4,697	19,825	593,500	893	14,088	596,800	153	537	3,130	593,200
GRGEN(PSQL)	30	760	27,715	24	1,180	406,000	–	–	–	96,486
GRGEN(SP)	< 1	8	79	< 1	5	64	< 1	< 1	5	99

One of the first graph rewrite tools is PROGRES and it is still amongst the most expressive ones [3]. As described by Zündorf [18], its matching algorithm is based on planned local search. A more contemporary tool is AGG, which also has the desirable property to rely closely on the theoretical foundations of the SPO approach [4]. The matching of AGG is done by reducing the problem to a constraint satisfaction problem [19]. To call FUJABA a graph rewrite tool is a kind of an understatement [5]. FUJABA is a tool for software visualization and two-way transformation based on UML. Some of its functionality relies on graph transformations. These parts can be utilized to perform general graph rewriting. The graph rewriting rules are programmed story diagrams in the sense of extended UML use case diagrams. Varró describes a technique for performing graph rewriting based on relational algebra [6]. Up to now, this tool is not accessible, but we have some example runs available [20].

The OPTIMIX system proposed by Uwe Assmann has a limited expressiveness [21]. It would be impossible to perform the benchmarks of our choice without significant simplifications. Therefore it is not included in our closer examination. But nevertheless OPTIMIX is interesting; because of its limitations it is possible to get some strong theoretical results, such as confluence and guaranteed termination. In general, this is not possible for the other tools mentioned above.

Dörr developed an idea for matching certain graphs in linear time [9]. His technique fails for graphs which contain edges that cause inevitably splitting of candidates. To our knowledge no actual tool was built using this approach. By defining a cost model, we extended this approach to all graphs, but had to sacrifice the linear runtime guarantee (see section 2.3). Independently Varro et al. proposed a quite similar method [22] which is not implemented, yet.

5   Performance

The benchmark uses various sizes of graphs and patterns as well as long and short transformation sequences. The example used as a benchmark by Varró was originally proposed to serve as distributed mutual exclusion algorithm. Varró has changed the algorithm slightly for benchmarking.

Our own measurements (for AGG and GRGEN) were carried out on an AMD Athlon XP 3000+ with 1GB main memory. Measurements by Varró (for PRO-GRES, FUJABA and VarróDB) were performed on a Intel Pentium 4 at 1.5 GHz

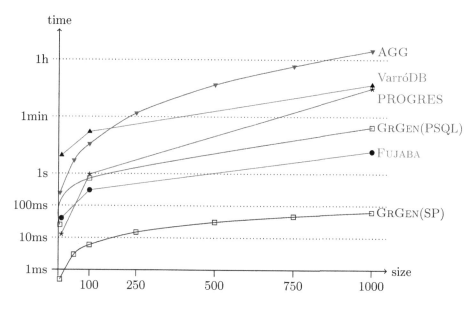

Fig. 4. Runtime of STS mutex benchmark (multiplicity optimizations off, parameter passing off, simultaneous execution off; for parameter details see [7])

with 768 MB main memory [20]. To reuse his results we multiplied Varró's figures by 0.68 which is the speed difference of both processors according to the SPEC organization [23].

Figure 4 shows the runtime of two GRGEN instances compared with the most prominent tools, namely AGG [4], FUJABA [5], PROGRES [3] and an approach presented by Varró [6], which we call VarróDB. GRGEN(SP) uses our most advanced graph engine, whereas GRGEN(PSQL) is based on a Postgres database for storing and matching graphs (see section 3). Further benchmark results, shown in table 2, support the overall impression. The other benchmarks proposed by Varró show analogous results and are omitted here (see [1]).

The memory usage of GRGEN(SP) for the largest mutex benchmark was below 1.6 MByte. In any benchmark we conducted GRGEN(SP) outperformed the next fastest tool at least by a factor of 40. Regarding the STS mutex benchmark GRGEN(SP) achieves even linear runtime in terms of benchmark size, i.e., the average runtime for a single rewrite rule is constant regardless the host graph size. This implies, that we have *reached* the speedup limit for the Varró benchmark; better tools can only lower the constant factor. The spread between GRGEN(SP) and the slowest tool is more than 6 orders of magnitude.

6 Conclusion

Graph rewriting has complex theoretical and practical aspects. We meet the computational challenge of *finding a match* with a heuristically optimizing approach based on search plans. The definition of the *rewrite semantics* closely follows the well-established SPO approach and provides some extensions.

We still have to answer the most important question: Can the user actually put the power of the theory to work? Therefore, let us consider what users might expect from GRGEN. The user wants to: define elements of a domain as graph elements, get expressive and concise rewrite specifications, get the results fast without excessive memory consumption, and easily integrate the graph rewriting into his applications.

GRGEN meets all those needs: In the meta model attributes and types can be defined both for nodes and edges. It is possible to check graphs against given connection assertions, but graphs not conforming to these assertions can also be processed. The specification language is expressive and concise. The type hierarchy defined by the meta model helps to express graph rewrite rules easily. GRGEN supports different rule application strategies: interactive application, regular graph rewrite sequences (RGS), and a low level selection by user supplied program code. An interactive environment for stepwise execution of graph rewrite rules and graph inspections is also provided. The performance of a rule application, especially of the potentially expensive pattern matching, is at least one order of magnitude faster than of any other tested system. The memory consumption of our search plan based graph engine is low, too. 10 million graph elements can be handled in 1 GB main memory. In other words: On average about 100 bytes were consumed per node or edge (without attributes assigned) including all administration overhead. The integration effort of the dynamically linked graph engines produced by GRGEN is small.

Thus, tool supported graph rewriting can be done both, fast and easy to use, based on the well established theoretical foundations of SPO built into the declarative graph rewrite language of GRGEN.

Acknowledgements. Thanks to all co-workers and students that helped during the design and implementation of GRGEN as well as the writing of this paper. Especially we want to thank Michael Beck, Dr. Markus Noga, Dr. Andreas Ludwig, Tom Gelhausen and the anonymous referees. We thank Gergely Varró for both his most influential work on the benchmarking of graph rewrite tools as well as allowing us to reuse his measurements. Finally, all this would not happened without the productive atmosphere and the generous support that Prof. Goos provides at his chair.

References

1. Geiß, R.: GRGEN. http://www.info.uni-karlsruhe.de/software.php/id=7 (2006)
2. Garey, M.R., Johnson, D.S.: Computers and Intractability; A Guide to the Theory of NP-Completeness. W. H. Freeman & Co., New York, NY, USA (1990)
3. Schürr, A.: The PROGRES Approach: Language and Environment. In: [24]. Volume 2. (1999) 487–550
4. Ermel, C., Rudolf, M., Taentzer, G.: The AGG Approach: Language and Environment. In: [24]. Volume 2. (1999) 551–603

5. Fujaba Developer Team: Fujaba-Homepage. http://www.fujaba.de/ (2005)
6. Varró, G., Friedl, K., Varró, D.: Graph Transformations in Relational Databases. In: Proc. GraBaTs 2004: Intl. Workshop on Graph Based Tools, Elsevier (2004)
7. Varró, G., Schürr, A., Varró, D.: Benchmarking for Graph Transformation. Technical report, Department of Computer Science and Information Theory, Budapest University of Technology and Economics (2005)
8. Batz, G.V.: Graphersetzung für eine Zwischendarstellung im Übersetzerbau. Master's thesis, Universität Karlsruhe (2005)
9. Dörr, H.: Efficient Graph Rewriting and its Implementation. Volume 922 of LNCS. Springer-Verlag New York, Inc., Secaucus, NJ, USA (1995)
10. Batz, G.V.: An Optimization Technique for Subgraph Matching Strategies. Technical Report 2006-7, Universität Karlsruhe, Fakultät für Informatik (2006)
11. Ehrig, H., Heckel, R., Korff, M., Löwe, M., Ribeiro, L., Wagner, A., Corradini, A.: Algebraic Approaches to Graph Transformation - Part II: Single Pushout Approach and Comparison with Double Pushout Approach. In: [24]. Volume 1. (1999) 247–312
12. Corradini, A., Montanari, U., Rossi, F., Ehrig, H., Heckel, R., Löwe, M.: Algebraic Approaches to Graph Transformation - Part I: Basic concepts and double pushout approach. In: [24]. Volume 1. (1999) 163–245
13. Ehrig, H., Ehrig, K., Prange, U., Taentzer, G.: Fundamentals of Algebraic Graph Transformation. Monographs in Theoretical Computer Science. Springer (2006)
14. Szalkowski, A.M.: Negative Anwendungsbedingungen für das suchprogramm-basierte Backend von GrGen (2005) Studienarbeit, Universität Karlsruhe.
15. Hack, S.: Graphersetzung für Optimierungen in der Codeerzeugung. Master's thesis, Universität Karlsruhe (2003)
16. Grund, D.: Negative Anwendungsbedingungen für den Graphersetzer GR-GEN(2004) Studienarbeit, Universität Karlsruhe.
17. Schürr, A.: Logic based programmed structure rewriting systems. Fundamenta Informaticae, Special Issues on Graph Transformations 26(3/4) (1996)
18. Zündorf, A.: Graph Pattern Matching in PROGRES. In: Proc. 5th. Int. Workshop on Graph-Grammars and their Application to Computer Science. Volume 1073 of LNCS., Springer (1996) 454–468
19. Rudolf, M.: Utilizing constraint satisfaction techniques for efficient graph pattern matching. In: TAGT'98: Selected papers from the 6th Intl. Workshop on Theory and Application of Graph Transformations. Volume 1764., LNCS (1998) 238–251
20. Varró, G.: Graph transformation benchmarks page. http://www.cs.bme.hu/~gervarro/benchmark/2.0/ (2005)
21. Assmann, U.: Graph rewrite systems for program optimization. ACM Trans. Program. Lang. Syst. 22(4) (2000) 583–637
22. Varró, G., Varró, D., Friedl, K.: Adaptive graph pattern matching for model transformations using model-sensitive search plans. In Karsai, G., Taentzer, G., eds.: Proc. of Int. Workshop on Graph and Model Transformation (GraMoT'05). Volume 152 of ENTCS., Tallinn, Estonia, Elsevier (2005) 191–205
23. Standard Performance Evaluation Corporation: All SPEC CPU2000 results published by SPEC page. http://www.spec.org/cpu2000/results/cpu2000.html (2005)
24. Rozenberg, G., ed.: Handbook of Graph Grammars and Computing by Graph Transformation. World Scientific (1999)

Realizing Graph Transformations by Pre- and Postconditions and Command Sequences

Fabian Büttner and Martin Gogolla

University of Bremen, Computer Science Department
Database Systems Group, D-28334 Bremen, Germany
{green, gogolla}@informatik.uni-bremen.de

Abstract. This paper studies two realizations of graph transformations which are based on a UML class diagram. The first realization achieves a representation in terms of descriptive pre- and postconditions. The second one yields an operationally executable command sequence in terms of basic commands for object and link creation, attribute modification, and object and link destruction. Our aim for realizing graph transformations in terms of target languages offering different views, i.e., descriptive or operational, is to take advantage of both views and to utilize the benefits which both views provide.

1 Introduction

This paper discusses *model behavior*. Under the notion *model* we understand a collection of UML descriptions [OMG04, RBJ05]. The paper focuses on a special kind of model *behavior*, namely behavior of operations. Behavior of operations is described graphically by a set of graph transformation rules having the aim of achieving understandable and intuitive but strictly formalizable characterizations. These behavior descriptions are *realized* in two different target languages. The first target language is given by OCL pre- and postconditions [WK03]. The second target language is determined by sequences of basic commands for object and link creation, attribute modification, and object and link destruction [OMG04, RBJ05].

We see several advantages in realizing the same source language, in our case graph transformations, in different target languages which possess quite different nature and complement each other: OCL pre- and postconditions have a *descriptive* nature, i.e., it is stated *what* properties have to hold before resp. after operation execution; the command sequences into which we translate have an *operational* nature, i.e., it is expressed *how* the operation is to be executed. Descriptive OCL pre- and postconditions can be analyzed with theorem prover like tools and approaches [BW02, GBR05]. Such tools and approaches aim to deduce new properties from the ones stated explicitly. Operational UML descriptions can be executed and are thus a basis for animating the behavior [OMG04, RBJ05]. Understanding the effect of a complex operation is supported by tracing the single execution steps. Deduction and reasoning as well as animation and prototypical execution is particularly useful is early software development phases.

A. Corradini et al. (Eds.): ICGT 2006, LNCS 4178, pp. 398–413, 2006.
c Springer-Verlag Berlin Heidelberg 2006

Realization in different target languages helps to better understand graph transformations insofar that the different target representations emphasize different aspects which are inherently and implicitly present. Realizing graph transformations in the two formalisms can help (1) to formally inspect properties of the graph transformations by checking the achieved OCL constraints with an OCL tool and (2) to understand and test the effects of the graph transformations by executing the achieved command sequences with a UML execution engine. Formal inspection subsumes proving properties of the graph transformation system as a whole.

Our starting point is a set of graph transformations which may be seen as a graphical description of operation behavior in terms of a UML *collaboration*. The UML Language Reference Manual [RBJ05, page 228] explains the notion collaboration as follows:

> A collaboration describes the context for an operation [...] in which the implementation of an operation [...] executes — this is, the arrangements of objects and links that exist when the execution begins, and the [objects and links] that are created or destroyed during execution.

Thus, a UML collaboration involves the objects and links existing before operation execution and the ones existing after operation execution. We distinguish between these two parts of a collaboration and denote a collaboration as a *graph transformation rule* possessing a left side showing the situation before operation execution and a right side presenting the scenario after execution. Rules are common instruments to describe steps within complex processes, and collaborations are the right mechanism to denote them in the UML. In our view, central aspects of graph transformations rules can be expressed in terms of UML diagrams. Utilizing this relationship helps to broaden the audience of graph transformations.

The structure of the rest of this paper is as follows. Section 2 forwards the basic idea of the paper by means of a simple example. Section 3 summarizes the language features in rules which our approach supports. Section 4 discusses the details of the first realization targeting OCL pre- and postconditions. Section 5 shows the details of the second realization targeting command sequences. Section 6 ends the paper with concluding remarks. Both realizations have been prototypically implemented in an extended version of USE [GBR05]. All examples have been checked and executed by the tool.

2   The Basic Idea

Figure 1 shows five example graph transformation rules. The rule fireManager is shown also in the textual rule representation we actually use in our implementation. The rules depend on the upper part of the UML class diagram in Fig. 2 (using the `worksFor` association for representing bi-directional access and the object-valued attribute `staffCar` for uni-directional access). They have a left and right side and name, and offer to employ OCL expressions: (A) In the

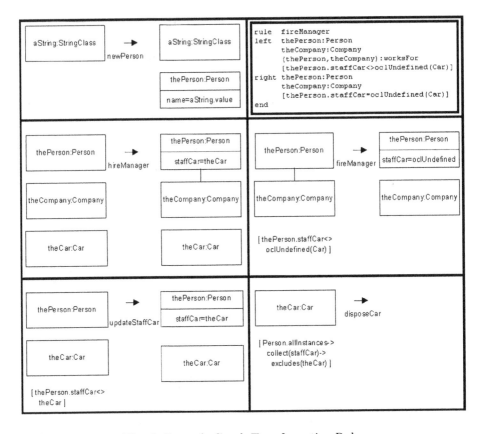

Fig. 1. Example Graph Transformation Rules

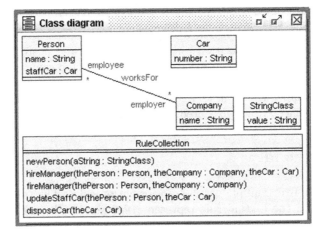

Fig. 2. Example Class Diagram

```
context newPerson(aString:StringClass)
pre   newPerson_pre: aString.isDefined
post newPerson_post: StringClass.allInstances->includes(aString) and
     Person.allInstances->exists( thePerson | thePerson.oclIsNew and
       thePerson.name=aString.value )

context hireManager(thePerson:Person, theCompany:Company, theCar:Car)
pre   hireManager_pre: thePerson.isDefined and theCompany.isDefined and
     theCar.isDefined and theCompany.employee->excludes(thePerson)
post hireManager_post: Person.allInstances->includes(thePerson) and
     Company.allInstances->includes(theCompany) and
     Car.allInstances->includes(theCar) and
     theCompany.employee->includes(thePerson) and
     thePerson.staffCar=theCar

context fireManager(thePerson:Person, theCompany:Company)
pre   fireManager_pre: thePerson.isDefined and theCompany.isDefined and
     theCompany.employee->includes(thePerson) and
     thePerson.staffCar<>oclUndefined(Car)
post fireManager_post: Person.allInstances->includes(thePerson) and
     Company.allInstances->includes(theCompany) and
     theCompany.employee->excludes(thePerson) and
     thePerson.staffCar=oclUndefined(Car)


-- newPerson(aString:StringClass)
-- assume parameter:Seq(OclAny)=Seq{aString}
let _aString = parameter->at(1);
openter rc newPerson(_aString); assign _thePerson := create Person;
set _thePerson.name := _aString.value;
opexit;

-- hireManager(thePerson:Person, theCompany:Company, theCar:Car)
-- assume parameter:Seq(OclAny)=Seq{thePerson,theCompany,theCar}
let _thePerson = parameter->at(1); let _theCompany = parameter->at(2);
let _theCar = parameter->at(3);
openter rc hireManager(_thePerson,_theCompany,_theCar);
insert(_thePerson,_theCompany) into worksFor;
set _thePerson.staffCar := _theCar;
opexit;

-- fireManager(thePerson:Person, theCompany:Company)
-- assume parameter:Seq(OclAny)=Seq{thePerson,theCompany}
let _thePerson = parameter->at(1); let _theCompany = parameter->at(2);
openter rc fireManager(_thePerson,_theCompany);
set _thePerson.staffCar := oclUndefined(Car);
delete(_thePerson,_theCompany) from worksFor;
opexit;
```

Fig. 3. Conditions and Commands Generated from Graph Transformations

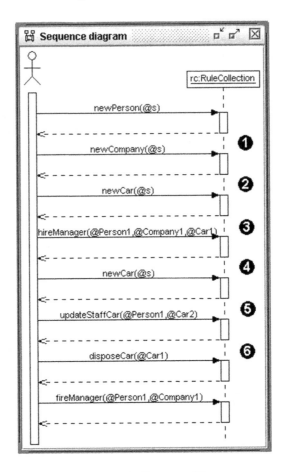

Fig. 4. Example Sequence Diagram

left side to restrict the rule applicability with a precondition, (B) in the object boxes to modify attributes with OCL expressions (like `aString.value` or `theCar`), and (C) in the right side to restrict applicability with a postcondition. This latter feature is not used in the example. The five rules are typical insofar that they cover atomic graph modifications: (A) Create an object, (B) create a link, (C) delete a link, (D) modify an attribute, and (E) delete an object.

Let us assume that each rule describes one operation. These generated operations are shown in the lower part of Fig. 2 in the class RuleCollection. The rule name determines the operation name, and the left side objects become the operation parameters. We have assigned the operations to a new class, but, if we indicate in the rule to which class the rule belongs, we could distribute the operations to given classes.

The central point we want to make is that we automatically generate an operation specification and an operation implementation from the graph transformations: The upper part of Fig. 3 shows the OCL pre- and postconditions

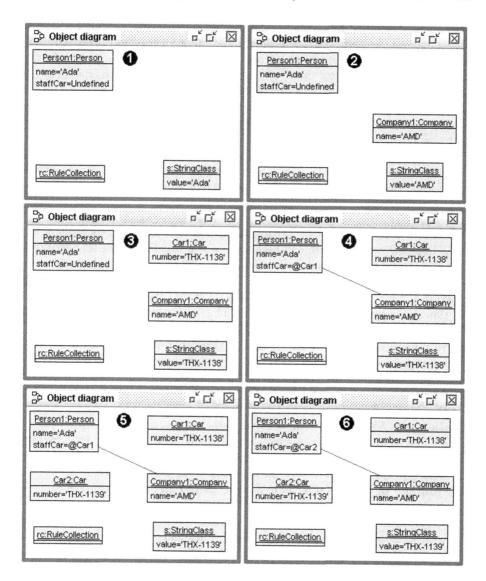

Fig. 5. Example Object Diagrams from Sequence Diagram

generated for the first three rules, and the lower part of Fig. 3 pictures the generated command sequences for executing the first three graph transformations. Details of the generation process will be discussed in the next section.

Figure 4 shows a sequence of rule applications resp. operation calls in form of a UML sequence diagram. The sequence diagram uses the operations `newCompany` and `newCar` not mentioned before, which are completely analogous to `newPerson`. Figure 5 traces the resulting working graph sequence for these rule applications with object diagrams. The sequence diagram can be regarded as an abbreviation of a command line protocol where the evaluation of all pre- and postconditions is given in full detail (which we do not show). The form of the arrows in the sequence diagram indicates that (A) the preconditions evaluate to true, i.e., it is allowed to apply the respective graph transformations at the redex determined by the actual parameter, and (B) the postconditions evaluate to true, i.e., the generated implementation with command sequences satisfies its specification. If a pre- or postcondition would have failed, the shape of the arrows would have been different indicating this failure.

To summarize: Our UML tool USE can be employed to execute the graph transformations and can give feedback on their properties. Let us now turn to some details of the conditions and commands in Fig. 3.

Variable 'parameter': Considering the command sequences, each operation is implemented by a command sequence which is called in the sequence diagram. Parameter passing currently works by letting the variable `parameter` hold the sequence of objects which are to be delivered as actual parameters.

Variable 'rc': In the example, we have decided that the new class RuleCollection is the container for the operations. Thus operation calls must be directed to an object `rc` of this class.

Parameters of the Operation: We have said that all objects of the left side become parameters of the operation. We might relax this: The necessity for being a parameter may be dropped, if the parameter is uniquely determined by another parameter, for example, because a *..1 multiplicity is present in the class diagram.

Data Type Valued Parameters: In the example, we had to introduce an auxiliary class StringClass which is essentially there only to pass a String-valued parameter to the operation. If the rule allows to explicitly identify data type valued parameters, such auxiliary classes can be avoided.

Advantage of Using OCL as a Target Language: The advantage that we see in our approach is that it provides full OCL support. For example, at any stage of the execution sequence, the current working graph can be inspected with arbitrary complex OCL queries. Additionally, OCL can be employed in the formulation of graph transformation for expressing rule application restrictions and attribute assignments. Traditional OCL invariants can be incorporated as well. Executing the graph transformation rules, displaying the result with object diagrams, and inspecting the resulting working graph with OCL gives support for checking whether the rules exactly perform the task which the rule designer expected from them.

3   Related Work

Rules have already been successfully applied to (meta-)modeling, for example, in [EHHS00] with a focus on UML 2.0 communication diagrams, but not on OCL on which we concentrate. Our mechanism of executing graph transformation with a UML engine supports checking and validating of graph transformations. We see our approach as an alternative mechanism to existing well tried graph transformation engines like AGG [dLT04], FUJABA [BGN+04], GREAT [KASS03], GROOVE [KR06] or PROGRES [SWZ96]. Our approach is similar to [VFV06] which realizes graph transformations operationally by translating them into basic database commands. [LSE05] proposes to specify contracts for software (pre- and postconditions) in a graphical way with graph transformations. The work concentrates on generating JML (Java Modeling Language) whereas we use OCL offering features not conceptually present in JML, e.g., `allInstances`. [Baa06] discusses in connection with OCL the so-called frame problem dealing with the parts of the system that should remain unchanged when a specific operation is executed. The work proposes a hybrid language with OCL and graph transformation features to handle that problem whereas we keep these two approaches distinct. A constraint language was already part of the graph transformation tool PROGRES. A nice comparison between that language and OCL can be found in [Sch01]. Due to this paper format an in-depth comparison is out of scope.

4   Language Features in Rules

Let us summarize how the rule language which we use for operation description looks like. We explain the language features with the schematic example in Fig. 6. We start with the simplified view that an operation is described by one rule

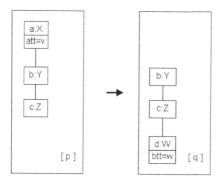

Fig. 6. Language Features in Rules

consisting of a pair of object diagrams where the left side of the rule shows the situation before operation execution and the right side the one after it.

– The connection between the two sides is established through common objects and links occurring in both sides (in Fig. 6 objects b and c and link (b,c)).
– Objects and links may only appear in the left side (in Fig. 6 object a and link (a,b)), whereas other objects and links only show up in the right side (in Fig. 6 object d and link (c,d)).

- The left and the right side may each contain one boolean OCL expression which allows to express an additional rule precondition (in Fig. 6 OCL expression p) and an additional rule postcondition (in Fig. 6 OCL expression q).
- The left side may contain attribute assertions permitting operation execution only if the given attribute values are present (in Fig. 6 attribute assertion att=v with att being an attribute and v being an OCL value expression).
- The right side may contain attribute assignments where the expressions to be assigned are evaluated in the state before operation execution (in Fig. 6 attribute assignment btt=w with btt being an attribute and w being an OCL value expression).

The object diagrams for the left and right side are not proper object diagrams, but they are generic object diagrams in the sense that the objects occurring in the diagram are substituted by concrete objects when the operation is executed and the parameters are passed.

Above we have assumed that one rule describes one operation. This may be generalized: One operation may be characterized by more than one rule without any difficulty if mutually exclusive preconditions are specified in the rules or if mutually exclusive object, link or attribute configurations are stated for different rules of a single operation in the left side. Then the operation can be seen as a larger case distinction which is reflected both in the pre- and postconditions and the command sequences. In future work, we want to allow more powerful operations by allowing for iteration and operation calling in rules as in [ZHG04, GBR05]. Rules would then have similar features like UML 1 collaboration diagrams.

5   Realization by OCL Pre- and Postconditions

The realization of rules by OCL pre- and postconditions is structured into 11 steps: (A) A single step for Initialization, (B) 3 steps handling objects (Object Creation, Object Preservation, Object Destruction), (C) 3 steps handling links (Link Creation, Link Preservation, Link Destruction), (D) 2 steps handling

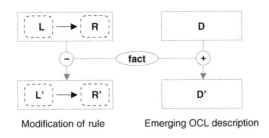

Modification of rule   Emerging OCL description

Fig. 7. Structure of Realization Steps for OCL

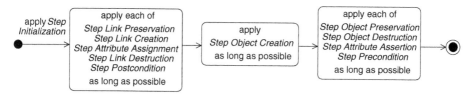

Fig. 8. Order of Realization Steps for OCL

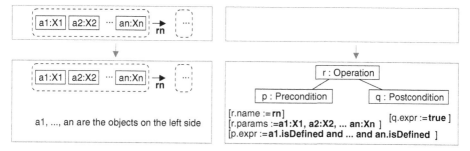

Fig. 9. Step Initialization

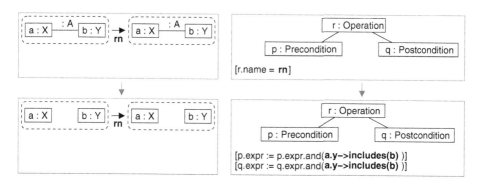

Fig. 10. Step Link Preservation

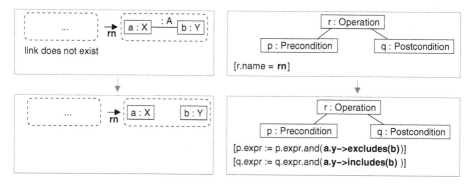

Fig. 11. Step Link Creation

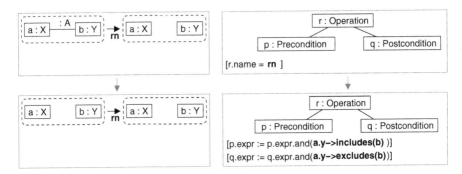

Fig. 12. Step Link Destruction

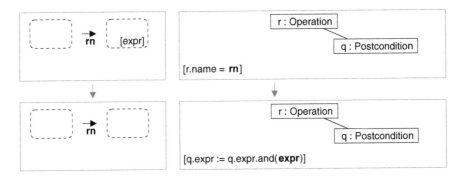

Fig. 13. Step Postcondition

attributes (Attribute Assertion, Attribute Assignment), and (E) 2 steps handling the additional pre- and postcondition (Precondition, Postcondition). Each step treats a particular aspect in the rule and, as depicted in Fig. 7, builds up the OCL description by removing a particular fact from the rule (L,R) and by adding this fact to the OCL description D. By this, one achieves an adapted rule (L',R') and a new OCL description D'. The process stops, if rules are not applicable any more. We show in the following figures the details of the central steps. We will refer to the process of modifying the rule as *rule adaption*.

The order in which the steps have to be applied is restricted as shown in Fig. 8. Within the respective block, the order is not significant, e.g., in the step Object Creation, it does not matter in which order the created objects of the rule are treated.

 - The step Initialization in Fig. 9 identifies the objects of the left side as parameters of a newly established operation having a precondition requiring the actual parameters to be different from undefined. The step does not adapt the rule (no rule adaption).
 - The step Link Preservation in Fig. 10 extends the already constructed pre- and postcondition by requirements for link existence at the start of the

operation as well as at the end of the operation. The step removes the considered link and thus indicates that this link has been completely handled.

- The step Link Creation in Fig. 11 guarantees in the precondition that the link is not existent and in the postcondition that the link exists. The step removes the link (rule adaption).

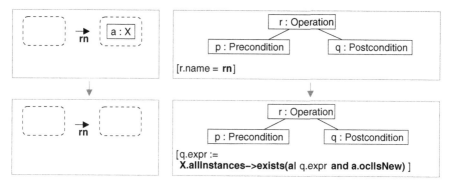

Fig. 14. Step Object Creation

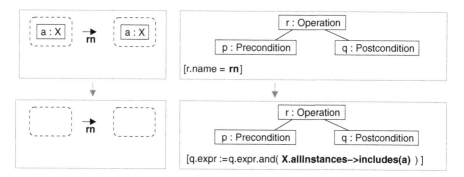

Fig. 15. Step Object Preservation

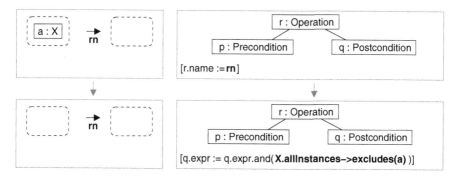

Fig. 16. Step Object Destruction

- The step Link Destruction in Fig. 12 guarantees in the precondition the existence of the link and in the postcondition the link's non-existence. The step removes the link (rule adaption).
- The step Postcondition in Fig. 13 adds the given OCL expression to the postcondition and removes the OCL expression.
- The step Object Creation in Fig. 14 guarantees in the postcondition that a new object is created and that the postcondition constructed so far is valid for this new object. The step removes the object (rule adaption).
- The step Object Preservation in Fig. 15 introduces the requirement that a preserved object still exists at postcondition time. Please recall, that all left side objects become operation parameters and are required in the initialization step to be defined, i.e., to be existent. The step removes the object (rule adaption).
- The step Object Destruction in Fig. 16 requires in the postcondition that the object does not belong to the current objects any more. The step removes the object (rule adaption).

We do not show the details of the steps Precondition, Attribute Assertion, and Attribute Assignment. Precondition works analogously to step Postcondition, Attribute Assertion is similar to step Precondition (because an assertion is basically a precondition), and Attribute Assignment resembles step Precondition (because an assignment is basically a postcondition).

6   Realization by Command Sequences

The realization of the rules by command sequences is structured into 5 steps as pictured in Fig. 17. The 5 steps and their order are as follows.

- Object Creation: The objects introduced by the right side of the rule are created.
- Link Creation: The links introduced by the right side of the rule are created by insertion into the link set of the association.
- Attribute Assignment: The attributes are modified according to the details given in the rule.
- Link Destruction: The links occurring only in the left side of the rule are destroyed by removing them from the link set of the association.
- Object Destruction: The objects occurring only the left side of the rule are destroyed.
- Steps *Link Creation* and *Link Destruction* may alternatively and equivalently be realized by modifying an object- and set-valued attribute on either side of the association (Steps 2b and 4b).

We do not cover handling of rule pre- or postconditions or attribute assertions. Basically, these features can be realized by allowing conditional execution of the commands already shown.

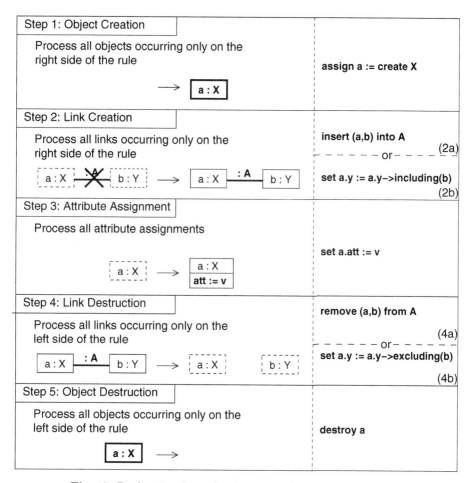

Fig. 17. Realization Steps for Command Sequence Generation

7   Conclusion

This paper proposes to translate graph transformations rules into two different formalisms: (A) A descriptive language expressing the effect of a rule in form of pre- and postconditions, and (B) a procedural language describing the effect of a rule in terms of basic imperative manipulation commands. The procedural realization can be employed to execute the graph transformations. The descriptive realization can be checked against the procedural one, i.e., the pre- and postconditions are evaluated during the execution. The pre- and postconditions can further be used for deduction and verification purposes. Both realizations give feedback to graph transformation designers for their development process. This feedback is in particular needed when the graph transformations work on a metamodel, e.g., the UML metamodel. We think that translating graph transformation into a standard software engineering language like UML including OCL broadens the audience of graph transformations, because standard UML and OCL tools are familiar for many developers. We have made double use of graph transformation rules: Firstly, we have used rules as the source language which we translate and realize in the two formalisms; secondly, we have described our translation in form of graph transformations.

We have used OCL as a vehicle, but we think that the basic idea behind the translation is independent of OCL. We could have used other languages like SQL achieving similar results.

Our approach enables us to automatically compute for a given set of rules and a given working graph, the redexes where rules could be applied. Multiplicities in the underlying class diagram can help to dramatically reduce the search space for the redexes. Due to space limitations we have not described this in detail. Other future work includes the following topics.

- There are several transformation alternatives to the decisions we have taken. For example, instead of using `oclIsNew` we could use explicitly `Class.allInstances->excludes` and `includes` in the pre- resp. postcondition. By this we would reduce the number of used OCL features. Another variation point concerns the parameters of the generated operations, which not necessarily have to be all objects of the left side. But the parameter list should be reduced only if the redex afterwards still uniquely determines the rule application.
- We want to improve the expressiveness of operation descriptions by allowing iteration and operation calling in the rules and plan to group rules into larger structuring units, so-called transformation units.
- A fundamental treatment of the relationship between the generated pre- and postcondition pair and the generated command sequence is needed, i.e., a study of the relationship between the descriptive and the operational view on rules. Such a study would ideally be inspired by the equivalence relationships between relational algebra and relational calculus from the database field.

References

[Baa06]   T. Baar. OCL and Graph-Transformations: A Symbiotic Alliance to Alleviate the Frame Problem. In MoDELS 2005 Satellite Events: International Workshops, Springer, Berlin, LNCS, pp. 20-31, 2006.

[BGN⁺04]   S. Burmester, H. Giese, J. Niere, M. Tichy, J.P. Wadsack, R. Wagner, L. Wendehals, and A. Zündorf. Tool integration at the meta-model level: the Fujaba approach. *STTT*, 6(3):203–218, 2004.

[BW02]   A.D. Brucker and B. Wolff. HOL-OCL: Experiences, consequences and design choices. In J.-M. Jézéquel, H. Hussmann, and S. Cook, editors, *Proc. 5th Int. Conf. Unified Modeling Language (UML'2002)*, 196–211. Springer, LNCS 2460, 2002.

[dLT04]   J. de Lara and G. Taentzer. Automated Model Transformation and Its Validation Using AToM 3 and AGG. In A.F. Blackwell, K. Marriott, and A. Shimojima, editors, *Diagrams*, LNCS 2980, 182–198. Springer, 2004.

[EHHS00]   G. Engels, J.H. Hausmann, R. Heckel, and S. Sauer. Dynamic meta modeling: A graphical approach to the operational semantics of behavioral diagrams in UML. In A. Evans, S. Kent, and B. Selic, editors, *Proc. 3rd Int. Conf. Unified Modeling Language (UML'2000)*, 323–337. Springer, LNCS 1939, 2000.

[GBR05]   M. Gogolla, J. Bohling, and M. Richters. Validating UML and OCL Models in USE by Automatic Snapshot Generation. *Journal on Software and System Modeling*, 4(4): 386-398, 2005.

[KASS03]   G. Karsai, A. Agrawal, F. Shi, and J. Sprinkle. On the Use of Graph Transformation in the Formal Specification of Model Interpreters. *Journal Universal Computer Science*, 9(11):1296–1321, 2003.

[KR06]   H. Kastenberg and A. Rensink. Model Checking Dynamic States in GROOVE. In A. Valmari, editor, *SPIN*, LNCS 3925, 299–305. Springer, 2006.

[LSE05]   M. Lohmann, S. Sauer, and G. Engels. Executable Visual Contracts. In IEEE Symp. Visual Languages and Human-Centric Computing (VL/HCC'05), 63-70, 2005.

[OMG04]   OMG, editor. *OMG Unified Modeling Language Specification, Version 2.0*. OMG, 2004.

[RBJ05]   J. Rumbaugh, G. Booch, and I. Jacobson. *The Unified Modeling Language Reference Manual, Second Edition*. Addison-Wesley, Reading, 2005.

[Sch01]   A. Schürr. Adding Graph Transformation Concepts to UML's Constraint Language OCL. In Proc. ETAPS Workshop UNIGRA, ENTCS, 44 (4), 403-410, 2001.

[SWZ96]   A. Schürr, A.J. Winter, and A. Zündorf. Developing Tools with the PROGRES Environment. In M. Nagl, editor, *IPSEN Book*, LNCS 1170, 356–369. Springer, 1996.

[VFV06]   G. Varró, K. Friedl, and D. Varró. Implementing a Graph Transformation Engine in Relational Databases. *Journal on Software and System Modeling*, 2006.

[WK03]   J. Warmer and A. Kleppe. *The Object Constraint Language: Precise Modeling with UML*. Addison-Wesley, 2003. 2nd Edition.

[ZHG04]   P. Ziemann, K. Hölscher, and M. Gogolla. From UML Models to Graph Transformation Systems. In M. Minas, editor, *Proc. Workshop Visual Languages and Formal Methods (VLFM'2004)*. ENTCS, 127(4), 17-33, 2004.

Heuristic Search for the Analysis
of Graph Transition Systems

Stefan Edelkamp[1], Shahid Jabbar[1], and Alberto Lluch Lafuente[2]

[1] Computer Science Department, University of Dortmund, Dortmund, Germany
{stefan.edelkamp, shahid.jabbar}@cs.uni-dortmund.de
[2] via del Giardino A 58, I-50053 Empoli, Italy
albertlluch@gmail.com

Abstract. Graphs are suitable modeling formalisms for software and hardware systems involving aspects such as communication, object orientation, concurrency, mobility and distribution. State spaces of such systems can be represented by graph transition systems, which are basically transition systems whose states and transitions represent graphs and graph morphisms. Heuristic search is a successful Artificial Intelligence technique for solving exploration problems implicitly present in games, planning, and formal verification. Heuristic search exploits information about the problem being solved to guide the exploration process. The main benefits are significant reductions in the search effort and the size of solutions. We propose the application of heuristic search for the analysis of graph transition systems. We define algorithms and heuristics and present experimental results.

1 Introduction

Graphs are a suitable formalism for software and hardware systems involving issues such as communication, object orientation, concurrency, distribution and mobility. The graphical nature of such systems appears explicitly in approaches like graph transformation systems [31] and implicitly in other modeling formalisms like algebras for communicating processes [26]. The properties of such systems mainly regard aspects such as temporal behavior and structural properties. They can be expressed, for instance, by logics used as a basis for a formal verification method, like model checking [5], whose success is mainly due to the ability to find and report errors.

Finding and reporting errors in model checking and many other analysis problems can be reduced to state space exploration problems. In most cases, the main drawback is the *state explosion problem*. In practice, the size of state spaces can be large enough (even infinite) to exhaust the available space and time resources. Heuristic search has been proposed as a solution in many fields, including model checking [13], planning [2] and games [24]. Basically, the idea is to apply algorithms that exploit the information about the problem being solved in order to guide the exploration process. The benefits are twofold: the search effort is reduced, for instance, errors are found faster and by consuming less memory, and

A. Corradini et al. (Eds.): ICGT 2006, LNCS 4178, pp. 414–429, 2006.

the solution quality is improved, i.e., counterexamples are shorter and thus may be more useful.

In cases like wide area networks with Quality of Service (QoS), one might not be interested in short paths, but in cheap or optimal ones according to a certain notion of *cost*. Typical examples are algebra defined on Reals, on Booleans, on Probabilities, or on any other system. To cover such a diversity, we generalize our approach by considering an abstract notion of cost.

Our work is mainly inspired by approaches to directed model checking [13], logics for graphs (like the *monadic second order logic* [8]), spatial logics used to reason about the behavior and structure of processes calculi [3] and graphs [4], approaches for the analysis of graph transformation systems [1,15,28,33], and cost-algebraic search algorithms [12,32].

The goal of our approach is to formalize a framework for the application of heuristic search in order to analyze structural properties of systems modeled by graph transition systems. We believe that our work additionally illustrates the benefits of applying heuristic search for state space exploration. Heuristic search is intended to reduce the analysis effort and, in addition, to deliver shorter solutions, which in our case means shorter paths in graph transition systems. Such paths might represent errors of a system or examples of interesting correct behaviours. It is worth saying that our approach offers no benefit if one is interested in exhaustively exploring a states space, like is usual when one needs to correctnes of a graph transition system.

Section 2 introduces a running example that is used along the paper to illustrate some of the concepts and methods. Section 3 defines our modeling formalism, namely graph transition systems. Section 4 defines the kind of properties we are interested in verifying. Section 5 summarizes the analysis algorithms and discusses their correctness. Section 6 proposes heuristics for the analysis of properties in graph transition systems. Abstraction is one of the most successful techniques in model checking. In Section 7, we discuss the role of abstraction to define useful heuristic estimates. Section 8 presents experimental results. Section 9 concludes the paper and outlines future research avenues.

2   The Arrow Distributed Directory Protocol

The *arrow distributed directory protocol* [9] is a solution to ensure exclusive access to mobile objects in a distributed system. The protocol induces a *distributed queue* structure on a distributed system. The distributed system is given as an undirected graph G, where vertices and edges respectively represent nodes and communication links. Costs are associated with the links in the usual way, and a mechanism for optimal routing is assumed.

The protocol works with a minimal spanning tree T of G. Each node has an arrow which either indicates the *direction* in which the object lies or is going to be. If a node owns the object or is requesting it, the arrow points to itself; we say that the node is a *terminal*. The directed graph induced by the arrows is called $\mathcal{L}$. The protocol works by propagating requests and updating arrows such

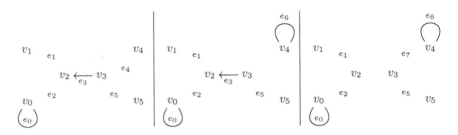

Fig. 1. Three states of the directory

that at any moment the paths induced by arrows, called *arrow paths*, lead to a terminal either owning the object or waiting for it.

Fig. 1 illustrates three states of a protocol instance with six nodes $v_0, \ldots, v_5$. For the sake of simplicity only $\mathcal{L}$ is depicted. The state on the left is the initial one: node v_0 has the object and all paths in $\mathcal{L}$ lead to it. The state on the right of the figure is the result of two steps: 1) node v_4 sends a request for the object through its arrow; and 2) v_3 processes it, making its arrow points to v_4. Request propagation should end by making all paths in $\mathcal{L}$ pointing towards v_4, where the object will be transfered once v_0 is finished with it. Each propagation step comprises of two transitions: deleting its out-going edge, and adding the new edge in the direction, where the request came from.

One could be interested in properties like: *Can a certain node v be terminal?* (Property 1), *Can a certain node v be terminal and all arrow paths end at v?* (Property 2), *Can some node be terminal?* (Property 3), *Can some node be terminal and all arrow paths end at it?* (Property 4).

3   Graph Transition Systems

This section presents our algebraic notion of costs. It shall be used as an abstraction of costs or weights associated to edges of graphs or transitions of transition systems. For a deeper treatment of the cost algebra, we refer to [12].

Definition 1. *A* cost algebra *is a 5-tuple* $\langle A, \times, \preceq, \mathbf{0}, \mathbf{1} \rangle$, *such that 1)* $\langle A, \times \rangle$ *is a monoid with* $\mathbf{1}$ *as identity element and* $\mathbf{0}$ *as its absorbing element, i.e.,* $a \times \mathbf{0} = \mathbf{0} \times a = \mathbf{0}$; *2)* $\preceq \subseteq A \times A$ *is a total ordering with* $\mathbf{0} = \bigsqcap A$ *and* $\mathbf{1} = \bigsqcup A$; *A is* isotone, *i.e.,* $a \preceq b$ *implies both* $a \times c \preceq b \times c$ *and 3)* $c \times a \preceq c \times b$ *for all* $a, b, c \in A$ [32].

In the rest of the paper $a \prec b$ abbreviates $a \preceq b$ and $a \neq b$. Moreover, $a \succeq b$ abbreviates $b \preceq a$, and $a \succ b$ abbreviates $a \succeq b$ and $a \neq b$.

Intuitively, A is the domain set of cost values, which is linearly ordered by $\preceq$ and has $\bigsqcup, \bigsqcap$ as least and greatest operations, and $\times$ is the operation used to cumulate values. Consider, for example, the following instances of cost algebras, typically used as cost or QoS formalisms: $\langle \{true, false\}, \wedge, \Rightarrow, false, true \rangle$ (Network and service availability), $\langle \mathbb{R}^+ \cup \{+\infty\}, +, \leq, +\infty, 0 \rangle$ (Price, propagation delay) or $\langle\ ^+\cup\{+\infty\},\min,\geq,0,+\infty\rangle$ (Bandwidth). In the rest of the paper, we consider a fixed cost algebra $\langle A,\times,\preceq,\mathbf{0},\mathbf{1}\rangle$.

Definition 2. *A graph G is a tuple $\langle V_G, E_G, src_G, tgt_G, \omega_G\rangle$ where V_G is a set of nodes, E_G is a set of edges, $src_G, tgt_G : E_G \rightarrow V_G$ are source and target functions, and $\omega_G : E_G \rightarrow A$ is a weighting function.*

Graphs usually have a particular start state $s_0^G \in V_G$, which we sometimes denote with s_0 if G is clear from the context.

Definition 3. *A path in a graph G is a alternating sequence of nodes and edges represented as $u_0 \overset{e_0}{\rightarrow} u_1 \ldots$ such that for each $i \geq 0$, we have $u_i \in V_G$, $e_i \in E_G$, $src_G(e_i) = u_i$ and $tgt_G(e_i) = u_{i+1}$, or, shortly $u_i \overset{e_i}{\rightarrow} u_{i+1}$.*

An initial path is a path starting at s_0^G. Finite paths are required to end at states. The length of a finite path p is denoted by $|p|$. The concatenation of two paths p, q is denoted by pq, where we require p to be finite and end at the initial state of q. The cost of a path is the cumulative cost of its edges. Formally,

Definition 4. *Let $p = u_0 \overset{e_0}{\rightarrow} \ldots \overset{e_{k-1}}{\rightarrow} u_k$ be a finite path in a graph G. The path cost $\omega_G(p)$ is $\omega_G(e) \times \omega_G(q)$ if $p = (u \overset{e}{\rightarrow} v)q$ and $\mathbf{1}$ otherwise.*

Let $\gamma(u)$ denote the set of all paths starting at node u. We shall use $\omega_G(u, V)$ to denote the cost of the optimal path starting at a node u and reaching a node v in a set $V \subseteq V_G$. For ease of notation, we write $\omega_G(u, \{v\})$ as $\omega_G(u, v)$.

Graph transition systems are suitable representations for software and hardware systems and extend traditional transition systems by relating states with graphs and transitions with partial graph morphisms. Intuitively, a partial graph morphism associated to a transition represents the relation between the graphs associated to the source and the target state of the transition, i.e., it models the merging, insertion, addition and renaming of graph items, where the cost of merged edges is the least one amongst the edges involved in the merging.

Definition 5. *A graph morphism $\psi : G_1 \rightarrow G_2$ is a pair of mappings $\psi_V : V_{G_1} \rightarrow V_{G_2}$, $\psi_E : E_{G_1} \rightarrow E_{G_2}$ such that we have $\psi_V \circ src_{G_1} = src_{G_2} \circ \psi_E$, $\psi_V \circ tgt_{G_1} = tgt_{G_2} \circ \psi_E$, and for each $e \in E_{G_2}$ such that $\{e' \mid \psi_E(e') = e\} = \emptyset$ we have, $\omega_{G_2}(e) = \bigsqcup\{\omega_{G_1}(e') \mid \psi_E(e') = e\}$. A graph morphism $\psi : G_1 \rightarrow G_2$ is called injective if so are ψ_V and ψ_E; identity if both ψ_V and ψ_E are identities, and isomorphism if both ψ_E and ψ_V are bijective. A graph G' is a subgraph of graph G, if $V_{G'} \subseteq V_G$ and $E_{G'} \subseteq E_G$, and the inclusions form a graph morphism. A partial graph morphism $\psi : G_1 \rightarrow G_2$ is a pair $\langle G_1', \psi_m\rangle$ where G_1' is a subgraph of G_1, and $\psi_m : G_1' \rightarrow G_2$ is a graph morphism.*

The composition of (partial) graph morphisms results in (partial) graph morphisms. Now, we extend transition systems with weights.

Definition 6. *A transition system is a graph $M = \langle S_M, T_M, in_M, out_M, \omega_M\rangle$ whose nodes and edges are called states and transitions, with in_M, out_M representing the source and target of an edge.*

Finally, we are ready to define graph transition systems, which are transition systems together with morphisms mapping states into graphs and transitions into partial graph morphisms.

Definition 7. *A graph transition system (GTS) is a pair $\langle M, g \rangle$, where M is a weighted transition system and $g : M \to \mathcal{U}(\mathbf{G}_p)$ is a graph morphism from M to the graph underlying $\mathbf{G}_p$, the category of graphs with partial graph morphisms. Therefore $g = \langle g^S, g^T \rangle$, and the component on states g^S maps each state $s \in S_M$ to a graph $g^S(s)$, while the component on transitions g^T maps each transitions $t \in T_M$ to a partial graph morphism $g^T(t) : g^S(in_M(t)) \Rightarrow g^S(out_M(t))$.*

In the rest of the paper, we shall consider a GTS $\langle M, g \rangle$ modeling the state space of our running example, where g maps states to $\mathcal{L}$, i.e., the graph induced by the arrows, and transitions to the corresponding partial graph morphisms. Consider Fig. 1, each of the three graphs depicted, say G_1, G_2 and G_3 corresponds to three states s_1, s_2, s_3, meaning that $g(s_1) = G_1$, $g(s_2) = G_2$ and $g(s_3) = G_3$. The figure illustrates a path $s_1 \xrightarrow{t_1} s_2 \xrightarrow{t_2} s_3$, where $g(t_1)$ is the identity restricted to all items but edge e_4. Similarly, $g(t_2)$ is the identity restricted to all items but edge e_3. Thus, in both transitions all other items are preserved (with their identity) except the edges mentioned.

4   Properties of Graph Transition Systems

The properties of a graph transition system can be expressed using different formalisms. One can use, for instance, a temporal graph logic like the ones proposed in [1,28], which combine temporal and graph logics. A similar alternative are spatial logics [3], which combine temporal and structural aspects. In graph transformation systems [7], one can use rules to find certain graphs: the goal might be to find a match for a certain transformation rule. For the sake of simplicity, however, we consider that the problem of satisfying or falsifying a property is reduced to the problem of finding a set of *goal states* characterized by a goal graph and the existence of an injective morphism.

Definition 8. *Given a GTS $\langle M, g \rangle$ and a graph G, the goal function $goal_G : S_M \to \{true, false\}$ is defined such that $goal_G(s) = true$ iff there is a partial injective graph morphism $\psi : G \to g(s)$.*

Intuitively, $goal_G$ maps a state s to *true*, if and only if G can be injectively matched with a subgraph of $g(s)$. It is worth mentioning that most graph transformations approaches consider injective rules, for which a match is precisely given by injective graph morphisms, and that the most prominent graph logic, namely the Monadic Second-Order (MSO) logic by Courcelle [8] and its first-order fragment (FO) can be used to express injective graph morphisms. The graph G will be called *goal graph*. It is of practical interest identifying particular cases of goal functions as the following goal types:

Fig. 2. Three graphs illustrating various goal criteria

1. ψ is an identity - the exact graph G is looked for. In our running example, this corresponds to Property 2 mentioned in Section 2. For instance, we look for the exact graph depicted to the left of Fig. 2.
2. ψ is a restricted identity - an exact subgraph of G is looked for. This is precisely Property 1. For instance, we look for a subgraph of the graph depicted to the left of Fig. 2. The graph in the center of Fig. 2 satisfies this.
3. ψ is an isomorphism - a graph isomorphic to G is looked for. This is precisely Property 4. For instance, we look for a graph isomorphic to the one depicted to the left of Fig. 2. The graph to the right of Fig. 2 satisfies this.
4. ψ is any injective graph morphism - we have the general case. This is precisely Property 3. For instance, we look for an injective match of the graph depicted in the center of Fig. 2. The graph to the right of Fig. 2 satisfies this.

Note that there is a type hierarchy, since goal type 1 is a subtype of goal types 2 and 3, which are subtypes of the most general goal type 4.

The computational complexity of the goal function varies according to the above cases. For goals of type 1 and 2, the computational efforts needed are just $O(|G|)$ and $O(|\psi(G)|)$, respectively. Unfortunately, for goal types 3 and 4, due to the search for isomorphisms, the complexity increases to a term exponential in $|G|$ for the graph isomorphism case and to a term exponential in $|\psi(G)|$ for the subgraph isomorphism case. The problem of graph isomorphism is not completely classified. It is expected not to be NP-complete [34].

Now we state the two analysis problems we consider.

Definition 9. *Given a GTS $\langle M, g \rangle$ and a graph G (the goal graph), the* reachability problem *consists of finding a state $s \in S_M$ such that* goal(s) *is true. The* optimality problem *of our approach consists of finding a finite initial path p ending at a state $s \in S_M$ such that* goal$_G(s)$ *is true and* $\omega(p) = \omega_M(s_0^M, S')$, *where* $S' = \{s \in S_M \mid goal_G(s) = true\}$.

For the sake of brevity, in the following, $\omega_M(s)$ abbreviates $\omega_M(s, S')$ with $S' = \{s \in S_M \mid goal_G(s) = true\}$, when $goal_G$ is clear from the context.

5   The Analysis of Graph Transition Systems

The two problems defined in the previous section can be solved with traditional graph exploration and shortest-path algorithms. For the reachability problem, for instance, one can use, amongst others, depth-first search, hill climbing, best-first search, Dijkstra's algorithm (and its simplest version breadth-first search) or A*. For the optimality problem, only the last two are suited.

Recall that Dijkstra's algorithm [6] maintains a set of nodes as search horizon and iteratively explores the (currently optimal) node in the horizon. A* [18] basically improves Dijkstra's algorithm by selecting the most *promising* node for expansion by considering not only the weight of the current optimal path to a node but also a heuristic estimate of its distance to the set of goal nodes. Contrarily, best-first search takes into account the heuristic only. For a deeper treatment of both algorithms, we refer to [6,18,27].

Dijkstra's algorithm and A* are traditionally defined over a simple instance of our cost algebra A, namely cost algebra $\langle \ ^+ \cup \{+\infty\}, +, \leq, +\infty, 0 \rangle$. Fortunately, the results that ensure the *admissibility* of Dijkstra's algorithm or A*, i.e., the fact that both algorithms correctly solve the optimality problem, have been generalized for the cost algebra [12]:

Proposition 1. *Dijkstra's algorithm solves the optimality problem.*

Definition 10. *Given a GTS $\langle M, g \rangle$ and a goal function* $goal_G$, *a heuristic* $h : S_M \to A$ *is* admissible, *if for all* $s \in S_M$ *we have* $h(s) \preceq \omega_M(s)$, *and have* $h(s) = 1$ *whenever* $goal_G(s)$; consistent, *if for each* $s \xrightarrow{t} s'$, *we have* $h(s) \preceq \omega(t) \times h(s')$.

A consistent heuristic is admissible if for all s such that $goal_G(s)$, $h(s) = 1$, even for our cost algebra [12].

It is worth saying that in some practical cases non-admissible strategies, like A* with non-admissible heuristics or best-first, find near-to-optimal solutions efficiently.

Proposition 2. *For an admissible heuristic A* solves the optimality problem.*

6   Heuristics for Graph Transition Systems

Now we propose various heuristics for the analysis of graph transition systems.

Items to Remove and Insert. Consider Fig. 1 and suppose we want to estimate the number of transitions necessary to transform the leftmost graph to the rightmost one. We need to remove e_3 and e_4, and to add e_6 and e_7. Since we know that each transition removes and adds at most one edge, we conclude that at least two transitions are necessary. We can generalize them as follows.

First, recall that partial graph morphisms are induced by system transitions. In the case of graph transformation systems, for instance, graph morphisms are induced by graph transformation rules, while in communication protocols by the operations of the processes. In most cases, such transitions are usually local and involve a few insertion/deletion/merging of items. We can thus determine, prior to the analysis, the number of items deleted and erased by graph morphisms.

Let n_m^i and n_m^d respectively be the maximum number of inserted and deleted nodes in any transition, and e_m^i and e_m^d respectively be the maximum number of inserted and deleted edges in any transition, where the merging of n items is interpreted as the deletion of $n-1$ items. Let G be the goal graph. In addition, let c_m be the least cost associated to transitions. On the other hand, for goals of type 2, only the number of items to add are relevant. When considering goals of type 3, we cannot rely on the identity of edges as in previous heuristics and have thus to base our heuristic on the number of items to be added or deleted. Finally, for type 4, only the number of items to be added is taken into account because we have use item identities.

Definition 11. *Heuristics $h_n^1, h_n^2, h_n^3, h_n^4$ are defined as follows:*

$$h_n^1(s) = c_m^{\max\{\lfloor |V_G \backslash V_{g(s)}|/n_m^i \rfloor, \lfloor |E_G \backslash E_{g(s)}|/e_m^i \rfloor, \lfloor |V_{g(s)} \backslash V_G|/n_m^d \rfloor, \lfloor |E_{g(s)} \backslash E_G|/e_m^d \rfloor\}}$$

$$h_n^2(s) = c_m^{\max\{\lfloor |V_G \backslash V_{g(s)}|/n_m^i \rfloor, \lfloor |E_G \backslash E_{g(s)}|/e_m^i \rfloor\}}$$

$$h_n^3(s) = c_m^{\max\{(\lfloor |V_G|-|V_{g(s)}|)/n_m^i \rfloor, (\lfloor |E_G|-|E_{g(s)}|)/e_m^i \rfloor, (\lfloor |V_{g(s)}|-|V_G|)/n_m^d \rfloor, (\lfloor |E_{g(s)}|-|E_G|)/e_m^d \rfloor\}}$$

$$h_n^4(s) = c_m^{\max\{(\lfloor |V_G|-|V_{g(s)}|)/n_m^i \rfloor, (\lfloor |E_G|-|E_{g(s)}|)/e_m^i \rfloor\}}$$

Proposition 3. *Heuristic h_n^1 (resp. h_n^2, h_n^3, h_n^4) is consistent and, for goals of type 1 (resp. 2,3,4), admissible.*

Isomorphism Heuristics. The main drawback of the previously presented heuristics for goals of type 4 is evident. If state graphs have more edges and nodes than the goal graph, the resulting heuristic is completely blind, i.e., it returns 1 for all states. Thus A* degenerates into Dijkstra and best-first into a random search. Thus, we propose functions inspired by heuristics to decide isomorphism or sub-graph isomorphism. For instance, if one has to decide whether two graphs are isomorphic one would check first whether the two graphs have the same number of items. If so, one could continue trying to match nodes with the same in- and out-degrees.

First, let $d_{in}(u)$ and $d_{out}(u)$ denote the in- and out-degree of a node u in a graph G, i.e., $d_{in}(u) = |\{e \in E_G \mid tgt(e) = u\}|$ and $d_{out}(u) = |\{e \in E_G \mid src(e) = u\}|$. Let further D_G be the set of pairs of in- and out-degrees of all nodes of G, i.e., $D_G = \bigcup_{u \, V_G} \langle in(u), out(u)\rangle$, and $\hat{D}_G$ be a vector with all elements of D_G ordered according to the first component of the tuples. Finally, let d_M denote the Manhattan distance between two vectors, i.e., $d_M(u,v) = \sum_i |u_i - v_i|$.

Definition 12. *Let G be the goal graph. We define h_c^4 as*

$$h_c^4(s) = c_m^{\sum_{i=0}^{\max\{|\hat{D}_G|,|\hat{D}_{g(s)}|\}} d_M(\hat{D}_G[i], \hat{D}_{g(s)}[i])},$$

where $\hat{D}_{G'}[i]$ is $\langle 0,0\rangle$ if $i \geq |\hat{D}_{G'}|$. In words, we compute for each graph G and $g(s)$ a node degree-ordered vector. Then we compute the Manhattan distances of elements in the same rank. Intuitively, we decide a match of nodes and establish how many in- and out-going edges have to be removed or inserted.

Note that if one graph has more nodes than the other, we consider that the graph with less nodes has extra nodes with no degree at all. If the goal type is 3, we can refine the heuristic by trying different matches of the two vectors, as formalized in the following heuristic:

Definition 13. *Let G be the goal graph. We define h_c^3 as*

$$h_c^3(s) = \begin{cases} h_c^4(s) & \text{if } |\hat{D}_G| \geq |\hat{D}_{g(s)}| \\ c_m \bigsqcup \bigsqcup_{j=0}^{|\hat{D}_G|-|\hat{D}_{g(s)}|} \sum_{i=0}^{|\hat{D}_G|} d_M(\hat{D}_G[i+j], \hat{D}_{g(s)}[i]) & \text{otherwise} \end{cases}$$

Obviously, none of the two heuristics presented in this section is consistent or admissible in general, and one could define other versions of the heuristics by changing some of the parameters used: the order criteria, the distance between vectors, etc. The idea of these heuristics is, indeed, to illustrate the wide variety of non-admissible heuristics one could define.

Formula-Based Heuristic. Based on the original formula-based heuristic [13] we define a heuristic that exploits the specification of goal states by graph formulae. The details on how to transform the goal function as a goal graph and a requirement on the injective morphism into a corresponding closed negation-free FO graph formula is out of the scope of the paper. A simple example for Property 3 is to find out if there is an edge with the same source and target, i.e., $\exists e.src(e) = tgt(e)$.

In addition to boolean connectives, FO ingredients include first-order node and edge quantifiers, and node and edge comparison. For a detailed description of the logic we refer to [8]. The idea of the formula-based heuristic is that each false predicate contributes to an increase to the value. In other words, FO formulae are interpreted over the domain of the cost algebra in the spirit of quantitative logics [20]. Thus, *true* is interpreted as **1**, *false* as c_m, disjunction as selection and conjunction as cumulation.

Definition 14. *Let G be a graph and f, g be closed negation-free FO formulae. The interpretation of FO formulae over the cost algebra is given by*

$$\begin{array}{ll} true^{\,G} = \mathbf{1} & false^{\,G} = c_m \\ f \vee g^{\,G} = f^{\,G} \sqcup g^{\,G} & f \wedge g^{\,G} = f^{\,G} \times g^{\,G} \\ \exists x.f^{\,G} = \bigsqcup_u{}_{V_G} f\{^u/_x\}^{\,G} & \forall x.f^{\,G} = \prod_u{}_{V_G} f\{^u/_x\}^{\,G} \\ \exists y.f^{\,G} = \bigsqcup_e{}_{E_G} f\{^e/_y\}^{\,G} & \forall y.f^{\,G} = \prod_e{}_{E_G} f\{^e/_y\}^{\,G} \\ u = u'^{\,G} = \mathbf{1} \text{ if } u = u', c_m \text{ otherwise} & e = e'^{\,G} = \mathbf{1} \text{ if } e = e', c_m \text{ otherwise} \end{array}$$

where $\prod$ denotes the iterated application of operator $\times$, x and y are node and edge variables, respectively, and u, u' and e, e' are node and edge constants.

Finally, we define the formula-based heuristic as the interpretation of the formula described over the cost algebra.

Definition 15. *Heuristic h_f is defined as $h_f(s) = f^{\,g(s)}$.*

The formula-based heuristic is neither consistent nor admissible in general as one transition can change the falsehood of more than one predicate.

Hamming Distance. The Hamming distance of two bit vectors is the number of vector indices on which the bits differ. As there are many different encodings of a graph, we choose a simple one based on the image of the state representation.

Definition 16. *If bin_G denotes the bit-vector representation of G, and if we interpret* false *as 0, and* true *as 1, we obtain*

$$h_h(s) = c_m^{bin_G} \quad bin_{g(s)} + \sum_{i=0}^{\min\{|bin_G|, |bin_{g(s)}|\}} bin_G[i] \quad bin_{g(s)}[i]$$

As more than one bit can change within one transition (e.g. the last one before reaching the goal) heuristic h_h is neither admissible nor consistent.

Tool-specific Heuristics. Finally, one can profit from specific heuristics available in the concrete tool that performs the analysis. For example, if the system is implemented and analyzed with HSF-SPIN [13], one can benefit from heuristics like the FSM Distance, which takes into account the finite automata representation of processes. If the Java Pathfinder [16] is used structural heuristics based on coverage metrics and interleavings are available. If planning tools like MIPS [11] are used, one can apply variants of the *relaxed planning heuristic* [21].

7 Abstraction and Heuristic Search

Abstraction is one of the most important issues to cope with large and infinite state spaces and to reduce the exploration efforts. Abstracted systems should be significantly smaller than the original one and while preserve some properties of concrete systems. The study of abstraction formalisms for graph transition systems is, however, out of the scope of this paper. We refer to [1] for an example of such a formalism. We assume that abstractions are available, state the properties necessary for abstractions to preserve our two problems (reachability and optimization) and propose how to use abstraction to define informed heuristics.

The preservation of the reachability problem means that the existence of an initial goal path in the concrete system must entail the existence of a corresponding initial goal path in the abstract system. Note that this does not mean the existence of *spurious* initial goal paths in the abstract system, i.e., abstract paths that do no not correspond to any concrete path. Similarly, the preservation of the optimization problem means that the cost of the optimal initial goal path in the concrete system should be greater than or equal to the cost of the optimal initial goal path in the abstract system.

Abstractions have been applied in combination with heuristic search both in model checking [14] and planning [10] approaches. The main idea is that the abstract system is explored in order to create a database that stores the exact distances from abstract states to the set of abstract goal states. The exact distance between abstract states is an admissible and consistent estimate of the distance between the corresponding concrete states. The distance database is thus used as heuristics for analyzing the concrete system.

We recall the notion of $\langle \alpha, \gamma \rangle$-simulations [25] typically used in model checking abstraction approaches.

Definition 17. *Let S, S' be two set of states. A* Galois connection *from 2^S to $2^{S'}$ is a pair of monotonic functions $\langle \alpha, \gamma \rangle$, with $\alpha : 2^S \to 2^{S'}$ (abstraction) and $\gamma : 2^{S'} \to 2^S$ (concretization) such that $\mathrm{id}_{2^S} \subseteq \gamma \circ \alpha$ and $\alpha \circ \gamma \subseteq \mathrm{id}_{2^{S'}}$.*

Definition 18. *Let $\langle M, g \rangle$ and $\langle M', g' \rangle$ be two GTSs and $\langle \alpha, \gamma \rangle$ be a Galois connection from 2^{S_M} to $2^{S_{M'}}$. We say that $\langle M', g' \rangle$ $\langle \alpha, \gamma \rangle$-simulates $\langle M, g \rangle$, written $\langle M, g \rangle \sqsubseteq_{\alpha, \gamma} \langle M', g' \rangle$, if $\alpha \circ \mathrm{pre} \circ \gamma \subseteq \mathrm{pre}'$, where $\mathrm{pre} : S_M \to S_M$ is defined by $\mathrm{pre}(S) = \{s \in S \mid s \xrightarrow{t} s'\}$ and pre' is defined similarly for $\langle M', g' \rangle$.*

We say that a simulation $\langle M, g \rangle \sqsubseteq_{\alpha, \gamma} \langle M', g' \rangle$ preserves a goal function $goal_G$ whenever $s' \in \alpha(s')$ implies $goal_G(s) \Rightarrow goal_G(s')$, and we call it *cost consistent* if for any transition $s_1 \xrightarrow{t} s_2$ in M there is a transition $s_1' \xrightarrow{t'} s_2'$ with $s_1' \in \alpha(s_1)$, $s_2' \in \alpha(s_2)$ and $\omega(t') \preceq \omega(t)$.

Proposition 4. *Let $\langle M, g \rangle$, $\langle M', g' \rangle$ be two GTSs such that $\langle M, g \rangle \sqsubseteq_{\alpha, \gamma} \langle M', g' \rangle$ and $goal_G$ is preserved. Then, there is a solution to the reachability problem in $\langle M', g' \rangle$ if there is a solution to the reachability problem in $\langle M, g \rangle$.*

As a consequence, if there is no solution to the reachability problem in the abstract graph transition system, there is no solution in the concrete system. Recall that the contrary is not true: there might be initial goal paths in the abstract system, but not in the concrete one. Such *spurious* solutions are usually eliminated by refining the abstractions [19]. Now we state that the optimality problem is preserved for cost consistent simulations.

Proposition 5. *Let $\langle M, g \rangle$, $\langle M', g' \rangle$ be two GTSs such that $\langle M, g \rangle \sqsubseteq_{\alpha, \gamma} \langle M', g' \rangle$ is cost consistent and $goal_G$ is preserved. Then $\omega_{M'}(s_0^{M'}) \preceq \omega_M(s_0^M)$.*

We now describe how to use abstraction to define informed heuristics.

Definition 19. *Let $\langle M, g \rangle$, $\langle M', g' \rangle$ be two GTSs such that $\langle M, g \rangle \sqsubseteq_{\alpha, \gamma} \langle M', g' \rangle$ is cost consistent and $goal_G$ is preserved. Heuristic h_a is defined as $h_a(s) = \omega_{M'}(s')$, for any $s' \in S_{M'}$ such that $s' \in \alpha(s)$.*

Proposition 6. *Heuristic h_a is consistent and admissible.*

When different abstractions are available, we can combine the different databases in various ways to obtain better heuristics. The first way is to trivially select the best value delivered by two heuristic databases, which trivially results in a consistent and admissible heuristic.

Definition 20. *Given two different abstraction database heuristics h_a and $h_{a'}$ we define $h_{a\ a'}$ as $h_{a\ a'}(s) = h_a(s) \sqcup h_{a'}(s)$.*

In some cases, however, it is possible to take their cumulative values using $\times$, which provides a much better guidance for the search process. The corresponding abstraction databases are called *disjoint*. Intuitively the idea is that each (non self-)transition in the concrete system either has a corresponding (non self-)transition in one of the abstracted systems but not in both.

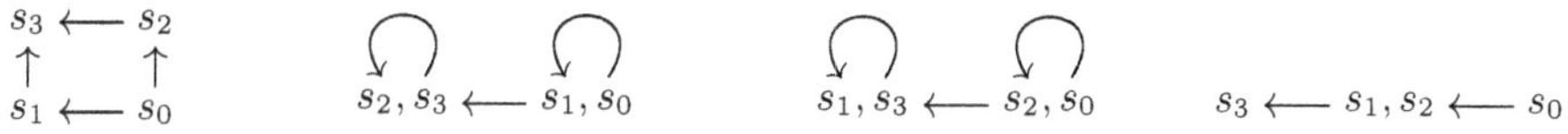

$$s_3 \longleftarrow s_2$$
$$\uparrow \qquad \uparrow$$
$$s_1 \longleftarrow s_0$$

$$\circlearrowleft \qquad \circlearrowleft$$
$$s_2, s_3 \longleftarrow s_1, s_0$$

$$\circlearrowleft \qquad \circlearrowleft$$
$$s_1, s_3 \longleftarrow s_2, s_0$$

$$\circlearrowleft$$
$$s_3 \longleftarrow s_1, s_2 \longleftarrow s_0$$

Fig. 3. A transition system (leftmost) with three different abstractions

Definition 21. *Let $\langle M, g \rangle$ be a GTS and $\langle M', g' \rangle$, $\langle M'', g'' \rangle$ be two abstracted GTS such that $\langle M, g \rangle \sqsubseteq_{\alpha, \gamma} \langle M', g' \rangle$, $\langle M, g \rangle \sqsubseteq_{\alpha', \gamma'} \langle M'', g'' \rangle$ are cost consistent and $goal_G$ is preserved by both simulations. We say that $\langle M', g' \rangle$, $\langle M'', g'' \rangle$ are* disjoint abstractions *whenever for any transition $s_1 \xrightarrow{t} s_2$ in M such that $s_1 \neq s_2$ either there is a transition $s_1' \xrightarrow{t} s_2'$ with $s_1' \neq s_2'$ and $s_1' \in \alpha(s_1)$, $s_2' \in \alpha(s_2)$, or $s_1'' \xrightarrow{t} s_2''$ with $s_1'' \neq s_2''$ and $s_1'' \in \alpha'(s_1)$, $s_2'' \in \alpha'(s_2)$.*

Fig. 3 depicts a concrete transition system (left) with three abstractions (given by node mergings). The center-left and center-right abstractions are mutually disjoint. However any of these together with the rightmost abstraction is not disjoint. For instance, the concrete transition from s_0 to s_3 in the leftmost graph has a corresponding abstract (non self-)transition in the center-left abstraction and in the rightmost one. As a result the distance from s_0 to s_3 would be estimated as 3 which is clearly not a lower bound.

Definition 22. *Given two different abstraction database heuristics h_a and $h_{a'}$, we define $h_{a\ a'}$ as $h_{a\ a'}(s) = h_a(s) \times h_{a'}(s)$.*

Proposition 7. *Let $\langle M, g \rangle$ be a GTS and $\langle M', g' \rangle$, $\langle M'', g'' \rangle$ be two disjoint abstracted GTS. Let $\langle M, g \rangle$ be a GTS and $\langle M', g' \rangle$, $\langle M'', g'' \rangle$ be two abstracted GTS such that $\langle M, g \rangle \sqsubseteq_{\alpha, \gamma} \langle M', g' \rangle$, $\langle M, g \rangle \sqsubseteq_{\alpha', \gamma'} \langle M'', g'' \rangle$ are cost consistent and $goal_G$ is preserved by both simulations. Let further h_a and $h_{a'}$ be the database heuristics constructed from $\langle M', g' \rangle$ and $\langle M'', g'' \rangle$, respectively. Then $h_{a\ a'}$ is consistent and admissible.*

8   Experimental Results

We validate our approach by presenting experimental results obtained with HSF-SPIN [13], a heuristic model checker compatible with the successful model checker SPIN [22]. The analysis we perform regards Property 2, i.e., *Can a certain node v_i be a terminal and no other requests are queued over the network?*. We have implemented the Arrow Distributed Directory Protocol in Promela, the specification language of both SPIN and HSF-SPIN. The implemented model allows for an easy definition of the minimal spanning tree underlying the protocol. A node is modeled as a non-deterministic process that can request an object, accept the request as a non-terminal node, accept the request as a terminal node, send the object if it has finished working on it, or receive the object sent directly over the network. We choose three different topologies: *star*, where all nodes are connected to one common node, *chain*, nodes forming a connected chain,

Table 1. Reachability experiments in the arrow distributed directory protocol

star	DJK	DFS	BF+h_n^1	BF+h_h	BF+h_f
expanded nodes	38,701	6,253	30	6,334	30
solution cost	20	134	58	32	58
chain	DJK	DFS	BF+h_n^1	BF+h_h	BF+h_f
expanded nodes	413,466	78,112	38	1,49	38
solution cost	28	118	74	74	74
tree	DJK	DFS	BF+h_n^1	BF+h_h	BF+h_f
expanded nodes	126,579	24,875	34	24,727	34
solution cost	24	126	66	44	66

and *tree*, where nodes are arranged in the form of a binary tree. Each instance consists of 10 nodes. In all our experiments, we set a memory bound of 512 MB.

The results in Table 1 correspond to the first phase of the goal-finding process, namely when one is interested in finding a goal state as quick as possible. The table shows the number of expanded nodes and solution length for Dijkstra's algorithm (DJK), depth-first search (DFS) and best-first search with heuristics h_n^1 (BF+h_n^1), h_h (BF+h_h) and h_f (BF+h_f).

As expected, Dijkstra's algorithm offers the optimal path to the desired state graph though requiring the greatest number of state expansions. Best-first offers the best performance in terms of node expansions with heuristics h_n^1 and h_f. It is worth mentioning that in this particular example h_f amounts to $(h_n^1)^5$. When applying the Hamming distance, the number of expanded nodes increases, still the solution length decreases.

Table 2 regards the second phase of the bug-finding process, namely when one is interested in finding optimal paths to a given goal state. The table shows the number of expanded nodes and solution length for Dijkstra's algorithm (DJK) and A* with heuristics h_n^1 (A*+h_n^1), h_h (A*+h_h) and h_f (A*+h_f). In addition, we used the Hamming distance applied to the whole state vector representation (A*+H_h), where s_b and s_d indicate that the goal state used for the heuristic was the one obtained with breadth- and depth-first search, respectively.

The first thing to observe is that h_n^1, being admissible, always delivers optimal paths and requires less search effort than Dijkstra's algorithm. The performance of the Hamming distance is not regular. Version h_h that takes into account the graph representation and $H_h(s_b)$ based on the whole bit-vector of the state obtained with Dijkstra's algorithm outperform heuristic h_n^1. On the other hand, the Hamming distance that takes into account the bit-vector representation of the state obtained with the depth-first search exploration shows poor performances by delivering non-optimal counterexamples and running out of memory. The reason is that the state vectors corresponding to states s_b and s_d are very different (though both representing the goal state). The bits representing data, not involving the goal graph items, result in rich information in one case and fuzzy in the other. Heuristic h_f performs the best, both in terms of the expanded nodes and the path length.

Table 2. Optimality experiments in the arrow distributed directory protocol

star	DJK	$A^*+H_h(s_d)$	$A^*+H_h(s_b)$	$A^*+h_n^1$	A^*+h_h	A^*+h_f
expanded nodes	38,701	o.m.	1,255	13,447	117	206
solution cost	20	o.m.	20	20	20	20
chain	DJK	$A^*+H_h(s_d)$	$A^*+H_h(s_b)$	$A^*+h_n^1$	A^*+h_h	A^*+h_f
expanded nodes	413,466	26,622	1,245	106,629	1,620	198
solution cost	28	42	28	28	28	28
tree	DJK	$A^*+H_h(s_d)$	$A^*+H_h(s_b)$	$A^*+h_n^1$	A^*+h_h	A^*+h_f
expanded nodes	126,579	o.m	1,481	33,720	6,197	224
solution cost	24	o.m.	24	24	24	24

9   Conclusion

We have presented an abstract approach for the analysis of graph transitions systems, which are traditional transition systems where states and transitions respectively represent graphs and partial graph morphisms. It is a useful formalism to represent the state space of systems involving graphs, like communication protocols, graph transformations, and visually described systems.

The analysis of such systems is reduced to exploration problems consisting of finding certain states reachable from the initial one. We analyze two problems: finding just one path and finding the optimal one, according to a certain notion of optimality. As algorithms, we propose the use of heuristic search. They use heuristic functions that lead the exploration to the set of goal states. We have proposed different such functions proving some of their properties. In addition, we have proposed the use of abstraction-based heuristics which exploit abstraction techniques in order to obtain informed heuristics.

We have illustrated our approach with a scenario in which one is interested in analyzing structural properties of communication protocols. As a concrete example we used the *arrow distributed directory protocol* [9], which ensures exclusive access to a mobile service in a distributed system. We implemented our approach in the heuristic model checker HSF-SPIN, an extension of the well-known model checker SPIN and presented promising preliminary experiments. In future work, we plan to realize a richer empirical evaluation of our approach, focusing on abstraction database heuristics and possibly profiting from existing approaches for the abstraction of graph transformation systems [1,30].

Acknowledgements. The first two authors thank DFG for continous support in the projects ED 74/2 and ED 74/3. Moreover, we thank R. Heckel and A. Rensink for discussion.

References

1. P. Baldan, A. Corradini, B. König, and B. König. Verifying a behavioural logic for graph transformation systems. In *CoMeta'03*, ENTCS, 2004.
2. B. Bonet and H. Geffner. Planning as heuristic search. *Artificial Intelligence*, 129(1–2):5–33, 2001.

3. L. Caires and L. Cardelli. A spatial logic for concurrency (part I). *Inf. Comput.*, 186(2):194–235, 2003.
4. L. Cardelli, P. Gardner, and G. Ghelli. A spatial logic for querying graphs. In *ICALP*, volume 2380, pages 597–610, 2002.
5. E. Clarke, O. Grumberg, and D. Peled. *Model Checking*. The MIT Press, 1999.
6. T. H. Cormen, C. E. Leiserson, R. L. Rivest, and C. Stein. *Introduction to Algorithms*. MIT Press, 2001.
7. A. Corradini, U. Montanari, F. Rossi, H. Ehrig, R. Heckel, and M. Löwe. *Algebraic approaches to graph transformation*, volume 1, chapter Basic concepts and double push-out approach. World Scientific, 1997.
8. B. Courcelle. *Handbook of graph grammars and computing by graph transformations*, volume 1 : Foundations, chapter 5, pages 313–400. World Scientific, 1997.
9. M. J. Demmer and M. Herlihy. The arrow distributed directory protocol. In *DISC*, pages 119 –133, 1998.
10. S. Edelkamp. Planning with pattern databases. In *ECP*, 2001. 13-24.
11. S. Edelkamp. Taming numbers and durations in the model checking integrated planning system. *JAIR*, 20:195–238, 2003.
12. S. Edelkamp, S. Jabbar, and A. Lluch Lafuente. Cost-algebraic heuristic search. In *AAAI*, pages 1362–1367, 2005.
13. S. Edelkamp, S. Leue, and A. Lluch Lafuente. Directed explicit-state model checking in the validation of communication protocols. *STTT*, 5(2-3):247–267, 2003.
14. S. Edelkamp and A. Lluch Lafuente. Abstraction databases in theory and model checking practice. In *ICAPS Workshop on Connecting Planning Theory with Practice*, 2004.
15. F. Gadducci and A. Lluch Lafuente. Graphical verification of a spatial logic for the pi-calculus. In *Graph Transformation for Verification and Concurrency*. ENTCS, 2005. to appear.
16. A. Groce and W. Visser. Model checking Java programs using structural heuristics. In *ISSTA*. ACM Press, 2002.
17. S. Gyapay, Á. Schmidt, and D. Varró. Joint optimization and reachability analysis in graph transformation systems with time. In *GT-VMT*, volume 109 of *ENTCS*, pages 137–147. Elsevier, 2004.
18. P. E. Hart, N. J. Nilsson, and B. Raphael. A formal basis for heuristic determination of minimum path cost. *IEEE Trans. on Systems Science and Cybernetics*, 4:100–107, 1968.
19. T. Henzinger, R. Jhala, R. Majumdar, and G. Sutre. Software verification with Blast. In *SPIN*, pages 235–239, 2003.
20. D. Hirsch, A. Lluch Lafuente, and E. Tuosto. A logic for application level QoS. In *Proceedings of the 3rd Workshop on Quantitative Aspects of Programming Languages*.
21. J. Hoffmann and B. Nebel. Fast plan generation through heuristic search. *JAIR*, 14:253–302, 2001.
22. G. Holzmann. *The Spin Model Checker: Primer and Reference Manual*. Addison-Wesley, 2003.
23. H. Kastenberg and A. Rensink. Model checking dynamic states in GROOVE. In *SPIN*, pages 299–305, 2006.
24. R. E. Korf. Depth-first iterative-deepening: An optimal admissible tree search. *Artificial Intelligence*, 27(1):97–109, 1985.
25. C. Loiseaux, S. Graf, J. Sifakis, A. Bouajjani, and S. Bensalem. Property preserving abstractions for the verification of concurrent systems. *Formal Methods in System Design*, 6:1–35, 1995.

26. R. Milner. *Communication and Concurrency.* Prentice Hall, 1989.
27. J. Pearl. *Heuristics.* Addison-Wesley, 1985.
28. A. Rensink. Towards model checking graph grammars. In *Automated Verification of Critical Systems*, Tech. Report DSSE-TR-2003, pages 150–160, 2003.
29. A. Rensink. Time and space issues in the generation of graph transition systems. In *GraBaTs*, volume 127 of *ENTCS*, pages 127–139. Elsevier, 2005.
30. A. Rensink and D. Distefano. Abstract graph transformation. In *SVV*, ENTCS, 2005. To appear.
31. G. Rozenberg, editor. *Handbook of graph grammars and computing by graph transformations.* World Scientific, 1997.
32. J. L. Sobrinho. Algebra and algorithms for QoS path computation and hop-by-hop routing in the internet. *IEEE/ACM Trans. Netw.*, 10(4):541–550, 2002.
33. D. Varrò. Automated formal verification of visual modeling languages by model checking. *Journal on Software and Systems Modeling*, 2003.
34. I. Wegener. *Komplexitätstheorie.* Springer, 2003. (in German).

Satisfiability of High-Level Conditions

Annegret Habel and Karl-Heinz Pennemann

Carl v. Ossietzky Universität Oldenburg, Germany[**]
{habel, pennemann}@informatik.uni-oldenburg.de

Abstract. In this paper, we consider high-level structures like graphs, Petri nets, and algebraic specifications and investigate two kinds of satisfiability of conditions and two kinds of rule matching over these structures. We show that, for weak adhesive HLR categories with class $\mathcal{A}$ of all morphisms and a class $\mathcal{M}$ of monomorphisms, strictly closed under decompositions, $\mathcal{A}$- and $\mathcal{M}$-satisfiability and $\mathcal{A}$- and $\mathcal{M}$-matching are expressively equivalent. The results are applied to the category of graphs, where $\mathcal{M}$ is the class of all injective graph morphisms.

1 Introduction

Conditions are most important for high-level systems in a large variety of application areas, especially in the area of safety-critical systems e.g. the specification of railroad control systems [8] and access control policies [11]. Conditions are properties on morphisms or objects which have to be satisfied.

Adhesive HLR systems, introduced in [6] are a new version of high-level replacement systems, combining HLR systems in the sense of [5] and adhesive categories [12]. Weak adhesive HLR categories consist of a category and a class $\mathcal{M}$ of monomorphisms and can be applied to all kinds of graphs, Petri nets, and algebraic specifications.

In this paper, we investigate rules and conditions on high-level structures and speak on $\mathcal{A}$-matching if the match of the rule is arbitrary and on $\mathcal{M}$-matching if the match is in $\mathcal{M}$. Accordingly, we speak on $\mathcal{A}$-satisfiability of a condition if the required morphisms are arbitrary and on $\mathcal{M}$-satisfiability if they are in $\mathcal{M}$. In the literature, nearly every combination of matching and satisfiability occurs. The different concepts have their advantages and disadvantages: $\mathcal{A}$-matching and $\mathcal{A}$-satisfiability allow a compact representation of a set of similar rules and conditions, but this might be a source of error. $\mathcal{M}$-matching and $\mathcal{M}$-satisfiability restrict the allowed morphisms, allow counting, are flexible and intuitive, but one may need sets of similar rules and conditions.

We systematically investigate the matching and satisfiability notions with the aim to find a simple transformation for switching from one notion to the other. We show how to transform conditions from $\mathcal{A}$- to $\mathcal{M}$-satisfiability such that a

[**] This work is supported by the German Research Foundation (DFG), grants GRK 1076/1 (Graduate School on Trustworthy Software Systems) and HA 2936/2 (Development of Correct Graph Transformation Systems).

A. Corradini et al. (Eds.): ICGT 2006, LNCS 4178, pp. 430–444, 2006.
© Springer-Verlag Berlin Heidelberg 2006

morphism $\mathcal{A}$-satisfies a condition if and only if it $\mathcal{M}$-satisfies the transformed condition and from $\mathcal{M}$- to $\mathcal{A}$-satisfiability such that a morphism $\mathcal{M}$-satisfies a condition if and only if it $\mathcal{A}$-satisfies the transformed condition, provided that the class $\mathcal{M}$ is strictly closed under decompositions. This is an important step for connecting conditions traditionally considered with $\mathcal{M}$-satisfiability with Rensink's graph predicates [13] and first order formulas, whose satisfiability notions have to be classified as $\mathcal{A}$-satisfiability: E.g., the semantics of formulas is concerned with arbitrary assignments of variables to values, i.e. an assignment is a not necessarily injective function from variables to the domain.

Moreover, we investigate $\mathcal{M}$-matching of rules in the framework of weak adhesive HLR categories and present two Simulation Theorems, saying that direct derivations can be simulated by direct derivations with $\mathcal{M}$-matching and vice versa. The transformations are illustrated by examples in the category of graphs where $\mathcal{M}$ is the class of all injective graph morphisms.

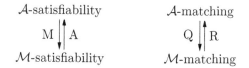

The paper is organized as follows. In Section 2, we recall the notions of conditions, rules, and direct derivations in the framework of weak adhesive HLR categories. In Section 3, we show that, for weak adhesive HLR categories with $\mathcal{M}$-initial object and $\mathcal{M}$ strictly closed under decompositions, $\mathcal{A}$-satisfiability and $\mathcal{M}$-satisfiability are expressively equivalent. In Section 4, we distinguish between $\mathcal{A}$-matching, where the match is allowed to be arbitrary, and $\mathcal{M}$-matching, where the match must be in $\mathcal{M}$, and show that, for weak adhesive HLR categories with $\mathcal{M}$ strictly closed under decompositions, $\mathcal{A}$-matching and $\mathcal{M}$-matching are expressively equivalent. A conclusion including further work is given in Section 5.

2   Conditions and Rules

In this section, we review the definitions of conditions and rules for high-level structures like graphs, Petri nets, and algebraic specifications. We use the framework of weak adhesive HLR categories. For a detailed introduction see [6,4].

Assumption. We assume that $\langle \mathcal{C}, \mathcal{M} \rangle$ is a weak adhesive HLR category.

Example 1. The category $\langle \mathbf{Graphs}, \mathcal{M} \rangle$ of directed graphs, where $\mathcal{M}$ is the class of all injective graph morphisms, is a weak adhesive HLR category.

Definition 1 (conditions). A *condition* over an object P is of the form $\exists a$ or $\exists(a, c)$ where $a: P \to C$ is a morphism and c is a condition over C. Moreover, Boolean formulas over conditions [over P] are conditions [over P]. A morphism $p: P \to G$ *satisfies* a condition $\exists a$ [$\exists(a, c)$] if there exists a morphism $q: C \to G$ in $\mathcal{M}$ with $q \circ a = p$ [satisfying c]. An object G *satisfies* a condition $\exists a$ [$\exists(a, c)$] if all morphisms $p: P \to G$ in $\mathcal{M}$ satisfy the condition.

$$P \xrightarrow{\ a\ } C$$
$$p \searrow \ \stackrel{=}{\ } \ \swarrow q$$
$$G$$

The satisfaction of conditions is extended onto Boolean conditions in the usual way. We write $p \models c$ or $p \models c$ [$G \models c$] to denote that morphism p [object G] satisfies c. Two conditions c and c' over P are *equivalent* on morphisms, denoted by $c \equiv c'$, if, for all morphisms $p \colon P \to G$, $p \models c$ if and only if $p \models c'$.

Remark 1. The conditions in Definition 1 correspond to nested constraints and application conditions in [8] and they subsume the previous notions of constraints and application conditions in [3].

In the definition, the required morphisms have to be in $\mathcal{M}$. We sometimes speak of $\mathcal{M}$-satisfiability. Besides $\mathcal{M}$-satisfiability, one may investigate $\mathcal{A}$-satisfiability, where the required morphisms are allowed to be arbitrary, i.e. in $\mathcal{A}$, where $\mathcal{A}$ denotes the class of all morphisms. The definition is obtained from the one of $\mathcal{M}$-satisfiability, by replacing all occurrences of $\mathcal{M}$ by $\mathcal{A}$ resp. deleting all occurrences of "in $\mathcal{M}$". We write $\models$ to denote $\mathcal{A}$-satisfiability.

Conditions of the form $\exists\, \mathrm{id}$ and $\neg \exists\, \mathrm{id}$ with identity $\mathrm{id} \colon P \to P$ are abbreviated by true and false, respectively. It turns out that, for every condition c, $\exists(\mathrm{id}, c) \equiv c$.

Example 2. In the category of graphs, the meaning of the following conditions for a graph morphism w.r.t. $\mathcal{M}$-satisfiability is:

$\neg\exists(\underset{1}{\bigcirc}\ \underset{2}{\bigcirc} \to \bigcirc\rightrightarrows\bigcirc)$ There do not exist parallel edges between the images of $1, 2$.

$\exists(\underset{1}{\bigcirc} \to S_n)$ The image of 1 has n outgoing edges to different nodes. S_n denotes a star with n outgoing edges.

$\neg\exists(\underset{1}{\bigcirc} \to S_{n+1})$ The image of 1 does not have $n+1$ outg. edges to diff. nodes.

$\exists(\underset{1}{\bigcirc}\ \underset{2}{\bigcirc} \to P_n)$ There is a simple path of length n connecting the im. of $1, 2$. P_n denotes a simple path of length n between the im. of $1, 2$.

$\mathcal{M}$-satisfiability restricts the kind of morphisms: no identification of nodes and edges is allowed. If one wants to have a condition c with arbitrary satisfiability, one has to add the so-called quotient conditions of c to the system. $\mathcal{M}$-satisfiability allows to express explicit counting such as the existence/nonexistence of n nodes or edges or a simple path of length n.

The meaning of the condition for a graph morphism w.r.t. $\mathcal{A}$-satisfiability is:

$\exists(\underset{1}{\bigcirc} \to \underset{1}{\bigcirc}\!\rightarrow\!\bigcirc)$ There exists an outgoing edge (proper edge or loop).

$\mathcal{A}$-satisfiability allows a compact representation of a set of similar conditions. On the other hand, $\mathcal{A}$- satisfiability may lead to misinterpretations of conditions.

The meaning of the condition $\exists(\bigcirc \rightarrow \bigcirc\!\!\underset{1}{\overset{\curvearrowright}{\bigcirc}})$ for a graph morphism w.r.t. $\mathcal{M}$-, $\mathcal{X}$- (class of all edge-injective graph morphisms), and $\mathcal{A}$-satisfiability is:

> There exist at least two proper outgoing edges to different nodes.
> There exist at least two outgoing edges (proper edges and/or loops).
> There exists at least one outgoing edge (proper edge or loop).

$\mathcal{A}$-satisfiability may be seen as the complement to $\mathcal{M}$-satisfiability. Both notions have their advantages as well as their disadvantages: E.g. $\mathcal{M}$-satisfiability allows explicit counting, but conditions get complex, if it is undesired to distinguish elements. For each notion there exist examples of properties, for which the conditions get complex when expressed by the other notion, so none is better than the other.

We consider rules with application conditions [3,8]. Examples and pointers to the literature can be found in [2,1,7].

Definition 2 (rules). A *plain rule* $q = \langle L \leftarrow K \rightarrow R \rangle$ consists of two morphisms in $\mathcal{M}$ with a common domain K. L is called the left-hand side, R the right-hand side, and K the interface. An *application condition* ac $= \langle \text{ac}_L, \text{ac}_R \rangle$ for q consists of two conditions over L and R, respectively. A *rule* $p = \langle q, \text{ac} \rangle$ consists of a plain rule q and an application condition ac for q.

$$
\begin{array}{ccccc}
L & \longleftarrow & K & \longrightarrow & R \\
m \downarrow & (1) & \downarrow & (2) & \downarrow m \\
G & \longleftarrow & D & \longrightarrow & H
\end{array}
$$

Given a plain rule q and a morphism $K \rightarrow D$, a *direct derivation* consists of two pushouts (1) and (2). We write $G \Rightarrow_{q,m,m^*} H$, $G \Rightarrow_q H$, or short $G \Rightarrow H$ and say that m is the *match* and m is the *comatch* of q in H. We speak of an $\mathcal{A}$-match ($\mathcal{A}$-matching) if m is arbitrary and of an $\mathcal{M}$-match ($\mathcal{M}$-matching) if m is in $\mathcal{M}$. Given a rule $p = \langle q, \text{ac} \rangle$, there is a *direct derivation* $G \Rightarrow_{p,m,m^*} H$ in $\langle X, Y \rangle$ with X, Y in $\{\mathcal{A}, \mathcal{M}\}$ if $G \Rightarrow_{q,m,m^*} H$, $m \in X$, $m \models_Y \text{ac}_L$, and $m \models_Y \text{ac}_R$. Given a set of rules $\mathcal{R}$, we write $G \Rightarrow_{\mathcal{R}} H$ if there is a rule p in $\mathcal{R}$ such that $G \Rightarrow_p H$.

Example 3. The meaning of the following rules w.r.t. $\mathcal{M}$-matching is:

Add $= \langle \bigcirc\bigcirc \leftarrow \bigcirc\bigcirc \rightarrow \bigcirc\!\!\rightarrow\!\!\bigcirc \rangle$ Addition of a proper edge.
Delete $= \langle \bigcirc\!\!\rightarrow\!\!\bigcirc \leftarrow \bigcirc\bigcirc \rightarrow \bigcirc\bigcirc \rangle$ Deletion of a proper edge.
Delete$_2$ $= \langle \bigcirc\!\!\rightrightarrows\!\!\bigcirc \leftarrow \bigcirc\bigcirc \rightarrow \bigcirc\bigcirc \rangle$ Deletion of two parallel proper edges.

where an edge is *proper* if its source and target are different.

$\mathcal{M}$-matching restricts the applicability of the rule: no identification of nodes and edges is allowed. The addition and deletion of a loop, for example, requires to explicitly consider the corresponding quotient rules of Add and Delete. Moreover, $\mathcal{M}$-matching allows explicit counting such as the deletion of n nodes or edges.

The meaning of the following rules w.r.t. $\mathcal{A}$-matching is:

Add $= \langle \bigcirc\bigcirc \leftarrow \bigcirc\bigcirc \rightarrow \bigcirc\!\rightarrow\!\bigcirc \rangle$ Addition of an edge (proper edge or loop).
Delete $= \langle \bigcirc\!\rightarrow\!\bigcirc \leftarrow \bigcirc\bigcirc \rightarrow \bigcirc\bigcirc \rangle$ Deletion of an edge (proper edge or loop).
Delete$_2 = \langle \bigcirc\!\rightleftarrows\!\bigcirc \leftarrow \bigcirc\bigcirc \rightarrow \bigcirc\bigcirc \rangle$ Deletion of two edges (proper or loops).

$\mathcal{A}$-matching allows a compact representation of a set of similar rules. Every rule represents a finite set of quotient rules; the quotient rules are implicitly in the system. This advantage may become a disadvantage whenever one forgets that the rules may be applied non-injective. Additionally, identifications of elements may only take place, if all elements involved are preserved. This corresponds to an additional implicit application condition and which is often forgotten.

In general, there are several cases where $\mathcal{M}$-matching is useful, e.g. no identification, several cases where $\mathcal{A}$-matching is desired, e.g. arbitrary identification of preserved elements, several important "mixed" cases, e.g. certain elements are possibly identified, others are not, and cases which are covered by neither $\mathcal{M}$- nor $\mathcal{A}$-matching, e.g. arbitrary identification of all elements.

3 $\mathcal{A}$-Satisfiability Versus $\mathcal{M}$-Satisfiability

In this section, we investigate the different satisfiability notions. One may ask whether $\mathcal{A}$-satisfiability and $\mathcal{M}$-satisfiability are expressively equivalent.

Assumption. We assume that $\langle \mathcal{C}, \mathcal{M} \rangle$ is a weak adhesive HLR category with epi-$\mathcal{M}$-factorizations, that is, for every morphism there is an epi-mono-factorization with monomorphism in $\mathcal{M}$.

There is a transformation from $\mathcal{A}$- to $\mathcal{M}$-satisfiability for morphisms. The construction is a quotient construction on conditions and similar to the transformation of constraints into application conditions in [3,8].

Theorem 1 (for morphisms: from $\mathcal{A}$- to $\mathcal{M}$-satisfiability).   *There is a transformation* M *on conditions such that, for every condition c over P and every morphism* $p\colon P \to G$, $p \models M(c) \Leftrightarrow p \models c$.

Construction. Let $M(c) = \vee_e \ \exists(e, M_e(c))$ where the disjunction $\vee_e$ ranges over all epimorphisms $e\colon P \to P'$ and, for every epimorphism $e\colon P \to P'$, the transformation M_e is defined inductively on the structure of the conditions:

$$M_e(\exists a) \quad = \vee_d \exists b$$
$$M_e(\exists(a,c)) = \vee_d \exists(b, M_f(c))$$

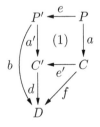

where (1) is the pushout of the morphisms $a\colon P \to C$ and $e\colon P \to P'$ leading to morphisms $a'\colon P' \to C'$ and $e'\colon C \to C'$ and the disjunction $\vee_d$ ranges over all epimorphisms $d\colon C' \to D$ such that $b = d \circ a'$ is in $\mathcal{M}$ and $f = d \circ e'$. For Boolean conditions, the transformations are extended in the usual way.

Proof. By structural induction, we show: For every condition c over P and every morphism $p: P \rightarrow G$, $p' \models M_e(c) \Leftrightarrow p \models c$ for some epi-$\mathcal{M}$-factorization $p = p' \circ e$ with epimorphism $e: P \rightarrow P'$ and monomorphism $p': P' \rightarrow G$ in $\mathcal{M}$. For conditions of the form $\exists(a, c)$ the statement is proved as follows: **If.** Let $p' \models M_f(\exists(a, c))$. Then there is some epimorphism $d: C' \rightarrow D$ with $b = d \circ a'$ in $\mathcal{M}$ and $f = d \circ e'$ such that $p' \models \exists(b, M_f(c))$. By definition of $\mathcal{M}$-satisfiability, there is some $q': D \rightarrow G$ in $\mathcal{M}$ such that $q' \circ b = p'$ and $q' \models M_e(c)$. Define $q = q' \circ f$. Then $q \circ a = p$ and, by inductive hypothesis, $q \models c$. Consequently, $p \models \exists(a, c)$.

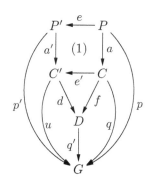

Only if. Let $p \models \exists(a, c)$. Then there is some $q: C \rightarrow G$ such that $q \circ a = p$ and $q \models c$. By the universal property of pushouts, there is some $u: C' \rightarrow G$ with $u \circ a' = p'$ and $u \circ e' = q$. Let $u = q' \circ d$ be an epi-$\mathcal{M}$-factorization of u with epimorphism d and monomorphism q' in $\mathcal{M}$. Then $q' \circ b = p'$. Since $\mathcal{M}$ is closed under decompositions, p' and q' in $\mathcal{M}$ imply b in $\mathcal{M}$. By inductive hypothesis, $q' \models M_f(c)$. Consequently, $p' \models \vee_d \exists(b, M_f(c)) = M_e(\exists(a, c))$. For conditions of the form $\exists a$, the proof is similar. For Boolean conditions, the statement follows from the definitions and the inductive hypothesis. Consequently, the statement holds for all conditions.

The statement implies that, for every morphism $p: P \rightarrow G$, $p \models M(c)$ iff $p' \models M_e(c)$ for some epi-$\mathcal{M}$-factorization $p = p' \circ e$ of p with epimorphism e and monomorphism p' in $\mathcal{M}$. This completes the proof.

Remark 2. The transformation M in Theorem 1 on conditions is very similar to the transformation of constraints into application conditions in [3,8] and the original paper of [10].

Example 4. In the category of graphs, the condition $c = \neg\exists(\underset{1}{\bigcirc} \rightarrow \underset{1}{\bigcirc}\text{--}\bigcirc)$ with meaning for graph morphisms w.r.t. $\mathcal{A}$-satisfiability "There does not exist an outgoing edge" is transformed into the condition $M(c)$ meaning w.r.t. $\mathcal{M}$-satisfiability "There does not exist a proper outgoing edge or a loop".

$$M(c) = M(\neg\exists(\underset{1}{\bigcirc} \rightarrow \underset{1}{\bigcirc}\text{--}\bigcirc)) = \neg M(\exists(\underset{1}{\bigcirc} \rightarrow \underset{1}{\bigcirc}\text{--}\bigcirc)) = \neg \vee_e \exists(e, M_e(c))$$
$$= \neg\exists(\mathrm{id}, M_{\mathrm{id}}(c)) = \neg\exists(\mathrm{id}, \exists(\underset{1}{\bigcirc} \rightarrow \underset{1}{\bigcirc}\text{--}\bigcirc) \vee \exists(\bigcirc \rightarrow \underset{1}{\overset{2}{\bigcirc}}))$$
$$\equiv \neg(\exists(\underset{1}{\bigcirc} \rightarrow \underset{1}{\bigcirc}\text{--}\bigcirc) \vee \exists(\bigcirc \rightarrow \underset{1}{\overset{2}{\bigcirc}}))$$

As corollary, we obtain a transformation from $\mathcal{A}$- to $\mathcal{M}$-satisfiability for objects.

Corollary 1 (for objects: from $\mathcal{A}$- to $\mathcal{M}$-satisfiability). There is a transformation M on conditions such that, for every condition c over P and every object G, $G \models \mathrm{M}(c) \Leftrightarrow G \models c$.

Proof. Let $\mathrm{M}(c) = \wedge_e \, M_e(c)$ where $\mathcal{E}$ denotes the set of all epimorphisms starting from P and $M_e(c)$ is as in the construction of Theorem 1. By the definition of $\mathcal{A}$-satisfiability of objects, epi-$\mathcal{M}$-factorization of morphisms, the proof of

Theorem 1, and the definition of $\mathcal{M}$-satisfiability of objects, we have $G \models c$ iff for all $p: P \to G$, $p \models c$ iff for all $p: P \to G$ with epi-$\mathcal{M}$-factorization $p = p' \circ e$, $p' \models \mathrm{M}_e(c)$ iff for all $p' \in \mathcal{M}$, $p' \models \wedge_e \mathrm{M}_e(c)$ iff $G \models \mathrm{M}(c)$.

Example 5. The condition $c = \exists(\bigcirc \to \bigcirc\!\!\to\!\!\bigcirc)$ with meaning for graphs w.r.t. $\mathcal{A}$-satisfiability "For every node, there exists an outgoing edge" is transformed into the condition $\mathrm{M}(c) = \exists(\underset{1}{\bigcirc} \to \underset{1}{\bigcirc}\!\!\to\!\!\bigcirc) \vee \exists(\bigcirc \to \bigcirc\!\!\!\!\!\circlearrowright)$ meaning w.r.t. $\mathcal{M}$-satisfiability "For every node, there exists a proper outgoing edge or a loop". The condition $c = \exists(\bigcirc\,\bigcirc \to \bigcirc\!\!\to\!\!\bigcirc)$ with meaning for graphs w.r.t. $\mathcal{A}$-satisfiability "For every pair of nodes, there exists a connecting edge" is transformed into the condition $\mathrm{M}(c) = \exists(\bigcirc\,\bigcirc \to \bigcirc\!\!\to\!\!\bigcirc) \wedge \exists(\bigcirc \to \bigcirc\!\!\!\!\!\circlearrowright)$ meaning w.r.t. $\mathcal{M}$-satisfiability "For every pair of distinct nodes, there exists a connecting edge and, for every node, there exists a loop".

There is a transformation from $\mathcal{M}$- to $\mathcal{A}$-satisfiability for morphisms, provided that $\mathcal{M}$ is strictly closed under decompositions, i.e. $g \circ f \in \mathcal{M}$ implies $f \in \mathcal{M}$.

Theorem 2 (for morphisms: from $\mathcal{M}$- to $\mathcal{A}$-satisfiability). *Let $\mathcal{M}$ be strictly closed under decompositions. Then there is a transformation A on conditions such that, for every condition c over an object P and every morphism $p: P \to G$, $p \models \mathrm{A}(c) \Leftrightarrow p \models c$.*

Construction. The transformation A is defined inductively on the structure of the conditions: For a morphism $a: P \to C$ and a condition c over C,

$$\mathrm{A}(\exists a) = \exists(a, \mathrm{inM}_C)$$
$$\mathrm{A}(\exists(a, c)) = \exists(a, \mathrm{inM}_C \wedge \mathrm{A}(c))$$

where $\mathrm{inM}_C = \neg\vee_e \exists e$ is a condition over C and the disjunction $\vee_e$ ranges over all epimorphisms $e: C \to C'$ not in $\mathcal{M}$. For Boolean conditions, the transformation is as usual.

Proof. The property "the match is in $\mathcal{M}$" can be expressed by an application condition inM_C: For every morphism $q: C \to G$, $q \in \mathcal{M} \Leftrightarrow q \models \mathrm{inM}_C$. This may be seen as follows: If $q \not\models \mathrm{inM}_C$, then $q \models \exists e$ for some epimorphism $e: C \to C'$ not in $\mathcal{M}$. Then there is some $q': C' \to G$ such that $q' \circ e = q$. Then q is not in $\mathcal{M}$. Otherwise, by the strict closure of $\mathcal{M}$ under decompositions, q in $\mathcal{M}$ would imply e in $\mathcal{M}$. If q is not in $\mathcal{M}$, we consider an epi-$\mathcal{M}$ factorization $q = q' \circ e$ of q with epimorphism e and monomorphism q' in $\mathcal{M}$. Then e is not in $\mathcal{M}$. Otherwise, by closure of $\mathcal{M}$ under compositions, e and q' in $\mathcal{M}$ would imply q in $\mathcal{M}$.

By structural induction, we show the statement of the theorem. For conditions of the form $\exists(a, c)$ we have the following. **Only if.** Let $p \models \exists(a, c)$. Then there is a morphism $q: C \to G$ in $\mathcal{M}$ such that $q \circ a = p$ and $q \models c$. By the inductive hypothesis and the application condition inM_C being equivalent to "match is in M", $q \models \mathrm{inM}_C$ and $q \models \mathrm{A}(c)$. Consequently, $p \models \exists(a, \mathrm{inM}_C \wedge \mathrm{A}(c)) = \mathrm{A}(\exists(a, c))$. **If.** Let $p \models \mathrm{A}(\exists(a, c))$. Then there is some $q: C \to G$ such that

$q \circ a = p$, $q \models$ inM$_C$, and $q \models$ A(c). By the inductive hypothesis, $q \in \mathcal{M}$ and $q \models c$. Thus, $p \models \exists(a, c)$. For conditions of the form $\exists a$ the proof is similar. For Boolean conditions, the statement follows from the definitions and the inductive hypothesis. This completes the inductive proof.

Example 6. The class $\mathcal{M}$ of injective graph morphisms is strictly closed under decompositions. The condition $c = \exists(\bigcirc\,\bigcirc \rightarrow \bigcirc\!\!\rightarrow\!\!\bigcirc)$ with meaning for graph morphisms w.r.t. $\mathcal{M}$-satisfiability "For the two nodes, there exists a connecting edge" is transformed into the condition $\mathrm{A}(c) = \exists(\bigcirc\,\bigcirc \rightarrow \bigcirc\!\!\rightarrow\!\!\bigcirc, \neg\exists(\bigcirc\!\!\rightarrow\!\!\bigcirc \rightarrow \bigcirc\!\!\text{⟲}))$ meaning w.r.t. $\mathcal{A}$-satisfiability "For the nodes, there exists a connecting edge and the endpoints are distinct".

There is a transformation from $\mathcal{M}$- to $\mathcal{A}$-satisfiability for objects, provided that $\mathcal{M}$ is strictly closed under decompositions and the category has an $\mathcal{M}$-initial object: In a category $\mathcal{C}$, an object I is *initial*, if for every object G in $\mathcal{C}$, there exists a unique initial morphism $i: I \rightarrow G$. In a weak adhesive category $\langle\mathcal{C}, \mathcal{M}\rangle$, an object I is $\mathcal{M}$-*initial* if I is initial in $\mathcal{C}$ and the initial morphisms are in $\mathcal{M}$.

Corollary 2 (for objects: from $\mathcal{M}$- to $\mathcal{A}$-satisfiability). For weak adhesive HLR categories with $\mathcal{M}$-initial object and $\mathcal{M}$ strictly closed under decompositions, there is a transformation A on conditions such that, for every condition c over an object P and every object G, $G \models$ A $(c) \Leftrightarrow G \models c$.

Proof. Let A $(c) = \forall(i, \text{inM}_P \Rightarrow \mathrm{A}(c))$ where $\forall(a, d)$ abbreviates the condition $\neg\exists(a, \neg d)$ and i denotes the unique morphism from the initial object I to P in $\mathcal{M}$. By the definition of $\mathcal{M}$-satisfiability, Theorem 2, the property of inM$_P$, and the definition of $\mathcal{A}$-satisfiability for objects, we have $G \models c$ iff for all $p: P \rightarrow G$ in $\mathcal{M}$, $p \models c$ iff for all $p: P \rightarrow G$, $p \models$ inM$_P$ implies $p \models$ A(c) iff for all $p: P \rightarrow G$, $p \models$ inM$_P \Rightarrow \mathrm{A}(c)$ iff for $j: I \rightarrow G$ in $\mathcal{M}$, $j \models \forall(i, \text{inM}_P \Rightarrow \mathrm{A}(c))$ iff $G \models \forall(i, \text{inM}_P \Rightarrow \mathrm{A}(c))$ iff $G \models$ A (c).

Example 7. The category $\langle\textbf{Graphs}, \mathcal{M}\rangle$ has an $\mathcal{M}$-initial object: the empty graph, denoted by $\emptyset$. The condition $c = \exists(\bigcirc\,\bigcirc \rightarrow \bigcirc\!\!\rightarrow\!\!\bigcirc)$ with the meaning for graph morphisms w.r.t. $\mathcal{M}$-satisfiability "Every pair of distinct nodes is connected by an edge" is transformed into the condition

$$\mathrm{A}\ (c) = \forall(\emptyset \rightarrow \bigcirc\,\bigcirc, \neg\exists(\bigcirc\,\bigcirc \rightarrow \bigcirc) \Rightarrow \exists(\bigcirc\,\bigcirc \rightarrow \bigcirc\!\!\rightarrow\!\!\bigcirc, \neg\exists(\bigcirc\!\!\rightarrow\!\!\bigcirc \rightarrow \bigcirc\!\!\text{⟲})))$$

meaning w.r.t. $\mathcal{A}$-satisfiability "For every pair of nodes, whenever the nodes are distinct, then there is a connecting edge (and the endpoints are distinct)".

By Theorems 1 and 2 and Corollaries 1 and 2 we obtain the following corollary.

Corollary 3. For weak adhesive HLR categories with epi-$\mathcal{M}$-factorizations, $\mathcal{M}$-initial object, and $\mathcal{M}$ strictly closed under decompositions, $\mathcal{A}$-satisfiability and $\mathcal{M}$-satisfiability are expressively equivalent.

Remark 3. The equivalence result is valid for nested constraints and application conditions [8]; it is not valid for plain constraints in the sense of [3].

Finally, we present two transformations of conditions over morphisms, i.e. given a morphism $e\colon P \to P'$, then the transformation transforms conditions over P into conditions over P'.

Lemma 1 (from $\mathcal{A}$- to $\mathcal{A}$-satisfiability). *For every (epi)morphism $e\colon P \to P'$, there is a transformation AA_e such that, for every condition c over P and every morphism $p'\colon P' \to G$, $p' \models \mathrm{AA}_e(c) \Leftrightarrow p' \circ e \models c$.*

Construction. For a morphism $e\colon P \to P'$, the transformation AA_e is defined inductively on the structure of the conditions:

$$\mathrm{AA}_e(\exists a) \quad = \exists a'$$
$$\mathrm{AA}_e(\exists(a,c)) = \exists(a', \mathrm{AA}_{e'}(c))$$

where (1) is the pushout of the morphisms a and e leading to the morphisms $a'\colon P' \to C'$ and $e'\colon C \to C'$. For Boolean conditions, the transformation is extended in the usual way.

Proof. By structural induction. For conditions of the form $\exists(a,c)$, the statement is proved as follows: **If.** Let $p' \models \mathrm{AA}_e(\exists(a,c))$. Then there is some $q'\colon C' \to G$ such that $q' \circ a' = p'$ and $q' \models \mathrm{AA}_{e'}(c)$. By inductive hypothesis, $q' \circ e' \models c$. Then $q' \circ e' \circ a = p' \circ e$. Consequently, $p' \circ e \models \exists(a,c)$.

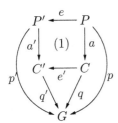

Only if. Let $p' \circ e \models \exists(a,c)$. Then there is some $q\colon C \to G$ such that $q \circ a = p' \circ e$ and $q \models c$. By the universal property of pushouts, there is a morphism $q'\colon C' \to G$ such that $q' \circ a' = p'$ and $q' \circ e' = q$. By inductive hypothesis, $q' \models \mathrm{AA}_{e'}(c)$. Consequently, $p' \models \exists(a', \mathrm{AA}_{e'}(c)) = \mathrm{AA}_e(\exists(a,c))$. For conditions of the form $\exists a$, the proof is similar.

For Boolean conditions, the statement follows from the definitions and the inductive hypothesis. Consequently, the statement holds for all conditions.

Example 8. Given the condition $c = \neg\exists(\bigcirc\bigcirc \to \bigcirc\!\!\to\!\!\bigcirc)$ with the meaning for graph morphisms w.r.t. $\mathcal{A}$-satisfiability "There does not exist a connecting edge" and the epimorphism $e\colon \bigcirc\bigcirc \to \bigcirc$, $\mathrm{AA}_e(c) = \neg\exists(\bigcirc \to \text{⊗})$ is the condition with the meaning "There does not exist an attached loop".

Lemma 2 (from $\mathcal{M}$- to $\mathcal{M}$-satisfiability). *For every epimorphism $e\colon P \to P'$, there is a transformation MM_e such that, for every condition c over P and every morphism $p'\colon P' \to G$ in $\mathcal{M}$, $p' \models \mathrm{MM}_e(c) \Leftrightarrow p' \circ e \models c$.*

Construction. For every epimorphism e, the transformation MM_e is defined inductively on the structure of the conditions: $\mathrm{MM}_e(\exists a) = \exists a'$ and $\mathrm{MM}_e(\exists(a,c)) = \exists(a',c)$ if e is the epimorphism of an epi-$\mathcal{M}$-factorization $e \circ a' = a$ and $\mathrm{MM}_e(\exists a) = \mathrm{MM}_e(\exists(a,c)) = \text{false}$ otherwise. For Boolean conditions, the transformation is extended in the usual way.

Proof. By structural induction. For conditions of the form $\exists(a,c)$, the statement is proved as follows: **If.** Let $p' \models \exists(a',c)$ for some epi-$\mathcal{M}$-factorization $a = a' \circ e$ of a. Then there is some $q' \colon C \to G$ in $\mathcal{M}$ such that $q' \circ a' = p'$ and $q' \models c$. Moreover, $q' \circ a = p' \circ e$. Consequently, $p' \circ e \models \exists(a,c)$. If e is not the epimorphism of an epi-$\mathcal{M}$-factorization of a, we have $p' \not\models \mathrm{MM}_e(\exists(a,c)) = \mathrm{false}$ and $p' \circ e \not\models \exists(a,c)$.

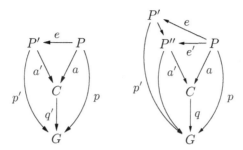

Only if. Let $p' \circ e \models \exists(a,c)$. Then there is some morphism $q \colon C \to G$ in $\mathcal{M}$ such that $q \circ a = p$ and $q \models c$ and some morphism $q' \colon P'' \to G$ in $\mathcal{M}$ with $q' = q \circ a'$. Whenever $a' \circ e'$ is an epi-$\mathcal{M}$-factorization of a, $q' \circ e'$ is an epi-$\mathcal{M}$-factorization of $p = p' \circ e$. By the uniqueness of epi-$\mathcal{M}$-factorizations, we have $e = e'$ and $p' = q'$ (up to isomorphism). Then $p' = q' \models \exists(a',c)$. For conditions of the form $\exists a$, the proof is similar. For Boolean conditions, the statement follows from the definitions and the inductive hypothesis. Consequently, the statement holds for all conditions.

Example 9. The condition $c = \exists(\bigcirc\bigcirc \to \bigotimes)$ with the meaning for graph morphisms w.r.t. $\mathcal{M}$-satisfiability "One of the nodes is connected by a loop" is transformed over the surjective graph morphism $e \colon \bigcirc\bigcirc \to \bigcirc$ into the condition $\mathrm{MM}_e(c) = \exists(\bigcirc \to \bigotimes)$ with the meaning "The node has an attached loop". The condition $c' = \exists(\bigcirc\bigcirc \to \bigcirc\!\!\rightarrow\!\!\bigcirc)$ with the meaning for graph morphisms w.r.t. $\mathcal{M}$-satisfiability "For the two nodes, there exists a connecting edge" is transformed into the condition $\mathrm{MM}_e(c') = \mathrm{false}$ with the meaning "Not satisfiable".

By Lemmas 1 and 2 and Theorems 1 and 2 we obtain the following corollary.

Corollary 4 (from $\mathcal{A}$- to $\mathcal{M}$ and $\mathcal{M}$- to $\mathcal{A}$-satisfiability). For every epimorphism $e \colon P \to P'$, there are transformations AM_e and MA_e such that, for every condition c over P and every morphism $p' \colon P' \to G$ in $\mathcal{M}$, $p' \models \mathrm{AM}_e(c) \Leftrightarrow p' \circ e \models c$ and $p' \models \mathrm{MA}_e(c) \Leftrightarrow p' \circ e \models c$.

Construction. For every epimorphism $e \colon P \to P'$, let $\mathrm{AM}_e = \mathrm{M} \circ \mathrm{AA}_e$ and $\mathrm{MA}_e = \mathrm{A} \circ \mathrm{MM}_e$.

4 $\mathcal{A}$-Matching Versus $\mathcal{M}$-Matching

The definition of a direct derivation allows $\mathcal{A}$-matching as well as $\mathcal{M}$-matching. One may ask whether $\mathcal{A}$-matching and $\mathcal{M}$-matching are expressively equivalent. We establish a Simulation Theorem saying that any direct derivation with arbitrary matching can be simulated by a direct derivation with $\mathcal{M}$-matching. This extends the Simulation Theorem in [7] to weak adhesive HLR categories with epi-$\mathcal{M}$-factorizations and makes use of so-called quotient rules and epi-$\mathcal{M}$-factorizations of morphisms and direct derivations.

Recall that a direct derivation $G \Rightarrow_{q,ac\ ,m,m^*} H$ in $\langle X, Y \rangle$ with X, Y in $\{\mathcal{A}, \mathcal{M}\}$ is a direct derivation $G \Rightarrow_{q,m,m^*} H$ with $m \in X$, $m \models_Y ac_L$, and $m \models_Y ac_R$.

Theorem 3 (from $\mathcal{A}$- to $\mathcal{M}$-matching). *For every $Y \in \{\mathcal{A}, \mathcal{M}\}$, there is a transformation Q_Y from rules into sets of rules such that, for every rule p,*

$$G \Rightarrow_p H \text{ in } \langle \mathcal{A}, Y \rangle \text{ if and only if } G \Rightarrow_{Q_Y(p)} H \text{ in } \langle \mathcal{M}, Y \rangle.$$

Construction. For a rule $q = \langle L \leftarrow K \rightarrow R \rangle$, the rule $q' = \langle L' \leftarrow K' \rightarrow R' \rangle$ is a *quotient rule* of q if there are two pushouts of the form

$$
\begin{array}{ccccc}
L & \longleftarrow & K & \longrightarrow & R \\
\downarrow e & (1) & \downarrow & (2) & \downarrow e \\
L' & \longleftarrow & K' & \longrightarrow & R'
\end{array}
$$

where the vertical morphisms are epimorphisms. For a rule $p = \langle q, ac \rangle$ with application condition, Q (p) is the set of rules $p' = \langle q', ac' \rangle$ where q' is a quotient rule of q, $ac'_L = AA_e(ac_L)$ and $ac'_R = AA_{e^*}(ac_R)$. Q (p) is the set of rules $p' = \langle q', ac' \rangle$ where q' is a quotient rule of q, $ac'_L = MM_e(ac_L)$, and $ac'_R = MM_{e^*}(ac_R)$.

Proof. Let $G \Rightarrow_{p',n,n^*} H$ be a direct derivation in $\langle \mathcal{M}, Y \rangle$ through $p' = \langle q', ac' \rangle$ in $Q_Y(p)$ with $q' = \langle L' \leftarrow K' \rightarrow R' \rangle$ and ac' as in the construction. Then the diagrams (1), (2), (1') and (2') in the figure below are pushouts. By the Composition Lemma of pushouts valid in every category, the composed diagrams (1)+(1') and (2)+(2') are pushouts as well. Hence, there is a direct derivation $G \Rightarrow_{q,m,m^*} H$ in $\langle \mathcal{A}, Y \rangle$. By assumption, $n \models_Y ac'_L$ and $n \models_Y ac'_R$. By Lemma 1 [2], $m \models_Y ac_L$ and $m \models_Y ac_R$. Thus, $G \Rightarrow_{p,m,m^*} H$ is a direct derivation in $\langle \mathcal{A}, Y \rangle$.

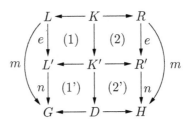

Vice versa, let $G \Rightarrow_{p,m,m^*} H$ be a direct derivation in $\langle \mathcal{A}, Y \rangle$ through $p = \langle q, ac \rangle$ with $q = \langle L \leftarrow K \rightarrow R \rangle$. Let $m = n \circ e$ an epi-$\mathcal{M}$ factorization of m with epimorphism $e \colon L \rightarrow L'$ and monomorphism $n \colon L' \rightarrow G$ in $\mathcal{M}$. Then there is a decomposition of the original diagrams into diagrams (1) and (1'), (2) and (2') as follows: Construct K' as a pullback object of $L' \rightarrow G \leftarrow D$ and denote the diagram by (1'). By the universal property of pullbacks, there is a unique morphism

$e': K \to K'$ such that $K \to K' \to D = K \to D$ and diagram (1) commutes. By the pushout-pullback decomposition, (1') and (1) are pushouts. Now construct R' as the pushout object of $K' \leftarrow K \to R$ and denote the diagram by (2). By the universal property of pushouts, there is a unique morphism $n : R' \to H$ such that $R \to R' \to H = R \to H$ and diagram (2') commutes. By the Decomposition Lemma for pushouts valid in every category, diagram (2') is a pushout. Since epimorphisms and $\mathcal{M}$-morphisms are closed under pullbacks and pushouts, the vertical morphisms e, e', and e are epimorphisms and the vertical morphisms n, n', and n are in $\mathcal{M}$. Then $q' = \langle L' \leftarrow K' \to R' \rangle$ is a quotient rule of q and $p' = \langle q', \mathrm{ac}' \rangle$ with $\mathrm{ac}' = \langle \mathrm{AA}_e(\mathrm{ac}_L), \mathrm{AA}_{e^*}(\mathrm{ac}_R) \rangle$ $[\langle \mathrm{MM}_e(\mathrm{ac}_L), \mathrm{MM}_{e^*}(\mathrm{ac}_R) \rangle]$ is a quotient rule or p in Q (p) [Q (p)]. By Lemma 1 [2], $G \Rightarrow_{p',n,n^*} H$ is a direct derivation in $\langle \mathcal{M}, \mathcal{A} \rangle$ $[\langle \mathcal{M}, \mathcal{M} \rangle]$. This completes the proof.

Remark 4. The construction and proof for the transformation Q of rules in Theorem 3 is very similar to transformation of right- to left application conditions in [3,8].

Example 10. Consider the rule $p = \langle q, \mathrm{ac} \rangle$ with $q = \langle \bigcirc\,\bigcirc \;\leftarrow\; \bigcirc\,\bigcirc \;\to\; \bigcirc\text{-}\bigcirc \rangle$ and $\mathrm{ac}_L = \neg\exists(\bigcirc\,\bigcirc \to \bigcirc\text{-}\bigcirc)$ with $\langle \mathcal{A}, Y \rangle$-meaning "Add a connecting edge, provided there does not exist one". Then $p' = \langle q', \mathrm{ac}'_L \rangle$ with $q' = \langle \bigcirc \;\leftarrow\; \bigcirc \;\to\; \reflectbox{$\otimes$} \rangle$ and $\mathrm{ac}'_L = \neg\exists(\bigcirc \to \reflectbox{$\otimes$})$ is a rule in Q (p) with $\langle \mathcal{M}, \mathcal{A} \rangle$-meaning "Add a loop at the node, provided there does not exist one". Furthermore, $p'' = \langle q', \mathrm{ac}''_L \rangle$ with $\mathrm{ac}''_L = \mathrm{false}$ is a rule in Q (p) with $\langle \mathcal{M}, \mathcal{M} \rangle$-meaning "Never add a loop at the node".

We can simulate $\mathcal{M}$-matching by $\mathcal{A}$-matching by using the application condition inM which is satisfied iff the match is in $\mathcal{M}$.

Theorem 4 (from $\mathcal{M}$- to $\mathcal{A}$-matching). *Let $\mathcal{M}$ be strictly closed under decompositions and $Y \in \{\mathcal{A}, \mathcal{M}\}$. Then there is a transformation R on rules such that, for every rule p, $G \Rightarrow_p H$ in $\langle \mathcal{M}, Y \rangle$ iff $G \Rightarrow_{\mathrm{R}(p)} H$ in $\langle \mathcal{A}, Y \rangle$.*

Construction. For every rule $p = \langle q, \mathrm{ac} \rangle$, let $\mathrm{R}(p) = \langle q, \mathrm{ac}' \rangle$ with

$$\mathrm{ac}' = \langle \mathrm{inM}_L \wedge \mathrm{ac}_L, \mathrm{inM}_R \wedge \mathrm{ac}_R \rangle.$$

Proof. By the proof of Theorem 2, the property "the match is in $\mathcal{M}$" can be expressed by the application condition inM. Thus, for every rule p, $G \Rightarrow_p H$ in $\langle \mathcal{M}, Y \rangle$ if and only if $G \Rightarrow_{\mathrm{R}(p)} H$ in $\langle \mathcal{A}, Y \rangle$.

Example 11. Consider the rule $p = \langle q, \mathrm{ac} \rangle$ with $q = \langle \bigcirc\,\bigcirc \;\leftarrow\; \bigcirc\,\bigcirc \;\to\; \bigcirc\text{-}\bigcirc \rangle$ and $\mathrm{ac}_L = \neg\exists(\bigcirc\,\bigcirc \to \bigcirc\text{-}\bigcirc)$ with $\langle \mathcal{M}, Y \rangle$-meaning "Add an edge between distinct nodes, provided there does not exist a connecting edge". Then $\mathrm{R}(p) = \langle q, \neg\exists(\bigcirc\,\bigcirc \to \bigcirc) \wedge \mathrm{ac}_L \rangle$ is a rule with $\langle \mathcal{A}, Y \rangle$-meaning "Add an edge between the nodes, provided the nodes are distinct and there does not exist a connecting edge".

As a consequence, we obtain transformations for all possible kinds of direct derivations.

Corollary 5. For weak adhesive HLR categories with epi-$\mathcal{M}$-factorizations and $\mathcal{M}$ strictly closed under decompositions and tuples $t = \langle X, Y, X', Y' \rangle \in \{\mathcal{A}, \mathcal{M}\}^4$, there is a transformation Q_t such that, for every rule p,

$$G \Rightarrow_p H \text{ in } \langle X, Y \rangle \text{ if and only if } G \Rightarrow_{Q_t(p)} H \text{ in } \langle X', Y' \rangle.$$

Proof. The transformation results follow directly from the transformation results on matching and satisfiability.

Finally, we obtain the following corollary.

Corollary 6. For weak adhesive HLR categories with epi-$\mathcal{M}$-factorizations and $\mathcal{M}$ strictly closed under decompositions, $\mathcal{A}$-matching and $\mathcal{M}$-matching are expressively equivalent.

5   Conclusion

We have shown that all notions of matching and satisfiability have their advantages and disadvantages and, for weak adhesive HLR categories with epi-$\mathcal{M}$-factorizations, $\mathcal{M}$-initial object, and $\mathcal{M}$ strictly closed under decompositions,

- $\mathcal{A}$-matching and $\mathcal{M}$-matching
- $\mathcal{A}$-satisfiability and $\mathcal{M}$-satisfiability

are expressively equivalent. The equivalence results are valid for nested constraints and application conditions in the sense of [8]; they are not valid for (basic) constraints in the sense of [3]. We have presented some transformation results for application conditions, plain rules, and rules with application conditions.

	For application conditions:	
M	from $\mathcal{A}$- to $\mathcal{M}$-satisfiability	Theorem 1
A	from $\mathcal{M}$- to $\mathcal{A}$-satisfiability	Theorem 2
AA_e	from $\mathcal{A}$- to $\mathcal{A}$-satisfiability	Lemma 1
AM_e	from $\mathcal{A}$- to $\mathcal{M}$-satisfiability	Corollary 4
MA_e	from $\mathcal{M}$- to $\mathcal{A}$-satisfiability	Corollary 4
MM_e	from $\mathcal{M}$- to $\mathcal{M}$-satisfiability	Lemma 2
	For plain rules:	
Q	from $\mathcal{A}$- to $\mathcal{M}$-matching	Theorem 3
R	from $\mathcal{M}$- to $\mathcal{A}$-matching	Theorem 4
	For rules:	
	from $\langle X, Y \rangle$ to $\langle X', Y' \rangle$	Corollary 5

This allows to switch from every combination of X-matching and Y-satisfiability to every other combination of X'-matching and Y'-satisfiability and may be the basis of a converter.

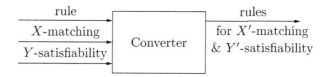

Summarizing, $\mathcal{A}$-matching as well as $\mathcal{M}$-matching are adequate matching notions and $\mathcal{A}$-satisfiability as well as $\mathcal{M}$-satisfiability are adequate satisfiability notions. Nevertheless, we propose to consider either $\mathcal{A}$-matching and $\mathcal{A}$-satisfiability or $\mathcal{M}$-matching and $\mathcal{M}$-satisfiability.

- The combination $\mathcal{A}$-matching and $\mathcal{M}$-satisfiability does not fit well together: Conditions of the form $c = \exists a$ or $\exists (a, c)$ are only satisfiable by morphisms with a certain degree of identifaction that depends on a. In fact, if a is an $\mathcal{M}$-morphism, $m \models c$ implies m in $\mathcal{M}$.
- In the case of $\mathcal{A}$-matching and $\mathcal{A}$-satisfiability, the transformation from constraints to application conditions is more simple and natural than the one for $\mathcal{A}$-matching and $\mathcal{M}$-satisfiability. The same holds for a transformation from application conditions to constraints. However, no direct transformation from right into left application conditions (L) is known, just the junction of A ∘ L ∘ M requiring that the category has an $\mathcal{M}$-initial object and $\mathcal{M}$ is strictly closed under decomposition.
- In the case of $\mathcal{M}$-matching and $\mathcal{M}$-satisfiability, the transformations from constraints to application conditions, from application conditions to constraints, and, consequently, the construction of weakest preconditions for high-level programs [9], are simpler and more intuitive than the one of $\mathcal{A}$-matching and $\mathcal{M}$-satisfiability.

Considering the above and the fact that $\mathcal{M}$-matching is the more explicit matching notion, our choice is $\mathcal{M}$-matching and $\mathcal{M}$-satisfiability.

Further topics will be the followings.

(1) Comparison of notions: A comparison of conditions – as considered in this paper – and first-order formulas on graphs and high-level structures.
(2) Extensions of the theory: The investigation of weak adhesive high-level replacement systems with merging similar to the investigation of graph replacement systems with merging as in [7].
(3) Implementation: A system for converting conditions and rules with one kind of matching and satisfiability into conditions and rules with the implemented kind of matching and satisfiability.

References

1. A. Corradini, U. Montanari, F. Rossi, H. Ehrig, R. Heckel, and M. Löwe. Algebraic approaches to graph transformation. Part I: Basic concepts and double pushout approach. In *Handbook of Graph Grammars and Computing by Graph Transformation*, volume 1, pages 163–245. World Scientific, 1997.

2. H. Ehrig. Introduction to the algebraic theory of graph grammars. In *Graph-Grammars and Their Application to Computer Science and Biology*, volume 73 of *Lecture Notes in Computer Science*, pages 1–69. Springer-Verlag, 1979.

3. H. Ehrig, K. Ehrig, A. Habel, and K.-H. Pennemann. Theory of constraints and application conditions: From graphs to high-level structures. *Fundamenta Informaticae*, 72, 2006.

4. H. Ehrig, K. Ehrig, U. Prange, and G. Taentzer. *Fundamentals of Algebraic Graph Transformation*. EATCS Monographs of Theoretical Computer Science. Springer-Verlag, Berlin, 2006.

5. H. Ehrig, A. Habel, H.-J. Kreowski, and F. Parisi-Presicce. Parallelism and concurrency in high level replacement systems. *Mathematical Structures in Computer Science*, 1:361–404, 1991.

6. H. Ehrig, A. Habel, J. Padberg, and U. Prange. Adhesive high-level replacement systems: A new categorical framework for graph transformation. *Fundamenta Informaticae*, 72, 2006.

7. A. Habel, J. Müller, and D. Plump. Double-pushout graph transformation revisited. *Mathematical Structures in Computer Science*, 11(5):637–688, 2001.

8. A. Habel and K.-H. Pennemann. Nested constraints and application conditions for high-level structures. In *Formal Methods in Software and System Modeling*, volume 3393 of *Lecture Notes in Computer Science*, pages 293–308. Springer-Verlag, 2005.

9. A. Habel, K.-H. Pennemann, and A. Rensink. Weakest preconditions for high-level programs. In *Proc. Int. Conference on Graph Transformation (ICGT 2006)*, this volume of *Lecture Notes in Computer Science*. Springer-Verlag, 2006.

10. R. Heckel and A. Wagner. Ensuring consistency of conditional graph grammars — a constructive approach. In *SEGRAGRA '95*, volume 2 of *Electronic Notes in Theoretical Computer Science*, pages 95–104, 1995.

11. M. Koch, L. V. Mancini, and F. Parisi-Presicce. Graph-based specification of access control policies. *Journal of Computer and System Sciences*, 71:1–33, 2005.

12. S. Lack and P. Sobociński. Adhesive categories. In *Proc. of Foundations of Software Science and Computation Structures (FOSSACS'04)*, volume 2987 of *Lecture Notes in Computer Science*, pages 273–288. Springer-Verlag, 2004.

13. A. Rensink. Representing first-order logic by graphs. In *Graph Transformations (ICGT'04)*, volume 3256 of *Lecture Notes in Computer Science*, pages 319–335. Springer-Verlag, 2004.

Weakest Preconditions for High-Level Programs

Annegret Habel[1], Karl-Heinz Pennemann[1], and Arend Rensink[2]

[1] University of Oldenburg, Germany[*]
{habel, pennemann}@informatik.uni-oldenburg.de
[2] University of Twente, Enschede, The Netherlands
rensink@cs.utwente.nl

Abstract. In proof theory, a standard method for showing the correctness of a program w.r.t. given pre- and postconditions is to construct a weakest precondition and to show that the precondition implies the weakest precondition. In this paper, graph programs in the sense of Habel and Plump 2001 are extended to programs over high-level rules with application conditions, a formal definition of weakest preconditions for high-level programs in the sense of Dijkstra 1975 is given, and a construction of weakest preconditions is presented.

1 Introduction

Graphs and related structures are associated with an accessible graphical representation. Transformation rules exploit this advantage, as they describe local change by relating a left- and a right-hand side. Nondeterministic choice, sequential composition and iteration give rise to rule-based programs [19].

Formal methods like verification with respect to a formal specification are important for the development of trustworthy systems. We use a graphical notion of conditions to specify valid objects as well as morphisms, e.g. matches for transformation rules. We distinguish the use of conditions by speaking of constraints in the first case, and application conditions for rules in the latter. Conditions seem to be adequate for describing requirements as well as for reasoning about the behavior of a system.

A well-known method for showing the correctness of a program with respect to a pre- and a postcondition (see e.g. [7,8]) is to construct a weakest precondition

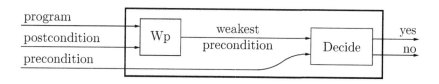

[*] This work is supported by the German Research Foundation (DFG), grants GRK 1076/1 (Graduate School on Trustworthy Software Systems) and HA 2936/2 (Development of Correct Graph Transformation Systems).

A. Corradini et al. (Eds.): ICGT 2006, LNCS 4178, pp. 445–460, 2006.

of the program relative to the postcondition and to prove that the precondition implies the weakest precondition.

In this paper, we use the framework of weak adhesive HLR categories to construct weakest preconditions for high-level rules and programs, using two known transformations from constraints to right application conditions, and from right to left application conditions, and additionally, a new transformation from application conditions to constraints.

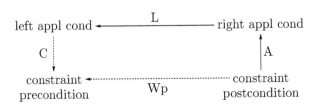

The paper is organized as follows. In Section 2, high-level conditions, rules and programs are defined and an access control for computer systems is introduced as a running example. In Section 3, two basic transformations of [16] are reviewed and, additionally, an essential transformation from application conditions into constraints is presented. In Section 4, weakest preconditions for high-level programs are formally defined and a transformation of programs and postconditions into weakest preconditions is given. In Section 5, related concepts and results are discussed. A conclusion including further work is given in Section 6. A long version of this paper including properties of weakest preconditions and detailed proofs of some results is available as a technical report [18].

2 Conditions and Programs

In this section, we will review the definitions of conditions, rules, and programs for high-level structures, e.g. graphs. We use the framework of weak adhesive HLR categories introduced as combination of HLR systems and adhesive categories. A detail introduction introduction can be found in [14,15]. As a running example, we consider a simple graph transformation system consisting of rules and programs. We demonstrate that programs are necessary extensions of rules for certain tasks and conditions can be used to describe a wide range of system properties, e.g. security properties.

Assumption. We assume that $\langle \mathcal{C}, \mathcal{M} \rangle$ is a weak adhesive HLR category with a category $\mathcal{C}$, a class $\mathcal{M}$ of monomorphisms, a $\mathcal{M}$-*initial* object, i.e. an object I in $\mathcal{C}$ such that there exists a unique morphism $I \rightarrow G$ in $\mathcal{M}$ for every object G in $\mathcal{C}$; binary coproducts and *epi-$\mathcal{M}$-factorization*, i.e. for every morphism there is an epi-mono-factorization with monomorphism in $\mathcal{M}$.

For illustration, we consider the category **Graph** of all directed, labeled graphs, which together with the class $\mathcal{M}$ of all injective graph morphisms constitutes a weak adhesive HLR category with binary coproducts and epi-$\mathcal{M}$-factorization and the empty graph $\emptyset$ as the $\mathcal{M}$-initial object.

Example 1 (access control graphs). In the following, we introduce state graphs of a simple access control for computer systems, which abstracts authentication and models user and session management in a simple way. We use this example solely for illustrative purposes. A more elaborated, role-based access control model is considered in [22]. The basic items of our model are users ⊛, sessions ◉, logs ◑, computer systems ▣, and directed edges between those items. An edge between a user and a system represents that the user has the right to access the system, i.e. establish a session with the system. Every user node is connected with one log, while an edge from a log to the system represents a failed (logged) login attempt. Every session is connected to a user and a system. The direction of the latter edge differentiates between sessions that have been proposed (an outgoing edge from a session node to a system) and sessions that have been established (an incoming edge to a session node from a system). Self-loops may occur in graphs during the execution of programs to select certain elements, but not beyond. An example of an access control graph is given in Figure 1.

Fig. 1. A state graph of the access control system

Conditions are nested constraints and application conditions in the sense of [16] generalizing the corresponding notions in [13] along the lines of [30].

Definition 1 (conditions). A *condition* over an object P is of the form $\exists a$ or $\exists (a, c)$, where $a: P \to C$ is a morphism and c is a condition over C. Moreover, Boolean formulas over conditions [over P] are conditions [over P]. Additionally, $\forall (a, c)$ abbreviates $\neg \exists (a, \neg c)$. A morphism $p: P \to G$ *satisfies* a condition $\exists a$ $[\exists (a, c)]$ over P if there exists a morphism $q: C \to G$ in $\mathcal{M}$ with $q \circ a = p$ [satisfying c]. An object G *satisfies* a condition $\exists a$ $[\exists (a, c)]$ if all morphisms $p: P \to G$ in $\mathcal{M}$ satisfy the condition. The satisfaction of conditions [over P] by objects [by morphisms with domain P] is extended onto Boolean conditions [over P] in the usual way. We write $p \models c$ $[G \models c]$ to denote that morphism p [object G] satisfies c. Two conditions c and c' over P are *equivalent* on objects, denoted by $c \equiv c'$, if, for all objects G, $G \models c$ if and only if $G \models c'$.

We allow infinite conjunctions and disjunctions of conditions. In the context of objects, conditions are also called *constraints*, in the context of rules, they are called *application conditions*. As the required morphisms of the semantics are to be in $\mathcal{M}$, we sometimes speak of $\mathcal{M}$-satisfiability as opposed to $\mathcal{A}$-satisfiability, where $\mathcal{A}$ is the class of all morphisms (see [17]).

Notation. For a morphism $a\colon P \to C$ in a condition, we just depict C, if P can be unambiguously inferred, i.e. for conditions over some left- or right-hand side and for constraints over the $\mathcal{M}$-initial object I. Note, that for every constraint over P, there is an equivalent constraint over I, i.e. $d \equiv \forall(I \to P, d)$, for $d = \exists a$ or $\exists(a, c)$ (see [18]).

Example 2 (access control conditions). Consider the access control graphs introduced in Example 1. Conditions allow to formulate statements on the graphs of the access control and can be combined to form more complex statements. The following conditions are over the empty graph:

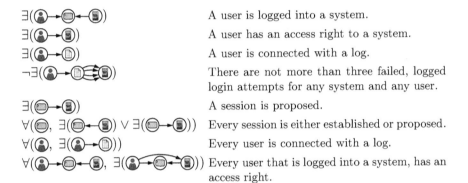

	A user is logged into a system.
	A user has an access right to a system.
	A user is connected with a log.
	There are not more than three failed, logged login attempts for any system and any user.
	A session is proposed.
	Every session is either established or proposed.
	Every user is connected with a log.
	Every user that is logged into a system, has an access right.

Fig. 2. Conditions on access control graphs

We consider rules with application conditions [13,16]. Examples and pointers to the literature can be found in [11,6].

Definition 2 (rules). A *plain rule* $p = \langle L \leftarrow K \to R \rangle$ consists of two morphisms in $\mathcal{M}$ with a common domain K. L is called the left-hand side, R the right-hand side, and K the interface. An *application condition* $\mathrm{ac} = \langle \mathrm{ac}_L, \mathrm{ac}_R \rangle$ for p consists of two application conditions over L and R, respectively. A *rule* $\hat{p} = \langle p, \mathrm{ac} \rangle$ consists of a plain rule p and an application condition ac for p.

$$
\begin{array}{ccccc}
L & \longleftarrow & K & \longrightarrow & R \\
{\scriptstyle m}\downarrow & (1) & \downarrow & (2) & \downarrow{\scriptstyle m^*} \\
G & \longleftarrow & D & \longrightarrow & H
\end{array}
$$

Given a plain rule p and a morphism $K \to D$, a *direct derivation* consists of two pushouts (1) and (2). We write $G \Rightarrow_{p,m,m^*} H$, $G \Rightarrow_p H$, or short $G \Rightarrow H$ and say that m is the *match* and m^{*} is the *comatch* of p in H. Given a rule

$\hat{p} = \langle p, \mathrm{ac} \rangle$ and a morphism $K \to D$, there is a *direct derivation* $G \Rightarrow_{\hat{p},m,m^*} H$ if $G \Rightarrow_{p,m,m^*} H$, $m \models \mathrm{ac}_L$, and $m \models \mathrm{ac}_R$. Let $\mathcal{A}$ be the class of all morphisms in $\mathcal{C}$. We distinguish between $\mathcal{A}$-*matching*, i.e. the general case, and $\mathcal{M}$-*matching*, i.e. if the match and the comatch are required to be in $\mathcal{M}$.

Notation. For the category **Graph**, we write $\langle L \Rightarrow R \rangle$ to abbreviate the rule $\langle L \leftarrow K \to R \rangle$, where K consists of all nodes common to L and R.

Example 3 (access control rules). Consider the access control graphs introduced in Example 1. The rules in Figure 3 are used to formalize the dynamic behavior of the access control system, i.e. are the basis of the access control programs.

Note, for every rule, every match is in $\mathcal{M}$. AddUser is a plain rule to introduce a user (and the associated log) to the system. Grant is a rule with application

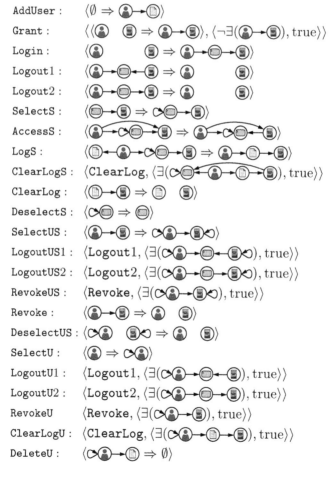

Fig. 3. Rules of the access control system

conditions: It grants a user the right to access a system, unless the user already has access. Login models a user proposing a session to a system, while Logout1 and Logout2 cancel an established or a proposed session, respectively. Rules with suffix S, US and U concern selected sessions (S), user and systems (US) and user (U) and are combined to programs in Figure 4. A description of each rule is given in [18].

We generalize the notions of programs on linear structures [7,8] and graph programs [19,27]) to high-level programs on rules.

Definition 3 (programs). *(High-level) Programs* are inductively defined:

(1) Skip and every rule p are programs.
(2) Every finite set S of programs is a program.
(3) Given programs P and Q, then $(P;Q)$, P^* and $P{\downarrow}$ are programs.

The *semantics* of a program P is a binary relation $P \subseteq C \times C$ on objects which is inductively defined as follows:

(1) Skip $= \{\langle G, G \rangle \mid G \in C\}$ and for every rule p, $p = \{\langle G, H \rangle \mid G \Rightarrow_p H\}$.
(2) For a finite set S of programs, $S = \cup_P P$.
(3) For programs P and Q, $(P;Q) = Q \circ P$, $P^* = P^*$ and

$$P{\downarrow} = \{\langle G, H \rangle \mid \langle G, H \rangle \in P^* \quad \text{and} \; \neg \exists M. \langle H, M \rangle \in P \}.$$

Programs according to (1) are *elementary*; a program according (2) describes the *nondeterministic choice* of a program; a program $(P;Q)$ is the *sequential composition* of P and Q, P^* is the *reflexive, transitive closure* of P, and $P{\downarrow}$ is the *iteration* of P as long as possible.

Example 4 (access control programs). Consider the access control graphs in Example 1. The dynamic part of the control system Control is the reflexive, transitive closure of the programs Control = {AddUser, Grant, Login, Logout, ProcessLogin, Revoke, DeleteUser}, depicted in Figure 3 and Figure 4, respectively. Logout cancels a session (established or proposed). ProcessLogin models

```
Logout       = {Logout1, Logout2}
ProcessLogin = SelectS; AccessS↓; LogS↓; ClearLogS↓; DeselectS↓
Revoke       = SelectUS; LogoutUS↓; RevokeUS; DeselectUS
LogoutU      = {LogoutU1, LogoutU2}
DeleteUser   = SelectU; LogoutU↓; RevokeU↓; ClearLogU↓; DeleteU
```

Fig. 4. Programs of the access control system

the reaction of a system towards a session proposal, which, dependent on the user's right, leads to an established session and the clearing of the user's log of failed attempts, or the denial and removal of that session and the logging of the failed attempt. Revoke removes a user's right to access a system, but not before closing the user's sessions to that system. Finally, DeleteUser is a program to delete a user and his/her associated log by canceling the user's sessions, by removing the user's access rights and by clearing the user's log.

Note, that there is no way to model certain actions like DeleteUser by a single rule, as a user, in principle, may have an arbitrary number of sessions or log entries. However, user deletion should be a transaction always applicable for every user.

Definition 4 (termination). A program P applied to an input object G *terminates* properly, if $\text{PDer}(P,G)$ is finite, i.e. $\exists k \in \quad . |\text{PDer}(P,G)| \leq k$, where $\text{PDer}(P,G)$ denotes the set of all partial derivations within the execution of a program P, starting with G (see [18]).

Remark 1. Execution of high-level programs requires backtracking, therefore the above definition of termination is more suitable than the classical one, i.e. the nonexistence of infinite derivations. This may be seen as follows: An infinite derivation implies infinitely many partial derivations. The other direction holds only if the number of matches is finite. By the uniqueness of pushouts, $\text{PDer}(p,G)$ is finite and there cannot be infinitely many derivations of finite length for any program P.

3   Basic Transformations of Conditions

In the following, we recall two known transformations from constraints to application conditions and from right- to left application conditions [21,13,16] and present a new transformation from application conditions to constraints. Combining these basic transformations, we obtain a transformation from a postcondition over the rule to a precondition. First, there is a transformation from constraints to application conditions such that a morphism satisfies the application condition if and only if the codomain satisfies the constraint.

Theorem 1 (transformation of constraints into application conditions). *There is a transformation* A *such that, for every constraint c and every rule $p = \langle L \leftarrow K \rightarrow R \rangle$, and all morphisms $m : R \rightarrow H$, $m \models A(p,c) \Leftrightarrow H \models c$.*

Second, there is a transformation from right to left application conditions such that a comatch satisfies an application condition if and only if the match satisfies the transformed application condition.

Theorem 2 (transformation of application conditions). *There is a transformation* L *such that, for every rule p, every right application condition ac for p, and all direct derivations $G \Rightarrow_{p,m,m^*} H$, $m \models L(p, \text{ac}) \Leftrightarrow m^* \models \text{ac}$.*

We consider a transformation of application conditions to constraints, which correspond to the universal closure of application conditions. For $\mathcal{A}$-matching however, the closure is over arbitrary morphisms and does not fit to the notion of $\mathcal{M}$-satisfiability. This is why a part of the application condition has to be transformed accordingly.

Theorem 3 (transformation of application conditions into constraints).
For weak adhesive HLR categories with $\mathcal{M}$-initial object, there is a transformation C *such that, for every application condition* ac *over* L *and for all objects* G,

$$G \models C(\mathrm{ac}) \Leftrightarrow \forall m\colon L \to G.\ m \models \mathrm{ac}$$

Construction. Define $C(\mathrm{ac}) = \bigwedge_{e\ E} \forall(e \circ i,\ C_e(\mathrm{ac}))$ where the junction ranges over all epimorphisms $e\colon L \to L'$ and $i\colon I \to L$ is the unique morphism from the $\mathcal{M}$-initial object to L. The transformation C_e is defined inductively on the structure of the conditions: $C_e(\exists a) = \exists a'$ and $C_e(\exists(a,c)) = \exists(a',c)$ if $a = a' \circ e$ is some epi-$\mathcal{M}$-factorization of a and $C_e(\exists a) = C_e(\exists(a,c)) = \mathrm{false}$ if there is no epi-$\mathcal{M}$-factorization of a with epimorphism e. For Boolean conditions, the transformation C_e is extended in the usual way.

Example 5. The application condition ac $= \neg\exists(\text{◎}{\to}\text{◎}) \wedge \neg\exists(\text{◎}{\leftarrow}\text{◎}) \wedge \neg\exists(\text{◎}⟲)$ over ◎ ◎ expresses that there is no edge between two given session nodes.

$$
\begin{aligned}
C(\mathrm{ac}) = \ & \forall(\text{◎ ◎},\ C_{\mathrm{id}}(\mathrm{ac})) \wedge \forall(\text{◎},\ C_e(\mathrm{ac})) \\
= \ & \forall(\text{◎ ◎},\ \neg C_{\mathrm{id}}(\exists(\text{◎}{\to}\text{◎})) \wedge \neg C_{\mathrm{id}}(\exists(\text{◎}{\leftarrow}\text{◎})) \wedge \neg C_{\mathrm{id}}(\exists(\text{◎}⟲))) \\
& \wedge \forall(\text{◎},\ \neg C_e(\exists(\text{◎}{\to}\text{◎})) \wedge \neg C_e(\exists(\text{◎}{\leftarrow}\text{◎})) \wedge \neg C_e(\exists(\text{◎}⟲))) \\
= \ & \forall(\text{◎ ◎},\ \mathrm{ac}) \wedge \forall(\text{◎},\ \neg\mathrm{false} \wedge \neg\mathrm{false} \wedge \neg\exists(\text{◎}⟲)) \\
\equiv \ & \forall(\text{◎ ◎},\ \neg\exists(\text{◎}{\to}\text{◎}) \wedge \neg\exists(\text{◎}{\leftarrow}\text{◎}) \wedge \mathrm{true}) \wedge \forall(\text{◎},\ \mathrm{true} \wedge \neg\exists(\text{◎}⟲)) \\
\equiv \ & \forall(\text{◎ ◎},\ \neg\exists(\text{◎}{\to}\text{◎}) \wedge \neg\exists(\text{◎}{\leftarrow}\text{◎})) \wedge \forall(\text{◎},\ \neg\exists(\text{◎}⟲))
\end{aligned}
$$

with id: ◎ ◎ → ◎ ◎ and e: ◎ ◎ → ◎.

Proof. In [17] is shown: For all $m'\colon L' \to G$ in $\mathcal{M}$ and all epimorphisms $e\colon L \to L'$,

$$m' \models C_e(\mathrm{ac}') \quad \Leftrightarrow \quad m' \circ e \models \mathrm{ac}' \qquad (*)$$

We show: $\forall m\colon L \to G$, $m \models \mathrm{ac}$ if and only if $G \models C(\mathrm{ac})$. "Only if". Assume $\forall m\colon L \to G$, $m \models \mathrm{ac}$. For $G \models C(\mathrm{ac})$ to hold, G has to satisfy $C_e(\mathrm{ac})$ for all epimorphisms $e\colon L \to L'$, i.e. for all epimorphisms $e\colon L \to L'$ and all morphisms $m'\colon L' \to G$ in $\mathcal{M}$ holds $m' \models C_e(\mathrm{ac})$. Given such morphisms e and m', define $m = m' \circ e$. By assumption, $m \models \mathrm{ac}$, and by $(*)$ we have $m' \models C_e(\mathrm{ac})$, hence $G \models C(\mathrm{ac})$. "If". Assume $G \models C(\mathrm{ac})$, i.e. G satisfies $C_e(\mathrm{ac})$ for all epimorphisms $e\colon L \to L'$, i.e. for all epimorphisms $e\colon L \to L'$ and all morphisms $m'\colon L' \to G$ in $\mathcal{M}$ holds $m' \models C_e(\mathrm{ac})$. Given an arbitrary morphism $m\colon L \to G$, consider the epi-$\mathcal{M}$-factorization $m' \circ e$. By assumption, $m' \models C_e(\mathrm{ac})$, and by $(*)$ we have $m \models \mathrm{ac}$.

Remark 2. The uniqueness of epi-$\mathcal{M}$-factorizations (up to isomorphism) follows immediately from the uniqueness of epi-mono-factorizations, as every $\mathcal{M}$-morphism is a monomorphism.

Remark 3. For weak adhesive HLR categories with $\mathcal{M}$-initial object and $\mathcal{M}$-matching, there is a simplified transformation C such that, for every application condition ac over L and for all objects G, $G \models C(\mathrm{ac}) \Leftrightarrow \forall m: L \to G \in \mathcal{M}.\ m \models \mathrm{ac}$. For an application condition ac over L and $i: I \to L$, let $C(\mathrm{ac}) = \forall(i, \mathrm{ac})$. For all $\mathcal{M}$-morphisms $m: L \to G$, $m \models \mathrm{ac}$ iff there exists an $\mathcal{M}$-morphism $p: I \to G$ such that for all $\mathcal{M}$-morphisms $m: L \to G$ holds $m \models \mathrm{ac}$ iff there exists an $\mathcal{M}$-morphism $p: I \to G$ such that for all $\mathcal{M}$-morphisms $m: L \to G$ with $p = m \circ i$ holds $m \models \mathrm{ac}$ iff $G \models \forall(i, \mathrm{ac})$ (Def.1).

Finally, the applicability of a rule can be expressed by a left application condition for the matching morphism.

Theorem 4 (applicability of a rule). *There is a transformation* Def *from rules into application conditions such that, for every rule p and every morphism $m: L \to G$,*

$$m \models \mathrm{Def}(p) \Leftrightarrow \exists H.G \Rightarrow_{p,m,m^*} H.$$

Construction. For a rule $p = \langle q, \mathrm{ac} \rangle$, let $\mathrm{Def}(p) = \mathrm{Appl}(q) \wedge \mathrm{ac}_L \wedge \mathrm{L}(p, \mathrm{ac}_R)$, where, for a rule $q = \langle L \xleftarrow{l} K \xrightarrow{r} R \rangle$, $\mathrm{Appl}(q) = \wedge_a\ _A \neg \exists a$ and the index set A ranges over all morphisms $a: L \to L'$ such that the pair $\langle l, a \rangle$ has no pushout complement and there is no decomposition $a = a'' \circ a'$ of a with proper morphism a'' in $\mathcal{M}$ (a'' not an isomorphism) such that $\langle l, a' \rangle$ has no pushout complement.

Example 6. An example of Appl is given below for $\texttt{DeleteSys} = \langle \circledcirc\!\!\!\!\times \leftarrow \emptyset \to \emptyset \rangle$. Intuitively, the application of $\texttt{DeleteSys}$ requires the absence of additional edges adjacent to the system node. Therefore, $\texttt{DeleteSys}$ may only be the last step in program deleting a system node. $\mathrm{Appl}(\texttt{DeleteSys})$ is a condition over $\circledcirc\!\!\!\!\times$.

$$\mathrm{Appl}(\texttt{DeleteSys}) = \quad \neg\exists(\circledcirc\!\!\to\!\!\circledcirc) \wedge \neg\exists(\circledcirc\!\!\leftarrow\!\!\circledcirc) \wedge \neg\exists(\circledcirc\!\!\to\!\!\bigcirc) \wedge \neg\exists(\circledcirc\!\!\leftarrow\!\!\bigcirc)$$
$$\wedge\ \neg\exists(\circledcirc\!\!\to\!\!\circledcirc) \wedge \neg\exists(\circledcirc\!\!\leftarrow\!\!\circledcirc) \wedge \neg\exists(\circledcirc\!\!\to\!\!\bigcirc) \wedge \neg\exists(\circledcirc\!\!\leftarrow\!\!\bigcirc)$$
$$\wedge\ \neg\exists(\times\!\!\circledcirc\!\!\times)$$

Proof. For plain rules, we show that, for every morphism $m: L \to G$,

$$m \models \mathrm{Appl}(q) \Leftrightarrow \exists H.G \Rightarrow_{q,m,m^*} H.$$

"Only if" Let $m \models \mathrm{Appl}(q)$. Assume there is no direct derivation $G \Rightarrow_{q,m,m^*} H$. Then the pair $\langle l, m \rangle$ has no pushout complement and there is a morphism $a: L \to L'$ such that $\langle l, a \rangle$ has no pushout complement and $m \models \exists a$. Then $m \not\models \mathrm{Appl}(q)$. A contradiction. Consequently, there is a direct derivation $G \Rightarrow_{q,m,m^*} H$.

"If" Let $G \Rightarrow_{q,m,m^*} H$. Then, for every morphism $a: L \to L'$, $m \models \exists a$ iff there is some $m': L' \to G$ in $\mathcal{M}$ such that $m' \circ a = m$. By the pushout-pullback decomposition, the pushout has a decomposition into two pushouts (1) and (2) and, in particular, $\langle l, a \rangle$ has a pushout complement. Consequently, for every morphism $a \in A$, $m \models \neg\exists a$, i.e. $m \models \mathrm{Appl}(q)$.

$$
\begin{array}{ccc}
L & \xleftarrow{\ l\ } & K \\
\downarrow & (1) & \vdots \\
L' & \dashleftarrow & K' \\
\downarrow & (2) & \vdots \\
G & \longleftarrow & D
\end{array}
$$

By the definition of Def and $\models$, Theorem 4, the statement above, and the definition of $\Rightarrow$, for every morphism $m: L \to G$, $m \models \mathrm{Def}(p)$ iff $m \models \mathrm{Appl}(q) \wedge$ $m \models \mathrm{ac}_L \wedge m \models \mathrm{L}(p, \mathrm{ac}_R)$ iff $\exists H.G \Rightarrow_{q,m,m^*} H \wedge m \models \mathrm{ac}_L \wedge m \models \mathrm{ac}_R$ iff $\exists H.G \Rightarrow_{p,m,m^*} H$. This completes the proof.

4   Weakest Preconditions

In the following, we define weakest preconditions for high-level programs similar to the ones for Dijkstra's guarded commands in [7,8], show how to construct weakest preconditions for high-level programs and demonstrate the use of weakest preconditions to reduce problems on programs, e.g. the invariance of conditions, onto tautology problems of conditions.

Definition 5 (weakest preconditions). For a program P relative to a condition d we define: A condition c is a *precondition*, if for all objects G satisfying c, (1) $\langle G, H \rangle \in P$ implies $H \models d$ for all H, (2) $\langle G, H \rangle \in P$ for some H, and (3) P terminates for G. A precondition c is a *weakest precondition*, denoted by $\mathrm{wp}(P, d)$, if for all other preconditions c' of P relative to d, c' implies c. A condition c is a *liberal precondition*, if for all objects $G \models c$ at least (1) is satisfied, and a *weakest liberal precondition*, denoted by $\mathrm{wlp}(P, d)$, if all other liberal preconditions c' of P relative to d imply c. A condition c is a *termination precondition*, if for all objects $G \models c$ properties (1) and (3) are satisfied, and a *weakest termination precondition*, denoted by $\mathrm{wtp}(P, d)$, if all other termination preconditions c' of P relative to d imply c.

The following fact points out a simple proof scheme for weakest preconditions.

Fact 1 (weakest preconditions). A condition c is a weakest precondition if, for all objects G, $G \models c$ if and only if properties (1)-(3) are satisfied.

For the construction of weakest preconditions, we make use of the fact that $\mathrm{wp}(P, d)$ is a conjunction of three properties and treat properties (1) and (3), and property (2) separately. We observe property (2) is equivalent to the negation of property (1) for $d = \neg\mathrm{true}$, hence we state:

Fact 2 (existence of results). $G \models \neg\mathrm{wlp}(P, \mathrm{false}) \Leftrightarrow$ property (2) is satisfied.

Assumption. We assume that $\langle \mathcal{C}, \mathcal{M} \rangle$ is a weak adhesive HLR category with finite number of matches, i.e. for every morphism $l: K \to L$ and every object G, there exist only a finite number of morphisms $m: L \to G$ s.t. $\langle l, m \rangle$ has a pushout complement.

Theorem 5 (weakest preconditions). *For weak adhesive HLR categories with finite number of matches, there are transformations* Wlp, Wtp *and* Wp *such that for every program P and every condition d,* $\mathrm{Wlp}(P, d)$ *is a weakest liberal precondition,* $\mathrm{Wtp}(P, d)$ *is a weakest termination precondition and* $\mathrm{Wp}(P, d)$ *is a weakest precondition of P relative to d.*

Construction. The transformations are defined inductively over the structure of programs. For every rule p, let

$$\mathrm{Wlp}(p, d) = \mathrm{Wtp}(p, d) = \mathrm{C}(\mathrm{Def}(p) \Rightarrow \mathrm{L}(p, \mathrm{A}(p, d))).$$

For any program P, $\mathrm{Wp}(P, d) = \mathrm{Wtp}(P, d) \wedge \neg\mathrm{Wlp}(P, \mathrm{false})$.
For any set $\mathcal{S}$ of programs and programs P, Q,

$$
\begin{aligned}
\mathrm{Wlp}(\mathtt{Skip}, d) &= d \\
\mathrm{Wlp}(\mathcal{S}, d) &= \textstyle\bigwedge_P \mathrm{Wlp}(P, d) \\
\mathrm{Wlp}((P; Q), d) &= \mathrm{Wlp}(P, \mathrm{Wlp}(Q, d)) \\
\mathrm{Wlp}(P^*, d) &= \textstyle\bigwedge_{i=0} \mathrm{Wlp}(P^i, d) \\
\mathrm{Wlp}(P{\downarrow}, d) &= \mathrm{Wlp}(P^*, \mathrm{Wlp}(P, \mathrm{false}) \Rightarrow d)
\end{aligned}
$$

$$
\begin{aligned}
\mathrm{Wtp}(\mathtt{Skip}, d) &= d \\
\mathrm{Wtp}(\mathcal{S}, d) &= \textstyle\bigwedge_P \mathrm{Wtp}(P, d) \\
\mathrm{Wtp}((P; Q), d) &= \mathrm{Wtp}(P, \mathrm{Wtp}(Q, d)) \\
\mathrm{Wtp}(P^*, d) &= \textstyle\bigwedge_{i=0} \mathrm{Wlp}(P^i, d \wedge \mathrm{Wtp}(P, \mathrm{true})) \wedge \bigvee_{k=0} \mathrm{Wlp}(P^{k+1}, \mathrm{false}) \\
\mathrm{Wtp}(P{\downarrow}, d) &= \mathrm{Wtp}(P^*, \mathrm{Wlp}(P, \mathrm{false}) \Rightarrow d)
\end{aligned}
$$

where for $i \geq 0$, P^i is inductively defined by $\mathtt{Skip}$ for $i = 0$ and by $P^{i+1} = (P^i; P)$.

Proof. We show $\mathrm{Wlp}(P, d) \equiv \mathrm{wlp}(P, d)$, $\mathrm{Wtp}(P, d) \equiv \mathrm{wtp}(P, d)$, and $\mathrm{Wp}(P, d) \equiv \mathrm{wp}(P, d)$. The first two proofs are done by induction over the structure of programs. First we consider elementary programs consisting of a single rule p. For all objects G, we have:

$$
\begin{aligned}
&G \models \mathrm{Wlp}(p, d) \\
\Leftrightarrow\ & G \models \mathrm{C}(\mathrm{Def}(p) \Rightarrow \mathrm{L}(p, \mathrm{A}(p, d))) && \text{(Def. Wlp)} \\
\Leftrightarrow\ & \forall L \overset{m}{\to} G.\ m \models (\mathrm{Def}(p) \Rightarrow \mathrm{L}(p, \mathrm{A}(p, d))) && \text{(Thm. 3)} \\
\Leftrightarrow\ & \forall L \overset{m}{\to} G.\ m \models \mathrm{Def}(p) \Rightarrow m \models \mathrm{L}(p, \mathrm{A}(p, d)) && \text{(Def. } \models) \\
\Leftrightarrow\ & \forall L \overset{m}{\to} G, R \overset{m^*}{\to} H.\ m \models \mathrm{Def}(p) \Rightarrow m \models \mathrm{A}(p, d) && \text{(Thm. 2)} \\
\Leftrightarrow\ & \forall L \overset{m}{\to} G, R \overset{m^*}{\to} H.\ (G \Rightarrow_{p,m,m^*} H) \Rightarrow H \models d && \text{(Thms. 4 \& 1)} \\
\Leftrightarrow\ & \forall H.\ \langle G, H \rangle \in\ p\ \Rightarrow H \models d && \text{(Def. } p) \\
\Leftrightarrow\ & G \models \mathrm{wlp}(p, d) && \text{(Def. wlp)}
\end{aligned}
$$

Thus, $\mathrm{Wlp}(p, d)$ is a weakest liberal precondition of p relative to d. Furthermore,

$G \models \mathrm{Wtp}(p, d)$ if and only if $G \models \mathrm{Wlp}(p, d)$, as every rule application terminates by the finiteness assumption and wtp reduces to wlp for single rules p. For composed programs, the statement follows by structural induction (see [18].)

For Wp, we now show for every program P, $\mathrm{Wp}(P, d) \equiv \mathrm{wp}(P, d)$: $\mathrm{Wp}(P, d)$ is defined as $\neg\mathrm{Wlp}(P, \mathrm{false}) \wedge \mathrm{Wtp}(P, d)$, which is, by the first two equations, equivalent to $\neg\mathrm{wlp}(P, \mathrm{false}) \wedge \mathrm{wtp}(P, d)$, which is equivalent to $\mathrm{wp}(P, d)$ (see [18].)

Example 7 (access control system). Consider the access control for computer systems, presented in Examples 1-4. For the system, one might want to ensure the validity of certain properties, e.g.:

(1) Always, every user logged into a system, has an access right to the system:
    secure implies wlp($\mathtt{Control}$, *secure*), where

$$secure = \forall(\text{⊙→⊡←⊡}, \exists(\text{⊙→⊡←⊡})).$$

(2) Every user can always be deleted: $\exists(\text{☺})$ implies wp(DeleteUser, true)

(3) Every user can always have his access right to a system revoked:
$\exists(\text{☺}\!\rightarrow\!\text{🖥})$ implies wp(Revoke, true)

By calculating weakest [liberal] preconditions, the problem to decide these properties can be reduced onto the tautology problem for conditions. The meaning of *secure* implies wlp(Control, *secure*) can be seen as follows: The constraint *secure* is an invariant, i.e. given a state satisfying *secure*, every next state will also satisfy *secure*. For a proof, we have to show *secure* implies Wlp(P, *secure*) for every program $P \in$ Control.

We give explicit proof of property (1) for the programs AddUser and Grant. For the program AddUser, *secure* implies Wlp(AddUser, *secure*), which can be proved as follows:

$$\text{A}(\text{AddUser}, secure) = \forall(\text{☺}\!\rightarrow\!\square \quad \text{☺}\!\rightarrow\!\square\!\leftarrow\!\blacksquare, \exists(\text{☺}\!\rightarrow\!\square \quad \text{☺}\!\rightarrow\!\square\!\leftarrow\!\blacksquare))$$
$$\wedge \forall(\square\!\leftarrow\!\text{☺}\!\rightarrow\!\square\!\leftarrow\!\blacksquare, \exists(\square\!\leftarrow\!\text{☺}\!\rightarrow\!\square\!\leftarrow\!\blacksquare))$$

$$\text{L}(\text{AddUser}, \text{A}(\text{AddUser}, secure)) = \forall(\text{☺}\!\rightarrow\!\square\!\leftarrow\!\blacksquare, \exists(\text{☺}\!\rightarrow\!\square\!\leftarrow\!\blacksquare)) = secure$$

$$\begin{aligned}
\text{Wlp}(\text{AddUser}, secure) &= \text{C}(\text{Def}(\text{AddUser}) \Rightarrow \text{L}(\text{AddUser}, \text{A}(\text{AddUser}, secure))) \\
&= \text{C}((\text{Appl}(\text{AddUser}) \wedge \text{true} \wedge \text{true}) \Rightarrow secure) \\
&\equiv \text{C}(\text{true} \Rightarrow secure) \equiv \text{C}(secure) \\
&= \forall(\emptyset, secure) \\
&\equiv secure
\end{aligned}$$

This is no surprise as we could also have argued that a newly added user cannot have an established session with a system, hence every application of AddUser preserves the satisfaction of *secure*. For the program Grant, *secure* implies Wlp(Grant, *secure*), even without the additional application condition.

$$\begin{aligned}
&\text{L}(\text{Grant}, \text{A}(\text{Grant}, secure)) \\
=\ &\forall(\text{☺} \quad \text{☺}\!\rightarrow\!\square\!\leftarrow\!\blacksquare \quad \blacksquare, \exists(\text{☺} \quad \text{☺}\!\rightarrow\!\square\!\leftarrow\!\blacksquare \quad \blacksquare)) \\
&\wedge \forall(\text{☺} \quad \text{☺}\!\rightarrow\!\square\!\leftarrow\!\blacksquare \quad , \exists(\text{☺} \quad \text{☺}\!\rightarrow\!\square\!\leftarrow\!\blacksquare \quad)) \\
&\wedge \forall(\quad \text{☺}\!\rightarrow\!\square\!\leftarrow\!\blacksquare \quad \blacksquare, \exists(\quad \text{☺}\!\rightarrow\!\square\!\leftarrow\!\blacksquare \quad \blacksquare)) \\
&\wedge \forall(\quad \text{☺}\!\rightarrow\!\square\!\leftarrow\!\blacksquare \quad , \exists(\quad \text{☺}\!\rightarrow\!\square\!\leftarrow\!\blacksquare \quad))
\end{aligned}$$

$$\begin{aligned}
&\text{Wlp}(\text{Grant}, secure) \\
&= \text{C}(\text{Def}(\text{Grant}) \Rightarrow \text{L}(\text{Grant}, \text{A}(\text{Grant}, secure))) \\
&= \text{C}((\text{Appl}(\text{Grant}) \wedge \neg\exists(\text{☺}\!\rightarrow\!\blacksquare) \wedge \text{true}) \Rightarrow \text{L}(\text{Grant}, \text{A}(\text{Grant}, secure))) \\
&\equiv \text{C}(\neg\exists(\text{☺}\!\rightarrow\!\blacksquare) \Rightarrow \text{L}(\text{Grant}, \text{A}(\text{Grant}, secure))) \\
&\text{if } \text{C}(\text{L}(\text{Grant}, \text{A}(\text{Grant}, secure))) \\
&\text{if } \text{L}(\text{Grant}, \text{A}(\text{Grant}, secure))
\end{aligned}$$

Note, *secure* implies L(Grant, A(Grant, *secure*)) and thus Wlp(Grant, *secure*). We also have *secure* implies Wlp(Login, *secure*), the proof of which is similar to *secure* implies Wlp(Grant, *secure*), as L(Grant, A(Grant, *secure*)) $\equiv$ L(Login, A(Login, *secure*)).

For Logout, ProcessLogin, Revoke and DeleteUser, we shall only sketch the proofs. It is easy to see that Logout1 as well as Logout2 preserve the satisfaction of *secure*, hence we can assume *secure* implies Wlp(Logout, *secure*). Concerning ProcessLogin, one can prove the invariance of *secure* for every used rule, i.e. SelectS, AccessS, LogS, ClearLogS and DeselectS, and moreover, for every subprogram of ProcessLogin. Intuitively the only interesting part is the proof of *secure* implies Wlp(AccessS, *secure*), while the validity of this claim is quite obvious. Concerning Revoke, one can show that LogoutUS↓ leaves no sessions for any selected user and system (see property (4)). As a consequence, RevokeUS will preserve the satisfaction of *secure*, as do all other parts of Revoke, hence *secure* implies Wlp(Revoke, *secure*). The proof of DeleteUser is similar.

(4) After execution of LogoutUS↓, there is no established session left for any selected user and system: wlp(LogoutUS↓, ¬∃(⟨graph⟩)) ≡ true.

(5) C(Appl(Logout1)) ∧ C(Appl(Logout2)) is an invariant for all programs P in Control, and all subprograms and rules of Revoke and DeleteUser.

One can show property (4) by using (5) as one observes Wlp(LogoutUS↓, *secure*) ≡ Wlp(LogoutU , true) ≡ true. Property (5) expresses that certain edges adjacent to a session node do not exist, while others have a multiplicity of at most 1. Proving property (5) for all rules used in Control is tedious, but nonetheless straightforward, as every subcondition may handled separately. Intuitively only subprograms and rules have to be considered that contain a session node, and moreover, that create or delete edges adjacent to session nodes.

5   Related Concepts

In this section we briefly review other work on using graph transformation for verification. Before we do so, however, we wish to point out one important global difference between this related work and the approach of this paper.

– The approach of this paper is based on the principle of *assertional reasoning*, and inherits both the advantage and the disadvantage of that principle. The advantage is that the approach is general where it can be made to apply, meaning that it provides a method to verify finite-state and infinite-state systems alike. The disadvantage is that finding invariants is hard and cannot be automated in general.

– Existing approaches are typically based on the principle of *model checking*, which essentially involves exhaustive exploration, either of the concrete states (which are often too numerous to cover completely) or on some level of abstraction (in which case the results become either unsound or incomplete). On the positive side, model checking is a push-button approach, meaning that it requires no human intervention.

In other words, there is a dividing line between the work in this paper and the related work reported below, which is parallel to the division between theorem proving and model checking in "mainstream" verification (see [20] for an early discussion). Since current wisdom holds that these approaches can actually be combined to join strengths (e.g., [5,25]), we expect that the same will turn out to hold in the context of graph transformation.

The first paper in which it was put forward that graph transformation systems can serve as a suitable specification formalism on the basis of which model checking can be performed was Varró [32]; this was followed up by [33] which describes a tool chain by which graph transformation systems are translated to Promela, and then model checked by SPIN. We pursued a similar approach independently in [28,29], though relying on dedicated (graph transformation-based) state space generation rather than an existing tool. The two strands were compared in [31]. Again independently, Dotti et al. [10,9] also describe a translation from a graph transformation-based specification formalism (which they call object-based graph grammars) to Promela.

Another model checking-related approach, based on the idea of McMillan unfoldings for Petri Nets (see [24]), has been pursued by Baldan, König et al. in, e.g., [2,1], and in combination with abstraction in [3,23]. The latter avoids the generation of complete (concrete) state spaces, at the price of being approximative, in other words, admitting either false positives (unsoundness) or false negatives (incompleteness) in the analysis. The (pure) model checking and abstraction-based techniques were briefly compared in [4].

Termination. In addition to the general verification methods discussed above, a lot of research has been carried out on more specific properties of graph grammars. Especially relevant in our context is the work on *termination* of graph grammars. This is known to be undecidable in general (see [26]), but under special circumstances may be shown to hold; for instance, Ehrig et al. discuss such a special case for *model transformation* in [12].

6   Conclusion

This paper extends graph programs to programs over high-level rules with application conditions, and defines weakest preconditions over high-level programs similar to the ones for Dijkstra's guarded commands in [7,8]. It presents transformations from application conditions to constraints, which, combined with two known transformations over constraints and application conditions, can be used to construct weakest preconditions for high-level rules as well as programs.

A known proof technique for showing the correctness of a program with respect to a pre- and a postcondition is to construct a weakest precondition and to show that the precondition implies the weakest precondition. We demonstrate the applicability of this method on our access control for computer systems.

Further topics could be the followings.

(1) Consideration of strongest postconditions.
(2) Comparison of notions: A comparison of conditions – as considered in this paper – and first-order formulas on graphs and high-level structures.
(3) Generalization of notions: The generalization of conditions to capture monadic second order properties.

(4) An investigation of the tautology problem for conditions with the aim to find a suitable class of conditions, for which the problem is decidable.

(5) Implementation: A system for computing/approximating weakest preconditions and for deciding/semideciding correctness of program specifications.

References

1. P. Baldan, A. Corradini, and B. König. Verifying finite-state graph grammars. In *Concurrency Theory*, volume 3170 of *LNCS*, pages 83–98. Springer, 2004.

2. P. Baldan and B. König. Approximating the behaviour of graph transformation systems. In *Graph Transformations (ICGT'02)*, volume 2505 of *LNCS*, pages 14–29. Springer, 2002.

3. P. Baldan, B. König, and B. König. A logic for analyzing abstractions of graph transformation systems. In *Static Analysis Symposium (SAS)*, volume 2694 of *LNCS*, pages 255–272. Springer, 2003.

4. P. Baldan, B. König, and A. Rensink. Graph grammar verification through abstraction. In B. König, U. Montanari, and P. Gardner, editors, *Graph Transformations and Process Algebras for Modeling Distributed and Mobile Systems*, number 04241 in Dagstuhl Seminar Proceedings, 2005.

5. E. M. Clarke, A. Biere, R. Raimi, and Y. Zhu. Bounded model checking using satisfiability solving. *Formal Methods in System Design*, 19(1):7–34, 2001.

6. A. Corradini, U. Montanari, F. Rossi, H. Ehrig, R. Heckel, and M. Löwe. Algebraic approaches to graph transformation. In *Handbook of Graph Grammars and Computing by Graph Trans.*, volume 1, pages 163–245. World Scientific, 1997.

7. E. W. Dijkstra. *A Discipline of Programming*. Prentice-Hall, 1976.

8. E. W. Dijkstra and C. S. Scholten. *Predicate Calculus and Program Semantics*. Springer, 1989.

9. O. M. dos Santos, F. L. Dotti, and L. Ribeiro. Verifying object-based graph grammars. *ENTCS*, 109:125–136, 2004.

10. F. L. Dotti, L. Foss, L. Ribeiro, and O. M. dos Santos. Verification of distributed object-based systems. In *Formal Methods for Open Object-Based Distributed Systems (FMOODS)*, volume 2884 of *LNCS*, pages 261–275. Springer, 2003.

11. H. Ehrig. Introduction to the algebraic theory of graph grammars. In *Graph-Grammars and Their Application to Computer Science and Biology*, volume 73 of *LNCS*, pages 1–69. Springer, 1979.

12. H. Ehrig, K. Ehrig, J. De Lara, G. Taentzer, D. Varró, and S. Varró-Gyapay. Termination criteria for model transformation. In *Proc. Fundamental Approaches to Software Engineering*, volume 2984 of *LNCS*, pages 214–228. Springer, 2005.

13. H. Ehrig, K. Ehrig, A. Habel, and K.-H. Pennemann. Theory of constraints and application conditions: From graphs to high-level structures. *Fundamenta Informaticae*, 72, 2006.

14. H. Ehrig, K. Ehrig, U. Prange, and G. Taentzer. *Fundamentals of Algebraic Graph Transformation*. EATCS Monographs of Theoretical Computer Science. Springer-Verlag, Berlin, 2006.

15. H. Ehrig, A. Habel, J. Padberg, and U. Prange. Adhesive high-level replacement systems: A new categorical framework for graph transformation. *Fundamenta Informaticae*, 72, 2006.

16. A. Habel and K.-H. Pennemann. Nested constraints and application conditions for high-level structures. In *Formal Methods in Software and System Modeling*, volume 3393 of *LNCS*, pages 293–308. Springer, 2005.

17. A. Habel and K.-H. Pennemann. Satisfiability of high-level conditions. In *Graph Transformations (ICGT'06)*, this volume of LNCS. Springer, 2006.
18. A. Habel, K.-H. Pennemann, and A. Rensink. Weakest preconditions for high-level programs: Long version. Technical Report 8/06, University of Oldenburg, 2006.
19. A. Habel and D. Plump. Computational completeness of programming languages based on graph transformation. In *Proc. Foundations of Software Science and Computation Structures*, volume 2030 of *LNCS*, pages 230–245. Springer, 2001.
20. J. Y. Halpern and M. Y. Vardi. Model checking vs. theorem proving: A manifesto. In J. Allen, R. Fikes, and E. Sandewall, editors, *Proc. International Conference on Principles of Knowledge Representation and Reasoning*, pages 325–334. Morgan Kaufmann Publishers, 1991.
21. R. Heckel and A. Wagner. Ensuring consistency of conditional graph grammars. In *SEGRAGRA'95*, volume 2 of *ENTCS*, pages 95–104, 1995.
22. M. Koch, L. V. Mancini, and F. Parisi-Presicce. Graph-based specification of access control policies. *Journal of Computer and System Sciences (JCSS)*, 71:1–33, 2005.
23. B. König and V. Kozioura. Counterexample-guided abstraction refinement for the analysis of graph transformation systems. In *Tools and Algorithms for the Construction and Analysis of Systems (TACAS)*, volume 3920 of *LNCS*, pages 197–211. Springer, 2006.
24. K. L. McMillan. Using unfoldings to avoid the state explosion problem in the verification of asynchronous circuits. In *Fourth Workshop on Computer-Aided Verification (CAV)*, volume 663 of *LNCS*, pages 164–174. Springer, 1992.
25. S. Owre, J. M. Rushby, and N. Shankar. PVS: A prototype verification system. In *11th International Conference on Automated Deduction (CADE)*, volume 607 of *LNCS*, pages 748–752. Springer, 1992.
26. D. Plump. Termination of graph rewriting is undecidable. *Fundamenta Informaticae*, 33(2):201–209, 1998.
27. D. Plump and S. Steinert. Towards graph programs for graph algorithms. In *Graph Transformations (ICGT'04)*, volume 3256 of *LNCS*, pages 128–143. Springer, 2004.
28. A. Rensink. Towards model checking graph grammars. In M. Leuschel, S. Gruner, and S. L. Presti, editors, *Workshop on Automated Verification of Critical Systems (AVoCS)*, Technical Report DSSE-TR-2003-2, pages 150–160. University of Southhampton, 2003.
29. A. Rensink. The GROOVE simulator: A tool for state space generation. In *Applications of Graph Transformations with Industrial Relevance (AGTIVE)*, volume 3062 of *LNCS*, page 485. Springer, 2004.
30. A. Rensink. Representing first-order logic by graphs. In *Graph Transformations (ICGT'04)*, volume 3256 of *LNCS*, pages 319–335. Springer, 2004.
31. A. Rensink, Á. Schmidt, and D. Varró. Model checking graph transformations: A comparison of two approaches. In *Graph Transformations (ICGT'04)*, volume 3256 of *LNCS*, pages 226–241. Springer, 2004.
32. D. Varró. Towards symbolic analysis of visual modeling languages. *ENTCS*, 72(3), 2003.
33. D. Varró. Automated formal verification of visual modeling languages by model checking. *Journal of Software and Systems Modelling*, 3(2):85–113, 2004.

Introductory Tutorial on Foundations and Applications of Graph Transformation

Reiko Heckel

University of Leicester, UK
reiko@mcs.le.ac.uk

Abstract. This tutorial is intended as a general introduction to graph transformation for scientists who are not familiar with the field. The tutorial will start with an informal introduction to the basic concepts like graph, rule, transformation, etc., discussing semantic choices like which notion of graph to use; how to put labels, attributes, or types; or what to do with dangling links during rewriting, etc., and mentioning different ways to formalise the basic concepts.

In the second part, the tutorial will give a survey of typical applications of graph transformation in software engineering, e.g., as a specification language and semantic model for concurrent and distributed systems, as a meta language for defining the syntax, semantics, and manipulation of diagrams, etc.

Finally, the tutorial will go into some details about the algebraic approach to graph transformation, its formal foundations and relevant theory and tools. This shall enable the participants to better appreciate the conference and its satellite events.

1 Motivation

Graphs and diagrams provide a simple and powerful approach to a variety of problems that are typical to computer science in general, and software engineering in particular. In fact, for most activities in the software process, a variety of visual notations have been proposed, including state diagrams, Structured Analysis, control flow graphs, architectural description languages, function block diagrams, and the UML family of languages. These notations produce models that can be easily seen as graphs and thus graph transformations are involved, either explicitly or behind the scenes, when specifying how these models should be built and interpreted, and how they evolve over time and are mapped to implementations. At the same time, graphs provide a universally adopted data structure, as well as a model for the topology of object-oriented, component-based and distributed systems. Computations in such systems are therefore naturally modelled as graph transformations, too.

A. Corradini et al. (Eds.): ICGT 2006, LNCS 4178, pp. 461–462, 2006.

2   Content

In this tutorial, we will introduce the basic concepts and approaches to graph transformation, demonstrate their application to software engineering problems, and provide a high-level survey of graph transformation theory and tools.

We start by introducing a simple set-theoretic presentation of the double-pushout approach [2] whose features are common to most graph transformation approaches and which provides the foundation for further elaboration. Then, we discuss alternatives and extensions, like *multi objects*, *programmed* transformations concerned with controlling the (otherwise non-deterministic) rewriting process, as well as *application conditions*, restricting the applicability of individual rules.

Typical applications of graph transformation to software engineering problems are presented in terms of examples. They include

- model and program transformation;
- syntax and semantics of visual languages;
- visual behaviour modelling and programming.

In particular, we distinguish between the use of graph transformation as a *modelling notation* (and semantic model) to reason on particular problems, like functional requirements or architectural reconfigurations of individual applications, and its use as a *meta language* to specify the syntax, semantics, and manipulation of visual modelling languages, like the UML.

The last part of the tutorial is dedicated to a survey on the algebraic approach to graph transformation, its formal foundations and relevant theory. This shall enable attendees to work their way through the relevant literature and to benefit from the presentations at the conference.

Previous versions of this tutorial together with accompanying papers have been presented in [1,3].

References

1. L. Baresi and R. Heckel. Tutorial introduction to graph transformation: A software engineering perspective. In A. Corradini, H. Ehrig, H.-J. Kreowski, and G. Rozenberg, editors, *Proc. of the First International Conference on Graph Transformation (ICGT 2002), Barcellona, Spain*, volume 2505 of *LNCS*, pages 402–429, Barcelona, Spain, October 2002. Springer-Verlag.
2. H. Ehrig, M. Pfender, and H.J. Schneider. Graph grammars: an algebraic approach. In *14th Annual IEEE Symposium on Switching and Automata Theory*, pages 167–180. IEEE, 1973.
3. R. Heckel. Graph transformation in a nutshell. In R. Heckel, editor, *Proceedings of the School on Foundations of Visual Modelling Techniques (FoVMT 2004) of the SegraVis Research Training Network*, volume 148 of *Electronic Notes in TCS*, pages 187–198. Elsevier, 2006.

Workshop on Graph Computation Models

Yves Métivier and Mohamed Mosbah

LaBRI, Université Bordeaux-1, France
{metivier, mosbah}@labri.fr

A variety of computation models have been developed using graphs and graph transformations. These include models for sequential, distributed, parallel or mobile computation. A graph may represent, in an abstract way, the underlying structure of a computer system, or it may stand for the computation steps running on such a system. In the former, the computation can be carried on the corresponding graph, implying a simplification of the complexity of the system. The aim of the workshop is to bring together researchers interested in all aspects of computation models based on graphs, and their applications. A particular emphasis will be made for models and tools describing general solutions. The workshop will include contributed papers, tutorials and tool demonstrations.

The tutorials will introduce many types of graph transformations and their use to study computation models. These graph transformations include graph relabeling systems, graph grammars, term graph rewritings. The computation models, on the other hand, will include mobile computing, programming, data transformations, concurrent and distributed computing. For instance, graph relabeling systems have been successfully used as a suitable tool for encoding distributed algorithms, for proving their correctness and for understanding their power. In this model, a network is represented by a graph whose vertices denote processors, and edges denote communication links. The local state of a processor (resp. link) is encoded by the label attached to the corresponding vertex (resp. edge). A rule is a local transformation of labels. A relabeling system is defined by a finite set of such rules. The application of the rules are asynchronous: there is no global clock available, and two conflict-free applications of rewriting rules may occur simultaneously, provided they do not attempt to modify the same local context in the host graph. Thus, the behaviour of the network is defined by its initial labeling and the rule base of the associated local rewriting calculus. Problems of interest in distributed computing include node election, node enumeration, spanning tree construction, termination detection, synchronisation, inter-node agreements, or local recognition of global properties. These studies rely on rule-based local computations on network graphs on the one hand, and the recognition and classification of certain initial network configurations on the other hand. The non-existence of deterministic distributed solutions to certain problems leads to propose also the investigation of probabilistic distributed algorithms, the formulation of which seems rather simple, but their analysis is difficult. Another important aspect is the relationship between the three principal paradigms of distributed computing — local computations, message passing, shared memory — and the comparison of their expressive powers. Similar questions arise where

A. Corradini et al. (Eds.): ICGT 2006, LNCS 4178, pp. 463–464, 2006.

those three paradigms are compared with mobile agent systems. The selected papers are still not yet known, but they will mainly be concerned with the applications of graph transformations as suitable and convenient tools to model some (computation) applications.

The workshop includes also a session of software and tool demonstrations based on graph computation models. They can be related to the tutorials or to the presented papers and can therefore be useful to illustrate the presented concepts. Tools can range from alpha-versions to fully developed products that are used in education, research or being prepared for commercialisation.

WORKSHOP CHAIRS AND ORGANISERS

- *Yves Métivier,*   University of Bordeaux 1
- *Mohamed Mosbah,*   University of Bordeaux 1

PROGRAMME COMMITTEE

- *Yves Métivier,*   University of Bordeaux 1
- *Mohamed Mosbah,*   University of Bordeaux 1
- *Annegret Habel,*   University of Oldenburg, Germany
- *Hans-Jörg Kreowski,*   University of Bremen, Germany
- *Detlef Plump,*   University of York, UK
- *Stefan Gruner,*   University Bordeaux 1, France

Workshop on Graph-Based Tools

Albert Zündorf[1] and Dániel Varró[2]

[1] University of Kassel
Department of Computer Science and Software Engineering
Software Engineering Research Group
`zuendorf@se.e-technik.uni-kassel.de`
[2] Budapest University of Technology and Economics
Department of Measurement and Information Systems
`varro@mit.bme.hu`

Abstract. The International Workshop on Graph Based Tools (GraBaTs 2006) is the third workshop of a series that serves as a forum for researchers and practitioners interested in the development and application of graph-based tools. Based upon mathematically solid underlying concepts, graph-based tools are frequently used in various application areas in software and systems engineering.

1   Motivation and History

Graphs are well-known means to capture structural aspects in various fields of computer science. Based upon mathematically solid underlying concepts, graph-based tools are frequently used in various application areas in software and systems engineering. Successful application areas include (but not limited to) compiler construction, constraint solving, CASE tool generation, software engineering, pattern recognition techniques, program analysis, software evolution, software visualization and animation, visual languages, and many more. A commonality in all these areas is that tools heavily rely on graphs as an underlying data structure.

The International Workshop on Graph Based Tools (GraBaTs 2006) is a forum for researchers and practitioners interested in the development and application of graph-based tools. The current event is already the third in a series which is traditionally organized bi-annually as a satellite event of the International Conference on Graph Transformation (ICGT) [3]. The first workshop on this topic [4] took place in 2002 in Barcelona, Spain, while the second workshop [5] was organized in 2004 in Rome, Italy.

Both events have demonstrated that the GraBaTs workshop is of special relevance for a conference on graph transformation. Frequently, the application of graph transformation technology requires the existence of reliable, user-friendly and efficient tool support.

Moreover, these tools are frequently built on top of basic services or frameworks provided by popular open development environments such as Eclipse [1] or NetBeans [2] which significantly reduce development efforts, and provide a professional look-and-feel of the tools thus facilitating industrial acceptance.

. A. Corradini et al. (Eds.): ICGT 2006, LNCS 4178, pp. 465–466, 2006.

2   Aims and Scope

This year, the GraBaTs workshop focused on attracting submissions mainly on the following topics:

- Tools for model-driven systems development, Meta CASE tools & generators;
- Tools for Visual languages: UML, Domain-specific languages, etc.
- Model transformation tools; Tool integration techniques; Animation and simulation tools
- Analysis tools (verification & validation, static analysis techniques, testing)
- Efficient algorithms (pattern matching, manipulation of large graph models)
- Case studies, empirical and experimental results on tool scalability, novel application areas

Furthermore, this year's workshop has a special focus on tool presentations organized as a separate session of the workshop. Authors were encouraged to submit tool papers which report on new features of existing tools or completely novel tools having graph-based foundations. The presentations of tools papers will contain a mandatory live demonstration part during the workshop.

The mission of the tool demonstration part is to provide an overview on the state-of-the-art of tools for the graph transformation community and to local participants as well.

3   Workshop Organization

The program committee of this workshop consists of Luciano Baresi, David Déharbe, Holger Giese, Gabor Karsai, Mark Minas, Arend Rensink, Andy Schürr, Gabriele Taentzer, Dániel Varró, Pieter Van Gorp, Hans Vangheluwe, Andreas Winter, and Albert Zündorf.

Altogether, 13 papers have been submitted for GraBaTs. More information about the workshop including its program and an electronic version of all accepted papers appearing the new electronic journal "Electronic Communications of EASST". can be found on the workshop web page:
`http://www.inf.mit.bme.hu/GRABATS2006`.

References

1. Eclipse framework. `http://www.eclipse.org`.
2. Netbeans software. `http://www.netbeans.org`.
3. A. Corradini, H. Ehrig, U. Montanari, L. Ribeiro, and G. Rozenberg (eds.). *Third International Conference on Graph Transformation (ICGT 2006)*, vol. 4178 of *Lecture Notes in Computer Science*. Springer Verlag, 2004. `http://www.dimap.ufrn.br/icgt2006/`
4. T. Mens, A. Schürr, and G. Taentzer (eds.). *First International Workshop on Graph Based Tools (GraBaTs 2002)*, vol. 72 of *ENTCS*. 2002.
5. T. Mens, A. Schürr, and G. Taentzer (eds.). *Second International Workshop on Graph Based Tools (GraBaTs 2004)*, vol. 127 of *ENTCS*. 2004.

Workshop on Petri Nets and Graph Transformations

Paolo Baldan[1], Hartmut Ehrig[2], Julia Padberg[2], and Grzegorz Rozenberg[3]

[1] Dipartimento di Informatica,
Università Ca' Foscari di Venezia, Italy
baldan@dsi.unive.it

[2] Institute for Software Technology and Theoretical Computer Science,
Technical University Berlin, Germany
{ehrig, padberg}@cs.tu-berlin.de

[3] Leiden Institute of Advanced Computer Science,
Universiteit Leiden, The Netherlands
rozenberg@liacs.nl

The *Workshop on Petri Nets and Graph Transformations*, which is currently at its second edition, is focussed on the mutual relationship between two prominent specification formalisms for concurrency and distribution, namely Petri nets and graph transformation systems. It belongs to the folklore that Petri nets can be seen as rewriting systems over (multi)sets, the rewriting rules being the transitions, and, as such, they can be seen as special graph transformation systems, acting over labelled discrete graphs. The basic notions of Petri nets like marking, enabling, firing, steps and step sequences can be naturally "translated" to corresponding notions of graph transformation systems. Due to this close correspondence there has been a mutual influence between the two fields, which has lead to a fruitful cross-fertilisation.

Several approaches to the concurrent semantics of graph transformation systems as well as techniques for their analysis and verification have been strongly influenced by the corresponding theories and constructions for Petri nets (see, e.g., [10]). For instance, the truly concurrent semantics of algebraic graph transformations presented in [3,2] can be seen as a generalisation to of the corresponding semantical constructions developed for Petri nets in [21,14]. Similarly, the concurrent semantics for EMS systems in [12] is partly inspired by the Goltz-Reisig process semantics for Petri nets. More recently, several approaches to the analysis and verification of graph transformation systems properties have been proposed (see, e.g., [18,4,20,6,17]) and also in this case the relation with Petri nets has been often a source of inspiration. In particular, some approaches are inspired by analogous techniques previously developed in the domain of Petri nets, e.g., based on invariants or on finite prefixes of the unfolding, and some others reduce the verification of a graph transformation systems to the analysis of a suitable abstraction expressed in the form of a Petri net.

Classical Petri nets models have been integrated with graph transformation systems in order to define rule-based changes in the Petri net structure. This can serve for a stepwise refinement of Petri net models, which leads from an abstract description of the system to the desired model. Alternatively, transformations

A. Corradini et al. (Eds.): ICGT 2006, LNCS 4178, pp. 467–469, 2006.
c Springer-Verlag Berlin Heidelberg 2006

over Petri nets can be used to define dynamically reconfiguring Petri nets, i.e., extended Petri net models where the standard behaviour, expressed by the token game over a fixed structure, is enriched with the possibility of altering the net structure (see, e.g., reconfigurable nets of [1] and high-level replacement systems applied to Petri nets in [16,7]).

As mentioned above, the theory of rewriting over categories of Petri falls into the realm of high-level replacement systems, a generalisation of graph transformation systems to general categories, the so-called called HLR categories [8], including, e.g., algebraic specifications. The HLR approach has been recently generalised with the introduction of adhesive categories [13] and adhesive HLR systems [9], which provide a quite elegant and general framework where (double-pushout) rewriting can be developed. The view of Petri nets as rewriting systems over adhesive categories [19] or as bigraphical reactive systems [15] has been recently used to automatically derive compositional behavioural equivalences for Petri nets.

As a further link between the two models, recall that graph transformation systems are also used for the development, the simulation, or animation of various types of Petri nets, e.g., via the the definition of visual languages and environments [5,11].

With the aim of favouring the cross-fertilisation and the exchange between the areas of Petri nets and of graph transformation, the workshop gathers researchers working in the field of low- and high-level Petri nets, and researchers working in the field of rewriting, including graph transformation, high-level replacement systems, rewriting systems over adhesive categories and rewriting logic. The contributions to the workshop will touch all the issues mentioned above: transfer of concepts and techniques from Petri nets to graph transformation, verification of graph transformation based on Petri net abstractions, theory and application of rewriting over Petri nets and encoding of (extensions) of Petri nets as rewrite theories.

References

1. E. Badouel, M. Llorens, and J. Oliver. Modeling concurrent systems: Reconfigurable nets. In H. R. Arabnia and Y. Mun, editors, *Proceedings of PDPTA'03*, volume 4, pages 1568–1574. CSREA Press, 2003.
2. P. Baldan. *Modelling concurrent computations: from contextual Petri nets to graph grammars*. PhD thesis, Department of Computer Science, University of Pisa, 2000. Available as technical report n. TD-1/00.
3. P. Baldan, A. Corradini, H. Ehrig, M. Löwe, U. Montanari, and F. Rossi. Concurrent Semantics of Algebraic Graph Transformation Systems. In Ehrig et al. [10].
4. P. Baldan, A. Corradini, and B. König. A static analysis technique for graph transformation systems. In K.G. Larsen and M. Nielsen, editors, *Proceedings of CONCUR'01*, volume 2154 of *LNCS*, pages 381–395. Springer Verlag, 2001.
5. R. Bardohl and C. Ermel. Scenario animation for visual behavior models: A generic approach. *Software and System Modeling*, 3(2):164–177, 2004.
6. F.L. Dottí, L. Foss, L. Ribeiro, and O. Marchi Santos. Verification of distributed object-based systems. In *Proceedings of FMOODS '03*, pages 261–275. Springer, 2003. LNCS 2884.

7. H. Ehrig, M. Gajewsky, and F. Parisi-Presicce. Replacement systems with applications to algebraic specifications and petri nets. In Ehrig et al. [10].
8. H. Ehrig, A. Habel, H.-J. Kreowski, and F. Parisi-Presicce. Parallelism and concurrency in High-Level Replacement Systems. *Mathematical Structures in Computer Science*, 1:361–404, 1991.
9. H. Ehrig, A. Habel, J. Padberg, and U. Prange. Adhesive high-level replacement categories and systems. In H. Ehrig, G. Engels, F. Parisi-Presicce, and G Rozenberg, editors, *Proceedings of ICGT'04*, volume 3256 of *LNCS*, pages 144–160. Springer Verlag, 2004.
10. H. Ehrig, J. Kreowski, U. Montanari, and G. Rozenberg, editors. *Handbook of Graph Grammars and Computing by Graph Transformation*, volume Volume III: Concurrency, Parallelism and Distribution. World Scientific, 1999.
11. C. Ermel and K. Ehrig. View transformation in visual environments applied to petri nets. In H. Ehrig, J. Padberg, and G. Rozenberg, editors, *Proceedings of PNGT'04*, volume 127 of *Electronic Notes in Theoretical Computer Science*, pages 61–86. Elsevier, 2005.
12. D. Janssens. ESM systems and the composition of their computations. In *Graph Transformations in Computer Science*, volume 776 of *LNCS*, pages 203–217. Springer Verlag, 1994.
13. S. Lack and P. Sobociński. Adhesive categories. In I. Walukiewicz, editor, *Proceedings of FoSSaCS'04*, volume 2987 of *LNCS*, pages 273–288. Springer Verlag, 2004.
14. J. Meseguer, U. Montanari, and V. Sassone. On the semantics of Place/Transition Petri nets. *Mathematical Structures in Computer Science*, 7:359–397, 1997.
15. R. Milner. Bigraphs for petri nets. In J. Desel, W. Reisig, and G. Rozenberg, editors, *Lectures on Concurrency and Petri Nets*, volume 3098 of *LNCS*, pages 686–701. Springer, 2003.
16. J. Padberg, H. Ehrig, and L. Ribeiro. High level replacement systems applied to algebraic high level net transformation systems. *Mathematical Structures in Computer Science*, 5(2):217–256, 1995.
17. J. Padberg and B. Enders. Rule invariants in graph transformation systems for analyzing safety-critical systems. In A. Corradini, H. Ehrig, H.-J. Kreowski, and G. Rozenberg, editors, *Proceedings of ICGT'02*, volume 2505 of *LNCS*, pages 334–350, 2002.
18. A. Rensink. Towards model checking graph grammars. In M. Leuschel, S. Gruner, and S. Lo Presti, editors, *Proceedings of the* 3rd *Workshop on Automated Verification of Critical Systems*, Technical Report DSSE–TR–2003–2, pages 150–160. University of Southampton, 2003.
19. V. Sassone and P. Sobocinski. A congruence for Petri nets. In *Proceedings of PNGT'04*, volume 127 of *Electronic Notes in Theoretical Computer Science*, pages 107–120. Elsevier Science, 2005.
20. Dániel Varró. Towards symbolic analysis of visual modelling languages. In P. Bottoni and M. Minas, editors, *Proc. GT-VMT 2002: International Workshop on Graph Transformation and Visual Modelling Techniques*, volume 72 of *Electronic Notes in Theoretical Computer Science*, pages 57–70. Elsevier, 2002.
21. G. Winskel. Event Structures. In *Petri Nets: Applications and Relationships to Other Models of Concurrency*, volume 255 of *LNCS*, pages 325–392. Springer Verlag, 1987.

3rd International Workshop on Software Evolution Through Transformations: Embracing Change

Jean-Marie Favré[1], Reiko Heckel[2], and Tom Mens[3]

[1] Universite Grenoble 1, France
Jean-Marie.Favre@imag.fr
[2] University of Leicester, UK
reiko@mcs.le.ac.uk
[3] Université de Mons-Hainaut, Belgium
tom.mens@umh.ac.be

Abstract. Transformation-based techniques such as refactoring, model transformation and model-driven development, architectural reconfiguration, etc. are at the heart of many software engineering activities, making it possible to cope with an ever changing environment.

This workshop provides a forum for discussing these techniques, their formal foundations and applications.

1 Motivation and Objectives

Since its birth as a discipline in the late 60ies Software Engineering had to cope with the breakdown of many of its original assumptions. Today we know that

- it is impossible to fix requirements up front;
- the design of the system is changing while it is being developed;
- the distinction between design time and run-time is increasingly blurred;
- a system's architecture will change or degrade while it is in use;
- technology will change more rapidly than it is possible to re-implement critical applications;

This recognition of lack of stability in software means that we have to cope with change, rather than defending against it. Processes, methods, languages, and tools have to be geared towards making change possible and cheap.

Transformations of development artifacts like specifications, designs, code, or run-time architectures are at the heart of many software engineering activities. Their systematic specification and implementation are the basis for a wide range of tools, from compilers and refactoring tools to model-driven CASE tools and formal verification environments. The workshop provides a forum for the discussion transformation-based techniques in software evolution.

A. Corradini et al. (Eds.): ICGT 2006, LNCS 4178, pp. 470–472, 2006.

2 Topics

Submissions to the workshop are based on a wide range of transformation formalisms like

- program transformation (over Java, C, or C++, etc.);
- model transformation (over UML and other visual languages);
- graph transformation;
- term rewriting;
- category theory, algebra, and logic;

discussing their application to software evolution activities like

- model-driven development;
- model and code refactoring, redesign and code optimisation;
- requirements evolution;
- reverse engineering, pattern detection, architecture recovery;
- architectural reconfiguration, self-organising or self-healing systems, service-oriented architectures;
- consistency management, co-evolution of models and code;
- merging of models, specifications, ontologies, etc;

The program will combine presentations of position and technical papers with discussions on selected topics. The nomination of papers for presentation is determined through a formal review process.

Accepted contributions will appear in the Electronic Communications of EASST, the European Association of Software Science and Technology. A preliminary version of the issue will be available at the workshop.

3 Program Committee

The following program committee is responsible for the selection of papers.

- Luciano Baresi, Politecnico di Milano, Italy
- Thaís Batista, Federal University of Rio Grande do Norte, Brazil
- Paulo Borba, Universidade Federal de Pernambuco, Recife, Brazil
- Artur Boronat, Universidad Politécnica de Valencia, Spain
- Christiano de Oliveira Braga, Universidad Complutense de Madrid, Spain
- Andrea Corradini, Università di Pisa, Italy
- Mohammad El-Ramly, University of Leicester, UK
- Jean-Marie Favre, Universite Grenoble 1, France [co-chair]
- Reiko Heckel, University of Leicester, UK [co-chair]
- Dirk Janssens, University of Antwerp, Belgium
- Tom Mens, Université de Mons-Hainaut, Belgium [co-chair]
- Anamaria Martins Moreira, Universidade Federal do Rio Grande do Norte, Natal, Brazil
- Leila Silva, Universidade Federal de Segipe, Brazil
- German Vega, Universite Grenoble 1, France

4   Acknowledgement

The workshop is supported by the European Research Training Network SegraVis on Syntactic and Semantic Integration of Visual Modelling Techniques, the Integrated Project Sensoria on Software Engineering for Service-Oriented Overlay Computers, and the ERCIM Working Group on *Software Evolution.*

Author Index

Lecture Notes in Computer Science

For information about Vols. 1–4063

please contact your bookseller or Springer